■ SECOND EDITION ■

# Fundamentals of Complex Analysis

## for Mathematics, Science, and Engineering

**E. B. SAFF**

*Distinguished Research
Professor of Mathematics
University of South Florida*

**A. D. SNIDER**

*Professor of Electrical
Engineering
University of South Florida*

*With an appendix by*
**LLOYD N. TREFETHEN**

*Associate Professor
of Computer Science
Cornell University*

Prentice Hall, Upper Saddle River, New Jersey 07458

*Library of Congress Cataloging-in-Publication Data*

Saff, E. B., (date)
   Fundamentals of complex analysis for mathematics, science, and
engineering / E. B. Saff, A. D. Snider; with an appendix by Lloyd N.
Trefethen.—2nd ed.
      p.    cm.
   Includes bibliographical references and index.
   ISBN 0-13-327461-6:
   1. Mathematical analysis.   2. Functions of complex variables.
I. Snider, Arthur David, (date).   II. Title.
QA300.S18     1993
515—dc20

92-33606
CIP

Acquisitions Editors: *Steven R. Conmy, Priscilla McGeehon*
Editorial/Production Supervision
   and Interior Design: *Kathleen M. Lafferty*
Copy Editor: *Linda Thompson*
Cover Designer: *Design Source*
Prepress Buyer: *Paula Massenaro*
Manufacturing Buyer: *Lori Bulwin*
Proofreader: *Bruce D. Colegrove*

Printed in the United States of America

20   19   18   17   16

ISBN 0-13-327461-6

Prentice-Hall International (UK) Limited, *London*
Prentice-Hall of Australia Pty. Limited, *Sydney*
Prentice Hall Canada Inc., *Toronto*
Prentice-Hall Hispanoamericana, S.A., *Mexico*
Prentice-Hall of India Private Limited, *New Delhi*
Prentice-Hall of Japan, Inc., *Tokyo*
Simon & Schuster Asia Pte. Ltd., *Singapore*
Editora Prentice-Hall do Brasil, Ltda., *Rio de Janeiro*

*To Loretta*
   *whose encouragement, typing, and coffee made it possible*

# Contents

# 5 Series Representations for Analytic Functions   181

# 6 Residue Theory   245

# 7 Conformal Mapping   301

# *8 The Transforms of Applied Mathematics    371

# Appendix I    Numerical Construction of Conformal Maps    430

# Appendix II    Table of Conformal Mappings    444

# Answers to Odd-Numbered Problems    450

# Index    465

# Preface

The *raison d'existence* for *Fundamentals of Complex Analysis* is our conviction that mathematics, science, and engineering undergraduates who have completed the calculus sequence are capable of understanding the basics of complex analysis and applying its methods to solve physical problems. Accordingly, we have addressed ourselves to this audience in our attempt to make the fundamentals of the subject more easily accessible to readers who have little inclination to wade through the rigors of the axiomatic approach. To accomplish this goal we have modeled the text after standard calculus books, both in level of exposition and layout, and have incorporated physical applications throughout the text so that the mathematical methodology will appear less sterile to the reader.

To be more specific about our mode of exposition, we begin by addressing the question most instructors ask first: To what extent is the book self-contained, i.e., which results are proved and which are merely stated? Let us say that we have elected to include all the proofs that reflect the spirit of analytic function theory and to omit most of those that involve deeper results from real analysis. For example, we do not prove the convergence of Riemann sums for complex integrals, nor do we prove the Cauchy criterion for convergence of a series, Goursat's generalization of Cauchy's theorem, or the Riemann mapping theorem. Moreover, in keeping with our philosophy of avoiding pedantics, we have shunned the ordered-pairs interpretations of complex numbers and retained the more intuitive approach (grounded in algebraic field extensions).

Cauchy's theorem is given two alternative presentations in Chapter 4. The first is based on the deformation of contours, or what is known to topologists as homotopy.

We have taken some pains to make this approach understandable and transparent to the novice because it is easy to visualize and to apply in specific situations. The second treatment interprets contour integrals in terms of line integrals and invokes Green's theorem to complete the argument. These parallel developments constitute the two parts of Section 4 in Chapter 4; either one may be read, and the other omitted, without disrupting the exposition (although it should not be difficult to discern our preference, from this paragraph).

Steady-state temperature patterns in two dimensions are, in our opinion, the most universally familiar instances of harmonic functions, so we have consistently chosen this interpretation for visualization of the theorems of analytic function theory. This and other applications to potential theory receive special emphasis in Chapter 7 on conformal mapping. We draw the distinction between direct methods, wherein a mapping must be constructed to solve a specific problem, and indirect methods that postulate a mapping and then investigate which problems it solves. In doing so we hope to dispel the impression, given in many older books, that all applications of the technique fall in the latter category. In this second edition many new wrinkles have been added that reflect the progress made in recent years on the numerical construction of conformal mappings. In fact, L. N. Trefethen has graciously contributed an appendix describing his computer implementation of the Schwarz-Christoffel transformation. A second appendix compiles a listing of some useful mappings having closed-form expressions.

Linear systems analysis is another application that recurs in the text. The basic ideas of frequency analysis are introduced in Chapter 3 following the study of the transcendental functions; Smith charts, circuit synthesis, and stability criteria are addressed at appropriate times; and the development culminates in Chapter 8 with the exposition of Fourier, Mellin, Laplace, Hilbert, and $z$-transforms. In keeping with the philosophy stated above, we offer proofs of the transform representations for those cases where analytic function theory applies.

Other features of the book are optional sections (marked *) that enrich either the mathematical development or the physical interpretation, a thorough coverage of the uses of residue theory in evaluating integrals, Summary and Suggested Reading sections at the end of each chapter, and a wealth of worked-out examples illustrating the theorems.

All the material in this book can be covered fairly leisurely in a two-semester course. One-semester courses can readily be tailored to the specific needs and interests of the class. At the University of South Florida both the math department and the engineering college offer courses based on the text, with complementary emphasis on the topics in later chapters.

For the edification of users of the first edition, let us point out the major innovations in the new edition:

- early introduction of Euler's formula to obviate the clumsy "cis" notation;
- applications of complex algebra in celestial mechanics and gear kinematics;
- incorporation of computational aspects of the subject, notably Trefethen's numerical implementation of the Schwarz-Christoffel transformation;
- exploitation of the thermal flow analogy for visualization of analytic function properties;

- extended applications to linear systems: circuit synthesis (positive real functions), stability, $z$- and Hilbert transforms, dispersion relations;
- introduction to Cauchy integrals and the Sokhotskyi-Plemelj formulas;
- greatly expanded problem sets;
- answers to odd-numbered problems now included.

We would like to thank Lloyd N. Trefethen, Deborah P. Levinson, and Randall Campbell-Wright, as well as the staff of Prentice Hall, for their contributions to the second edition. We also thank the reviewers of this edition:

Sam H. Davis,
   Rice University
Frederic Dias, Worcester
   Polytechnic Institute
James A. Hummel,
   University of Maryland

William Perry,
   Texas A&M University
Calvin Wilcox,
   University of Utah

The efforts of Kenneth Hoffman, Ken Gross, Jack Britton, James Gard, and David Horowitz in polishing the first edition continue, of course, to be in force. Finally we wish to acknowledge our dissertation directors, Joseph L. Walsh and Paul Garabedian, for their mentorship in our mathematical growth.

*E. B. Saff*
*A. D. Snider*

# :1:

# Complex Numbers

## 1.1 THE ALGEBRA OF COMPLEX NUMBERS

To achieve a proper perspective for studying the system of complex numbers, let us begin by briefly reviewing the construction of the various numbers used in computation.

We start with the rational numbers. These are ratios of integers and are written in the form $m/n$, $n \neq 0$, with the stipulation that all rationals of the form $n/n$ are equal to 1. The arithmetic operations of addition, subtraction, multiplication, and division with these numbers can always be performed in a finite number of steps, and the results are, again, rational numbers. Furthermore, there are certain simple rules concerning the order in which the computations can proceed. These are the familiar commutative, associative, and distributive laws:

*Commutative Law of Addition*

$$a + b = b + a$$

*Commutative Law of Multiplication*

$$ab = ba$$

*Associative Law of Addition*

$$a + (b + c) = (a + b) + c$$

*Associative Law of Multiplication*

$$a(bc) = (ab)c$$

1

*Distributive Law*

$$(a + b)c = ac + bc,$$

for any rationals $a$, $b$, and $c$.

Notice that the rationals are the only numbers we would ever need, to solve equations of the form

$$ax + b = 0.$$

The solution, for nonzero $a$, is $x = -b/a$, and since this is the ratio of two rationals, it is itself rational.

However, if we try to solve quadratic equations in the rational system, we find that some of them have no solution; for example, the simple equation

$$x^2 = 2 \tag{1}$$

cannot be satisfied by any rational number (see Prob. 29 at the end of this section). Therefore, to get a more satisfactory number system, we extend the concept of "number" by appending to the rationals a new symbol, mnemonically written as $\sqrt{2}$, which is defined to be a solution of Eq. (1). Our revised concept of a number is now an expression in the standard form

$$a + b\sqrt{2}, \tag{2}$$

where $a$ and $b$ are rationals. Addition and subtraction are performed according to

$$(a + b\sqrt{2}) \pm (c + d\sqrt{2}) = (a \pm c) + (b \pm d)\sqrt{2}. \tag{3}$$

Multiplication is defined via the distributive law with the proviso that the square of the symbol $\sqrt{2}$ can always be replaced by the rational number 2. Thus we have

$$(a + b\sqrt{2})(c + d\sqrt{2}) = (ac + 2bd) + (bc + ad)\sqrt{2}. \tag{4}$$

Finally, using the well-known process of *rationalizing the denominator* we can put the quotient of any two of these new numbers into the standard form

$$\frac{a + b\sqrt{2}}{c + d\sqrt{2}} = \frac{a + b\sqrt{2}}{c + d\sqrt{2}} \frac{c - d\sqrt{2}}{c - d\sqrt{2}} = \frac{ac - 2bd}{c^2 - 2d^2} + \frac{bc - ad}{c^2 - 2d^2}\sqrt{2}. \tag{5}$$

This procedure of "calculating with radicals" should be very familiar to the reader, and the resulting arithmetic system can easily be shown to contain no inconsistencies and to satisfy the commutative, associative, and distributive laws. However, observe that the symbol $\sqrt{2}$ has not been absorbed by the rational numbers painlessly. Indeed, in the standard form (2) and in the algorithms (3), (4), and (5) its presence stands out like a sore thumb. Actually, we are only using the symbol $\sqrt{2}$ to "hold a place" while we compute around it using the rational components, except for those occasional opportunities when it occurs squared and we are temporarily relieved of having to carry it. So the inclusion of $\sqrt{2}$ as a number is a somewhat artificial process, devised solely so that we might have a richer system in which we can solve the equation $x^2 = 2$.

With this in mind, let us jump to the stage where we have appended all the real numbers to our system. Some of them, such as $\sqrt[3]{17}$, arise as solutions of more complicated equations, while others, such as $\pi$ and $e$, come from certain limit processes. Each irrational is absorbed in a somewhat artificial manner, but once again the resulting conglomerate of numbers and arithmetic operations contains no inconsistencies and satisfies the commutative, associative, and distributive laws.[†]

At this point we observe that we still cannot solve the equation

$$x^2 = -1. \tag{6}$$

But now our experience suggests that we can expand our number system once again by appending a symbol for a solution to Eq. (6); customarily the symbol used is $i$.[‡] Next we imitate the model of expressions (2) through (5) (pertaining to $\sqrt{2}$) and thereby generalize our concept of number as follows:

---

**Definition 1.**    A **complex number** is an expression of the form $a + bi$, where $a$ and $b$ are real numbers. Two complex numbers $a + bi$ and $c + di$ are said to be equal ($a + bi = c + di$) if and only if $a = c$ and $b = d$.

---

*The operations of addition and subtraction of complex numbers are given by*

$$(a + bi) \pm (c + di) := (a \pm c) + (b \pm d)i,$$

where the symbol $:=$ means "is defined to be."

In accordance with the distributive law and the proviso that $i^2 = -1$, we postulate the following:

*The multiplication of two complex numbers is defined by*

$$(a + bi)(c + di) := (ac - bd) + (bc + ad)i.$$

To compute the quotient of two complex numbers, we again "rationalize the denominator":

$$\frac{a + bi}{c + di} = \frac{a + bi}{c + di} \frac{c - di}{c - di} = \frac{ac + bd}{c^2 + d^2} + \frac{bc - ad}{c^2 + d^2} i.$$

Thus we formally postulate the following:

*The division of complex numbers is given by*

$$\frac{a + bi}{c + di} := \frac{ac + bd}{c^2 + d^2} + \frac{bc - ad}{c^2 + d^2} i \qquad (\text{if } c^2 + d^2 \neq 0).$$

These are rules for computing in the complex number system. The usual algebraic properties (commutativity, associativity, etc.) are easy to verify and appear as exercises.

---

[†] The algebraic aspects of extending a number field are discussed in Ref. [7] at the end of this chapter.

[‡] Engineers often use the letter $j$.

**Example 1**

Find the quotient

$$\frac{(6 + 2i) - (1 + 3i)}{(-1 + i) - 2}.$$

**Solution.**

$$\frac{(6 + 2i) - (1 + 3i)}{(-1 + i) - 2} = \frac{5 - i}{-3 + i} = \frac{(5 - i)}{(-3 + i)} \frac{(-3 - i)}{(-3 - i)}$$

$$= \frac{-15 - 1 - 5i + 3i}{9 + 1}$$

$$= -\frac{8}{5} - \frac{1}{5}i. \quad \blacksquare^{\dagger}$$

Historically, $i$ was considered as an "imaginary" number because of the blatant impossibility of solving Eq. (6) with any of the numbers at hand. With the perspective we have developed, we can see that this label could also be applied to the numbers $\sqrt{2}$ or $\sqrt[4]{17}$; like them, $i$ is simply one more symbol appended to a given number system to create a richer system. Nonetheless, tradition dictates the following designations:

---

**Definition 2.**    The **real part** of the complex number $a + bi$ is the (real) number $a$; its **imaginary part** is the (real) number $b$. If $a$ is zero, the number is said to be a **pure imaginary number**.

---

For convenience we customarily use a single letter, usually $z$, to denote a complex number. Its real and imaginary parts are then written $\operatorname{Re} z$ and $\operatorname{Im} z$, respectively. With this notation we have $z = \operatorname{Re} z + i \operatorname{Im} z$.

Observe that the equation $z_1 = z_2$ holds if and only if $\operatorname{Re} z_1 = \operatorname{Re} z_2$ and $\operatorname{Im} z_1 = \operatorname{Im} z_2$. Thus any equation involving complex numbers can be interpreted as a pair of real equations.

The set of all complex numbers is sometimes denoted as **C**. Unlike the real number system, there is no natural ordering for the elements of **C**; it is meaningless, for example, to ask whether $2 + 3i$ is greater than or less than $3 + 2i$. (See Prob. 30.)

**EXERCISES 1.1**

**1.** Verify that $-i$ is also a root of Eq. (6).

**2.** Verify the commutative, associative, and distributive laws for complex numbers.

---

$^{\dagger}$ A slug marks the end of solutions or proofs throughout the text.

3. Notice that 0 and 1 retain their "identity" properties as complex numbers; i.e., $0 + z = z$ and $1 \cdot z = z$ when $z$ is complex.
   (a) Verify that complex subtraction is the inverse of complex addition (i.e., $z_3 = z_2 - z_1$ if and only if $z_3 + z_1 = z_2$).
   (b) Verify that complex division, as given in the text, is the inverse of complex multiplication (i.e., if $z_2 \neq 0$, then $z_3 = z_1/z_2$ if and only if $z_3 z_2 = z_1$).
4. Prove that if $z_1 z_2 = 0$, then $z_1 = 0$ or $z_2 = 0$.

*In Problems 5–13, write the number in the form a + bi.*

5. (a) $-3\left(\dfrac{i}{2}\right)$        (b) $(8 + i) - (5 + i)$        (c) $\dfrac{2}{i}$

6. (a) $(-1 + i)^2$        (b) $\dfrac{2 - i}{\frac{1}{3}}$        (c) $i(\pi - 4i)$

7. (a) $\dfrac{8i - 1}{i}$        (b) $\dfrac{-1 + 5i}{2 + 3i}$        (c) $\dfrac{3}{i} + \dfrac{i}{3}$

8. $\dfrac{(8 + 2i) - (1 - i)}{(2 + i)^2}$        9. $\dfrac{2 + 3i}{1 + 2i} - \dfrac{8 + i}{6 - i}$        10. $\left[\dfrac{2 + i}{6i - (1 - 2i)}\right]^2$

11. $i^3(i + 1)^2$        12. $(2 + i)(-1 - i)(3 - 2i)$        13. $((3 - i)^2 - 3)i$

14. Show that $\text{Re}(iz) = -\text{Im}\, z$ for every complex number $z$.
15. Let $k$ be an integer. Show that

$$i^{4k} = 1, \qquad i^{4k+1} = i, \qquad i^{4k+2} = -1, \qquad i^{4k+3} = -i.$$

16. Use the result of Problem 15 to find
    (a) $i^7$        (b) $i^{62}$        (c) $i^{-202}$        (d) $i^{-4321}$
17. Use the result of Problem 15 to evaluate

$$3i^{11} + 6i^3 + \frac{8}{i^{20}} + i^{-1}.$$

18. Show that the complex number $z = -1 + i$ satisfies the equation $z^2 + 2z + 2 = 0$.
19. Write the complex equation $z^3 + 5z^2 = z + 3i$ as two real equations.
20. Solve each of the following equations for $z$.

    (a) $iz = 4 - zi$        (b) $\dfrac{z}{1 - z} = 1 - 5i$

    (c) $(2 - i)z + 8z^2 = 0$        (d) $z^2 + 16 = 0$

21. The complex numbers $z_1, z_2$ satisfy the system of equations

$$(1 - i)z_1 + 3z_2 = 2 - 3i,$$

$$iz_1 + (1 + 2i)z_2 = 1.$$

Find $z_1, z_2$.
22. Find all solutions to the equation $z^4 - 16 = 0$.
23. Let $z$ be a complex number such that $\text{Re}\, z > 0$. Prove that $\text{Re}(1/z) > 0$.
24. Let $z$ be a complex number such that $\text{Im}\, z > 0$. Prove that $\text{Im}(1/z) < 0$.
25. Let $z_1, z_2$ be two complex numbers such that $z_1 + z_2$ and $z_1 z_2$ are each negative real numbers. Prove that $z_1$ and $z_2$ must be real numbers.

**26.** Verify that $\text{Re}\left(\sum_{j=1}^{n} z_j\right) = \sum_{j=1}^{n} \text{Re} \, z_j$ and that $\text{Im}\left(\sum_{j=1}^{n} z_j\right) = \sum_{j=1}^{n} \text{Im} \, z_j$. [The real (imaginary) part of the sum is the sum of the real (imaginary) parts.] Formulate, and then *disprove*, the corresponding conjectures for multiplication.

**27.** Prove the *binomial theorem* for complex numbers:

$$(z_1 + z_2)^n = z_1^n + \binom{n}{1} z_1^{n-1} z_2 + \cdots + \binom{n}{k} z_1^{n-k} z_2^k + \cdots + z_2^n,$$

where $n$ is a positive integer, and

$$\binom{n}{k} := \frac{n!}{k!(n-k)!}.$$

**28.** Use the binomial theorem (Problem 27) to compute $(2 - i)^5$.

**29.** Prove that there is no rational number $x$ that satisfies $x^2 = 2$. [HINT: Show that if $p/q$ were a solution, where $p$ and $q$ are integers, then 2 would have to divide both $p$ and $q$. This contradicts the fact that such a ratio can always be written without common divisors.]

**30.** The definition of the order relation denoted by $>$ in the real number system is based upon the existence of a subset $\mathscr{P}$ (the positive reals) having the following properties:

  (i) For any number $\alpha \neq 0$, either $\alpha$ or $-\alpha$ (but not both) belongs to $\mathscr{P}$.
  (ii) If $\alpha$ and $\beta$ belong to $\mathscr{P}$, so does $\alpha + \beta$.
  (iii) If $\alpha$ and $\beta$ belong to $\mathscr{P}$, so does $\alpha \cdot \beta$.

When such a set $\mathscr{P}$ exists we write $\alpha > \beta$ if and only if $\alpha - \beta$ belongs to $\mathscr{P}$.[†] Prove that the *complex* number system does not possess a nonempty subset $\mathscr{P}$ having properties (i), (ii), and (iii). [HINT: Argue that neither $i$ nor $-i$ could belong to such a set $\mathscr{P}$.]

**31.** Write a computer program for calculating sums, differences, products, and quotients of complex numbers. The input and output parameters should be the corresponding real and imaginary parts.

**32.** The straightforward method of computing the product $(a + bi)(c + di) = (ac - bd) + i(bc + ad)$ requires four (real) multiplications (and two signed additions). On most computers multiplication is far more time consuming than addition. Devise an algorithm for computing $(a + bi)(c + di)$ with only three multiplications (at the cost of extra additions). [HINT: Start with $(a + b)(c + d)$.]

## 1.2 POINT REPRESENTATION OF COMPLEX NUMBERS; ABSOLUTE VALUE AND COMPLEX CONJUGATES

It is presumed that the reader is familiar with the Cartesian coordinate system (Fig. 1.1) which establishes a one-to-one correspondence between points in the $xy$-plane and ordered pairs of real numbers. The ordered pair $(-2, 3)$, for example, corresponds to that point $P$ that lies two units to the left of the $y$-axis and three units above the $x$-axis.

The Cartesian coordinate system suggests a convenient way to represent complex numbers as points in the $xy$-plane; namely, to each complex number $a + bi$ we associate that point in the $xy$-plane which has the coordinates $(a, b)$. The complex

---

[†] On computers this is, in fact, the method by which the statement $\alpha > \beta$ is tested.

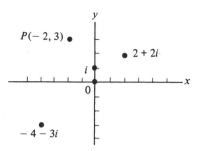

**Figure 1.1**  Cartesian coordinate system.

number $-2 + 3i$ is therefore represented by the point $P$ in Fig. 1.1. Also shown in Fig. 1.1 are the points that represent the complex numbers $0$, $i$, $2 + 2i$, and $-4 - 3i$.

When the $xy$-plane is used for the purpose of describing complex numbers it is referred to as the *complex plane* or *z-plane*.[†] Since each point on the $x$-axis represents a real number, this axis is called the *real axis*. Analogously, the $y$-axis is called the *imaginary axis* for it represents the pure imaginary numbers.

Hereafter, we shall refer to the point that represents the complex number $z$ as simply *the point* $z$; i.e., the point $z = a + bi$ is the point with coordinates $(a, b)$.

**Example 1**

Suppose that $n$ particles with masses $m_1, m_2, \ldots, m_n$ are located at the respective points $z_1, z_2, \ldots, z_n$ in the complex plane. Show that the center of mass of the system is the point

$$\hat{z} = \frac{m_1 z_1 + m_2 z_2 + \cdots + m_n z_n}{m_1 + m_2 + \cdots + m_n}.$$

**Solution.**  Write $z_1 = x_1 + y_1 i$, $z_2 = x_2 + y_2 i, \ldots, z_n = x_n + y_n i$, and let $M$ be the total mass $\sum_{k=1}^{n} m_k$. Presumably the reader will recall that the center of mass of the given system is the point with coordinates $(\hat{x}, \hat{y})$, where

$$\hat{x} = \frac{\sum_{k=1}^{n} m_k x_k}{M}, \qquad \hat{y} = \frac{\sum_{k=1}^{n} m_k y_k}{M}.$$

But clearly $\hat{x}$ and $\hat{y}$ are, respectively, the real and imaginary parts of the complex number $(\sum_{k=1}^{n} m_k z_k)/M = \hat{z}$.  ∎

*Absolute Value.*  By the Pythagorean theorem the distance from the point $z = a + bi$ to the origin is given by $\sqrt{a^2 + b^2}$. Special notation for this distance is given in

---

**Definition 3.**  The **absolute value** or **modulus** of the number $z = a + bi$ is denoted by $|z|$ and is given by

$$|z| := \sqrt{a^2 + b^2}.$$

---

[†] The representation of complex numbers in the plane was proposed (independently) by Caspar Wessel and Jean Pierre Argand around 1800. The term *Argand diagram* is sometimes used.

In particular,

$$|0| = 0, \qquad \left|\frac{i}{2}\right| = \frac{1}{2}, \qquad |3 - 4i| = \sqrt{9 + 16} = 5.$$

The reader should note that $|z|$ is always a nonnegative real number and that the *only* complex number whose modulus is zero is the number 0.

Let $z_1 = a_1 + b_1 i$ and $z_2 = a_2 + b_2 i$. Then

$$|z_1 - z_2| = |(a_1 - a_2) + (b_1 - b_2)i| = \sqrt{(a_1 - a_2)^2 + (b_1 - b_2)^2},$$

which is the distance between the points with coordinates $(a_1, b_1)$ and $(a_2, b_2)$ (see Fig. 1.2). Hence *the distance between the points $z_1$ and $z_2$ is given by $|z_1 - z_2|$*. This fact is useful in describing certain curves in the plane. Consider, for example, the set of all numbers $z$ that satisfy the equation

$$|z - z_0| = r, \tag{1}$$

where $z_0$ is a fixed complex number and $r$ is a fixed positive real number. This set consists of all points $z$ whose distance from $z_0$ is $r$. Consequently Eq. (1) is the equation of a circle.

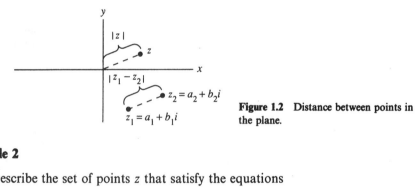

**Figure 1.2**   Distance between points in the plane.

## Example 2

Describe the set of points $z$ that satisfy the equations

**(a)** $|z + 2| = |z - 1|,$ **(b)** $|z - 1| = \text{Re } z + 1.$

**Solution.**   **(a)** A point $z$ satisfies Eq. (a) if and only if it is equidistant from the points $-2$ and 1. Hence Eq. (a) is the equation of the perpendicular bisector of the line segment joining $-2$ and 1; i.e., Eq. (a) describes the line $x = -\frac{1}{2}$ (see Fig. 1.3).

A more routine method for solving Eq. (a) is to set $z = x + iy$ in the equation and perform the algebra:

$$|z + 2| = |z - 1|,$$
$$|x + iy + 2| = |x + iy - 1|,$$
$$(x + 2)^2 + y^2 = (x - 1)^2 + y^2,$$
$$4x + 4 = -2x + 1,$$
$$x = -\tfrac{1}{2}.$$

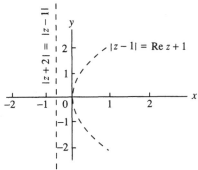

**Figure 1.3**  Graphs for Example 2.

**(b)** The geometric interpretation of Eq. (b) is less obvious, so we proceed directly with the mechanical approach and derive $\sqrt{(x-1)^2 + y^2} = x + 1$, or $y^2 = 4x$, which describes a parabola (see Fig. 1.3).  ■

*Complex Conjugates.*    The reflection of the point $z = a + bi$ in the real axis is the point $a - bi$ (see Fig. 1.4). As we shall see, the relationship between $a + bi$ and $a - bi$ will play a significant role in the theory of complex variables. We introduce special notation for this concept in the next definition.

---

**Definition 4.**   The **complex conjugate** of the number $z = a + bi$ is denoted by $\bar{z}$ and is given by

$$\bar{z} := a - bi.$$

---

Thus,

$$\overline{-1 + 5i} = -1 - 5i, \qquad \overline{\pi - i} = \pi + i, \qquad \overline{8} = 8.$$

It follows from Definition 4 that $z = \bar{z}$ if and only if $z$ is a real number. Also it is clear that the conjugate of the sum (difference) of two complex numbers is equal to the sum (difference) of their conjugates; i.e.,

$$\overline{z_1 \pm z_2} = \overline{z_1} \pm \overline{z_2}.$$

Perhaps not so obvious is the analogous property for multiplication.

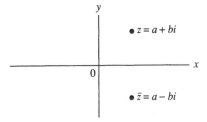

**Figure 1.4**  Complex conjugates.

### Example 3

Prove that the conjugate of the product of two complex numbers is equal to the product of the conjugates of these numbers.

**Solution.**   It is required to verify that

$$\overline{(z_1 z_2)} = \bar{z}_1 \bar{z}_2. \tag{2}$$

Write $z_1 = a_1 + b_1 i$, $z_2 = a_2 + b_2 i$. Then

$$\overline{(z_1 z_2)} = \overline{a_1 a_2 - b_1 b_2 + (a_1 b_2 + a_2 b_1)i}$$
$$= a_1 a_2 - b_1 b_2 - (a_1 b_2 + a_2 b_1)i.$$

On the other hand,

$$\bar{z}_1 \bar{z}_2 = (a_1 - b_1 i)(a_2 - b_2 i) = a_1 a_2 - b_1 b_2 - a_1 b_2 i - a_2 b_1 i$$
$$= a_1 a_2 - b_1 b_2 - (a_1 b_2 + a_2 b_1)i.$$

Thus Eq. (2) holds.   ■

There is another, possibly more enlightening, way to see Eq. (2). Notice that when we represent a complex number in terms of two real numbers and the symbol $i$, as in $z = a + bi$, then the action of conjugation is equivalent to changing the sign of the $i$ term. Now recall the role that $i$ plays in computations; it merely holds a place while we compute around it, replacing its square by $-1$ whenever it arises. Except for these occurrences $i$ is never really absorbed into the computations; we could just as well call it $j$, $\lambda$, $\sqrt{-1}$, or any other symbol whose square we agree to replace by $-1$. *In fact, without affecting the validity of the calculation, we could replace it throughout by the symbol* $(-i)$, *since the square of the latter is also* $-1$. Thus, for instance, if in the expression $(a_1 + b_1 i)(a_2 + b_2 i)$ we replace $i$ by $-i$ and then multiply, the only thing different about the product will be the appearance of $-i$ instead of $i$. But expressed in terms of conjugation, this is precisely the statement of Example 3.[†]

In addition to Eq. (2) the following properties can be seen:

$$\overline{\left(\frac{z_1}{z_2}\right)} = \frac{\bar{z}_1}{\bar{z}_2} \qquad (z_2 \neq 0); \tag{3}$$

$$\operatorname{Re} z = \frac{z + \bar{z}}{2}; \tag{4}$$

$$\operatorname{Im} z = \frac{z - \bar{z}}{2i}; \tag{5}$$

$$\bar{\bar{z}} = z. \tag{6}$$

Property (4) demonstrates that the sum of a complex number and its conjugate

---

[†] By the same token we should be able to replace $\sqrt{2}$ by $-\sqrt{2}$ in $(3 + 2\sqrt{2})(4 - 3\sqrt{2})$ *either before or after multiplying* and obtain the same result. (Try it.)

is real, whereas (5) shows that the difference is (pure) imaginary. Property (6) means that the conjugate of the conjugate of a complex number is the number itself.

It is clear from Definition 4 that

$$|z| = |\bar{z}|;$$

i.e., the points $z$ and $\bar{z}$ are equidistant from the origin. Furthermore, since

$$z\bar{z} = (a + bi)(a - bi) = a^2 + b^2,$$

we have

$$z\bar{z} = |z|^2. \tag{7}$$

This is a useful fact to remember: *The square of the modulus of a complex number equals the number times its conjugate.*

Observe that we have already employed complex conjugates in Sec. 1.1 in the process of rationalizing the denominator for the division algorithm. Thus, for instance, if $z_1$ and $z_2$ are complex numbers, then we can rewrite the ratio $z_1/z_2$ with a real denominator as

$$\frac{z_1}{z_2} = \frac{z_1 \bar{z}_2}{z_2 \bar{z}_2} = \frac{z_1 \bar{z}_2}{|z_2|^2}. \tag{8}$$

## EXERCISES 1.2

1. Show that the point $(z_1 + z_2)/2$ is the midpoint of the line segment joining $z_1$ and $z_2$.
2. Given four particles of masses 2, 1, 3, and 5 located at the respective points $1 + i$, $-3i$, $1 - 2i$, and $-6$, find the center of mass of this system.
3. Which of the points $i$, $2 - i$, and $-3$ is farthest from the origin?
4. Let $z = 3 - 2i$. Plot the points $z$, $-z$, $\bar{z}$, $-\bar{z}$, and $1/z$ in the complex plane.
5. Show that the points 1, $-1/2 + i\sqrt{3}/2$, and $-1/2 - i\sqrt{3}/2$ are the vertices of an equilateral triangle.
6. Show that the points $3 + i$, 6, and $4 + 4i$ are the vertices of a right triangle.
7. Describe the set of points $z$ in the complex plane that satisfy each of the following.
   (a) $\text{Im } z = -2$          (b) $|z - 1 + i| = 3$
   (c) $|2z - i| = 4$             (d) $|z - 1| = |z + i|$
   (e) $|z| = \text{Re } z + 2$      (f) $|z - 1| + |z + 1| = 7$
   (g) $|z| = 3|z - 1|$           (h) $\text{Re } z \geq 4$
   (i) $|z - i| < 2$             (j) $|z| > 6$
8. Show, both analytically and graphically, that $|z - 1| = |\bar{z} - 1|$.
9. Show that if $r$ is a nonnegative real number, then $|rz| = r|z|$.
10. Prove that $|\text{Re } z| \leq |z|$ and $|\text{Im } z| \leq |z|$.
11. Prove that if $|z| = \text{Re } z$, then $z$ is a nonnegative real number.
12. Verify properties (3), (4), and (5).
13. Prove that if $(\bar{z})^2 = z^2$, then $z$ is either real or pure imaginary.

**14.** Prove that $|z_1 z_2| = |z_1| |z_2|$. [HINT: Use Eqs. (7) and (2) to show that $|z_1 z_2|^2 = |z_1|^2 |z_2|^2$.]

**15.** Prove that $(\bar{z})^k = \overline{(z^k)}$ for every integer $k$ (provided $z \neq 0$ when $k$ is negative).

**16.** Prove that if $|z| = 1$ $(z \neq 1)$, then $\mathrm{Re}[1/(1 - z)] = \frac{1}{2}$.

**17.** Let $a_1, a_2, \ldots, a_n$ be real constants. Show that if $z_0$ is a root of the equation
$z^n + a_1 z^{n-1} + \cdots + a_n = 0$, then so is $\bar{z}_0$.

**18.** Use the familiar formula for the roots of a quadratic polynomial to give another proof of the statement in Prob. 17 for the case $n = 2$.

## 1.3 VECTORS AND POLAR FORMS

With each point $z$ in the complex plane we can associate a *vector*, namely, the directed line segment from the origin to the point $z$. Recall that vectors are characterized by length and direction, and that a given vector remains unchanged under translation. Thus the vector determined by $z = 1 + i$ is the same as the vector from the point $2 + i$ to the point $3 + 2i$ (see Fig. 1.5). Note that every vector parallel to the real axis corresponds to a real number, while those parallel to the imaginary axis represent pure imaginary numbers. Observe, also, that the length of the vector associated with $z$ is $|z|$.

Let $\mathbf{v}_1$ and $\mathbf{v}_2$ denote the vectors determined by the points $z_1$ and $z_2$, respectively. The vector sum $\mathbf{v} = \mathbf{v}_1 + \mathbf{v}_2$ is given by the parallelogram law, which is illustrated in Fig. 1.6. If $z_1 = x_1 + iy_1$ and $z_2 = x_2 + iy_2$, then the terminal point of the vector $\mathbf{v}$ in Fig. 1.6 has the coordinates $(x_1 + x_2, y_1 + y_2)$; i.e., it corresponds to the point $z_1 + z_2$. Thus we see that *the correspondence between complex numbers and planar vectors carries over to the operation of addition.*

Hereafter, the vector determined by the point $z$ will be simply called *the vector $z$.*

Recall the geometric fact that the length of any side of a triangle is less than or equal to the sum of the lengths of the other two sides. If we apply this theorem to the triangle in Fig. 1.6 with vertices 0, $z_1$, and $z_1 + z_2$, we deduce a very important law relating the magnitudes of complex numbers and their sum:

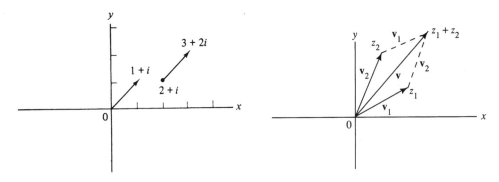

**Figure 1.5**  Complex numbers as vectors.

**Figure 1.6**  Vector addition.

---

**The Triangle Inequality.**  For any two complex numbers $z_1$ and $z_2$ we have

$$|z_1 + z_2| \le |z_1| + |z_2|. \qquad (1)$$

---

The triangle inequality can easily be extended to more than two complex numbers, as requested in Prob. 22.

The vector $z_2 - z_1$, when added to the vector $z_1$, obviously yields the vector $z_2$. Thus $z_2 - z_1$ can be represented as the directed line segment from $z_1$ to $z_2$ (see Fig. 1.7). Applying the geometric theorem to the triangle in Fig. 1.7, we deduce another form of the triangle inequality:

$$|z_2| \le |z_1| + |z_2 - z_1|$$

or

$$|z_2| - |z_1| \le |z_2 - z_1|. \qquad (2)$$

Inequality (2) states that the difference in the lengths of any two sides of a triangle is no greater than the length of the third side.

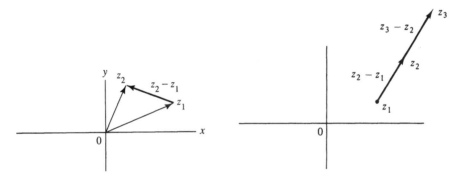

**Figure 1.7**  Vector subtraction.     **Figure 1.8**  Collinear points.

**Example 1**

Prove that the three distinct points $z_1$, $z_2$, and $z_3$ lie on the same straight line if and only if $z_3 - z_2 = c(z_2 - z_1)$ for some real number $c$.

**Solution.**  Recall that two vectors are parallel if and only if one is a (real) scalar multiple of the other. In the language of complex numbers, this says that $z$ is parallel to $w$ if and only if $z = cw$, where $c$ is real. From Fig. 1.8 we see that the condition that the points $z_1$, $z_2$, and $z_3$ be collinear is equivalent to the statement that the vector $z_3 - z_2$ is parallel to the vector $z_2 - z_1$. Using our characterization of parallelism, the conclusion follows immediately.  ∎

There is another set of parameters that characterize the vector from the origin to the point $z$ (other, that is, than the real and imaginary parts of $z$) which more intimately reflects its interpretation as an object with magnitude and direction. These

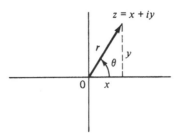

**Figure 1.9**   Polar coordinates.

are the polar coordinates, $r$ and $\theta$, of the point $z$. The coordinate $r$ is the distance from the origin to $z$, and $\theta$ is the angle of inclination of the vector $z$, measured positively in a counterclockwise sense from the positive real axis (and thus negative when measured clockwise) (see Fig. 1.9). *We shall always measure angles in radians in this book*; the use of degree measure is fine for visualization purposes, but it becomes quite treacherous in any discipline where calculus is involved. Notice that $r$ is the modulus, or absolute value, of $z$ and is never negative: $r = |z|$.

From Fig. 1.9 we readily derive the equations expressing the rectangular (or *Cartesian*) coordinates $(x, y)$ in terms of the polar coordinates $(r, \theta)$:

$$x = r \cos \theta, \qquad y = r \sin \theta. \tag{3}$$

On the other hand, the expressions for $(r, \theta)$ in terms of $(x, y)$ contain some minor but troublesome complications. Indeed the coordinate $r$ is given, unambiguously, by

$$r = \sqrt{x^2 + y^2} = |z|. \tag{4}$$

However, observe that although it is certainly true that $\tan \theta = y/x$, the natural conclusion

$$\theta = \tan^{-1}\left(\frac{y}{x}\right)$$

is *invalid* for points $z$ in the second and third quadrants (since the standard interpretation of the arctangent function places its range in the first and fourth quadrants). One *can* state that the angle is uniquely determined by the pair of equations

$$\cos \theta = \frac{x}{|z|}, \qquad \sin \theta = \frac{y}{|z|}, \tag{5}$$

but in practice we usually compute $\tan^{-1}(y/x)$ and adjust for the quadrant problem by adding or subtracting $\pi$ (radians) when appropriate (see Prob. 14).

The nuisance aspects of $\theta$ do not end here, however. Even using Eqs. (5) one can, because of its identification as an angle, determine $\theta$ only up to an integer multiple of $2\pi$. To accommodate this feature we shall call the value of any of these angles an *argument*, or *phase*, of $z$, denoted

$$\arg z.$$

Thus if $\theta_0$ qualifies as a value of arg $z$, then so do

$$\theta_0 \pm 2\pi, \; \theta_0 \pm 4\pi, \; \theta_0 \pm 6\pi, \ldots,$$

and every value of arg $z$ must be one of these.[†] In particular, the values of arg $i$ are

$$\frac{\pi}{2}, \frac{\pi}{2} \pm 2\pi, \frac{\pi}{2} \pm 4\pi, \ldots$$

and we write

$$\arg i = \frac{\pi}{2} + 2k\pi \qquad (k = 0, \pm 1, \pm 2, \ldots).$$

It is convenient to have a notation for some *definite* value of arg $z$. Notice that any half-open interval of length $2\pi$ will contain one and only one value of the argument. By specifying such an interval we say that we have selected a particular *branch* of arg $z$. Figure 1.10 on page 16 illustrates three possible branch selections. The first diagram (Fig. 1.10(a)) depicts the branch that selects the value of arg $z$ from the interval $(-\pi, \pi]$; it is known as the *principal value of the argument* and is denoted Arg $z$ (with capital A). The principal value is most commonly used in complex arithmetic computer codes; it is inherently discontinuous, jumping by $2\pi$ as $z$ crosses the negative real axis. This line of discontinuities is known as the *branch cut*.

Of course, *any* branch of arg $z$ must have a jump of $2\pi$ somewhere. The branch depicted in Fig. 1.10(b) is discontinuous on the *positive* real axis, taking values from the interval $(0, 2\pi]$. The branch in Fig. 1.10(c) has the same branch cut but selects values from the interval $(2\pi, 4\pi]$.

The notation $\arg_\tau z$ is used for the branch of arg $z$ taking values from the interval $(\tau, \tau + 2\pi]$. Thus $\arg_{-\pi} z$ is the principal value Arg $z$, and the branches depicted in Fig. 1.10(b) and 1.10(c), respectively, are $\arg_0 z$ and $\arg_{2\pi} z$. Note that arg $0$ cannot be sensibly defined for any branch.

With these conventions in hand, one can now write $z = x + iy$ in the *polar form* [Recall Eq. (3)]

$$z = x + iy = r(\cos \theta + i \sin \theta) = r \operatorname{cis} \theta, \tag{6}$$

where we abbreviate the "cosine plus $i$ sine" operator as cis.

**Example 2**

Find $\arg(1 + \sqrt{3}i)$ and write $1 + \sqrt{3}i$ in polar form.

**Solution.**   Note that $r = |1 + \sqrt{3}i| = 2$ and that the equations $\cos \theta = \frac{1}{2}$, $\sin \theta = \sqrt{3}/2$ are satisfied by $\theta = \pi/3$. Hence $\arg(1 + \sqrt{3}i) = \pi/3 + 2k\pi$, $k = 0, \pm 1, \pm 2, \ldots$ [in particular, $\operatorname{Arg}(1 + \sqrt{3}i) = \pi/3$]. The polar form of $1 + \sqrt{3}i$ is $2(\cos \pi/3 + i \sin \pi/3) = 2 \operatorname{cis} \pi/3$.  ■

In many circumstances one of the forms $x + iy$ or $r \operatorname{cis} \theta$ may be more suitable than the other. The rectangular form, for example, is very convenient for addition or subtraction, whereas the polar form can be a monstrosity (see Prob. 21). On the

---

[†] An alternative way to express arg $z$ is to write it as the *set*

$$\arg z = \{\theta_0 + 2k\pi : k = 0, \pm 1, \pm 2, \ldots\}.$$

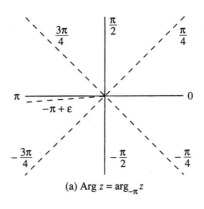

(a) Arg $z = \arg_{-\pi} z$

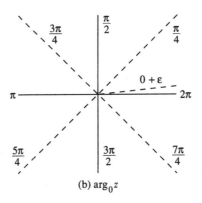

(b) $\arg_0 z$

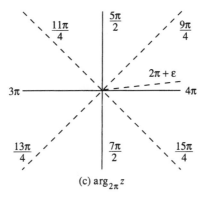

(c) $\arg_{2\pi} z$

**Figure 1.10** Branches of $\arg z$.

other hand, the polar form lends a very interesting geometric interpretation to the process of multiplication. If we let

$$z_1 = r_1(\cos\theta_1 + i\sin\theta_1), \qquad z_2 = r_2(\cos\theta_2 + i\sin\theta_2),$$

then we compute

$$z_1 z_2 = r_1 r_2[(\cos\theta_1\cos\theta_2 - \sin\theta_1\sin\theta_2) + i(\sin\theta_1\cos\theta_2 + \cos\theta_1\sin\theta_2)],$$

and so

$$z_1 z_2 = r_1 r_2[\cos(\theta_1 + \theta_2) + i\sin(\theta_1 + \theta_2)]. \tag{7}$$

The abbreviated version of (7) reads as follows:

$$z_1 z_2 = (r_1 \operatorname{cis}\theta_1)(r_2 \operatorname{cis}\theta_2) = (r_1 r_2)\operatorname{cis}(\theta_1 + \theta_2)$$

and we see that

*The modulus of the product is the product of the moduli:*

$$|z_1 z_2| = |z_1||z_2| \quad (= r_1 r_2); \tag{8}$$

*The argument of the product is the sum of the arguments:*

$$\arg z_1 z_2 = \arg z_1 + \arg z_2. \tag{9}$$

(To be precise, the ambiguous Eq. (9) is to be interpreted as saying that if particular values are assigned to any pair of terms therein, then one can find a value for the third term that satisfies the identity.)

Geometrically, the vector $z_1 z_2$ has length equal to the product of the lengths of the vectors $z_1$ and $z_2$ and has angle equal to the sum of the angles of the vectors $z_1$ and $z_2$ (see Fig. 1.11). For instance, since the vector $i$ has length 1 and angle $\pi/2$, it follows that the vector $iz$ can be obtained by rotating the vector $z$ through a right angle in the counterclockwise direction.

Observing that division is the inverse operation to multiplication, we are led to the following equations:

$$\frac{z_1}{z_2} = \frac{r_1}{r_2}[\cos(\theta_1 - \theta_2) + i\sin(\theta_1 - \theta_2)] = \frac{r_1}{r_2}\operatorname{cis}(\theta_1 - \theta_2), \tag{10}$$

$$\arg\left(\frac{z_1}{z_2}\right) = \arg z_1 - \arg z_2, \tag{11}$$

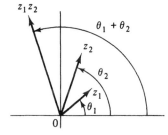

**Figure 1.11**   Geometric interpretation of the product.

and

$$\left|\frac{z_1}{z_2}\right| = \frac{|z_1|}{|z_2|}. \tag{12}$$

Equation (10) can be proved in a manner similar to Eq. (7), and Eqs. (11) and (12) follow immediately. Geometrically, the vector $z_1/z_2$ has length equal to the quotient of the lengths of the vectors $z_1$ and $z_2$ and has angle equal to the difference of the angles of the vectors $z_1$ and $z_2$.

**Example 3**

Write the quotient $(1 + i)/(\sqrt{3} - i)$ in polar form.

**Solution.** The polar forms for $(1 + i)$ and $(\sqrt{3} - i)$ are

$$1 + i = |1 + i| \operatorname{cis}(\arg(1 + i)) = \sqrt{2} \operatorname{cis}(\pi/4),$$
$$\sqrt{3} - i = 2 \operatorname{cis}(-\pi/6).$$

Hence, from Eq. (10), we have

$$\frac{1 + i}{\sqrt{3} - i} = \frac{\sqrt{2}}{2} \operatorname{cis}\left[\frac{\pi}{4} - \left(-\frac{\pi}{6}\right)\right] = \frac{\sqrt{2}}{2} \operatorname{cis}\frac{5\pi}{12}. \quad \blacksquare$$

**Example 4**

Prove that the line $l$ through the points $z_1$ and $z_2$ is perpendicular to the line $L$ through the points $z_3$ and $z_4$ if and only if

$$\arg\frac{z_1 - z_2}{z_3 - z_4} = \frac{\pi}{2} + 2k\pi \qquad (k = 0, \pm 1, \pm 2, \dots) \tag{13}$$

or

$$\arg\frac{z_1 - z_2}{z_3 - z_4} = \frac{3\pi}{2} + 2k\pi \qquad (k = 0, \pm 1, \pm 2, \dots). \tag{14}$$

**Solution.** Note that the lines $l$ and $L$ are perpendicular if and only if the vectors $z_1 - z_2$ and $z_3 - z_4$ are perpendicular. Let $\theta_0$ be the particular value of $\arg[(z_1 - z_2)/(z_3 - z_4)]$ that satisfies $0 \le \theta_0 < 2\pi$. Since

$$\arg\frac{z_1 - z_2}{z_3 - z_4} = \arg(z_1 - z_2) - \arg(z_3 - z_4),$$

we deduce that $\theta_0$ is the positive angle with initial side $z_3 - z_4$ and terminal side $z_1 - z_2$ (see Fig. 1.12). Thus the lines $l$ and $L$ are perpendicular if and only if $\theta_0 = \pi/2$ or $\theta_0 = 3\pi/2$, i.e., if and only if Eq. (13) or (14) holds. $\quad \blacksquare$

Recall that, geometrically, the vector $\bar{z}$ is the reflection in the real axis of the vector $z$ (see Fig. 1.13). Hence we see that *the argument of the conjugate of a complex number is the negative of the argument of the number*; i.e.,

$$\arg \bar{z} = -\arg z. \tag{15}$$

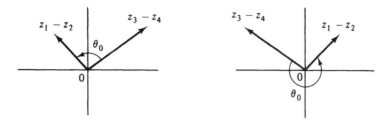

**Figure 1.12**  Perpendicular vectors.

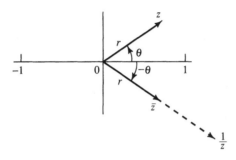

**Figure 1.13**  The argument of the conjugate and the reciprocal.

In fact, as a special case of Eq. (11) we also have

$$\arg \frac{1}{z} = -\arg z.$$

Thus $\bar{z}$ and $z^{-1}$ have the same argument and represent parallel vectors (see Fig. 1.13).

## EXERCISES 1.3

1. Let $z_1 = 2 - i$ and $z_2 = 1 + i$. Use the parallelogram law to construct each of the following vectors.
   **(a)** $z_1 + z_2$  **(b)** $z_1 - z_2$  **(c)** $2z_1 - 3z_2$
2. Show that $|z_1 z_2 z_3| = |z_1| |z_2| |z_3|$.
3. Translate the following geometric theorem into the language of complex numbers: The sum of the squares of the lengths of the diagonals of a parallelogram is equal to the sum of the squares of its sides. (See Fig. 1.6.)
4. Show that for any integer $k$, $|z^k| = |z|^k$ (provided $z \neq 0$ when $k$ is negative).
5. Find the following.
   **(a)** $\left| \dfrac{1 + 2i}{-2 - i} \right|$  **(b)** $|(\overline{1 + i})(2 - 3i)(4i - 3)|$
   **(c)** $\left| \dfrac{i(2 + i)^3}{(1 - i)^2} \right|$  **(d)** $\left| \dfrac{(\pi + i)^{100}}{(\pi - i)^{100}} \right|$
6. Draw each of the following vectors.
   **(a)** $7 \operatorname{cis}\left(\dfrac{3\pi}{4}\right)$  **(b)** $4 \operatorname{cis}\left(\dfrac{-\pi}{6}\right)$

**(c)** $\text{cis}\left(\dfrac{3\pi}{4}\right)$

**(d)** $3\,\text{cis}\left(\dfrac{27\pi}{4}\right)$

7. Find the argument of each of the following complex numbers and write each in polar form.
  **(a)** $-1/2$
  **(b)** $-3 + 3i$
  **(c)** $-\pi i$
  **(d)** $-2\sqrt{3} - 2i$
  **(e)** $(1 - i)(-\sqrt{3} + i)$
  **(f)** $(\sqrt{3} - i)^2$
  **(g)** $\dfrac{-1 + \sqrt{3}i}{2 + 2i}$
  **(h)** $\dfrac{-\sqrt{7}(1 + i)}{\sqrt{3} + i}$

8. Show geometrically that the nonzero complex numbers $z_1$ and $z_2$ satisfy $|z_1 + z_2| = |z_1| + |z_2|$ if and only if they have the same argument.

9. Given the vector $z$, interpret geometrically the vector $(\cos \phi + i \sin \phi)z$.

10. Show that $\arg z_1 z_2 z_3 = \arg z_1 + \arg z_2 + \arg z_3$.

11. Show that $\arg z_1 \overline{z_2} = \arg z_1 - \arg z_2$.

12. Find the following.
  **(a)** $\text{Arg}(-6 - 6i)$
  **(b)** $\text{Arg}(-\pi)$
  **(c)** $\text{Arg}(10i)$
  **(d)** $\text{Arg}(\sqrt{3} - i)$

13. Decide which of the following statements are always true.
  **(a)** $\text{Arg } z_1 z_2 = \text{Arg } z_1 + \text{Arg } z_2$ if $z_1 \neq 0$, $z_2 \neq 0$.
  **(b)** $\text{Arg } \bar{z} = -\text{Arg } z$ if $z$ not a real number.
  **(c)** $\text{Arg}(z_1/z_2) = \text{Arg } z_1 - \text{Arg } z_2$ if $z_1 \neq 0$, $z_2 \neq 0$.
  **(d)** $\arg z = \text{Arg } z + 2\pi k$, $k = 0, \pm 1, \pm 2, \dots$, if $z \neq 0$.

14. Show that a correct formula for $\arg z$ can be computed using the form

$$\arg z = \begin{cases} \tan^{-1}\left(\dfrac{y}{x}\right) + \left(\dfrac{\pi}{2}\right)[1 - \text{sgn}(x)] & \text{if } x \neq 0, \\[2mm] \left(\dfrac{\pi}{2}\right)\text{sgn}(y) & \text{if } x = 0 \text{ and } y \neq 0, \\[2mm] \text{undefined} & \text{if } x = y = 0, \end{cases}$$

where the BASIC "signum" function is specified by

$$\text{sgn}(\eta) := \begin{cases} +1 & \text{if } \eta > 0, \\ 0 & \text{if } \eta = 0, \\ -1 & \text{if } \eta < 0. \end{cases}$$

Show also that the expression $\text{sgn}(y) \cos^{-1}(x/\sqrt{x^2 + y^2})$ equals $\text{Arg } z$ at its points of continuity.

15. Prove that $|z_1 - z_2| \leq |z_1| + |z_2|$.

16. Prove that $||z_1| - |z_2|| \leq |z_1 - z_2|$.

17. Show that the vector $z_1$ is parallel to the vector $z_2$ if and only if $\text{Im}(z_1 \bar{z}_2) = 0$.

18. Show that every point $z$ on the line through the distinct points $z_1$ and $z_2$ is of the form $z = z_1 + c(z_2 - z_1)$, where $c$ is a real number. What can be said about the value of $c$ if $z$ also lies strictly between $z_1$ and $z_2$?

19. Prove that $\arg z_1 = \arg z_2$ if and only if $z_1 = cz_2$, where $c$ is a positive real number.

20. Let $z_1, z_2$, and $z_3$ be distinct points and let $\phi$ be a particular value of $\arg[(z_3 - z_1)/(z_2 - z_1)]$. Prove that

$$|z_3 - z_2|^2 = |z_3 - z_1|^2 + |z_2 - z_1|^2 - 2|z_3 - z_1||z_2 - z_1| \cos \phi.$$

[HINT: Consider the triangle with vertices $z_1, z_2, z_3$.]

**21.** If $r \text{ cis } \theta = r_1 \text{ cis } \theta_1 + r_2 \text{ cis } \theta_2$, determine $r$ and $\theta$ in terms of $r_1$, $r_2$, $\theta_1$, and $\theta_2$. Check your answer by applying the law of cosines.

**22.** Use mathematical induction to prove the *generalized triangle inequality*:

$$\left| \sum_{k=1}^{n} z_k \right| \leq \sum_{k=1}^{n} |z_k|.$$

**23.** Let $m_1$, $m_2$, and $m_3$ be three positive real numbers and let $z_1$, $z_2$, and $z_3$ be three complex numbers, each of modulus less than or equal to 1. Use the generalized triangle inequality (Prob. 22) to prove that

$$\left| \frac{m_1 z_1 + m_2 z_2 + m_3 z_3}{m_1 + m_2 + m_3} \right| \leq 1,$$

and give a physical interpretation of the inequality.

**24.** Write computer programs for converting between rectangular and polar coordinates (using the principal value of the argument).

## 1.4 THE COMPLEX EXPONENTIAL

The familiar exponential function $f(x) = e^x$ has a natural and extremely useful extension to the complex plane. Indeed the complex function $e^z$ provides a basic tool for the application of complex variables to electrical circuits, control systems, wave propagation, and time-invariant physical systems in general.

To find a suitable definition for $e^z$ when $z = x + iy$, we want to preserve the basic identities satisfied by the real function $e^x$. So first of all we postulate that the multiplicative property should persist:

$$e^{z_1} e^{z_2} = e^{z_1 + z_2}. \tag{1}$$

This simplifies matters considerably, since Eq. (1) enables the decomposition

$$e^z = e^{x+iy} = e^x e^{iy} \tag{2}$$

and we see that to define $e^z$, we need only specify $e^{iy}$ (in other words we will be able to exponentiate complex numbers once we discover how to exponentiate imaginary ones).

Next we propose that the differentiation law

$$\frac{de^z}{dz} = e^z \tag{3}$$

be preserved. Differentiation with respect to a complex variable $z = x + iy$ is a very profound and, at this stage, ambiguous operation; indeed Chapter 2 is devoted to a painstaking study of this concept (and the rest of the book is dedicated to exploring its consequences). But, thanks to the factorization displayed in Eq. (2) we need only consider (for the moment) a special case of Eq. (3)—namely,

$$\frac{de^{iy}}{d(iy)} = e^{iy}$$

or, equivalently (by the chain rule),

$$\frac{de^{iy}}{dy} = ie^{iy}. \tag{4}$$

The consequences of postulating Eq. (4) become more apparent if we differentiate again:

$$\frac{d^2e^{iy}}{dy^2} = \frac{d}{dy}(ie^{iy}) = i^2e^{iy} = -e^{iy};$$

in other words, the function $g(y) := e^{iy}$ satisfies the differential equation

$$\frac{d^2g}{dy^2} = -g. \tag{5}$$

Now observe that any function of the form

$$A \cos y + B \sin y \qquad (A, B \text{ constants})$$

satisfies Eq. (5). In fact, from the theory of differential equations it is known that every solution of Eq. (5) must have this form. Hence we can write

$$g(y) = A \cos y + B \sin y. \tag{6}$$

To evaluate $A$ and $B$ we use the conditions that

$$g(0) = e^{i0} = e^0 = 1 = A \cos 0 + B \sin 0$$

and

$$\frac{dg}{dy}(0) = ig(0) = i = -A \sin 0 + B \cos 0.$$

Thus $A = 1$ and $B = i$, leading us to the identification

$$\boxed{e^{iy} = \cos y + i \sin y.} \tag{7}$$

Equation (7) is known as *Euler's equation*. Combining Eqs. (7) and (2) we formulate the following.

---

**Definition 5.**  If $z = x + iy$, then $e^z$ is defined to be the complex number

$$e^z := e^x(\cos y + i \sin y). \tag{8}$$

---

It is not difficult to verify directly that $e^z$, as defined above, satisfies the usual algebraic properties of the exponential function—in particular, the multiplicative identity (1) and the associated division rule

$$\frac{e^{z_1}}{e^{z_2}} = e^{z_1 - z_2} \tag{9}$$

(see Prob. 15). In Sec. 2.4 we will obtain further confirmation that we have made the "right choice" by showing that Definition 5 produces a function that has the extremely desirable property of *analyticity*. Another confirmation is exhibited in the following example.

**Example 1**

Show that Euler's equation is formally consistent with the usual Taylor series expansions

$$e^x = 1 + x + \frac{x^2}{2!} + \frac{x^3}{3!} + \frac{x^4}{4!} + \frac{x^5}{5!} + \cdots,$$

$$\cos x = 1 - \frac{x^2}{2!} + \frac{x^4}{4!} - \cdots,$$

$$\sin x = x - \frac{x^3}{3!} + \frac{x^5}{5!} - \cdots.$$

**Solution.** We shall study series representations of complex functions in full detail in Chapter 5. For now we ignore questions of convergence, etc., and simply substitute $x = iy$ into the exponential series:

$$e^{iy} = 1 + iy + \frac{(iy)^2}{2!} + \frac{(iy)^3}{3!} + \frac{(iy)^4}{4!} + \frac{(iy)^5}{5!} + \cdots$$

$$= \left(1 - \frac{y^2}{2!} + \frac{y^4}{4!} - \cdots\right) + i\left(y - \frac{y^3}{3!} + \frac{y^5}{5!} - \cdots\right)$$

$$= \cos y + i \sin y. \quad \blacksquare$$

Euler's equation (7) enables us to write the polar form (Sec. 1.3) of a complex number as

$$z = r \text{ cis } \theta = r(\cos \theta + i \sin \theta) = re^{i\theta}.$$

Thus we can (and do) drop the awkward "cis" artifice and use, as the standard polar representation,

$$z = re^{i\theta} = |z|e^{i \arg z}. \tag{10}$$

In particular, notice the following identities:

$$e^{i0} = e^{2\pi i} = e^{-2\pi i} = e^{4\pi i} = e^{-4\pi i} = \cdots = 1,$$

$$\boxed{e^{\pi i} = -1,}^{\dagger}$$

$$e^{(\pi/2)i} = i, \qquad e^{(-\pi/2)i} = -i.$$

---

† Students of mathematics have often marveled at this identity. The constant $e$ comes from calculus, $\pi$ comes from geometry, and $i$ comes from algebra—and the combination $e^{\pi i}$ gives $-1$, the basic unit for generating the arithmetic system from the counting numbers (cardinals)!

Observe also that $|e^{i \arg z}| = 1$ and that Euler's equation leads to the following representations of the customary trigonometric functions:

$$\cos \theta = \operatorname{Re} e^{i\theta} = \frac{e^{i\theta} + e^{-i\theta}}{2}, \tag{11}$$

$$\sin \theta = \operatorname{Im} e^{i\theta} = \frac{e^{i\theta} - e^{-i\theta}}{2i}. \tag{12}$$

The rules derived in Sec. 1.3 for multiplying and dividing complex numbers in polar form now find very natural expressions:

$$z_1 z_2 = (r_1 e^{i\theta_1})(r_2 e^{i\theta_2}) = (r_1 r_2) e^{i(\theta_1 + \theta_2)}, \tag{13}$$

$$\frac{z_1}{z_2} = \frac{r_1 e^{i\theta_1}}{r_2 e^{i\theta_2}} = \left(\frac{r_1}{r_2}\right) e^{i(\theta_1 - \theta_2)}, \tag{14}$$

and complex conjugation of $z = re^{i\theta}$ is accomplished by changing the sign of $i$ in the exponent:

$$\bar{z} = re^{-i\theta}. \tag{15}$$

### Example 2

Compute (a) $(1 + i)/(\sqrt{3} - i)$ and (b) $(1 + i)^{24}$.

**Solution.** (a) This quotient was evaluated using the cis operator in Example 3 of Sec. 1.3; using the exponential the calculations take the form

$$1 + i = \sqrt{2}\,\operatorname{cis}(\pi/4) = \sqrt{2}e^{i\pi/4}, \qquad \sqrt{3} - i = 2\,\operatorname{cis}(-\pi/6) = 2e^{-i\pi/6},$$

and, therefore,

$$\frac{1 + i}{\sqrt{3} - i} = \frac{\sqrt{2}e^{i\pi/4}}{2e^{-i\pi/6}} = \frac{\sqrt{2}}{2}\,e^{i5\pi/12}.$$

(b) The exponential forms become

$$(1 + i)^{24} = (\sqrt{2}e^{i\pi/4})^{24} = (\sqrt{2})^{24}e^{i24\pi/4} = 2^{12}e^{i6\pi} = 2^{12}. \quad\blacksquare$$

In the solution to part (b) we glossed over the justification for the identity $(e^{i\pi/4})^{24} = e^{i24\pi/4}$. Actually, a careful scrutiny yields much more—a powerful formula involving trigonometric functions which we describe in the next example.

### Example 3

Prove *De Moivre's formula:*

$$\boxed{(\cos\theta + i\sin\theta)^n = \cos n\theta + i\sin n\theta, \qquad n = 1, 2, 3, \ldots .} \tag{16}$$

**Solution.**  By the multiplicative property, Eq. (1),

$$(e^{i\theta})^n = \underbrace{e^{i\theta}e^{i\theta}\cdots e^{i\theta}}_{(n \text{ times})} = e^{i\theta + i\theta + \cdots + i\theta} = e^{in\theta}.$$

Now applying Euler's formula (7) to the first and last members of this equation string, we deduce (16).  ∎

De Moivre's formula can be a convenient tool for deducing multiple-angle trigonometric identities, as is illustrated by the following example. (See also Probs. 12 and 20.)

**Example 4**

Express $\cos 3\theta$ in terms of $\cos \theta$ and $\sin \theta$.

**Solution.**  By Eq. (16) (with $n = 3$) we have

$$\cos 3\theta = \text{Re}(\cos 3\theta + i \sin 3\theta) = \text{Re}(\cos \theta + i \sin \theta)^3. \qquad (17)$$

According to the binomial formula,

$$(a + b)^3 = a^3 + 3a^2b + 3ab^2 + b^3.$$

Thus, making the obvious identifications $a = \cos \theta$, $b = i \sin \theta$ in Eq. (17), we deduce

$$\cos 3\theta = \text{Re}[\cos^3 \theta + 3 \cos^2 \theta (i \sin \theta) + 3 \cos \theta (-\sin^2 \theta) - i \sin^3 \theta]$$

$$= \cos^3 \theta - 3 \cos \theta \sin^2 \theta.  \quad \blacksquare$$

## EXERCISES 1.4

*In Problems 1 and 2 write each of the given numbers in the form $a + bi$.*

1. (a) $e^{-i\pi/4}$ 

   (b) $\dfrac{e^{1+i3\pi}}{e^{-1+i\pi/2}}$ 

   (c) $e^{e^i}$

2. (a) $\dfrac{e^{3i} - e^{-3i}}{2i}$ 

   (b) $2e^{3+i\pi/6}$ 

   (c) $e^z$, where $z = 4e^{i\pi/3}$

*In Problems 3 and 4 write each of the given numbers in the polar form $re^{i\theta}$.*

3. (a) $\dfrac{1-i}{3}$ 

   (b) $-8\pi(1 + \sqrt{3}i)$ 

   (c) $(1 + i)^6$

4. (a) $\left(\cos \dfrac{2\pi}{9} + i \sin \dfrac{2\pi}{9}\right)^3$ 

   (b) $\dfrac{2 + 2i}{-\sqrt{3} + i}$ 

   (c) $\dfrac{2i}{3e^{4+i}}$

5. Show that $|e^{x+iy}| = e^x$ and $\arg e^{x+iy} = y + 2k\pi$ $(k = 0, \pm1, \pm2, \ldots)$.

6. Show that, for real $\theta$,

   (a) $\tan \theta = \dfrac{e^{i\theta} - e^{-i\theta}}{i(e^{i\theta} + e^{-i\theta})}$

   (b) $\csc \theta = \dfrac{2}{e^{i(\theta - \pi/2)} - e^{-i(\theta + \pi/2)}}$

7. Show that $e^z = e^{z+2\pi i}$ for all $z$. (*The exponential function is periodic with period* $2\pi i$.)

8. Show that, for all $z$,

   (a) $e^{z+\pi i} = -e^z$

   (b) $\overline{e^z} = e^{\bar{z}}$

9. Show that $(e^z)^n = e^{nz}$ for any integer $n$.

10. Show that $|e^z| \leq 1$ if $\operatorname{Re} z \leq 0$.

11. Determine which of the following properties of the real exponential function remain true for the complex exponential function (i.e., for $x$ replaced by $z$).

    (a) $e^x$ is never zero.

    (b) $e^x$ is a one-to-one function.

    (c) $e^x$ is defined for all $x$.

    (d) $e^{-x} = 1/e^x$.

12. Use De Moivre's formula together with the binomial formula to derive the following identities.

    (a) $\sin 3\theta = 3 \cos^2 \theta \sin \theta - \sin^3 \theta$

    (b) $\sin 4\theta = 4 \cos^3 \theta \sin \theta - 4 \cos \theta \sin^3 \theta$

13. Show how the following trigonometric identities follow from Eqs. (11) and (12).

    (a) $\sin^2 \theta + \cos^2 \theta = 1$

    (b) $\cos(\theta_1 + \theta_2) = \cos \theta_1 \cos \theta_2 - \sin \theta_1 \sin \theta_2$

14. Does De Moivre's formula hold for negative integers $n$?

15. (a) Show that the multiplicative law (1) follows from Definition 5.

    (b) Show that the division rule (9) follows from Definition 5.

16. Let $z$ $(\neq 0)$ have the polar representation $z = re^{i\theta}$. Show that $\exp(\ln r + i\theta) = z$.[†]

17. Show that the function $z(t) = e^{it}$, $0 \leq t \leq 2\pi$, describes the unit circle $|z| = 1$ traversed in the counterclockwise direction (as $t$ increases from 0 to $2\pi$). Then describe each of the following curves.

    (a) $z(t) = 3e^{it}$,    $0 \leq t \leq 2\pi$

    (b) $z(t) = 2e^{it} + i$,    $0 \leq t \leq 2\pi$

    (c) $z(t) = 2e^{i2\pi t}$,    $0 \leq t \leq \frac{1}{2}$

    (d) $z(t) = 3e^{-it} + 2 - i$,    $0 \leq t \leq 2\pi$

18. Sketch the curves that are given for $0 \leq t \leq 2\pi$ by

    (a) $z(t) = e^{(1+i)t}$

    (b) $z(t) = e^{(1-i)t}$

    (c) $z(t) = e^{(-1+i)t}$

    (d) $z(t) = e^{(-1-i)t}$

19. Let $n$ be a positive integer greater than 2. Show that the points $e^{2\pi ik/n}$, $k = 0, 1, \ldots, n - 1$, form the vertices of a regular polygon.

20. Prove that if $z \neq 1$, then

$$1 + z + z^2 + \cdots + z^n = \frac{z^{n+1} - 1}{z - 1}.$$

Use this result and De Moivre's formula to establish the following identities.

   (a) $1 + \cos \theta + \cos 2\theta + \cdots + \cos n\theta = \dfrac{1}{2} + \dfrac{\sin[(n + \frac{1}{2})\theta]}{2 \sin\left(\dfrac{\theta}{2}\right)}$

---

[†] As a convenience in printing we sometimes write $\exp(z)$ instead of $e^z$.

**(b)** $\sin \theta + \sin 2\theta + \cdots + \sin n\theta = \dfrac{\sin\left(\dfrac{n\theta}{2}\right) \sin\left((n+1)\dfrac{\theta}{2}\right)}{\sin\left(\dfrac{\theta}{2}\right)}$, where $0 < \theta < 2\pi$.

**21.** Prove that if $n$ is a positive integer, then

$$\left| \frac{\sin(n\theta/2)}{\sin(\theta/2)} \right| \le n \qquad (\theta \ne 0, \ \pm 2\pi, \ \pm 4\pi, \ldots).$$

[HINT: Argue first that if $z = e^{i\theta}$, then the left-hand side equals $|(1 - z^n)/(1 - z)|$.]

## 1.5 POWERS AND ROOTS

In this section we shall derive formulas for the $n$th power and the $m$th roots of a complex number.

Let $z = re^{i\theta} = r(\cos \theta + i \sin \theta)$ be the polar form of the complex number $z$. By taking $z_1 = z_2 = z$ in Eq. (13) of Sec. 1.4, we obtain the formula

$$z^2 = r^2 e^{i2\theta}.$$

Since $z^3 = zz^2$, we can apply the identity a second time to deduce that

$$z^3 = r^3 e^{i3\theta}.$$

Continuing in this manner we arrive at the formula for the $n$th power of $z$:

$$z^n = r^n e^{in\theta} = r^n(\cos n\theta + i \sin n\theta). \tag{1}$$

Clearly this is just an extension of De Moivre's formula, discussed in Example 3 of Sec. 1.4.

Equation (1) is an appealing formula for raising a complex number to a positive integer power. It is easy to see that the identity is also valid for negative integers $n$ (see Prob. 2). The question arises whether the formula will work for $n = 1/m$, so that $\zeta = z^{1/m}$ is an $m$th root of $z$ satisfying

$$\zeta^m = z. \tag{2}$$

Certainly if we define

$$\zeta = \sqrt[m]{r} \, e^{i\theta/m} \tag{3}$$

(where $\sqrt[m]{r}$ denotes the customary, positive, $m$th root), we compute a complex number $\zeta$ satisfying (2) [as is easily seen by applying (1)]. But the matter is more complicated than this; the number 1, for instance, has *two* square roots: 1 and $-1$. And each of these has, in turn, two square roots—generating *four* fourth roots of 1, namely, 1, $-1$, $i$, and $-i$.

To see how the additional roots fit into the scheme of things, let's work out the polar description of the equation $\zeta^4 = 1$ for each of these numbers:

$$1^4 = (1e^{i0})^4 = 1^4 e^{i0} = 1,$$

$$i^4 = (1e^{i\pi/2})^4 = 1^4 e^{i2\pi} = 1,$$

$$(-1)^4 = (1e^{i\pi})^4 = 1^4 e^{i4\pi} = 1,$$
$$(-i)^4 = (1e^{i3\pi/2})^4 = 1^4 e^{i6\pi} = 1.$$

It is instructive to trace the consecutive powers of these roots in the Argand diagram. Thus Fig. 1.14 shows that $i$, $i^2$, $i^3$, and $i^4$ complete one revolution before landing on 1; $(-1)$, $(-1)^2$, $(-1)^3$, and $(-1)^4$ go around twice; the powers of $(-i)$ go around three times counterclockwise, and of course 1, $1^2$, $1^3$, and $1^4$ never move.

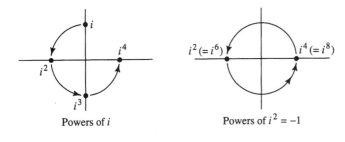

Powers of $i$ 

Powers of $i^2 = -1$

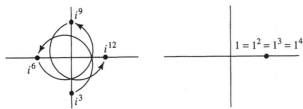

Powers of $i^3 = -i$

Powers of $i^4 = 1$

**Figure 1.14** Successive powers of the fourth roots of unity.

Clearly, the multiplicity of roots is tied to the ambiguity in representing 1, in polar form, as $e^{i0}$, $e^{i2\pi}$, $e^{i4\pi}$, etc. Thus to compute *all* the $m$th roots of a number $z$, we must apply formula (3) to *every* polar representation of $z$. For the cube roots of unity, for example, we would have

| Polar representation of 1 | Application of (3) |
|---|---|
| $\vdots$ | $\vdots$ |
| $1 = e^{-i6\pi}$ | $1^{1/3} = e^{-i6\pi/3} = 1$ |
| $1 = e^{-i4\pi}$ | $1^{1/3} = e^{-i4\pi/3} = -\dfrac{1}{2} + i\dfrac{\sqrt{3}}{2}$ |
| $1 = e^{-i2\pi}$ | $1^{1/3} = e^{-i2\pi/3} = -\dfrac{1}{2} - i\dfrac{\sqrt{3}}{2}$ |
| $1 = e^{i0}$ | $1^{1/3} = e^{i0/3} = 1$ |
| $1 = e^{i2\pi}$ | $1^{1/3} = e^{i2\pi/3} = -\dfrac{1}{2} + i\dfrac{\sqrt{3}}{2}$ |
| $1 = e^{i4\pi}$ | $1^{1/3} = e^{i4\pi/3} = -\dfrac{1}{2} - i\dfrac{\sqrt{3}}{2}$ |
| $1 = e^{i6\pi}$ | $1^{1/3} = e^{i6\pi/3} = 1$ |
| $\vdots$ | $\vdots$ |

Obviously the roots recur in sets of three, since $e^{i2\pi m_1/3} = e^{i2\pi m_2/3}$ whenever $m_1 - m_2 = 3$.

Generalizing, we can see that *there are exactly m distinct mth roots of unity, denoted by* $1^{1/m}$, *and they are given by*

$$1^{1/m} = e^{i2k\pi/m} = \cos\frac{2k\pi}{m} + i\sin\frac{2k\pi}{m} \qquad (k = 0, 1, 2, \ldots, m-1). \tag{4}$$

The arguments of these roots are $2\pi/m$ radians apart, and the roots themselves form the vertices of a regular polygon (Fig. 1.15).

Taking $k = 1$ in (4) we obtain the root[†]

$$\omega_m := e^{i2\pi/m} = \cos\frac{2\pi}{m} + i\sin\frac{2\pi}{m},$$

and it is easy to see that the complete set of roots can be displayed as

$$1, \omega_m, \omega_m^2, \ldots, \omega_m^{m-1}.$$

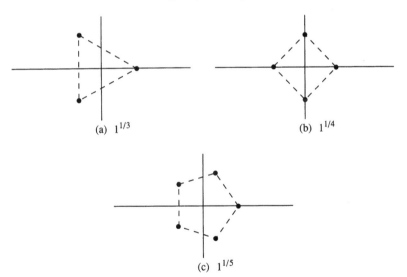

(a) $1^{1/3}$                    (b) $1^{1/4}$

(c) $1^{1/5}$

**Figure 1.15**   Regular polygons formed by the roots of unity.

**Example 1**

Prove that

$$1 + \omega_m + \omega_m^2 + \cdots + \omega_m^{m-1} = 0. \tag{5}$$

---

[†] A number $w$ is said to be a *primitive mth root of unity* if $w^m$ equals 1 but $w^k \neq 1$ for $k = 1, 2, \ldots,$ $m-1$. Clearly, $\omega_m$ is a primitive root.

**Solution.** This result is obvious from a physical point of view, since, by symmetry, the center of mass $(1 + \omega_m + \omega_m^2 + \cdots + \omega_m^{m-1})/m$ of the system of $m$ unit masses located at the $m$th roots of unity must be at the origin (see Fig. 1.15).

To give an algebraic proof we simply note that

$$(\omega_m - 1)(1 + \omega_m + \omega_m^2 + \cdots + \omega_m^{m-1}) = \omega_m^m - 1 = 0.$$

Since $\omega_m \neq 1$, Eq. (5) follows.  ∎

To obtain the $m$th roots of an *arbitrary* (nonzero) complex number $z = re^{i\theta}$, we generalize the idea displayed by (4) and, reasoning similarly, conclude that *the $m$ distinct $m$th roots of $z$ are given by*

$$z^{1/m} = \sqrt[m]{|z|}\,e^{i(\theta + 2k\pi)/m} \qquad (k = 0, 1, 2, \ldots, m - 1). \tag{6}$$

Equivalently, we can form these roots by taking any single one such as (3) and multiplying by the $m$th roots of unity.

### Example 2

Find all the cube roots of $\sqrt{2} + i\sqrt{2}$.

**Solution.** The polar form for $\sqrt{2} + i\sqrt{2}$ is

$$\sqrt{2} + i\sqrt{2} = 2e^{i\pi/4}.$$

Putting $|z| = 2$, $\theta = \pi/4$, and $m = 3$ into Eq. (6), we obtain

$$(\sqrt{2} + i\sqrt{2})^{1/3} = \sqrt[3]{2}\,e^{i(\pi/12 + 2k\pi/3)} \qquad (k = 0, 1, 2).$$

Hence $\sqrt[3]{2}(\cos \pi/12 + i \sin \pi/12)$, $\sqrt[3]{2}(\cos 3\pi/4 + i \sin 3\pi/4)$, and $\sqrt[3]{2}(\cos 17\pi/12 + i \sin 17\pi/12)$ are the cube roots of $\sqrt{2} + i\sqrt{2}$.  ∎

### Example 3

Let $a$, $b$, and $c$ be *complex* constants and let $a \neq 0$. Prove that the solutions of the equation

$$az^2 + bz + c = 0 \tag{7}$$

are given by the (usual) quadratic formula

$$z = \frac{-b \pm \sqrt{b^2 - 4ac}}{2a}, \tag{8}$$

where $\sqrt{b^2 - 4ac}$ denotes one of the values of $(b^2 - 4ac)^{1/2}$.

**Solution.** By completing the square, Eq. (7) can be written in the equivalent form

$$(2az + b)^2 = b^2 - 4ac.$$

Hence

$$2az + b = (b^2 - 4ac)^{1/2} = \pm\sqrt{b^2 - 4ac},$$

which is equivalent to Eq. (8).  ∎

## EXERCISES 1.5

1. Prove identity (1) by using induction.
2. Show that formula (1) also holds for negative integers $n$.
3. Let $n$ be a positive integer. Prove that $\arg z^n = n \operatorname{Arg} z + 2k\pi$, $k = 0, \pm 1, \pm 2, \ldots$, for $z \neq 0$.
4. Use identity (1) to show that
   (a) $(\sqrt{3} - i)^7 = -64\sqrt{3} + i64$         (b) $(1 + i)^{95} = 2^{47}(1 - i)$
5. Find all the values of the following.
   (a) $(-16)^{1/4}$              (b) $1^{1/5}$              (c) $i^{1/4}$

   (d) $(1 - \sqrt{3}i)^{1/3}$    (e) $(i - 1)^{1/2}$       (f) $\left(\dfrac{2i}{1+i}\right)^{1/6}$
6. Describe how to construct geometrically the fifth roots of $z_0$ if
   (a) $z_0 = -1$                 (b) $z_0 = i$              (c) $z_0 = 1 + i$
7. Solve each of the following equations.
   (a) $2z^2 + z + 3 = 0$
   (b) $z^2 - (3 - 2i)z + 1 - 3i = 0$
   (c) $z^2 - 2z + i = 0$
8. Let $a$, $b$, and $c$ be real numbers and let $a \neq 0$. Show that the equation $az^2 + bz + c = 0$ has
   (a) two real solutions if $b^2 - 4ac > 0$.
   (b) two nonreal conjugate solutions if $b^2 - 4ac < 0$.
9. Solve the equation $z^3 - 3z^2 + 6z - 4 = 0$.
10. Find all four roots of the equation $z^4 + 1 = 0$ and use them to deduce the factorization $z^4 + 1 = (z^2 - \sqrt{2}z + 1)(z^2 + \sqrt{2}z + 1)$.
11. Solve the equation $(z + 1)^5 = z^5$.
12. Show that the $n$ points $z_0^{1/n}$ form the vertices of a regular $n$-sided polygon inscribed in the circle of radius $\sqrt[n]{|z_0|}$ about the origin.
13. Show that $\omega_3 = (-1 + \sqrt{3}i)/2$ and that $\omega_4 = i$. Use these values to verify identity (5) for the special cases $n = 3$ and $n = 4$.
14. Let $m$ and $n$ be positive integers which have no common factor. Prove that the set of numbers $(z^{1/n})^m$ is the same as the set of numbers $(z^m)^{1/n}$. We denote this common set of numbers by $z^{m/n}$. Show that

$$z^{m/n} = \sqrt[n]{|z|^m}\left[\cos\frac{m}{n}(\theta + 2k\pi) + i\sin\frac{m}{n}(\theta + 2k\pi)\right] \qquad (k = 0, 1, 2, \ldots, n - 1).$$

15. Use the result of Prob. 14 to find all the values of $(1 - i)^{3/2}$.
16. Figure 1.15(c) shows the root $\omega_5$ raised to the first, second, third, fourth, and fifth powers. Using the same diagram, label the points which correspond to the first five powers of $\omega_5^2$. Do the same for the powers of $\omega_5^3$ and $\omega_5^4$.
17. Let $m$ be a fixed positive integer and let $l$ be an integer that is not divisible by $m$. Prove the following generalization of Eq. (5):

$$1 + \omega_m^l + \omega_m^{2l} + \cdots + \omega_m^{(m-1)l} = 0.$$

18. Show that if $\alpha$ and $\beta$ are $n$th and $m$th roots of unity, respectively, then the product $\alpha\beta$ is a $k$th root of unity for some integer $k$.

**19.** Write a computer program for solving the quadratic equation

$$az^2 + bz + c = 0, \qquad a \neq 0.$$

Use as inputs the real and imaginary parts of $a$, $b$, $c$ and print the solutions in both rectangular and polar form.

## 1.6 PLANAR SETS

In this section we shall give some basic definitions concerning point sets in the plane. The terminology is helpful when we wish to express precise conditions for the validity of the advanced theorems of complex analysis.

The set of all points that satisfy the inequality

$$|z - z_0| < \rho,$$

where $\rho$ is a positive real number, is called an *open disk* or *neighborhood of $z_0$*. This set consists of all the points that lie inside the circle of radius $\rho$ about $z_0$. In particular, the solution sets of the inequalities

$$|z - 2| < 3, \qquad |z + i| < \frac{1}{2}, \qquad |z| < 8$$

are neighborhoods of the respective points 2, $-i$, and 0. We shall make frequent reference to the neighborhood $|z| < 1$, which is called the *open unit disk*.

A point $z_0$ which lies in a set $S$ is called an *interior point of $S$* if there is some neighborhood of $z_0$ which is completely contained in $S$. For example, if $S$ is the right half-plane Re $z > 0$ and $z_0 = .01$, then $z_0$ is an interior point of $S$ because $S$ contains the neighborhood $|z - z_0| < .01$ (see Fig. 1.16).

If every point of a set $S$ is an interior point of $S$, we say that $S$ is an *open set*. Any neighborhood is an open set (Prob. 1). Each of the following inequalities also describes an open set: (a) $\rho_1 < |z - z_0| < \rho_2$, (b) $|z - 3| > 2$, (c) Im $z > 0$, and (d) $1 < $ Re $z < 2$. These sets are sketched in Fig. 1.17. Note that the solution set $T$ of the inequality $|z - 3| \geq 2$ is *not* an open set since no point on the circle $|z - 3| = 2$ is an interior point of $T$. Note also that an open interval of the real axis is *not* an open set since it contains no open disk.

Let $w_1, w_2, \ldots, w_{n+1}$ be $n + 1$ points in the plane. For each $k = 1, 2, \ldots, n$, let $l_k$ denote the line segment joining $w_k$ to $w_{k+1}$. Then the successive line segments $l_1$, $l_2, \ldots, l_n$ form a continuous chain known as a *polygonal path* that joins $w_1$ to $w_{n+1}$.

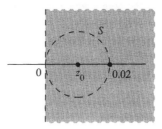

**Figure 1.16**  Interior point.

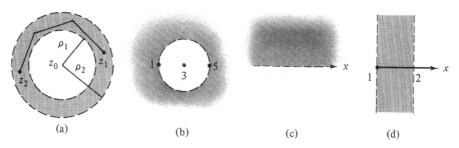

(a)                    (b)              (c)              (d)

**Figure 1.17**   Open sets.

An open set $S$ is said to be *connected* if every pair of points $z_1$, $z_2$ in $S$ can be joined by a polygonal path that lies entirely in $S$ [see Fig. 1.17(a)]. Roughly speaking, this means that $S$ consists of a "single piece." Each of the sets in Fig. 1.17 is connected. The set consisting of all those points in the plane that do not lie on the circle $|z| = 1$ is an example of an open set that is not connected; indeed, if $z_1$ is a point inside the circle and $z_2$ is a point outside, then every polygonal path that joins $z_1$ and $z_2$ must intersect the circle.

We call an open connected set a *domain*. Therefore, all the sets in Fig. 1.17 are domains.

In the calculus of functions of a single real variable a useful and familiar fact is that, on an interval, the vanishing of the derivative implies that the function is identically constant. We now present an extension of this result to functions of two real variables which underscores the importance of the notion of a domain.

---

**Theorem 1.**   Suppose $u(x, y)$ is a real-valued function defined in a domain $D$. If the first partial derivatives of $u$ satisfy

$$\frac{\partial u}{\partial x} = \frac{\partial u}{\partial y} = 0 \tag{1}$$

at all points of $D$, then $u \equiv$ constant in $D$.

---

*Proof.*   Notice that the assumption $\partial u/\partial x = 0$ implies that $u$ remains constant along any horizontal line segment contained in $D$; indeed, on such a segment, $u$ is a function of a single variable (namely, $x$) whose derivative vanishes. Similarly, the assumption $\partial u/\partial y = 0$ means that $u$ is constant along any vertical line segment that lies in $D$. Putting these facts together we see that $u$ remains unchanged along any polygonal path in $D$ that has all its segments parallel to the coordinate axes. Now from the definition of connectedness we know that any pair of points in $D$ can be joined by some polygonal path lying entirely in $D$. The catch is that this path may have some segments that are neither vertical nor horizontal. However, it turns out from topological considerations (see Prob. 21) that any such segment can be replaced by a chain of small horizontal and vertical segments lying in $D$ (see Fig. 1.18). Thus, Theorem 1 follows.   ∎

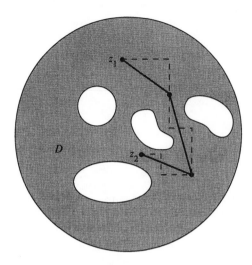

**Figure 1.18** Polygonal path with vertical and horizontal segments.

For the reader who prefers to avoid such topological arguments, an alternative proof of Theorem 1 can be given with the aid of the chain rule (see Prob. 22).

What is crucial for Theorem 1 is the connectedness property of domains; in fact, the theorem is no longer true if $D$ is merely assumed to be an open set (see Prob. 19).

We now continue with our discussion of planar sets. A point $z_0$ is said to be a *boundary point* of a set $S$ if every neighborhood of $z_0$ contains at least one point of $S$ and at least one point not in $S$. The set of all boundary points of $S$ is called the *boundary* or *frontier* of $S$. The boundaries of the sets in Fig. 1.17 are as follows: (a) the two circles $|z - z_0| = \rho_1$ and $|z - z_0| = \rho_2$, (b) the circle $|z - 3| = 2$, (c) the real axis, and (d) the two lines $\operatorname{Re} z = 1$ and $\operatorname{Re} z = 2$. Since each point of a domain $D$ is an interior point of $D$, it follows that a domain cannot contain any of its boundary points.

A set $S$ is said to be *closed* if it contains all of its boundary points. (As requested in Prob. 13, the reader should verify that $S$ being closed is equivalent to $\mathbf{C} \backslash S$ being open.) The set described by the inequality $0 < |z| \le 1$ is not closed since it does not contain the boundary point 0. The set of points $z$ that satisfy the inequality

$$|z - z_0| \le \rho \qquad (\rho > 0)$$

is a closed set, for it contains its boundary $|z - z_0| = \rho$. Therefore, we call this set a *closed disk*.

A set of points $S$ is said to be *bounded* if there exists a positive real number $R$ such that $|z| < R$ for every $z$ in $S$. In other words, $S$ is bounded if it is contained in some neighborhood of the origin. An *unbounded* set is one that is not bounded. Of the sets in Fig. 1.17 only (a) is bounded. A set that is both closed and bounded is said to be *compact*.

A *region* is a domain together with some, none, or all of its boundary points. In particular, every domain is a region.

## EXERCISES 1.6

1. Prove that the neighborhood $|z - z_0| < \rho$ is an open set. [HINT: Show that if $z_1$ belongs to the neighborhood, then so do all points $z$ that satisfy $|z - z_1| < R$, where $R = \rho - |z_1 - z_0|$.]

*Problems 2–8 refer to the sets described by the following inequalities:*

(a) $|z - 1 + i| \le 3$   (b) $|\text{Arg } z| < \dfrac{\pi}{4}$

(c) $0 < |z - 2| < 3$   (d) $-1 < \text{Im } z \le 1$

(e) $|z| \ge 2$   (f) $(\text{Re } z)^2 > 1$

2. Sketch each of the given sets.
3. Which of the given sets are open?
4. Which of the given sets are domains?
5. Which of the given sets are bounded?
6. Describe the boundary of each of the given sets.
7. Which of the given sets are regions?
8. Which of the given sets are closed regions?

9. Prove that any set consisting of finitely many points is bounded.
10. Prove that the closed disk $|z - z_0| \le \rho$ is bounded.
11. Let $S$ be the set consisting of the points $1, \frac{1}{2}, \frac{1}{3}, \dots$. What is the boundary of $S$?
12. Let $z_0$ be a point of the set $S$. Prove that if $z_0$ is not an interior point of $S$, then $z_0$ must be a boundary point of $S$.
13. Let $S$ be a subset of $\mathbf{C}$. Prove that $S$ is closed if and only if its complement $\mathbf{C} \backslash S$ is an open set.
14. A point $z_0$ is said to be an *accumulation point* of a set $S$ if every neighborhood of $z_0$ contains infinitely many points of the set $S$. Prove that a closed region contains all its accumulation points.

*Problems 15–18 refer to the following definitions: Let $S$ and $T$ be sets. The set consisting of all points belonging to $S$ or $T$ or both $S$ and $T$ is called the* union *of $S$ and $T$ and is denoted by $S \cup T$. The set consisting of all points belonging to both $S$ and $T$ is called the* intersection *of $S$ and $T$ and is denoted by $S \cap T$.*

15. Let $S$ and $T$ be the sets described by $|z + 1| < 2$ and $|z - i| < 1$, respectively. Sketch the sets $S \cup T$ and $S \cap T$.
16. If $S$ and $T$ are open sets, prove that $S \cup T$ is an open set.
17. If $S$ and $T$ are domains, is $S \cap T$ necessarily a domain?
18. Prove that if $S$ and $T$ are domains which have at least one point in common, then $S \cup T$ is a domain.
19. Let

$$u(x, y) := \begin{cases} 1 & \text{for } |z| < 1, \\ 0 & \text{for } |z| > 2. \end{cases}$$

Show that $\partial u/\partial x = \partial u/\partial y = 0$ in the open set

$$D := \{z: |z| < 1\} \cup \{z: |z| > 2\},$$

but $u$ is not constant in $D$. Why doesn't this contradict Theorem 1?

**20.** Suppose $u(x, y)$ is a real-valued function defined in a domain $D$. If

$$\frac{\partial u}{\partial x} = y \quad \text{and} \quad \frac{\partial u}{\partial y} = x$$

at all points of $D$, prove that $u(x, y) = xy + c$ for some constant $c$.

**21.** Let $D$ be a domain and $l$ be a closed line segment lying in $D$. A theorem in elementary topology states that $l$ (and, in fact, any compact subset of $D$) can be covered by a finite number of open disks that lie in $D$ and have their centers on $l$. Use this fact to prove that any two points in a domain $D$ can be joined by a polygonal path in $D$ having all its segments parallel to the coordinate axes.

**22.** Prove Theorem 1 by completing the following steps.

**(a)** Show that any line segment can be parametrized by

$$x = at + b, \qquad y = ct + d,$$

where $a$, $b$, $c$, and $d$ are real constants and $t$ ranges between 0 and 1. Hence the values of $u$ along a line segment lying in $D$ are given by

$$U(t) := u(at + b, ct + d), \qquad 0 \le t \le 1.$$

**(b)** Use assumption (1) of Theorem 1 and the chain rule to show that $dU/dt = 0$ for $0 \le t \le 1$ and thereby conclude that $u$ is constant on any line segment in $D$.

**(c)** By appealing to the definition of connectedness, argue that $u$ must have the same value at any two points of $D$.

## *1.7 SOME APPLICATIONS OF COMPLEX VARIABLES IN MECHANICS

One can take advantage of the interpretation of a complex number $z$ as a vector to solve certain two-dimensional problems in physics and engineering mechanics. In this section we shall illustrate this with brief analyses of satellite orbits.

The motion of a satellite around the earth is governed by Newton's second law—force equals mass times acceleration—and Newton's law of gravitation. The latter states that the force on the satellite is directed toward the center of the earth, is proportional to the product of the mass of the earth $m_e$ and the mass of the satellite $m_s$, and is inversely proportional to the square of the distance between the satellite and the earth's center. If we assume that the earth is stationary[†] and is located at the origin of the orbital plane, then this force $F$ can be written in terms of the satellite's position $z$ as

$$F = \frac{Gm_e m_s}{|z|^2}\left(\frac{-z}{|z|}\right)$$

---

[†] Corrections for the earth's motion are discussed in Ref. [8].

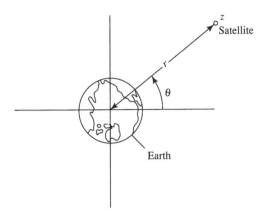

**Figure 1.19**   Satellite coordinates.

with $G$ a constant (see Fig. 1.19). The equation of motion is then

$$-\frac{Gm_e m_s}{|z|^3} z = m_s \frac{d^2 z}{dt^2}. \tag{1}$$

As is clear from Eq. (1), the mass of the satellite does not figure into the motion, so we shall take $m_s = 1$ hereafter.

Rewriting (1) in polar form with the substitution $K = Gm_e$, we have

$$-\frac{K}{r^3} re^{i\theta} = \frac{d^2}{dt^2} (re^{i\theta}) = (\ddot{r} + 2i\dot{r}\dot{\theta} + ir\ddot{\theta} - r\dot{\theta}^2)e^{i\theta}, \tag{2}$$

where the dot superscript denotes differentiation with respect to time. Canceling $e^{i\theta}$ in Eq. (2) and taking the imaginary part results in

$$2\dot{r}\dot{\theta} + r\ddot{\theta} = 0. \tag{3}$$

By multiplying Eq. (3) by $r$ we learn that the quantity $r^2\dot{\theta}$ is *conserved* (constant) throughout the motion, and we denote this constant by $l$:

$$\frac{dl}{dt} = \frac{d}{dt}(r^2\dot{\theta}) = 0. \tag{4}$$

Since $r\dot{\theta}$ is the component of the velocity $\dot{z}$ perpendicular to the vector $z$, the quantity $l$ can be identified as the angular momentum.

The analysis of the real part of (2) (with $e^{i\theta}$ canceled) is somewhat more complicated. First we eliminate $\dot{\theta}$ in favor of $l$ to obtain a differential equation with only one unknown ($r$):

$$-\frac{K}{r^2} = \ddot{r} - r\dot{\theta}^2 = \ddot{r} - \frac{l^2}{r^3}. \tag{5}$$

This equation is simplified if we change the independent parameter from time to angle $\theta$. The chain rule tells us[†]

---

[†] The case $l = 0$ requires $\theta = $ constant. Such orbits are of no interest to us, since they correspond to the satellite dropping straight down.

$$\frac{d}{dt} = \frac{d\theta}{dt}\frac{d}{d\theta} = \frac{l}{r^2}\frac{d}{d\theta} \tag{6}$$

and we have the fortunate circumstance that

$$\frac{dr}{dt} = \frac{l}{r^2}\frac{dr}{d\theta} = -l\frac{d(r^{-1})}{d\theta}. \tag{7}$$

Using (7) in (5) and applying (6) again for the second derivative, we deduce

$$\frac{-K}{r^2} = \frac{l}{r^2}\frac{d}{d\theta}\left(-l\frac{d(r^{-1})}{d\theta}\right) - \frac{l^2}{r^3}$$

or, equivalently,

$$\frac{d^2}{d\theta^2}(r^{-1}) + (r^{-1}) = \frac{K}{l^2}. \tag{8}$$

This linear second-order differential equation with constant coefficients is easily solved, and its solutions can be parametrized in terms of the constants of integration $\varepsilon$ and $\theta_p$ as

$$r^{-1} = \frac{K}{l^2}[1 + \varepsilon \cos(\theta - \theta_p)]$$

or, equivalently,

$$r = \frac{l^2/K}{1 + \varepsilon \cos(\theta - \theta_p)}. \tag{9}$$

Equation (9) is the polar form of the equation of the general *conic section*, and we invite the reader to refer to any calculus text for confirmation of the following classifications (see Fig. 1.20):

For $\varepsilon > 1$ the orbit is a hyperbola.
For $\varepsilon = 1$ the orbit is a parabola.
For $\varepsilon < 1$ the orbit is an ellipse.
For $\varepsilon = 0$ the orbit is a circle.

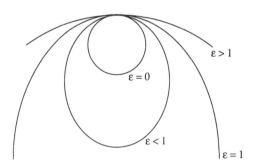

$\varepsilon > 1$

$\varepsilon = 0$

$\varepsilon < 1$

$\varepsilon = 1$   **Figure 1.20**   Eccentricities of orbits.

Here $\varepsilon$ is known as the *eccentricity* of the orbit and $\theta_p$ is the angle at which $r$ is minimal; this point is called the *perigee*. Further details concerning these *Keplerian orbits* are provided in the exercises.

## EXERCISES 1.7

1. **(a)** Show, using Eq. (5), that the quantity

$$E := \dot{r}^2/2 + l^2/2r^2 - K/r$$

   is constant during the satellite's motion.
   **(b)** Prove:

$$\frac{|\dot{z}|^2}{2} = \frac{\dot{r}^2}{2} + \frac{l^2}{2r^2}.$$

   (This quantity is the *kinetic energy* of the satellite. The quantity $-K/r$ in part (a) is the *potential energy*, and the constant $E$ is the *total energy*.)
2. Derive the expression for the eccentricity $\varepsilon$ in terms of the total energy $E$ of Prob. 1 and the angular momentum $l$:

$$\varepsilon = (1 + 2El^2/K^2)^{1/2}.$$

   [HINT: Evaluate Eq. (9) at perigee, $\theta = \theta_p$.]
3. Communications satellites are placed in *synchronous orbits*, i.e., circular orbits whose angular velocity $\dot{\theta}$ matches the rotation of the earth. What elevation and velocity must a communications satellite possess? Take $G = 6.67 \times 10^{-11}$ m$^3$/kg·s$^2$, $m_e = 5.98 \times 10^{24}$ kg, and earth radius $= 6.37 \times 10^6$ m.
4. This problem demonstrates how complex notation can simplify the kinematic analysis of planar mechanisms.
   Consider the crank-and-piston linkage depicted in Fig. 1.21. The crank arm $a$ rotates about the fixed point $O$ while the piston arm $c$ executes horizontal motion. (If this were a gasoline engine, combustion forces would drive the piston and the connecting arm $b$ would transform this energy into a rotation of the crankshaft.) For engineering analysis it is important to be able to relate the crankshaft's angular coordinates—position, velocity, and acceleration—to the corresponding linear coordinates for the piston. Although this calculation can be carried out using vector analysis, the following complex variable technique is more "automatic."
   Let the crankshaft pivot $O$ lie at the origin of the coordinate system, and let $z$ be the complex number giving the location of the base of the piston rod, as depicted in Fig. 1.21,

$$z = l + id,$$

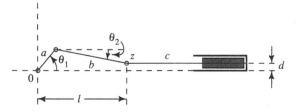

**Figure 1.21**   Crank-and-piston linkage.

where $l$ gives the piston's (linear) excursion and $d$ is a fixed offset. The crank arm is described by $ae^{i\theta_1}$ and the connecting arm by $be^{i\theta_2}$ ($\theta_2$ is negative in Fig. 1.21). Exploit the obvious identity

$$ae^{i\theta_1} + be^{i\theta_2} = z = l + id$$

to derive the expression relating the piston position to the crankshaft angle:

$$l = a \cos \theta_1 + b \cos \left[ \sin^{-1} \left( \frac{d - a \sin \theta_1}{b} \right) \right].$$

5. Suppose the mechanism in Prob. 4 has the dimensions

$$a = 0.1 \text{ m}, \qquad b = 0.2 \text{ m}, \qquad d = 0.1 \text{ m}$$

and the crankshaft rotates at a uniform velocity of 2 rad/s. Compute the position and velocity of the piston when $\theta_1 = \pi$.

6. For the linkage illustrated in Fig. 1.22, use complex variables to outline a scheme for expressing the angular position, velocity, and acceleration of arm $c$ in terms of those of arm $a$. (You needn't work out the equations.)

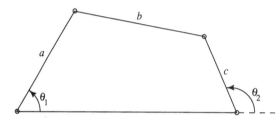

**Figure 1.22**   Linkage in Prob. 6.

7. A *central field* in celestial mechanics is a force field for which the force vector on a satellite is always directed toward the origin. Show that angular momentum is always conserved in a central field.

## SUMMARY

The complex number system is an extension of the real number system and consists of all expressions of the form $a + bi$, where $a$ and $b$ are real and $i^2 = -1$. The operations of addition, subtraction, multiplication, and division with complex numbers are performed in a manner analogous to "computing with radicals." Geometrically, complex numbers can be represented by points or vectors in the plane. Thus certain theorems from geometry, such as the triangle inequality, can be translated into the language of complex numbers. Associated with a complex number $z = a + bi$ are its absolute value, given by $|z| = \sqrt{a^2 + b^2}$, and its complex conjugate, given by $\bar{z} = a - bi$. The former is the distance from the point $z$ to the origin, while the latter is the reflection of the point $z$ in the $x$-axis. The numbers $z$, $\bar{z}$, and $|z|$ are related by $z\bar{z} = |z|^2$.

Every nonzero complex number $z$ can be written in the polar form $z = r(\cos \theta + i \sin \theta)$, where $r = |z|$ and $\theta$ is the angle of inclination of the vector $z$. Any of the equivalent angles $\theta + 2k\pi$, $k = 0, \pm 1, \pm 2, \dots$, is called the argument of $z$ (arg $z$). The polar form is useful in finding powers and roots of $z$.

For $z = x + iy$, the complex exponential $e^z$ is defined by $e^z = e^x(\cos y + i \sin y)$. In particular, if $\theta$ is real, Euler's equation states that $e^{i\theta} = \cos \theta + i \sin \theta$. Moreover, the polar form of a complex number can be written simply as $z = re^{i\theta}$.

Special terminology is used in describing point sets in the plane. Important is the concept of a domain $D$. Such a set is characterized by two properties: (i) each point $z$ of $D$ is the center of an open disk completely contained in $D$; (ii) each pair of points $z_1$ and $z_2$ in $D$ can be joined by a polygonal path that lies entirely in $D$. If some of the boundary points are adjoined to a domain, the resulting set is called a region.

## SUGGESTED READING

### *Introductory Level*

[1] Boas, R. P. *Invitation to Complex Analysis*. Random House/Birkhäuser Math Series, New York, 1987.

[2] Churchill, R. V., and Brown, J. W. *Complex Variables and Applications*, 5th ed. McGraw-Hill Book Company, New York, 1990.

[3] Levinson, Norman, and Redheffer, Raymond M. *Complex Variables*. Holden-Day, Inc., San Francisco, 1970.

### *Advanced Level*

[4] Ahlfors, L. V. *Complex Analysis*, 3rd ed. McGraw-Hill Book Company, New York, 1979.

[5] Hille, E. *Analytic Function Theory*, Vol. I, 2nd ed. Chelsea, New York, 1973.

[6] Nehari, Z. *Conformal Mapping*. Dover Publishing, New York, 1975.

### *Extending Number Fields*

[7] Fraleigh, John B. *A First Course in Abstract Algebra*, 4th ed. Addison-Wesley Publishing Company, Reading, Mass., 1989.

### *Mechanics*

[8] Goldstein, Herbert. *Classical Mechanics*, 2nd ed. Addison-Wesley Publishing Company, Reading, Mass., 1980.

[9] Martin, George H. *Kinematics and Dynamics of Machines*, 2nd ed. McGraw-Hill Book Company, New York, 1982.

# :2:

# Analytic Functions

## 2.1 FUNCTIONS OF A COMPLEX VARIABLE

The concept of a complex number $z$ was introduced in Chapter 1 in order to solve certain algebraic equations. We shall now study functions $f(z)$ defined on these complex variables. Our objective is to mimic the concepts, theorems, and mathematical structure of calculus; we want to differentiate and integrate $f(z)$. The notion of a derivative is far more subtle in the complex case because of the intrinsically two-dimensional nature of the complex variable, and the exposition of this point will consume all of Chapter 2. The payoff is enormous, however, and the remainder of the book will be devoted to developing the mathematical consequences and demonstrating their applications to physical problems.

Let us begin with a careful review of the basics. Recall that a *function* $f$ is a rule that assigns to each element in a set $A$ one and only one element in a set $B$. If $f$ assigns the value $b$ to the element $a$ in $A$, we write

$$b = f(a)$$

and call $b$ the *image* of $a$ under $f$. The set $A$ is the *domain of definition* of $f$ (even if $A$ is not a domain in the sense of Chapter 1), and the set of all images $f(a)$ is the *range* of $f$. We sometimes refer to $f$ as a *mapping* of $A$ into $B$.

Here we are concerned with complex-valued functions of a complex variable, so that the domains of definition and the ranges are subsets of the complex numbers. If $f(z)$ is expressed by a formula such as

$$f(z) = \frac{z^2 - 1}{z^2 + 1},$$

then, unless stated otherwise, we take the domain of $f$ to be the set of all $z$ for which the formula is well defined. (Thus the domain for this $f$ comprises all $z$ except for $\pm i$.)

If $w$ denotes the value of the function $f$ at the point $z$, we then write $w = f(z)$. Just as $z$ decomposes into real and imaginary parts as $z = x + iy$, the real and imaginary parts of $w$ are each (real-valued) functions of $z$ or, equivalently, of $x$ and $y$, and so we customarily write

$$w = u(x, y) + iv(x, y),$$

with $u$ and $v$ denoting the real and imaginary parts, respectively, of $w$. Thus a complex-valued function of a complex variable is, in essence, a pair of real functions of two real variables.

**Example 1**

Write the function $w = f(z) = z^2 + 2z$ in the form $w = u(x, y) + iv(x, y)$.

**Solution.**   Setting $z = x + iy$ we obtain

$$w = f(z) = (x + iy)^2 + 2(x + iy) = x^2 - y^2 + i2xy + 2x + i2y.$$

Hence $w = (x^2 - y^2 + 2x) + i(2xy + 2y)$ is the desired form.   ∎

Unfortunately, it is generally impossible to draw the graph of a complex function; to display two real functions of two real variables graphically would require four dimensions. We can, however, visualize some of the properties of a complex function $w = f(z)$ by sketching its domain of definition in the $z$-plane and its range in the $w$-plane, and depicting the relationship as in Fig. 2.1.

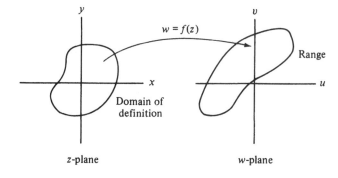

z-plane                    w-plane

**Figure 2.1**   Representation of a complex function.

**Example 2**

Describe the range of the function $f(z) = x^2 + 2i$ defined on the closed unit disk $|z| \leq 1$.

**Solution.**   We have $u(x, y) = x^2$ and $v(x, y) = 2$. Thus as $z$ varies over the closed unit disk, $u$ varies between 0 and 1, and $v$ is constant. The range is therefore the line segment from $w = 2i$ to $w = 1 + 2i$.   ∎

**Example 3**

Describe the function $f(z) = z^3$ for $z$ in the semidisk given by $|z| \leq 2$, Im $z \geq 0$ [see Fig. 2.2(a)].

**Solution.** From Sec. 1.5 we know that the points $z$ in the sector of the semidisk from Arg $z = 0$ to Arg $z = 2\pi/3$, when cubed, cover the entire disk $|w| \leq 8$. The cubes of the remaining $z$-points also fall in this disk, overlapping it in the upper half-plane, as depicted in Fig. 2.2(b). ■

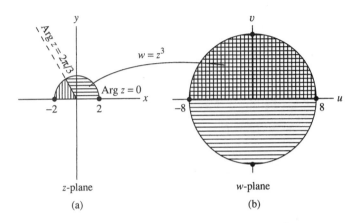

**Figure 2.2**    Mapping of a semidisk under $f(z) = z^3$.

# EXERCISES 2.1

1. Write each of the following functions in the form $w = u(x, y) + iv(x, y)$.

    (a) $f(z) = 3z^2 + 5z + i + 1$

    (b) $g(z) = \dfrac{1}{z}$

    (c) $h(z) = \dfrac{z + i}{z^2 + 1}$

    (d) $q(z) = \dfrac{2z^2 + 3}{|z - 1|}$

    (e) $F(z) = e^{3z}$

    (f) $G(z) = e^z + e^{-z}$

2. Find the domain of definition of each of the functions in Prob. 1.

3. Describe the range of each of the following functions.

    (a) $f(z) = z + 5$ for Re $z > 0$

    (b) $g(z) = z^2$ for $z$ in the first quadrant, Re $z \geq 0$, Im $z \geq 0$

    (c) $h(z) = \dfrac{1}{z}$   for $0 < |z| \leq 1$

    (d) $p(z) = -2z^3$ for $z$ in the quarter-disk $|z| < 1$, $0 < $ Arg $z < \dfrac{\pi}{2}$

4. The mapping $w = g(z) = 1/z$ is called an *inversion*.

    (a) Show that $g$ maps the circle $|z| = r$ onto the circle $|w| = 1/r$.

    (b) Show that $g$ maps the ray Arg $z = \theta_0$, $-\pi < \theta_0 < \pi$, onto the ray Arg $w = -\theta_0$.

**5.** For the complex exponential function $f(z) = e^z$ defined in Sec. 1.4:
   **(a)** Describe the domain of definition and the range.
   **(b)** Show that $f(-z) = 1/f(z)$.
   **(c)** Describe the image of the vertical line Re $z = 1$.
   **(d)** Describe the image of the horizontal line Im $z = \pi/4$.
   **(e)** Describe the image of the infinite strip $0 \leq$ Im $z \leq \pi/4$.

**6.** The *Joukowski mapping* is defined by

$$w = J(z) = \frac{1}{2}\left(z + \frac{1}{z}\right).$$

Show that

**(a)** $J(z) = J\left(\dfrac{1}{z}\right).$

**(b)** $J$ maps the unit circle $|z| = 1$ onto the real interval $[-1, 1]$.
**(c)** $J$ maps the circle $|z| = r$ $(r > 0, r \neq 1)$ onto the ellipse

$$\frac{u^2}{\left[\frac{1}{2}\left(r + \frac{1}{r}\right)\right]^2} + \frac{v^2}{\left[\frac{1}{2}\left(r - \frac{1}{r}\right)\right]^2} = 1,$$

which has foci at $\pm 1$.

**7.** A function of the form $F(z) = z + c$, where $c$ is a complex constant, generates a *translation mapping*. Sketch the image of the semidisk $|z| \leq 2$, Im $z \geq 0$, [see Fig. 2.2(a)] under $F$ when
**(a)** $c = 3$; **(b)** $c = 2i$; **(c)** $c = -1 - i$.

**8.** A function of the form $G(z) = e^{i\phi}z$, where $\phi$ is a real constant, generates a *rotation mapping*. Sketch the image of the semidisk $|z| \leq 2$, Im $z \geq 0$ [see Fig. 2.2(a)] under $G$ when
**(a)** $\phi = \pi/4$; **(b)** $\phi = -\pi/4$; **(c)** $\phi = 3\pi/4$.

**9.** A function of the form $H(z) = \rho z$, where $\rho$ is a positive real constant, generates a *magnification mapping* when $\rho > 1$ and a *reduction mapping* when $\rho < 1$. Sketch the image of the semidisk $|z| \leq 2$, Im $z \geq 0$ [see Fig. 2.2(a)] under $H$ when **(a)** $\rho = 3$; **(b)** $\rho = \frac{1}{2}$.

**10.** Let $F(z) = z + i$, $G(z) = e^{i\pi/4}z$, and $H(z) = z/2$. Sketch the image of the semidisk $|z| \leq 2$, Im $z \geq 0$ [see Fig. 2.2(a)] under each of the following composite mappings:
**(a)** $G(F(z))$     **(b)** $G(H(z))$     **(c)** $H(F(z))$     **(d)** $F(G(H(z)))$

**11.** Let $F(z) = z - 3$, $G(z) = -iz$, and $H(z) = 2z$. Sketch the image of the circle $|z| = 1$ under each of the following composite mappings:
**(a)** $G(F(z))$     **(b)** $G(H(z))$     **(c)** $H(F(z))$     **(d)** $F(G(H(z)))$

**12.** A function of the form $f(z) = az + b$, where $a$ and $b$ are complex constants, is called a *linear transformation*. Show that every linear transformation can be expressed as the composition of a magnification (or reduction; Prob. 9), a rotation (Prob. 8), and a translation (Prob. 7). [HINT: Write $a$ in polar form.]

**13.** (*Electric Field*) A uniformly charged infinite rod, standing perpendicular to the $z$-plane at the point $z_0$, generates an electric field at every point in the plane. The intensity of this field varies inversely as the distance from $z_0$ to the point and is directed along the line from $z_0$ to the point.
   **(a)** Show that the (vector) field at the point $z$ is given by the function $F(z) = 1/(\bar{z} - \bar{z}_0)$, in appropriate units.
   **(b)** If three such rods are located at the points $1 + i$, $-1 + i$, and $0$, find the positions of equilibrium (i.e., the points where the vector sum of the fields is zero).

## 2.2 LIMITS AND CONTINUITY

As we observed in Chapter 1, the definition of absolute value can be used to designate the distance between two complex numbers. Having a concept of distance, we can proceed to introduce the notions of limit and continuity.

Informally, when we have an infinite sequence of complex numbers $z_1$, $z_2$, $z_3$, ..., we say that the number $z_0$ is the limit of the sequence if the $z_n$ eventually (i.e., for large enough $n$) stay arbitrarily close to $z_0$. More precisely, we state

---

**Definition 1.** A sequence of complex numbers $\{z_n\}_1^\infty$ is said to have the **limit** $z_0$ or to **converge** to $z_0$, and we write

$$\lim_{n \to \infty} z_n = z_0$$

or, equivalently,

$$z_n \to z_0 \quad \text{as} \quad n \to \infty,$$

if for any $\varepsilon > 0$ there exists an integer $N$ such that $|z_n - z_0| < \varepsilon$ for all $n > N$.

---

Geometrically, this means that each term $z_n$, for $n > N$, lies in the open disk of radius $\varepsilon$ about $z_0$ (see Fig. 2.3).

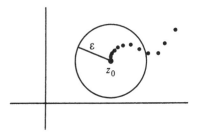

**Figure 2.3** A convergent sequence.

A related concept is the limit of a complex-valued function $f(z)$. Roughly speaking, we say that the number $w_0$ is the limit of the function $f(z)$ as $z$ approaches $z_0$, if $f(z)$ stays close to $w_0$ whenever $z$ is sufficiently near $z_0$. In precise terms we give

---

**Definition 2.** Let $f$ be a function defined in some neighborhood of $z_0$, with the possible exception of the point $z_0$ itself. We say that the **limit of** $f(z)$ **as** $z$ **approaches** $z_0$ **is the number** $w_0$ and write

$$\lim_{z \to z_0} f(z) = w_0$$

or, equivalently,

$$f(z) \to w_0 \quad \text{as} \quad z \to z_0,$$

if for any $\varepsilon > 0$ there exists a positive number $\delta$ such that

$$|f(z) - w_0| < \varepsilon \quad \text{whenever} \quad 0 < |z - z_0| < \delta.$$

---

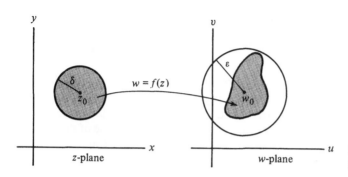

**Figure 2.4** Mapping property of a function with limit $w_0$ as $z \to z_0$.

Geometrically, this says that *any* neighborhood of $w_0$ contains all the values assumed by $f$ in some full neighborhood of $z_0$, except possibly the value $f(z_0)$; see Fig. 2.4.

**Example 1**

Prove that $\lim_{z \to i} z^2 = -1$.

**Solution.**  We must show that for given $\varepsilon > 0$ there is a positive number $\delta$ such that

$$|z^2 - (-1)| < \varepsilon \quad \text{whenever} \quad 0 < |z - i| < \delta.$$

Since

$$z^2 - (-1) = z^2 + 1 = (z - i)(z + i) = (z - i)(z - i + 2i),$$

it follows from the properties of absolute value derived in Sec. 1.3 (in particular, the triangle inequality) that

$$|z^2 - (-1)| = |z - i||z - i + 2i| \le |z - i|(|z - i| + 2). \tag{1}$$

Thus to ensure that the left-hand member of (1) is less than $\varepsilon$, we merely have to insist that $z$ lie in a neighborhood of $i$ whose radius $\delta$ is less than each of the numbers 1 and $\varepsilon/3$, for then the right-hand member of (1) will satisfy

$$|z - i|(|z - i| + 2) < \frac{\varepsilon}{3}(1 + 2) = \varepsilon. \quad \blacksquare$$

There is an obvious relation between the limit of a function and the limit of a sequence; namely, if $\lim_{z \to z_0} f(z) = w_0$, then for every sequence $\{z_n\}_1^\infty$ converging to $z_0$ $(z_n \ne z_0)$ the sequence $\{f(z_n)\}_1^\infty$ converges to $w_0$. The converse of this statement is also valid and is left as an exercise.

The condition of continuity is expressed in

---

**Definition 3.**  Let $f$ be a function defined in a neighborhood of $z_0$. Then $f$ is **continuous at $z_0$** if

$$\lim_{z \to z_0} f(z) = f(z_0).$$

---

In other words, for $f$ to be continuous at $z_0$, it must have a limiting value at $z_0$, and this limiting value must be $f(z_0)$.

A function $f$ is said to be *continuous on a set S* if it is continuous at each point of $S$.

Clearly the definitions of this section are direct analogues of concepts introduced in elementary calculus. In fact, one can show that $f(z)$ approaches a limit precisely when its real and imaginary parts approach limits (see Prob. 16); similarly, the continuity of the latter functions is equivalent to the continuity of $f$. Because of the analogy, many of the familiar theorems on real sequences, limits, and continuity remain valid in the complex case. Two such theorems are stated here.

---

**Theorem 1.**   If $\lim_{z \to z_0} f(z) = A$ and $\lim_{z \to z_0} g(z) = B$, then

(i) $\lim_{z \to z_0} (f(z) \pm g(z)) = A \pm B$,

(ii) $\lim_{z \to z_0} f(z)g(z) = AB$,

(iii) $\lim_{z \to z_0} \dfrac{f(z)}{g(z)} = \dfrac{A}{B}$    if $B \neq 0$.

---

**Theorem 2.**   If $f(z)$ and $g(z)$ are continuous at $z_0$, then so are $f(z) \pm g(z)$ and $f(z)g(z)$. The quotient $f(z)/g(z)$ is also continuous at $z_0$ provided $g(z_0) \neq 0$.

---

(Theorem 2 is, in fact, an immediate consequence of Theorem 1.)

One can easily verify that the constant functions as well as the function $f(z) = z$ are continuous on the whole plane C. Thus from Theorem 2 we deduce that the *polynomial functions* in $z$, i.e., functions of the form

$$a_0 + a_1 z + a_2 z^2 + \cdots + a_n z^n,$$

where the $a_i$ are constants, are also continuous on the whole plane. *Rational functions* in $z$, which are defined as quotients of polynomials, i.e.,

$$\frac{a_0 + a_1 z + \cdots + a_n z^n}{b_0 + b_1 z + \cdots + b_m z^m},$$

are therefore continuous at each point where the denominator does not vanish. These considerations provide a much simpler solution for problems such as Example 1, as we illustrate next.

**Example 2**

Find the limits, as $z \to 2i$, of the functions $f_1(z) = z^2 - 2z + 1$, $f_2(z) = (z + 2i)/z$, and $f_3(z) = (z^2 + 4)/z(z - 2i)$.

**Solution.**  Since $f_1(z)$ and $f_2(z)$ are continuous at $z = 2i$, we simply evaluate them there, i.e.,

$$\lim_{z \to 2i} f_1(z) = f_1(2i) = (2i)^2 - 2(2i) + 1 = -3 - 4i,$$

$$\lim_{z \to 2i} f_2(z) = f_2(2i) = \frac{2i + 2i}{2i} = 2.$$

The function $f_3(z)$ is not continuous at $z = 2i$ because it is not defined there (the denominator vanishes). However, for $z \neq 2i$ and $z \neq 0$ we have

$$f_3(z) = \frac{(z + 2i)(z - 2i)}{z(z - 2i)} = \frac{z + 2i}{z} = f_2(z),$$

and so

$$\lim_{z \to 2i} f_3(z) = \lim_{z \to 2i} f_2(z) = 2. \quad \blacksquare$$

Note that in the preceding example the discontinuity of $f_3(z)$ at $z = 2i$ can be removed by suitably defining the function at this point [set $f_3(2i) = 2$]. In general, if a function can be defined or redefined at a single point $z_0$ so as to be continuous there, we say that this function has a *removable discontinuity* at $z_0$.

In closing this section, we wish to emphasize an important distinction between the concepts of limit in the (one-dimensional) real and complex cases. For the latter situation, observe that a sequence $\{z_n\}_1^\infty$ may approach a limit $z_0$ from *any* direction in the plane, or even along a spiral, etc. Thus the manner in which a sequence of numbers approaches its limit can be much more complicated in the complex case.

## EXERCISES 2.2

1. Sketch the first five terms of the sequence $(i/2)^n$, $n = 1, 2, 3, \ldots$, and then describe the convergence of this sequence.
2. Sketch the first five terms of the sequence $(2i)^n$, $n = 1, 2, 3, \ldots$, and then describe the divergence of this sequence.
3. Using Definition 1, prove that $\lim_{n \to \infty} z_n = 0$ if and only if $\lim_{n \to \infty} |z_n| = 0$.
4. Prove that if $|z_0| < 1$, then $z_0^n \to 0$ as $n \to \infty$. Also prove that if $|z_0| > 1$, then the sequence $z_0^n$ diverges.
5. Decide whether each of the following sequences converges, and if so, find its limit.

   **(a)** $z_n = \dfrac{i}{n}$        **(b)** $z_n = i(-1)^n$        **(c)** $z_n = \text{Arg}\left(-1 + \dfrac{i}{n}\right)$

   **(d)** $z_n = \dfrac{n(2 + i)}{n + 1}$        **(e)** $z_n = \left(\dfrac{1 - i}{4}\right)^n$        **(f)** $z_n = \exp\left(\dfrac{2n\pi i}{5}\right)$

6. Use Definition 2 to prove that $\lim_{z \to 1 + i} (6z - 4) = 2 + 6i$.

7. Use Definition 2 to prove that $\lim_{z \to -i} 1/z = i$.

8. Use Theorem 1 to prove Theorem 2.

9. Find each of the following limits.

(a) $\displaystyle\lim_{z \to 2+3i} (z - 5i)^2$

(b) $\displaystyle\lim_{z \to 2} \frac{z^2 + 3}{iz}$

(c) $\displaystyle\lim_{z \to 3i} \frac{z^2 + 9}{z - 3i}$

(d) $\displaystyle\lim_{z \to i} \frac{z^2 + 1}{z^4 - 1}$

(e) $\displaystyle\lim_{\Delta z \to 0} \frac{(z_0 + \Delta z)^2 - z_0^2}{\Delta z}$

(f) $\displaystyle\lim_{z \to 1+2i} |z^2 - 1|$

10. Show that the function Arg $z$ is discontinuous at each point on the nonpositive real axis.

11. Let $f(z)$ be defined by

$$f(z) = \begin{cases} \dfrac{2z}{z + 1} & \text{if } z \neq 0, \\ 1 & \text{if } z = 0. \end{cases}$$

At which points does $f(z)$ have a limit, and at which points is it continuous? Which of the discontinuities of $f(z)$ are removable?

12. Prove that the function $g(z) = \bar{z}$ is continuous on the whole plane.

13. Prove that if $f(z)$ is continuous at $z_0$, then so are the functions $\overline{f(z)}$, Re $f(z)$, Im $f(z)$, and $|f(z)|$. [HINT: To show that $|f(z)|$ is continuous at $z_0$ use inequality (2) on page 13.]

14. Let $g$ be a function defined in a neighborhood of $z_0$ and let $f$ be a function defined in a neighborhood of the point $g(z_0)$. Show that if $g$ is continuous at $z_0$ and $f$ is continuous at $g(z_0)$, then the composite function $f(g(z))$ is continuous at $z_0$.

15. Let $f(z) = [x^2/(x^2 + y^2)] + 2i$. Does $f$ have a limit at $z = 0$? [HINT: Investigate $\{f(z_n)\}$ for sequences $\{z_n\}$ approaching 0 along the real and imaginary axes separately.]

16. Let $f(z) = u(x, y) + iv(x, y)$, $z_0 = x_0 + iy_0$, and $w_0 = u_0 + iv_0$. Prove that

$$\lim_{z \to z_0} f(z) = w_0$$

if, and only if,

$$\lim_{\substack{x \to x_0 \\ y \to y_0}} u(x, y) = u_0 \quad \text{and} \quad \lim_{\substack{x \to x_0 \\ y \to y_0}} v(x, y) = v_0.$$

[HINT: Use the triangle inequality and the facts that $|\text{Re } w| \leq |w|$, $|\text{Im } w| \leq |w|$.]

17. Use Prob. 16 to find $\lim_{z \to 1-i} [x/(x^2 + 3y)] + ixy$.

18. Use Prob. 16 to prove that $f(z) = e^z$ is continuous on the whole plane **C**.

19. Find each of the following limits:

(a) $\displaystyle\lim_{z \to 0} e^z$

(b) $\displaystyle\lim_{z \to 2\pi i} (e^z - e^{-z})$

(c) $\displaystyle\lim_{z \to \pi i/2} (z + 1)e^z$

(d) $\displaystyle\lim_{z \to -\pi i} \exp\left(\frac{z^2 + \pi^2}{z + \pi i}\right)$

20. Show that if $\lim_{n \to \infty} f(z_n) = w_0$ for every sequence $\{z_n\}_1^\infty$ converging to $z_0$ ($z_n \neq z_0$), then $\lim_{z \to z_0} f(z) = w_0$. [HINT: Show that if this were not true, then one could construct a sequence $\{z_n\}_1^\infty$ violating the hypothesis.]

## 2.3 ANALYTICITY

Now that we have a secure notion of functions of a complex variable, we are ready to turn to the main topic of this book—the theory of *analytic* functions. Before we proceed with the rigorous exposition, however, it will prove useful for the reader's perspective if we give an informal preview of what it is we want to achieve.

So far we have viewed a complex function of a complex variable, $f(z)$, as nothing more than an arbitrary mapping from the $xy$-plane to the $uv$-plane. We have individual names for the real and imaginary parts of $z$ ($x$ and $y$, respectively) and for the real and imaginary parts of $f$ ($u$ and $v$); and *any* pair $u(x, y)$ and $v(x, y)$ of two-variable functions gives us a complex function ($u + iv$) in this sense. But notice that there is something special about the pair

$$u_1(x, y) = x^2 - y^2, \qquad v_1(x, y) = 2xy,$$

as opposed to (say)

$$u_2(x, y) = x^2 - y^2, \qquad v_2(x, y) = 3xy;$$

namely, the complex function $u_1 + iv_1$ treats $z = x + iy$ as a single "unit," because it equals $x^2 - y^2 + i2xy = (x + iy)^2$ and thus it respects the complex structure of $z = x + iy$. However (apparently, at least), the formulation of $u_2 + iv_2$ requires us to break apart the real and imaginary parts of $z$.

In (real) calculus we don't deal with functions that look at a number like $3 + 4\sqrt{2}$ and square the 3 but cube the 4! The interesting calculus functions treat the number as an indivisible module. We seek to classify the complex functions that behave this same way with regard to their complex argument. Thus we want to admit functions such as

$$z = x + iy \qquad \text{(admissible)},$$
$$z^2 = x^2 - y^2 + i2xy \qquad \text{(admissible)},$$
$$z^3 = x^3 - 3xy^2 + i(3x^2y - y^3) \qquad \text{(admissible)},$$
$$\frac{1}{z} = \frac{x}{x^2 + y^2} - i\frac{y}{x^2 + y^2} \qquad \text{(admissible)},$$

and their basic arithmetic combinations (sums, products, quotients, powers, and roots) but ban such functions as

$$\text{Re } z = x \qquad \text{(inadmissible)},$$
$$\text{Im } z = y \qquad \text{(inadmissible)},$$
$$x^2 - y^2 + i3xy \qquad \text{(inadmissible)}.$$

Notice that we will have to ban the conjugate function $\bar{z}$, because if we admit it we will open the gate to $x \left[= (z + \bar{z})/2\right]$ and $y \left[= (z - \bar{z})/2i\right]$:

$$\bar{z} = x - iy \qquad \text{(inadmissible)}.$$

Similarly, admitting the modulus $|z|$ would be a mistake as well, since $\bar{z} = |z|^2/z$:

$$|z| \quad \text{(inadmissible).}$$

One could criticize our "inadmissible" classification of $u_2 + iv_2 = x^2 - y^2 + i3xy$, because we have not yet *proved* that it cannot be written in terms of $z$ alone. The following computation is instructive: we set

$$x = (z + \bar{z})/2, \qquad y = (z - \bar{z})/2i \tag{1}$$

in $u_2 + iv_2$ and obtain, after some algebra,

$$u_2 + iv_2 = x^2 - y^2 + i3xy = \frac{(z + \bar{z})^2}{4} - \frac{(z - \bar{z})^2}{(-4)} + i3 \frac{(z + \bar{z})}{2} \frac{(z - \bar{z})}{2i}$$

$$= \frac{5}{4}z^2 - \frac{1}{4}\bar{z}^2.$$

Now we see that if we admit $u_2 + iv_2$, we would have to admit $\bar{z}^2$ [since it equals $5z^2 - 4(u_2 + iv_2)$] and its undesirable square root $\bar{z}$.

## Example 1

Express the following functions in terms of $z$ and $\bar{z}$:

$$f_1(z) = \frac{x - 1 - iy}{(x - 1)^2 + y^2}, \qquad f_2(z) = x^2 + y^2 + 3x + 1 + i3y.$$

**Solution.**   Using relations (1) we obtain

$$f_1(z) = \frac{\dfrac{z + \bar{z}}{2} - 1 - i\dfrac{z - \bar{z}}{2i}}{\left(\dfrac{z + \bar{z}}{2} - 1\right)^2 + \left(\dfrac{z - \bar{z}}{2i}\right)^2}$$

$$= \frac{\bar{z} - 1}{z\bar{z} - z - \bar{z} + 1} = \frac{1}{z - 1},$$

$$f_2(z) = \frac{(z + \bar{z})^2}{4} + \frac{(z - \bar{z})^2}{4i^2} + 3\left(\frac{z + \bar{z}}{2}\right) + 1 + i3\left(\frac{z - \bar{z}}{2i}\right)$$

$$= z\bar{z} + 3z + 1. \quad \blacksquare$$

Clearly we will want to accept $f_1$ as admissible, but the presence of $\bar{z}$ in $f_2$ disqualifies it. However, this procedure—the disqualification of functions with $\bar{z}$ in their formulas—does not lead to a workable criterion. For instance, who among us would recognize that the function

$$\frac{z^2\bar{z}^2 + z^2 + \bar{z}^2 - 2\bar{z}z^2 - 2\bar{z} + 1}{10\bar{z} + z\bar{z}^2 - 2z\bar{z} - 5\bar{z}^2 + z - 5}$$

has a canceling common factor of $(\bar{z} - 1)^2$ in its numerator and denominator and thus is admissible?

The function $e^z$ is even more vexing. The definition we have adopted separates the real and imaginary parts of $z$:

$$e^z = e^x(\cos y + i \sin y).  \qquad (2)$$

But recall (Example 1, Sec. 1.4, page 23) how this definition was motivated. We postulated a "formula" for $e^z$ based on the Taylor expansion for $e^x$,

$$e^z = 1 + z + \frac{z^2}{2!} + \frac{z^3}{3!} + \cdots  \qquad (3)$$

and used it to construct $e^{iy}$. Thus (2) was "derived" from expression (3)—which apparently respects the complex structure of $z$. We'll postpone deciding on the admissibility of this function until Sec. 2.4.

Over the next four chapters we shall see that the criterion we are seeking—the test that will distinguish the admissible functions from the others—can be expressed simply in terms of *differentiability*. The following definition is a straightforward extension of the definition in the real case and appears innocuous enough.

---

**Definition 4.**   Let $f$ be a complex-valued function defined in a neighborhood of $z_0$. Then the **derivative** of $f$ at $z_0$ is given by

$$\frac{df}{dz}(z_0) \equiv f'(z_0) := \lim_{\Delta z \to 0} \frac{f(z_0 + \Delta z) - f(z_0)}{\Delta z},$$

provided this limit exists. (Such an $f$ is said to be *differentiable at $z_0$*.)

---

The catch here is that $\Delta z$ is a complex number, so it can approach zero in many different ways (from the right, from below, along a spiral, etc.); but the difference quotient must tend to a *unique* limit $f'(z_0)$ independent of the manner in which $\Delta z \to 0$. Let us see why this notion disqualifies $\bar{z}$.

## Example 2

Show that $\bar{z}$ is nowhere differentiable.

**Solution.**   The difference quotient for $f(z) = \bar{z}$ takes the form

$$\frac{f(z_0 + \Delta z) - f(z_0)}{\Delta z} = \frac{\overline{(z_0 + \Delta z)} - \overline{z_0}}{\Delta z} = \frac{\overline{\Delta z}}{\Delta z}.$$

Now if $\Delta z \to 0$ through real values, then $\Delta z = \Delta x$ (see Fig. 2.5 on page 54) and $\overline{\Delta z} = \Delta z$, so the difference quotient is 1. On the other hand, if $\Delta z \to 0$ from above, then $\Delta z = i \Delta y$ and $\overline{\Delta z} = -\Delta z$, so the quotient is $-1$. Consequently there is no way of assigning a unique value to the derivative of $\bar{z}$ at any point. Hence $\bar{z}$ is not differentiable.  ∎

A similar analysis demonstrates that neither $x$, $y$, nor $|z|$ is differentiable (see Prob. 4).

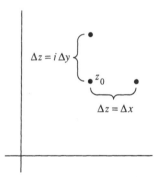

**Figure 2.5** Horizontal and vertical approach to zero of $\Delta z$.

Let us reassure ourselves that the elementary functions such as sums, products, and quotients of powers of $z$ are differentiable.

**Example 3**

Show that, for any positive integer $n$,

$$\frac{d}{dz} z^n = nz^{n-1}. \tag{4}$$

**Solution.**    Using the binomial theorem (Prob. 27 in Exercises 1.1) we find

$$\frac{(z + \Delta z)^n - z^n}{\Delta z} = \frac{nz^{n-1}\,\Delta z + \dfrac{n(n-1)}{2}z^{n-2}(\Delta z)^2 + \cdots + (\Delta z)^n}{\Delta z}.$$

Thus

$$\frac{d}{dz} z^n = \lim_{\Delta z \to 0} \frac{(z + \Delta z)^n - z^n}{\Delta z} = nz^{n-1}. \quad\blacksquare$$

Notice that the proof was just the same as for the real-variable case. In fact the validity of any of the following rules can be proven from Definition 4 by mimicking the corresponding proof from elementary calculus.

---

**Theorem 3.**    If $f$ and $g$ are differentiable at $z_0$, then

$$(f \pm g)'(z_0) = f'(z_0) \pm g'(z_0), \tag{5}$$

$$(cf)'(z_0) = cf'(z_0) \qquad \text{(for any constant } c\text{)}, \tag{6}$$

$$(fg)'(z_0) = f(z_0)g'(z_0) + f'(z_0)g(z_0), \tag{7}$$

$$\left(\frac{f}{g}\right)'(z_0) = \frac{g(z_0)f'(z_0) - f(z_0)g'(z_0)}{g(z_0)^2} \qquad \text{if } g(z_0) \neq 0, \tag{8}$$

and

$$\frac{d}{dz} f(g(z)) = f'(g(z))g'(z). \tag{9}$$

---

Notice that Eq. (9) is the *chain rule*. Furthermore, differentiability implies continuity, as in the real case (see Prob. 3).

It follows from Example 3 and rules (5) and (6) that any polynomial in $z$,

$$P(z) = a_n z^n + a_{n-1} z^{n-1} + \cdots + a_1 z + a_0,$$

is differentiable in the whole plane and that its derivative is given by

$$P'(z) = na_n z^{n-1} + (n-1)a_{n-1} z^{n-2} + \cdots + a_1.$$

Consequently, from rule (8), any rational function of $z$ is differentiable at every point in its domain of definition. We see then that for purposes of differentiation, polynomial and rational functions in $z$ can be treated as if $z$ were a real variable.

**Example 4**

Compute the derivative of

$$f(z) = \left(\frac{z^2 - 1}{z^2 + 1}\right)^{100}.$$

**Solution.** Unless $z = \pm i$ (where the denominator is zero), the usual calculus rules apply. Thus

$$f'(z) = 100\left(\frac{z^2 - 1}{z^2 + 1}\right)^{99} \frac{(z^2 + 1)2z - (z^2 - 1)2z}{(z^2 + 1)^2} = 400z \frac{(z^2 - 1)^{99}}{(z^2 + 1)^{101}}. \quad\blacksquare$$

As we demonstrate in Prob. 10, it is possible for a complex function to be differentiable solely at isolated points. Of course, this also occurs in real analysis. Such functions are treated there as exceptional cases, while the general theorems usually apply only to functions differentiable over open intervals of the real line. By analogy, then, we distinguish a special class of complex functions in

---

**Definition 5.**   A complex-valued function $f(z)$ is said to be **analytic**[†] on an open set $G$ if it has a derivative at every point of $G$.

---

We emphasize that analyticity is a property defined over open sets, while differentiability could conceivably hold at one point only. Occasionally, however, we shall use the abbreviated phrase "*$f(z)$ is analytic at the point $z_0$*" to mean that $f(z)$ is analytic in some neighborhood of $z_0$. Thus we can say that a rational function of $z$ is analytic at every point for which its denominator is nonzero. If $f(z)$ is analytic on the whole complex plane, then it is said to be *entire*. For example, all polynomial functions of $z$ are entire.

As we shall see in the next few chapters, analyticity is the criterion that we have been seeking for functions that respect the complex structure of the variable $z$. In fact, Sec. 5.2 will demonstrate that all analytic functions can be written in terms of $z$ alone (no $x$, $y$, or $\bar{z}$).

---

[†] Some authors use the words *holomorphic* or *regular* instead of analytic.

When a function is given in terms of real and imaginary parts as $u(x, y) + iv(x, y)$, it may be very tedious to apply the definition to determine if $f$ is analytic. In the next section we will establish a test that is easier to use. Also, we will verify the analyticity of $e^z$. We will have no further occasion for using the substitution method based on Eqs. (1).

## EXERCISES 2.3

1. Let $f(z)$ be defined in a neighborhood of $z_0$. Show that finding $\lim_{\Delta z \to 0} [f(z_0 + \Delta z) - f(z_0)]/\Delta z$ is equivalent to finding $\lim_{z \to z_0} [f(z) - f(z_0)]/(z - z_0)$.

2. Prove that if $f(z)$ is differentiable at $z_0$, then

$$f(z) = f(z_0) + f'(z_0)(z - z_0) + \lambda(z)(z - z_0),$$

   where $\lambda(z) \to 0$ as $z \to z_0$.

3. Prove that if $f(z)$ is differentiable at $z_0$, then it is continuous at $z_0$. [HINT: Use the result of Prob. 2.]

4. Using Definition 4, prove that each of the following functions is nowhere differentiable.
   (a) Re $z$          (b) Im $z$          (c) $|z|$

5. Prove rules (5) and (7).

6. Prove that formula (4) is also valid for negative integers $n$.

7. Use rules (5)–(9) to find the derivatives of the following functions.
   (a) $f(z) = 6z^3 + 8z^2 + iz + 10$          (b) $f(z) = (z^2 - 3i)^{-6}$

   (c) $f(z) = \dfrac{z^2 - 9}{iz^3 + 2z + \pi}$          (d) $f(z) = \dfrac{(z + 2)^3}{(z^2 + iz + 1)^4}$

   (e) $f(z) = 6i(z^3 - 1)^4(z^2 + iz)^{100}$

8. (*Geometric Interpretation of $f'$*) Suppose that $f$ is analytic at $z_0$ and $f'(z_0) \neq 0$. Show that

$$\lim_{z \to z_0} \frac{|f(z) - f(z_0)|}{|z - z_0|} = |f'(z_0)|$$

   and

$$\lim_{z \to z_0} \{\arg[f(z) - f(z_0)] - \arg(z - z_0)\} = \arg f'(z_0).$$

   Thus, on setting $w = f(z)$ and $w_0 = f(z_0)$ we see that for $z$ near $z_0$, the mapping $f$ dilates distances by the factor $|f'(z_0)|$:

$$|w - w_0| \approx |f'(z_0)| \times |z - z_0|.$$

   Also, $f$ rotates vectors emanating from $z_0$ by an angle of arg $f'(z_0)$:

$$\arg(w - w_0) \approx \arg(z - z_0) + \arg f'(z_0).$$

   In other words, for $z$ near $z_0$ the mapping $w = f(z)$ behaves like the linear transformation

$$w = f(z_0) + f'(z_0)(z - z_0)$$

$$= c + e^{i\phi}\rho(z - z_0).$$

(See Prob. 12, Exercises 2.1.)

9. For each of the following determine the points at which the function is not analytic.

   (a) $\dfrac{1}{z - 2 + 3i}$

   (b) $\dfrac{iz^3 + 2z}{z^2 + 1}$

   (c) $\dfrac{3z - 1}{z^2 + z + 4}$

   (d) $z^2(2z^2 - 3z + 1)^{-2}$

10. Let $f(z) = |z|^2$. Use Definition 4 to show that $f$ is differentiable at $z = 0$ but is not differentiable at any other point. HINT: Write

$$\frac{|z_0 + \Delta z|^2 - |z_0|^2}{\Delta z} = \frac{(z_0 + \Delta z)(\bar{z}_0 + \overline{\Delta z}) - z_0 \bar{z}_0}{\Delta z}$$

$$= \bar{z}_0 + \overline{\Delta z} + z_0 \frac{\overline{\Delta z}}{\Delta z}.$$

11. Discuss the analyticity of each of the following functions.

   (a) $8\bar{z} + i$

   (b) $\dfrac{z}{\bar{z} + 2}$

   (c) $\dfrac{z^3 + 2z + i}{z - 5}$

   (d) $x^2 - y^2 + 2xyi$

   (e) $x^2 + y^2 + y - 2 + ix$

   (f) $\left(x + \dfrac{x}{x^2 + y^2}\right) + i\left(y - \dfrac{y}{x^2 + y^2}\right)$

   (g) $|z|^2 + 2z$

   (h) $\dfrac{|z| + z}{2}$

12. Let $P(z) = (z - z_1)(z - z_2) \cdots (z - z_n)$. Show by induction on $n$ that

$$\frac{P'(z)}{P(z)} = \frac{1}{z - z_1} + \frac{1}{z - z_2} + \cdots + \frac{1}{z - z_n}.$$

   [NOTE: $P'(z)/P(z)$ is called the *logarithmic derivative* of $P(z)$.]

13. Let $f(z)$ and $g(z)$ be entire functions. Decide which of the following statements are always true.

   (a) $f(z)^3$ is entire.

   (b) $f(z)g(z)$ is entire.

   (c) $f(z)/g(z)$ is entire.

   (d) $5f(z) + ig(z)$ is entire.

   (e) $f(1/z)$ is entire.

   (f) $g(z^2 + 2)$ is entire.

   (g) $f(g(z))$ is entire.

14. Prove *L'Hospital's rule*: If $f(z)$ and $g(z)$ are analytic at $z_0$ and $f(z_0) = g(z_0) = 0$, but $g'(z_0) \neq 0$, then

$$\lim_{z \to z_0} \frac{f(z)}{g(z)} = \frac{f'(z_0)}{g'(z_0)}.$$

   HINT: Write

$$\frac{f(z)}{g(z)} = \frac{f(z) - f(z_0)}{z - z_0} \bigg/ \frac{g(z) - g(z_0)}{z - z_0}.$$

15. Use L'Hospital's rule to find $\lim_{z \to i} (1 + z^6)/(1 + z^{10})$.

16. Let $f(z) = z^3 + 1$, and let $z_1 = (-1 + \sqrt{3}i)/2$, $z_2 = (-1 - \sqrt{3}i)/2$. Show that there is no point $w$ on the line segment from $z_1$ to $z_2$ such that

$$f(z_2) - f(z_1) = f'(w)(z_2 - z_1).$$

   This shows that the mean-value theorem of calculus does not extend to complex functions.

17. Let $F(z) = f(z)g(z)h(z)$, where $f$, $g$, and $h$ are each differentiable at $z_0$. Prove that

$$F'(z_0) = f'(z_0)g(z_0)h(z_0) + f(z_0)g'(z_0)h(z_0) + f(z_0)g(z_0)h'(z_0).$$

## 2.4 THE CAUCHY-RIEMANN EQUATIONS

The property of analyticity for a function indicates some type of connection between its real and imaginary parts. The precise expression of this kinship is easily derived, as we shall see shortly, by letting $\Delta z$ approach zero from the right and from above in Definition 4. In this section we shall explore the nature of this relationship.

If the function $f(z) = u(x, y) + iv(x, y)$ is differentiable at $z_0 = x_0 + iy_0$, then the limit

$$f'(z_0) = \lim_{\Delta z \to 0} \frac{f(z_0 + \Delta z) - f(z_0)}{\Delta z}$$

can be computed by allowing $\Delta z \ (= \Delta x + i \, \Delta y)$ to approach zero from any convenient direction in the complex plane. If it approaches horizontally, then $\Delta z = \Delta x$, and we obtain

$$f'(z_0) = \lim_{\Delta x \to 0} \frac{u(x_0 + \Delta x, y_0) + iv(x_0 + \Delta x, y_0) - u(x_0, y_0) - iv(x_0, y_0)}{\Delta x}$$

$$= \lim_{\Delta x \to 0} \left[ \frac{u(x_0 + \Delta x, y_0) - u(x_0, y_0)}{\Delta x} \right] + i \lim_{\Delta x \to 0} \left[ \frac{v(x_0 + \Delta x, y_0) - v(x_0, y_0)}{\Delta x} \right].$$

(It may be helpful to consult Fig. 2.5 again.) Since the limits of the bracketed expressions are just the first partial derivatives of $u$ and $v$ with respect to $x$, we deduce that

$$f'(z_0) = \frac{\partial u}{\partial x} (x_0, y_0) + i \frac{\partial v}{\partial x} (x_0, y_0). \tag{1}$$

On the other hand, if $\Delta z$ approaches zero vertically, then $\Delta z = i \, \Delta y$ and we have

$$f'(z_0) = \lim_{\Delta y \to 0} \left[ \frac{u(x_0, y_0 + \Delta y) - u(x_0, y_0)}{i \, \Delta y} \right] + i \lim_{\Delta y \to 0} \left[ \frac{v(x_0, y_0 + \Delta y) - v(x_0, y_0)}{i \, \Delta y} \right].$$

Hence

$$f'(z_0) = -i \frac{\partial u}{\partial y} (x_0, y_0) + \frac{\partial v}{\partial y} (x_0, y_0). \tag{2}$$

But the right-hand members of Eqs. (1) and (2) are equal to the same complex number $f'(z_0)$, so by equating real and imaginary parts we see that the equations

$$\boxed{\frac{\partial u}{\partial x} = \frac{\partial v}{\partial y}, \quad \frac{\partial u}{\partial y} = -\frac{\partial v}{\partial x}} \tag{3}$$

must hold at $z_0 = x_0 + iy_0$. Equations (3) are called the *Cauchy-Riemann* equations. We have thus established

> **Theorem 4.** A necessary condition for a function $f(z) = u(x, y) + iv(x, y)$ to be differentiable at a point $z_0$ is that the Cauchy-Riemann equations hold at $z_0$.
>
> Consequently, if $f$ is analytic in an open set $G$, then the Cauchy-Riemann equations must hold at every point of $G$.

**Example 1**

Show that the function $f(z) = (x^2 + y) + i(y^2 - x)$ is not analytic at any point.

**Solution.** Since $u(x, y) = x^2 + y$ and $v(x, y) = y^2 - x$, we have

$$\frac{\partial u}{\partial x} = 2x, \qquad \frac{\partial v}{\partial y} = 2y,$$

$$\frac{\partial u}{\partial y} = 1, \qquad \frac{\partial v}{\partial x} = -1.$$

Hence the Cauchy-Riemann equations are simultaneously satisfied only on the line $x = y$ and therefore in no open disk. Thus, by Theorem 4 the function $f(z)$ is nowhere analytic.  ∎

To be mathematically precise, we point out that the Cauchy-Riemann equations alone are *not* sufficient to ensure differentiability; one needs the additional hypothesis of continuity of the first partial derivatives of $u$ and $v$. The complete story is given in the following theorem.

> **Theorem 5.** Let $f(z) = u(x, y) + iv(x, y)$ be defined in some open set $G$ containing the point $z_0$. If the first partial derivatives of $u$ and $v$ exist in $G$, are continuous at $z_0$, and satisfy the Cauchy-Riemann equations at $z_0$, then $f$ is differentiable at $z_0$.
>
> Consequently, if the first partial derivatives are continuous and satisfy the Cauchy-Riemann equations at all points of $G$, then $f$ is analytic in $G$.

*Proof.* The difference quotient for $f$ at $z_0$ can be written in the form

$$\frac{f(z_0 + \Delta z) - f(z_0)}{\Delta z}$$

$$= \frac{[u(x_0 + \Delta x, y_0 + \Delta y) - u(x_0, y_0)] + i[v(x_0 + \Delta x, y_0 + \Delta y) - v(x_0, y_0)]}{\Delta x + i \, \Delta y} \tag{4}$$

where $z_0 = x_0 + iy_0$ and $\Delta z = \Delta x + i \, \Delta y$. The above expressions are well defined if $|\Delta z|$ is so small that the closed disk with center $z_0$ and radius $|\Delta z|$ lies entirely in $G$. Let us rewrite the difference

$$u(x_0 + \Delta x, y_0 + \Delta y) - u(x_0, y_0)$$

as

$$[u(x_0 + \Delta x, y_0 + \Delta y) - u(x_0, y_0 + \Delta y)] + [u(x_0, y_0 + \Delta y) - u(x_0, y_0)]. \qquad \textbf{(5)}$$

Because the partial derivatives exist in $G$, the mean-value theorem says that there is a number $x^*$ between $x_0$ and $x_0 + \Delta x$ such that

$$u(x_0 + \Delta x, y_0 + \Delta y) - u(x_0, y_0 + \Delta y) = \Delta x \frac{\partial u}{\partial x}(x^*, y_0 + \Delta y).$$

Furthermore, since the partial derivatives are continuous at $(x_0, y_0)$, we can write

$$\frac{\partial u}{\partial x}(x^*, y_0 + \Delta y) = \frac{\partial u}{\partial x}(x_0, y_0) + \varepsilon_1,$$

where the function $\varepsilon_1 \to 0$ as $x^* \to x_0$ and $\Delta y \to 0$ (in particular, as $\Delta z \to 0$). Thus the first bracketed expression in (5) can be written as

$$u(x_0 + \Delta x, y_0 + \Delta y) - u(x_0, y_0 + \Delta y) = \Delta x \left[ \frac{\partial u}{\partial x}(x_0, y_0) + \varepsilon_1 \right].$$

The second bracketed expression in (5) is treated similarly, introducing the function $\varepsilon_2$. Then working the same strategy for the $v$-difference in Eq. (4), we ultimately have

$$\frac{f(z_0 + \Delta z) - f(z_0)}{\Delta z} = \frac{\Delta x \left[ \dfrac{\partial u}{\partial x} + \varepsilon_1 + i\dfrac{\partial v}{\partial x} + i\varepsilon_3 \right] + \Delta y \left[ \dfrac{\partial u}{\partial y} + \varepsilon_2 + i\dfrac{\partial v}{\partial y} + i\varepsilon_4 \right]}{\Delta x + i\,\Delta y}$$

where each partial derivative is evaluated at $(x_0, y_0)$ and where each $\varepsilon_i \to 0$ as $\Delta z \to 0$. Now we use the Cauchy-Riemann equations to express the difference quotient as

$$\frac{\Delta x \left[ \dfrac{\partial u}{\partial x} + i\dfrac{\partial v}{\partial x} \right] + i\,\Delta y \left[ \dfrac{\partial u}{\partial x} + i\dfrac{\partial v}{\partial x} \right]}{\Delta x + i\,\Delta y} + \frac{\lambda}{\Delta x + i\,\Delta y}, \qquad \textbf{(6)}$$

where $\lambda := \Delta x(\varepsilon_1 + i\varepsilon_3) + \Delta y(\varepsilon_2 + i\varepsilon_4)$. Since

$$\left| \frac{\lambda}{\Delta x + i\,\Delta y} \right| \leq \left| \frac{\Delta x}{\Delta x + i\,\Delta y} \right| |\varepsilon_1 + i\varepsilon_3| + \left| \frac{\Delta y}{\Delta x + i\,\Delta y} \right| |\varepsilon_2 + i\varepsilon_4|$$

$$\leq |\varepsilon_1 + i\varepsilon_3| + |\varepsilon_2 + i\varepsilon_4|,$$

we see that the last term in (6) approaches zero as $\Delta z \to 0$, and so

$$\lim_{\Delta z \to 0} \frac{f(z_0 + \Delta z) - f(z_0)}{\Delta z} = \frac{\partial u}{\partial x}(x_0, y_0) + i\frac{\partial v}{\partial x}(x_0, y_0);$$

i.e., $f'(z_0)$ exists. ∎

It follows from Theorem 5 that the nowhere analytic function $f(z)$ of Example 1 is, nonetheless, differentiable at each point on the line $x = y$.

As promised in Sec. 1.4, we now offer one last vindication of our definition of the complex exponential by demonstrating its analyticity.

**Example 2**

Prove that the function $f(z) = e^z = e^x \cos y + i e^x \sin y$ is entire, and find its derivative.

**Solution.**  Since $\partial u/\partial x = e^x \cos y$, $\partial v/\partial y = e^x \cos y$, $\partial u/\partial y = -e^x \sin y$, and $\partial v/\partial x = e^x \sin y$, the first partial derivatives are continuous and satisfy the Cauchy-Riemann equations at every point in the plane. Hence $f(z)$ is entire. From Eq. (1) we see that

$$f'(z) = \frac{\partial u}{\partial x} + i \frac{\partial v}{\partial x} = e^x \cos y + i e^x \sin y.$$

Not surprisingly, $f'(z) = f(z)$.  ■

As a further application of these techniques, let us prove the following theorem whose analogue in the real case is well known.

---

**Theorem 6.**  If $f(z)$ is analytic in a domain $D$ and if $f'(z) = 0$ everywhere in $D$, then $f(z)$ is constant in $D$.

---

Before we proceed with the proof, we observe that the *connectedness* property of the domain is essential. Indeed, if $f(z)$ is defined by

$$f(z) = \begin{cases} 0 & \text{if } |z| < 1, \\ 1 & \text{if } |z| > 2, \end{cases}$$

then $f$ is analytic and $f'(z) = 0$ on its *domain of definition* (which is not a *domain!*), yet $f$ is not constant.

*Proof of Theorem 6.*   Since $f'(z) = 0$ in $D$, we see from Eqs. (1) and (2) that all the first partial derivatives of $u$ and $v$ vanish in $D$; that is,

$$\frac{\partial u}{\partial x} = \frac{\partial u}{\partial y} = \frac{\partial v}{\partial x} = \frac{\partial v}{\partial y} = 0.$$

Thus, by Theorem 1 in Sec. 1.6 (see page 33), we have $u = $ constant and $v = $ constant in $D$. Consequently, $f = u + iv$ is also constant in $D$.  ■

Using the preceding theorem and the Cauchy-Riemann equations, one can show that an analytic function $f(z)$ must be constant when any one of the following conditions hold in the domain $D$:

$$\text{Re } f(z) \text{ is constant.} \tag{7}$$

$$\text{Im } f(z) \text{ is constant} \tag{8}$$
[this follows from (7) by considering $if(z)$].

$$|f(z)| \text{ is constant.} \tag{9}$$

The proofs are left as problems.

## EXERCISES 2.4

1. Use the Cauchy-Riemann equations to show that the following functions are nowhere differentiable.
   **(a)** $w = \bar{z}$          **(b)** $w = \text{Re } z$          **(c)** $w = 2y - ix$

2. Show that $h(z) = x^3 + 3xy^2 - 3x + i(y^3 + 3x^2y - 3y)$ is differentiable on the coordinate axes but is nowhere analytic.

3. Use Theorem 5 to show that $g(z) = 3x^2 + 2x - 3y^2 - 1 + i(6xy + 2y)$ is entire. Write this function in terms of $z$.

4. Let
$$f(z) = \begin{cases} \dfrac{x^{4/3}y^{5/3} + ix^{5/3}y^{4/3}}{x^2 + y^2} & \text{if } z \neq 0, \\ 0 & \text{if } z = 0. \end{cases}$$

Show that the Cauchy-Riemann equations hold at $z = 0$ but that $f$ is not differentiable at this point. [HINT: Consider the difference quotient $f(\Delta z)/\Delta z$ for $\Delta z \to 0$ along the real axis and along the line $y = x$.]

5. Show that the function $f(z) = e^{x^2 - y^2}[\cos(2xy) + i \sin(2xy)]$ is entire, and find its derivative.

6. If $u$ and $v$ are expressed in terms of polar coordinates $(r, \theta)$, show that the Cauchy-Riemann equations can be written in the form
$$\frac{\partial u}{\partial r} = \frac{1}{r}\frac{\partial v}{\partial \theta}, \qquad \frac{\partial v}{\partial r} = -\frac{1}{r}\frac{\partial u}{\partial \theta}.$$

7. Show that if two analytic functions have the same derivative throughout a domain $D$, then they differ only by an additive constant.

8. Show that if $f$ is analytic in a domain $D$ and $\text{Re } f(z)$ is constant in $D$, then $f(z)$ must be constant in $D$.

9. Show, by contradiction, that the function $F(z) = |z^2 - z|$ is nowhere analytic because of condition (8).

10. Show that if $f(z)$ is analytic and real-valued in a domain $D$, then $f(z)$ is constant in $D$.

11. Suppose that $f(z)$ and $\overline{f(z)}$ are analytic in a domain $D$. Show that $f(z)$ is constant in $D$.

12. Show that if $f$ is analytic in a domain $D$ and $|f(z)|$ is constant in $D$, then the function $f(z)$ is constant in $D$. [HINT: $|f|^2$ is constant, so $\partial|f|^2/\partial x = \partial|f|^2/\partial y = 0$ throughout $D$. Using these two relations and the Cauchy-Riemann equations, deduce that $f'(z) = 0$.]

13. Given that $f(z)$ and $|f(z)|$ are each analytic in a domain $D$, prove that $f(z)$ is constant in $D$.

14. Show that if the analytic function $w = f(z)$ maps a domain $D$ onto a portion of a line, then $f$ must be constant throughout $D$.

15. The *Jacobian* of a mapping
$$u = u(x, y), \qquad v = v(x, y)$$

from the $xy$-plane to the $uv$-plane is defined to be the determinant
$$J(x_0, y_0) := \begin{vmatrix} \dfrac{\partial u}{\partial x} & \dfrac{\partial u}{\partial y} \\ \dfrac{\partial v}{\partial x} & \dfrac{\partial v}{\partial y} \end{vmatrix},$$

where the partial derivatives are all evaluated at $(x_0, y_0)$. Show that if $f = u + iv$ is analytic at $z_0 = x_0 + iy_0$, then $J(x_0, y_0) = |f'(z_0)|^2$.

16. The notion of analyticity as discussed in the preceding section requires that $f(x, y) = u(x, y) + iv(x, y)$ can be written in terms of $(x + iy)$ alone, without using $\bar{z} = (x - iy)$.[†] To make this concept more explicit, we introduce the change of variables

$$\begin{cases} \xi = x + iy \\ \eta = x - iy \end{cases} \quad \text{or, equivalently,} \quad \begin{cases} x = \dfrac{\xi + \eta}{2} \\ y = \dfrac{\xi - \eta}{2i} \end{cases}$$

producing the function

$$\tilde{f}(\xi, \eta) := f(x(\xi, \eta), y(\xi, \eta)).$$

(a) Using the chain rule show formally that

$$\frac{\partial \tilde{f}}{\partial \xi} = \frac{1}{2}\left(\frac{\partial u}{\partial x} + \frac{\partial v}{\partial y}\right) + \frac{i}{2}\left(\frac{\partial v}{\partial x} - \frac{\partial u}{\partial y}\right),$$

$$\frac{\partial \tilde{f}}{\partial \eta} = \frac{1}{2}\left(\frac{\partial u}{\partial x} - \frac{\partial v}{\partial y}\right) + \frac{i}{2}\left(\frac{\partial u}{\partial y} + \frac{\partial v}{\partial x}\right).$$

(b) Since $\eta$ is the same as $\bar{z}$, the statement "$f$ is independent of $\bar{z}$" is equivalent to

$$\frac{\partial \tilde{f}}{\partial \eta} = \frac{\partial \tilde{f}}{\partial \bar{z}} = 0.$$

Show that this condition is the same as the Cauchy-Riemann equations for $f$.

## 2.5 HARMONIC FUNCTIONS

Solutions of the two-dimensional *Laplace equation*

$$\Delta \phi := \frac{\partial^2 \phi}{\partial x^2} + \frac{\partial^2 \phi}{\partial y^2} = 0 \qquad (1)$$

are among the most important functions in mathematical physics. The electrostatic potential solves Eq. (1) in two-dimensional free space, as does the scalar magnetostatic potential; the corresponding field in any direction is given by the directional derivative of $\phi(x, y)$. Two-dimensional fluid flow problems are described by such functions under certain idealized conditions, and $\phi$ can also be interpreted as the displacement of a membrane stretched across a loop of wire, if the loop is nearly flat. In the next section we shall discuss equilibrium temperature distributions as models for solutions to (1).

One of the most important applications of analytic function theory to applied mathematics is the abundance of solutions of Eq. (1) that it supplies. We shall adopt the following standard terminology for these solutions.

---

[†] That is, "z-bar" is barred!

**Definition 6.**   A real-valued function $\phi(x, y)$ is said to be **harmonic** in a domain $D$ if all its second-order partial derivatives are continuous in $D$ and if at each point of $D$, $\phi$ satisfies Eq. (1).

The sources of these harmonic functions are the real and imaginary parts of analytic functions, as we prove in the next theorem.

**Theorem 7.**   If $f(z) = u(x, y) + iv(x, y)$ is analytic in a domain $D$, then each of the functions $u(x, y)$ and $v(x, y)$ is harmonic in $D$.

*Proof.*   In a later chapter we shall show that the real and imaginary parts of any analytic function have continuous partial derivatives of all orders. Assuming this fact, we recall from elementary calculus that under such conditions mixed partial derivatives can be taken in any order; i.e.,

$$\frac{\partial}{\partial y}\frac{\partial u}{\partial x} = \frac{\partial}{\partial x}\frac{\partial u}{\partial y}. \tag{2}$$

Using the Cauchy-Riemann equations for the first derivatives, we transform Eq. (2) into

$$\frac{\partial^2 v}{\partial y^2} = -\frac{\partial^2 v}{\partial x^2}.$$

which is equivalent to Eq. (1). Thus $v$ is harmonic in $D$, and a similar computation proves that $u$ is also.   ∎

Conversely, if we are given a function $u(x, y)$ harmonic in, say, an open disk, then we can find another harmonic function $v(x, y)$ so that $u + iv$ is an analytic function of $z$ in the disk. Such a function $v$ is called a *harmonic conjugate* of $u$. The construction of $v$ is effected by exploiting the Cauchy-Riemann equations, as we illustrate in the following example.

**Example 1**

Construct an analytic function whose real part is $u(x, y) = x^3 - 3xy^2 + y$.

**Solution.**   First we verify that

$$\frac{\partial^2 u}{\partial x^2} + \frac{\partial^2 u}{\partial y^2} = 6x - 6x = 0,$$

and so $u$ is harmonic in the whole plane. Now we have to find a mate, $v(x, y)$, for $u$ such that the Cauchy-Riemann equations are satisfied. Thus we must have

$$\frac{\partial v}{\partial y} = \frac{\partial u}{\partial x} = 3x^2 - 3y^2 \tag{3}$$

and

$$\frac{\partial v}{\partial x} = -\frac{\partial u}{\partial y} = 6xy - 1. \tag{4}$$

If we hold $x$ constant and integrate Eq. (3) with respect to $y$, we get

$$v(x, y) = 3x^2y - y^3 + \text{constant},$$

but the "constant" could conceivably be any differentiable function of $x$; it need only be independent of $y$. Therefore, we write

$$v(x, y) = 3x^2y - y^3 + \psi(x).$$

We can find $\psi(x)$ by plugging this last expression into Eq. (4);

$$\frac{\partial v}{\partial x} = 6xy + \psi'(x) = 6xy - 1. \tag{5}$$

This yields $\psi'(x) \equiv -1$, and so $\psi(x) = -x + a$, where $a$ is some (genuine) constant. It follows that a harmonic conjugate of $u(x, y)$ is given by

$$v(x, y) = 3x^2y - y^3 - x + a,$$

and the analytic function

$$f(z) = x^3 - 3xy^2 + y + i(3x^2y - y^3 - x + a),$$

which we recognize as $z^3 - i(z - a)$, solves the problem.  ∎

This procedure will always work for any $u(x, y)$ harmonic in a *disk*, as is shown in Prob. 19.[†] Thus we can learn a great deal about analytic functions by studying harmonic functions and vice versa.

The harmonic functions forming the real and imaginary parts of an analytic function $f(z)$ each generate a family of curves in the $xy$-plane, namely, the *level curves*, or *isotimic curves*

$$u(x, y) = \text{constant} \tag{6}$$

and

$$v(x, y) = \text{constant}. \tag{7}$$

If $u$ is interpreted as an electrostatic potential, then the curves (6) are the *equipotentials*. If $u$ is temperature, (6) describes the *isotherms*.

For the function $f(z) = z^2 = x^2 - y^2 + i2xy$, the level curves $u = x^2 - y^2 = $ constant are hyperbolas asymptotic to the lines $y = \pm x$, as shown in Fig. 2.6(a). The curves $v = 2xy = $ constant are also hyperbolas, asymptotic to the coordinate axes; see Fig. 2.6(b).

Notice that if the two families of curves are superimposed as in Fig. 2.6(c) they appear to intersect at right angles. The same effect occurs with the level curves for

---

[†] Also see Prob. 20 for an example of a function harmonic in a *punctured* disk that does not have a harmonic conjugate in the punctured disk.

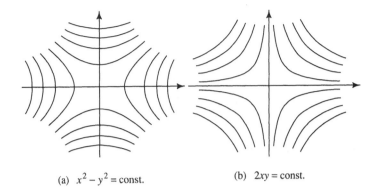

(a) $x^2 - y^2 = $ const.

(b) $2xy = $ const.

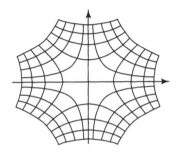

(c) Orthogonal level curves:
Re($z^2$) = const. and Im($z^2$) = const.

**Figure 2.6** Level curves of real and imaginary parts of $z^2$.

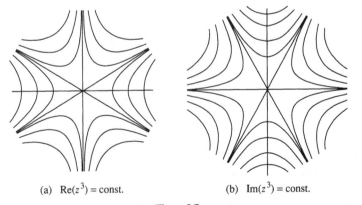

(a) Re($z^3$) = const.

(b) Im($z^3$) = const.

**Figure 2.7**

the analytic functions $z^3$ (Fig. 2.7), $1/z$ (Fig. 2.8), and $e^z$ (Fig. 2.9). This is no accident; the level curves of the real and imaginary parts of an analytic function $f(z)$ will always intersect at right angles—unless $f'(z) = 0$ at the point of intersection. This can be seen from the Cauchy-Riemann equations as follows.

Recall that the vector with components $[\partial u/\partial x, \partial u/\partial y]$ is the *gradient* of $u$ and is normal to the level curves of $u$. Similarly, $[\partial v/\partial x, \partial v/\partial y]$ is normal to the level

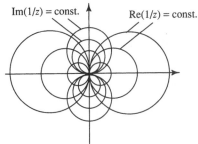

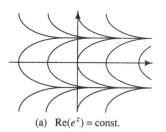

Figure 2.8   Level curves:
Re(1/z) = const., Im(1/z) = const.

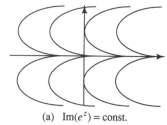

(a)  Re($e^z$) = const.                                  (a)  Im($e^z$) = const.

**Figure 2.9**

curves of $v$. The scalar (dot) product of these gradient vectors is

$$\frac{\partial u}{\partial x}\frac{\partial v}{\partial x} + \frac{\partial u}{\partial y}\frac{\partial v}{\partial y} = \frac{\partial v}{\partial y}\frac{\partial v}{\partial x} - \frac{\partial v}{\partial x}\frac{\partial v}{\partial y} = 0$$

by the Cauchy-Riemann equations. Thus *if these gradients are nonzero*, they are perpendicular, and hence so are the level curves. As a rule, *level curves of conjugate harmonic functions intersect at right angles.*

The following examples illustrate how analytic function theory can be used to solve Laplace's equation in regions whose boundaries are identifiable as level curves.

**Example 2**

Find a function $\phi(x, y)$ that is harmonic in the region of the right half-plane between the curves $x^2 - y^2 = 2$ and $x^2 - y^2 = 4$ and takes the value 3 on the left edge and the value 7 on the right edge (Fig. 2.10 on page 68).

**Solution.**   We recognize $x^2 - y^2$ as the real part of $z^2$, so the boundary curves are level curves of a known harmonic function. To meet the specified boundary conditions, we add some flexibility by considering

$$\phi(x, y) = A(x^2 - y^2) + B = \text{Re}(Az^2 + B), \qquad A, B \text{ real},$$

and adjust $A$ and $B$ accordingly. When $x^2 - y^2 = 2$, $\phi = 3$;

$$A(2) + B = 3.$$

When $x^2 - y^2 = 4$, $\phi = 7$;

$$A(4) + B = 7.$$

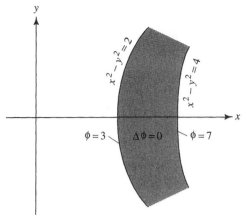

**Figure 2.10**  Laplace's equation for the region of Example 2.

Solving for $A$ and $B$ we find the solution to be

$$\phi(x, y) = 2(x^2 - y^2) - 1. \quad \blacksquare$$

In Chapter 7 we shall consider more profound examples of this idea.

## EXERCISES 2.5

1. Verify directly that the real and imaginary parts of the following analytic functions satisfy Laplace's equation.

   **(a)** $f(z) = z^2 + 2z + 1$           **(b)** $g(z) = \dfrac{1}{z}$

   **(c)** $h(z) = e^z$

2. Find the most general harmonic polynomial of the form $ax^2 + bxy + cy^2$.

3. Verify that each given function $u$ is harmonic (in the region where it is defined) and then find a harmonic conjugate of $u$.

   **(a)** $u = y$                **(b)** $u = e^x \sin y$              **(c)** $u = xy - x + y$

   **(d)** $u = \sin x \cosh y$    **(e)** $u = \ln |z|$ for Re $z > 0$     **(f)** $u = \mathrm{Im}(e^{z^2})$

4. Show that if $v(x, y)$ is a harmonic conjugate of $u(x, y)$ in a domain $D$, then *every* harmonic conjugate of $u(x, y)$ in $D$ must be of the form $v(x, y) + a$, where $a$ is a real constant.

5. Show that if $v$ is a harmonic conjugate for $u$, then $-u$ is a harmonic conjugate for $v$.

6. Show that if $v$ is a harmonic conjugate of $u$ in a domain $D$, then $uv$ is harmonic in $D$.

7. Find a function $\phi(x, y)$ that is harmonic in the infinite vertical strip $\{z : -1 \le \mathrm{Re}\ z \le 3\}$ and takes the value 0 on the left edge and the value 4 on the right edge.

8. Suppose that the functions $u$ and $v$ are harmonic in a domain $D$.

   **(a)** Is the sum $u + v$ necessarily harmonic in $D$?

   **(b)** Is the product $uv$ necessarily harmonic in $D$?

   **(c)** Is $\partial u / \partial x$ harmonic in $D$? (You may use the fact—which we will prove in Chapter 4— that harmonic functions have continuous partial derivatives of all orders.)

9. Find a function $\phi(x, y)$ that is harmonic in the region of the first quadrant between the curves $xy = 2$ and $xy = 4$ and takes the value 1 on the lower edge and the value 3 on the upper edge. [HINT: Begin by considering $z^2$.]

10. Show that in polar coordinates $(r, \theta)$ Laplace's equation becomes

$$\frac{\partial^2 \phi}{\partial r^2} + \frac{1}{r}\frac{\partial \phi}{\partial r} + \frac{1}{r^2}\frac{\partial^2 \phi}{\partial \theta^2} = 0.$$

11. Let $f(z) = z + 1/z$. Show that the level curve Im $f(z) = 0$ consists of the real axis (excluding $z = 0$) and the circle $|z| = 1$. [The level curves Im $f(z) = $ constant can be interpreted as streamlines for fluid flow around a cylindrical obstacle.]

12. Prove that if $r$ and $\theta$ are polar coordinates, then the functions $r^n \cos n\theta$ and $r^n \sin n\theta$, where $n$ is an integer, are harmonic as functions of $x$ and $y$. [HINT: Recall De Moivre's formula.]

13. Find a function harmonic inside the wedge bounded by the nonnegative $x$-axis and the half-line $y = x$ ($x \geq 0$) that goes to zero on these sides but is not identically zero. [HINT: See Prob. 12.] The level curves for this function can be interpreted as streamlines for a fluid flowing inside this wedge, under certain idealized conditions.

14. Suppose that $f(z)$ is analytic and nonzero in a domain $D$. Prove that $\ln |f(z)|$ is harmonic in $D$.

15. Find a function harmonic outside the circle $|z| = 3$ that goes to zero on $|z| = 3$ but is not identically zero.

16. Show that if $\phi(x, y)$ and $\psi(x, y)$ are harmonic, then $u$ and $v$ defined by

$$u(x, y) = \phi_x \phi_y + \psi_x \psi_y$$

and

$$v(x, y) = \tfrac{1}{2}(\phi_x^2 + \psi_x^2 - \phi_y^2 - \psi_y^2)$$

satisfy the Cauchy-Riemann equations.

17. Find a function $\phi(x, y)$ harmonic in the upper half-plane Im $z > 0$ and continuous on Im $z \geq 0$ such that
   **(a)** $\phi(x, 0) = x^2 + 5x + 1$ for all $x$.
   **(b)** $\phi(x, 0) = 2x^3/(x^2 + 4)$ for all $x$.
   [HINT: $\phi = \text{Re}[2z^3/(z^2 + 4)]$ won't work because the function $2z^3/(z^2 + 4)$ is not analytic at $z = 2i$ in the upper half-plane. Instead, write

$$\frac{2x^3}{x^2 + 4} = \frac{x^2}{x - 2i} + \frac{x^2}{x + 2i} = 2\,\text{Re}\,\frac{x^2}{x + 2i},$$

which suggests the proper choice for $\phi$.]

18. Show that if $\phi(x, y)$ is harmonic, then $\phi_x - i\phi_y$ is analytic. (You may assume that $\phi$ has continuous partial derivatives of all orders.)

19. By tracing the steps in Example 1, show that every function $u(x, y)$ harmonic in a disk has a harmonic conjugate $v(x, y)$. [HINT: The only difficulty which could occur is in the step corresponding to Eq. (5), where in order to find $\psi'(x)$ we must be certain that all appearances of the variable $y$ cancel. Show that this is guaranteed because $u$ is harmonic.]

20. Show that although $u = \ln |z|$ is harmonic in the complex plane except at $z = 0$ (i.e., in the domain $\mathbb{C}\backslash\{0\}$), $u$ does *not* have a harmonic conjugate $v$ throughout $\mathbb{C}\backslash\{0\}$. In other words, show that there is no function $v$ such that $\ln |z| + iv(z)$ is analytic in $\mathbb{C}\backslash\{0\}$. [HINT: Show that if $\ln |z| + iv(z)$ is analytic in $\mathbb{C}\backslash\{0\}$, then $v(z) = \text{Arg } z + a$ except along the nonpositive real axis.]

## *2.6 STEADY-STATE TEMPERATURE AS A HARMONIC FUNCTION

It is useful to have a familiar physical model for harmonic functions as an aid in visualizing and remembering their properties. The equilibrium temperatures in a slab, as we shall see, fill this role nicely.

Figure 2.11 depicts a uniform slab of a thermally conducting material, such as a copper plate or a ceramic substrate for microelectronic circuitry. It has constant thickness, so its top and bottom surfaces lie parallel to the $xy$-plane. We assume that these surfaces are also insulated, and no heat flows in the vertical direction. As a result the equilibrium temperature $T$ is a function of $x$ and $y$;

$$T = T(x, y).$$

This temperature distribution is maintained by heat sources (or sinks) and insulation placed around the edges, so that the isotherms appear as illustrated in Fig. 2.12.

Now once the temperature has reached equilibrium, $T(x, y)$ will be a harmonic function:

$$\Delta T(x, y) = \frac{\partial^2 T}{\partial x^2} + \frac{\partial^2 T}{\partial y^2} = 0. \tag{1}$$

The physical reason for this is as follows. Focusing attention on a small square in the slab as depicted in Fig. 2.13, we call upon Fourier's law of heat conduction, which states that the rate at which heat flows through each side of the square is proportional

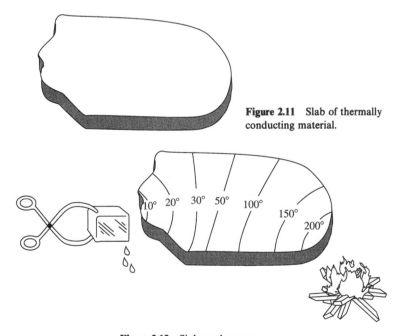

**Figure 2.11**   Slab of thermally conducting material.

**Figure 2.12**   Sinks and sources.

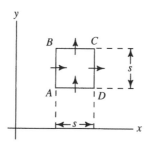

**Figure 2.13**  Heat flow.

to the rate of change of temperature in the direction of the flow. Thus the flow through *AB* and *CD* is proportional to $\partial T/\partial x$, and that through *BC* and *AD* is proportional to $\partial T/\partial y$. (In fact, the constant of proportionality, which depends on the cross-section area and the material, is negative, since heat flows from hot to cold!) The heat flows are depicted as *entering* the square through *AB* and *AD* and *exiting* through *BC* and *CD*. Therefore, the net *outflow* of heat is proportional to

$$\frac{\partial T}{\partial x}\bigg|_{CD} - \frac{\partial T}{\partial x}\bigg|_{AB} + \frac{\partial T}{\partial y}\bigg|_{BC} - \frac{\partial T}{\partial y}\bigg|_{AD}.$$

For small dimensions *s* the difference in the first derivatives can be approximated by the second derivative, and the net outflux is proportional to

$$\frac{\partial^2 T}{\partial x^2} s + \frac{\partial^2 T}{\partial y^2} s. \tag{2}$$

At equilibrium the temperature has settled; the square has finished cooling down (or heating up), and the net outflux will be zero. Dividing expression (2) by *s*, then, we conclude that $T(x, y)$ satisfies (1).

The fact that harmonic functions arise as temperature distributions permits us to anticipate some of their mathematical properties. For example, look at the iso-thermal curves in Fig. 2.14. They indicate a "hot spot" in the interior of the slab. This cannot occur at equilibrium, because heat would flow away from the hot spot and it would cool down. Of course, this pattern *could* be maintained by an external source underneath the slab, but we have precluded this by assuming that such sources are located only on the edge. We conclude that the temperature distribution can never exhibit such an *interior* maximum. The rigorous formulation and generalization of this observation to harmonic functions is identified in Chapter 4 as the *maximum principle*, which says that a harmonic function cannot take its maximum in the interior of a region, except in the trivial case when it is constant throughout.

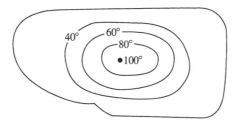

**Figure 2.14**  Isotherms.

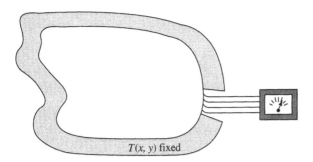

**Figure 2.15**   Adjustable boundary temperatures.

As another example consider the following experiment. The edges of the slab in Fig. 2.15 are maintained at fixed temperatures by external heat sources, except for a small section along which *we* can control the temperature to our liking (using some type of adjustable furnace). Then, on physical grounds we would expect to be able, by turning up the furnace sufficiently, to raise the temperature of an arbitrary interior point to any specified value—although, of course, we couldn't guarantee to replicate a whole *pattern* of temperatures across the slab.[†] Our expectation is premised on the intuitive feeling that the interior temperatures are completely determined by the edge temperature distribution. This is actually an instance of the *boundary value property* of harmonic functions, and in fact in Chapter 4 we shall study *Poisson's formulas*, which express the explicit relationship between the interior and boundary values of such functions for certain geometries.

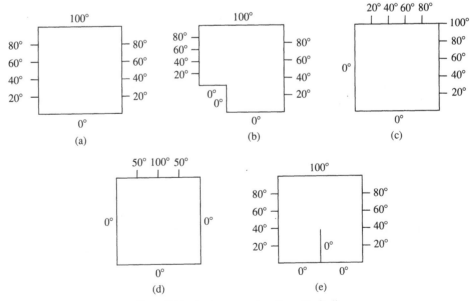

**Figure 2.16**   Isotherm constructions (Prob. 1).

---

[†] Note also that the thermodynamic reality of a zero absolute temperature inhibits our ability to *cool* interior points arbitrarily.

## EXERCISES 2.6

1. Using only your physical intuition, sketch the family of isotherms that you would expect to see at equilibrium for slabs with edge temperatures maintained as shown in Fig. 2.16.
2. Sketch the isotherms for the edge-temperature distribution in Fig. 2.17. Does this configuration violate the maximum principle?
3. Sketch the isotherms for the edge-temperature distribution in Fig. 2.18. Does this configuration violate the maximum principle?

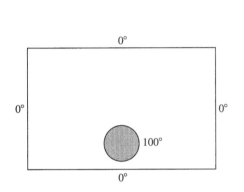

**Figure 2.17**  Isotherm construction (Prob. 2).

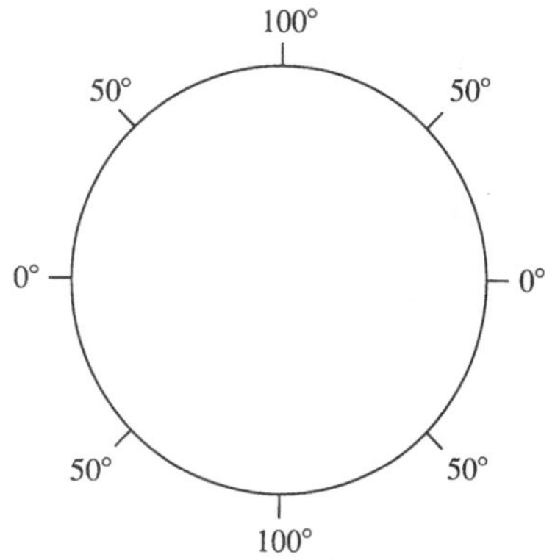

**Figure 2.18**  Isotherm construction (Prob. 3).

## SUMMARY

A complex-valued function $f$ of a complex variable $z = x + iy$ can be considered as a pair of real functions of two real variables in accordance with $f(z) = u(x, y) + iv(x, y)$. The definitions of limit, continuity, and derivative for such functions are direct analogues of the corresponding concepts introduced in calculus, but the greater freedom of $z$ to vary in two dimensions lends added strength to these conditions. In particular, the existence of a derivative, defined as the limit of $[f(z + \Delta z) - f(z)]/\Delta z$ as $\Delta z \to 0$, implies a strong relationship between the functions $u$ and $v$, namely, the Cauchy-Riemann equations

$$\frac{\partial u}{\partial x} = \frac{\partial v}{\partial y}, \quad \frac{\partial u}{\partial y} = -\frac{\partial v}{\partial x}.$$

If the function $f$ is differentiable in an open set, it is said to be analytic. This property can be established by showing that the first-order partial derivatives of $u$ and $v$ are continuous and satisfy the Cauchy-Riemann equations on the open set. Analyticity of a function $f$ is the mathematical expression of the intuitive condition that $f$ respects the complex structure of $z$; i.e., $f$ can be computed using $x$ and $y$ only in the combination $(x + iy)$. If $f$ is given in terms of $z$ alone, the basic techniques of calculus can be used to find its derivative.

The real and imaginary parts of an analytic function are harmonic; i.e., they satisfy Laplace's equation

$$\frac{\partial^2 \phi}{\partial x^2} + \frac{\partial^2 \phi}{\partial y^2} = 0,$$

and their second-order partial derivatives are continuous. Furthermore, the level curves of the real part intersect those of the imaginary part orthogonally. Given a harmonic function $u(x, y)$ it is possible to construct another harmonic function $v(x, y)$ so that $u(x, y) + iv(x, y)$ is analytic in some domain; such a function $v$ is called a harmonic conjugate of $u$. Harmonic functions can be physically interpreted as equilibrium temperature distributions.

## SUGGESTED READING

In addition to the references following Chapter 1, the following texts may be helpful for special topics:

### *Harmonic Functions*

[1] Davis, H. F., and Snider, A. D. *Introduction to Vector Analysis*, 6th ed. Wm. C. Brown, Dubuque, Iowa, 1991.

[2] Hayman, W. K., and Kennedy, P. B. *Subharmonic Functions*, Vol. I, London Math. Soc. Monographs. Academic Press, London, 1976.

[3] Hille, E. *Analytic Function Theory*, Vol. II. Chelsea, New York, 1973.

# :3:

# Elementary Functions

## 3.1 THE EXPONENTIAL, TRIGONOMETRIC, AND HYPERBOLIC FUNCTIONS

The complex exponential function $e^z$ plays a prominent role in analytic function theory, not only because of its own important properties but because it is used to define the complex trigonometric and hyperbolic functions. Recall the definition from Chapter 1: If $z = x + iy$,

$$\boxed{e^z = e^x(\cos y + i \sin y).} \tag{1}$$

As a consequence of Example 2 in Sec. 2.4, we know that $e^z$ is an *entire* function and that

$$\frac{d}{dz} e^z = e^z.$$

The polar components of $e^z$ are readily derived from (1):

$$|e^z| = e^x, \tag{2}$$

$$\arg e^z = y + 2k\pi \qquad (k = 0, \pm 1, \pm 2, \ldots). \tag{3}$$

It follows that $e^z$ is never zero. However, $e^z$ does assume every other complex value (see Prob. 4).

Recall that a function $f$ is *one-to-one* on a set $S$ if the equation $f(z_1) = f(z_2)$, where $z_1$ and $z_2$ are in $S$, implies that $z_1 = z_2$. As is shown in calculus, the exponential

**75**

function is one-to-one on the real axis. However it is *not* one-to-one on the complex plane. In fact, we have the following.

---

**Theorem 1.**

(i) A  necessary  and  sufficient  condition  that

$$e^z = 1$$

is  that  $z = 2k\pi i$,  where  $k$  is  an  integer.

(ii) A  necessary  and  sufficient  condition  that

$$e^{z_1} = e^{z_2}$$

is  that  $z_1 = z_2 + 2k\pi i$,  where  $k$  is  an  integer.

---

*Proof of* (i).    First suppose that $e^z = 1$, with $z = x + iy$. Then we must have

$$|e^z| = |e^{x+iy}| = e^x = 1,$$

and so $x = 0$. This implies that

$$e^z = e^{iy} = \cos y + i \sin y = 1,$$

or, equivalently,

$$\cos y = 1, \qquad \sin y = 0.$$

These two simultaneous equations are clearly satisfied only when $y = 2k\pi$ for some integer $k$; i.e., $z = 2k\pi i$.

Conversely, if $z = 2k\pi i$, where $k$ is an integer, then

$$e^z = e^{2k\pi i} = e^0(\cos 2k\pi + i \sin 2k\pi) = 1.$$

*Proof of* (ii). It follows from the division rule that

$$e^{z_1} = e^{z_2} \quad \text{if, and only if,} \quad e^{z_1 - z_2} = 1.$$

But, by part (i), the last equation holds precisely when $z_1 - z_2 = 2k\pi i$, where $k$ is an integer. ■

One important consequence of Theorem 1 is the fact that $e^z$ is periodic. In general, a function $f$ is said to be *periodic* in a domain $D$ if there exists a nonzero constant $\lambda$ such that the equation $f(z + \lambda) = f(z)$ holds for every $z$ in $D$. Any constant $\lambda$ with this property is called a *period* of $f$. Since, for all $z$,

$$e^{z + 2\pi i} = e^z,$$

we see that $e^z$ is periodic with complex period $2\pi i$. Consequently, if we divide up the $z$-plane into the infinite horizontal strips

$$S_n := \{x + iy \mid -\infty < x < \infty, (2n - 1)\pi < y \le (2n + 1)\pi\} \qquad (n = 0, \pm 1, \pm 2, \ldots),$$

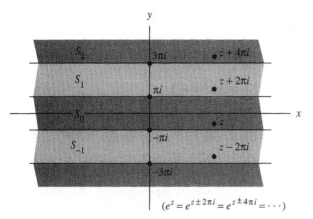

$$(e^z = e^{z \pm 2\pi i} = e^{z \pm 4\pi i} = \cdots)$$    **Figure 3.1**   Fundamental regions for $e^z$.

as shown in Fig. 3.1, then $e^z$ will behave in the same manner on each strip. Furthermore, from part (ii) of Theorem 1, it follows that $e^z$ is one-to-one on each strip $S_n$. For these reasons any one of these strips is called a *fundamental region* for $e^z$.

From the identity

$$e^{iy} = \cos y + i \sin y,$$

and its obvious consequence

$$e^{-iy} = \cos y - i \sin y,$$

we deduce, by subtracting and adding these equations, that

$$\sin y = \frac{e^{iy} - e^{-iy}}{2i}, \qquad \cos y = \frac{e^{iy} + e^{-iy}}{2}.$$

These real variable formulas suggest the following extensions of the trigonometric functions to complex "angles."

---

**Definition 1.**   Given any complex number $z$, we define

$$\sin z := \frac{e^{iz} - e^{-iz}}{2i}, \qquad \cos z := \frac{e^{iz} + e^{-iz}}{2}.$$

---

Since $e^{iz}$ and $e^{-iz}$ are entire functions, so are $\sin z$ and $\cos z$. In fact,

$$\frac{d}{dz} \sin z = \frac{d}{dz}\left(\frac{e^{iz} - e^{-iz}}{2i}\right) = \frac{1}{2i}(ie^{iz} - (-i)e^{-iz}) = \cos z, \qquad \textbf{(4)}$$

and, similarly,

$$\frac{d}{dz} \cos z = -\sin z. \qquad \textbf{(5)}$$

We recognize that Eqs. (4) and (5) agree with the familiar formulas derived in calculus.

Some further identities that remain valid in the complex case are listed below:

$$\sin(z + 2\pi) = \sin z, \qquad \cos(z + 2\pi) = \cos z. \tag{6}$$

$$\sin(-z) = -\sin z, \qquad \cos(-z) = \cos z. \tag{7}$$

$$\sin^2 z + \cos^2 z = 1. \tag{8}$$

$$\sin(z_1 \pm z_2) = \sin z_1 \cos z_2 \pm \sin z_2 \cos z_1. \tag{9}$$

$$\cos(z_1 \pm z_2) = \cos z_1 \cos z_2 \mp \sin z_1 \sin z_2. \tag{10}$$

$$\sin 2z = 2 \sin z \cos z, \qquad \cos 2z = \cos^2 z - \sin^2 z. \tag{11}$$

The proofs of these identities follow directly from the properties of the exponential function and are left to the exercises. Notice that Eqs. (6) imply that $\sin z$ and $\cos z$ are both periodic with period $2\pi$.

**Example 1**

> Prove that $\sin z = 0$ if, and only if, $z = k\pi$, where $k$ is an integer.
>
> **Solution.** If $z = k\pi$, then clearly $\sin z = 0$. Now suppose, conversely, that $\sin z = 0$. Then we have
>
> $$\frac{e^{iz} - e^{-iz}}{2i} = 0,$$
>
> or, equivalently,
>
> $$e^{iz} = e^{-iz}.$$
>
> By Theorem 1(ii) it follows that
>
> $$iz = -iz + 2k\pi i,$$
>
> which implies that $z = k\pi$ for some integer $k$.  ∎

Thus the only zeros of $\sin z$ are its real zeros. The same is true of the function $\cos z$; i.e.,

$$\cos z = 0 \quad \text{if and only if} \quad z = \frac{\pi}{2} + k\pi,$$

where $k$ is an integer.

The other four complex trigonometric functions are defined by

$$\tan z := \frac{\sin z}{\cos z}, \qquad \cot z := \frac{\cos z}{\sin z}, \qquad \sec z := \frac{1}{\cos z}, \qquad \csc z := \frac{1}{\sin z}.$$

Notice that the functions $\cot z$ and $\csc z$ are analytic except at the zeros of $\sin z$, i.e., the points $z = k\pi$, whereas the functions $\tan z$ and $\sec z$ are analytic except at the points $z = \pi/2 + k\pi$, where $k$ is any integer. Furthermore, the usual rules for differentiation remain valid for these functions:

$$\frac{d}{dz} \tan z = \sec^2 z, \qquad \frac{d}{dz} \sec z = \sec z \tan z,$$

$$\frac{d}{dz} \cot z = -\csc^2 z, \qquad \frac{d}{dz} \csc z = -\csc z \cot z.$$

The preceding discussion has emphasized the similarity between the real trigonometric functions and their complex extensions. However, this analogy should not be carried too far. For example, the real cosine function is bounded by 1, i.e.,

$$|\cos x| \le 1, \qquad \text{for all real } x,$$

but

$$|\cos (iy)| = \left| \frac{e^{-y} + e^{y}}{2} \right| = |\cosh y|,$$

which is unbounded and, in fact, is never less than 1! 
    The complex hyperbolic functions are defined by a natural extension of their definitions in the real case:

---

**Definition 2.** For any complex number $z$ we define

$$\sinh z := \frac{e^z - e^{-z}}{2}, \qquad \cosh z := \frac{e^z + e^{-z}}{2}. \qquad (12)$$

---

Notice that the functions (12) are entire and satisfy

$$\frac{d}{dz} \sinh z = \cosh z, \qquad \frac{d}{dz} \cosh z = \sinh z. \qquad (13)$$

By using the trigonometric identities and the readily verified formulas

$$\sinh z = -i \sin iz, \qquad \cosh z = \cos iz, \qquad (14)$$

one can easily show that the familiar hyperbolic identities are valid in the complex case. The four remaining complex hyperbolic functions are given by

$$\tanh z := \frac{\sinh z}{\cosh z}, \qquad \coth z := \frac{\cosh z}{\sinh z}, \qquad \text{sech } z := \frac{1}{\cosh z}, \qquad \text{csch } z := \frac{1}{\sinh z}.$$

## EXERCISES 3.1

1. Show that $e^z = (1 + i)/\sqrt{2}$ if and only if $z = (\pi/4 + 2k\pi)i$, $k = 0, \pm 1, \pm 2, \ldots$.
2. Let $f(z) = e^z - (1 + z + z^2/2 + z^3/6)$. Show that $f^{(3)}(0) = 0$.
3. Find the sum $\sum_{k=0}^{100} e^{kz}$.

**4.** Let $\omega$ ($\neq 0$) have the polar representation $\omega = re^{i\theta}$. Show that $\exp(\ln r + i\theta) = \omega$.

**5.** Write each of the following numbers in the form $a + bi$.

(a) $\exp\left(2 + \dfrac{i\pi}{4}\right)$

(b) $\dfrac{\exp(1 + i3\pi)}{\exp(-1 + i\pi/2)}$

(c) $\sin(2i)$

(d) $\cos(1 - i)$

(e) $\sinh(1 + \pi i)$

(f) $\cosh\left(\dfrac{i\pi}{2}\right)$

**6.** Establish the trigonometric identities (8) and (9).

**7.** Show that the formula $e^{iz} = \cos z + i \sin z$ holds for all complex numbers $z$.

**8.** Verify the differentiation formulas (13).

**9.** Find $dw/dz$ for each of the following.

(a) $w = \exp(\pi z^2)$

(b) $w = \cos(2z) + i \sin\left(\dfrac{1}{z}\right)$

(c) $w = \exp[\sin(2z)]$

(d) $w = \tan^3 z$

(e) $w = [\sinh z + 1]^2$

(f) $w = \tanh z$

**10.** Explain why the function $f(z) = \sin(z^2) + e^{-z} + iz$ is entire.

**11.** Explain why the function $\text{Re}\left(\dfrac{\cos z}{e^z}\right)$ is harmonic in the whole plane.

**12.** Establish the following hyperbolic identities by using the relations (14) and the corresponding trigonometric identities.

(a) $\cosh^2 z - \sinh^2 z = 1$

(b) $\sinh(z_1 + z_2) = \sinh z_1 \cosh z_2 + \cosh z_1 \sinh z_2$

(c) $\cosh(z_1 + z_2) = \cosh z_1 \cosh z_2 + \sinh z_1 \sinh z_2$

**13.** Show the following.

(a) $\sin(x + iy) = \sin x \cosh y + i \cos x \sinh y$

(b) $\cos(x + iy) = \cos x \cosh y - i \sin x \sinh y$

**14.** Prove the following.

(a) $e^{iz}$ is periodic with period $2\pi$.

(b) $\tan z$ is periodic with period $\pi$.

(c) $\sinh z$ and $\cosh z$ are both periodic with period $2\pi i$.

(d) $\tanh z$ is periodic with period $\pi i$.

**15.** Prove that $\cos z = 0$ if and only if $z = \pi/2 + k\pi$, where $k$ is an integer.

**16.** Verify the identity

$$\cos z_2 - \cos z_1 = 2 \sin \frac{z_1 + z_2}{2} \sin \frac{z_1 - z_2}{2},$$

and use it to show that $\cos z_1 = \cos z_2$ if and only if $z_2 = \pm z_1 + 2k\pi$, where $k$ is an integer.

**17.** Find all numbers $z$ such that

(a) $e^{4z} = 1$

(b) $e^{iz} = 3$

(c) $\cos z = i \sin z$

**18.** Prove the following.

(a) $\lim\limits_{z \to 0} \dfrac{\sin z}{z} = 1$

(b) $\lim\limits_{z \to 0} \dfrac{\cos z - 1}{z} = 0$

[HINT: Use the fact that $f'(0) = \lim_{z \to 0} [f(z) - f(0)]/z$.]

**19.** Prove that the function $e^z$ is one-to-one on any open disk of radius $\pi$.

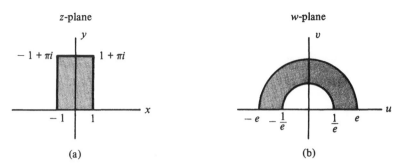

**Figure 3.2**   Mapping of a rectangle under $w = e^z$.

20. Show that the function $w = e^z$ maps the shaded rectangle in Fig. 3.2(a) one-to-one onto the semi-annulus in Fig. 3.2(b).

21. Below is an outline of an alternative proof that $\sin^2 z + \cos^2 z = 1$ for all $z$. Justify each step in the proof.
    (a) The function $f(z) = \sin^2 z + \cos^2 z$ is entire.
    (b) $f'(z) = 0$ for all $z$.
    (c) $f(z)$ is a constant function.
    (d) $f(0) = 1$.
    (e) $f(z) = 1$ for all $z$.

22. Using only real arithmetic operations, write a computer program whose input $(x, y)$ is the real and imaginary parts of $z = x + iy$ and whose output is the real and imaginary parts of (a) $e^z$, (b) $\sin z$, and (c) $\cosh z$.

## 3.2 THE LOGARITHMIC FUNCTION

In discussing a correspondence $w = f(z)$ we have used the word "function" to mean that $f$ assigns a single value $w$ to each permissible value of $z$. Sometimes this fact is emphasized by saying "$f$ is a single-valued function." Of course there are equations that do not define single-valued functions—for example, $w = \arg z$ and $w = z^{1/2}$. Indeed, for each nonzero $z$ there are an infinite number of distinct values of $\arg z$ and two distinct values of $z^{1/2}$. In general, if for some values of $z$ there corresponds more than one value of $w = f(z)$, then we say that $w = f(z)$ is a *multiple-valued function*. We commonly obtain multiple-valued functions by taking the inverses of single-valued functions that are not one-to-one. This is, in fact, how we obtain the complex logarithmic function log $z$.

So we want to define[†] log $z$ as the inverse of the exponential function; i.e.,

$$w = \log z \quad \text{if} \quad z = e^w. \tag{1}$$

Since $e^w$ is never zero, we presume that $z \neq 0$. To find log $z$ explicitly, let us write $z$ in polar form as $z = re^{i\theta}$ and write $w$ in standard form as $w = u + iv$. Then the equa-

---

† All logarithms in this text are taken to the base $e$ and are hereafter abbreviated log or Log. The notations ln and Ln are subsequently not used.

tion $z = e^w$ becomes

$$re^{i\theta} = e^{u+iv} = e^u e^{iv}. \tag{2}$$

Taking magnitudes of both sides of (2) we deduce that $r = e^u$, or that $u$ is the ordinary (real) logarithm of $r$:

$$u = \text{Log } r.$$

(We capitalize Log here to distinguish the natural logarithmic function of real variables.) The equality of the remaining factors in Eq. (2), namely, $e^{i\theta} = e^{iv}$, identifies $v$ as the (multiple-valued) polar angle $\theta = \arg z$:

$$v = \arg z = \theta.$$

Thus $w = \log z$ is also a multiple-valued function. The explicit definition is as follows.

---

**Definition 3.** If $z \neq 0$, then we define $\log z$ to be any of the infinitely many values

$$\log z := \text{Log } |z| + i \arg z$$
$$= \text{Log } |z| + i \text{ Arg } z + i2k\pi \qquad (k = 0, \pm1, \pm2, \ldots). \tag{3}$$

---

The multiple-valuedness of $\log z$ simply reflects the fact that the imaginary part of the logarithm is the polar angle $\theta$; the real part is single-valued. As examples, consider

$$\log 3 = \text{Log } 3 + i \arg 3 = (1.098 \ldots) + i2k\pi,$$

$$\log(-1) = \text{Log } 1 + i \arg(-1) = i(2k + 1)\pi,$$

$$\log(1 + i) = \text{Log } |1 + i| + i \arg(1 + i)$$

$$= \frac{1}{2} \text{Log } 2 + i\left(\frac{\pi}{4} + 2k\pi\right),$$

where $k = 0, \pm1, \pm2, \ldots$.

The familiar properties of the real logarithmic function extend to the complex case, but the precise statements of these extensions are complicated by the fact that $\log z$ is multiple-valued. For example, if $z \neq 0$, then we have

$$z = e^{\log z},$$

but

$$\log e^z = z + 2k\pi i \qquad (k = 0, \pm1, \pm2, \ldots).$$

Furthermore, using the representation (3) and the equations

$$\arg z_1 z_2 = \arg z_1 + \arg z_2,$$

$$\arg\left(\frac{z_1}{z_2}\right) = \arg z_1 - \arg z_2, \tag{4}$$

one can readily verify that

$$\log z_1 z_2 = \log z_1 + \log z_2, \tag{5}$$

and that

$$\log\left(\frac{z_1}{z_2}\right) = \log z_1 - \log z_2. \tag{6}$$

As with Eqs. (4), we must interpret Eqs. (5) and (6) to mean that if particular values are assigned to any two of their terms, then one can find a value of the third term so that the equation is satisfied. For example, if $z_1 = z_2 = -1$ and we select $\pi i$ to be the value of $\log z_1$ and $\log z_2$, then Eq. (5) will be satisfied if we use the particular value $2\pi i$ for $\log(z_1 z_2) = \log 1$.

Recall that in Sec. 1.3 we used the notion of a *branch cut* to resolve the ambiguity in the designation of the polar angle $\theta = \arg z$. We took Arg $z$ to be the principal value of arg $z$—the value in the interval $(-\pi, \pi]$—which jumps by $2\pi$ as $z$ crosses the branch cut along the negative real axis. Other branches $\arg_\tau z$ resulted from restricting the values of arg $z$ to $(\tau, \tau + 2\pi]$ and shifting the $2\pi$-discontinuities to the ray $\theta = \tau$.

Clearly the same artifice will generate single-valued *branches* of log $z$. The *principal value of the logarithm* Log $z$ is the value inherited from the principal value of the argument:

$$\boxed{\operatorname{Log} z := \operatorname{Log} |z| + i \operatorname{Arg} z.} \tag{7}$$

[Notice that we can use the same convention—Log with a capital L—for the principal value as for the usual (real) value, since Arg $z = 0$ if $z$ is positive real.] Log $z$ also inherits, from Arg $z$, the discontinuities along the branch cut; it jumps by $2\pi i$ as $z$ crosses the negative real axis. However, at all points off the nonpositive real axis, Log $z$ *is* continuous, and this fact enables us to prove the next theorem.

---

**Theorem 2.** The function Log $z$ is analytic in the domain $D^*$ consisting of all points of the complex plane except those lying on the nonpositive real axis (see Fig. 3.3). Furthermore,

$$\frac{d}{dz}\operatorname{Log} z = \frac{1}{z}, \qquad \text{for } z \text{ in } D^*. \tag{8}$$

---

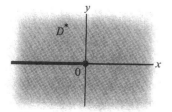

**Figure 3.3**  Analyticity domain for Log $z$.

*Proof.* Let us set $w = \text{Log } z$. Our goal is to prove that, for $z_0$ in $D^*$ and $w_0 = \text{Log } z_0$, the limit of the difference quotient[†]

$$\lim_{z \to z_0} \frac{w - w_0}{z - z_0} \tag{9}$$

exists and equals $1/z_0$. We are guided in this endeavor by the knowledge that $z = e^w$ and that the exponential function *is* analytic so that

$$\lim_{w \to w_0} \frac{z - z_0}{w - w_0} = \frac{dz}{dw}\bigg|_{w = w_0} = e^{w_0} = z_0.$$

Observe that we will have accomplished our goal if we can show that

$$\lim_{z \to z_0} \frac{w - w_0}{z - z_0} = \lim_{w \to w_0} \frac{1}{\dfrac{z - z_0}{w - w_0}}, \tag{10}$$

because the limit on the right exists and equals $1/z_0$. But (10) will follow from the trivial identity

$$\frac{w - w_0}{z - z_0} = \frac{1}{\dfrac{z - z_0}{w - w_0}}, \tag{11}$$

provided we show that

(a) As $z$ approaches $z_0$, $w$ must approach $w_0$, and
(b) For $z \neq z_0$, $w$ will not coincide with $w_0$ [so that the terms in Eq. (10) are meaningful].

Condition (a) follows from the continuity, in $D^*$, of $w = \text{Log } z$. Condition (b) is even more immediate; if $w$ coincided with $w_0$, $z$ would have to equal $z_0$, since $z = e^w$. Thus $w = \text{Log } z$ is differentiable at every point in $D^*$ and hence is analytic there.[‡] ∎

---

**Corollary 1.**   The function Arg $z$ is harmonic in the domain $D^*$ of Theorem 2.

---

**Corollary 2.**   The real function Log $|z|$ is harmonic in the entire plane with the exception of the origin. (See Prob. 8.)

---

[†] It is convenient here to use form (9) instead of the usual form $\lim_{\Delta z \to 0} [\text{Log}(z_0 + \Delta z) - \text{Log } z_0]/\Delta z$. The equivalence of these limits can be seen by putting $\Delta z = z - z_0$.

[‡] Advanced readers will observe that the same proof could be applied to *any* function $f(z)$ that is analytic and one-to-one around $z_0$ and for which $f'(z_0) \neq 0$, to conclude that the *inverse function* is analytic and has derivative $1/f'(z_0)$.

**Example 1**

Determine the domain of analyticity for the function $f(z):= \text{Log}(3z - i)$. Compute $f'(z)$.

**Solution.** Since $f$ is the composition of Log with the function $g(z) = 3z - i$, the chain rule asserts that $f$ will be differentiable at each point $z$ for which $3z - i$ lies in the domain $D^*$ of Theorem 2. Thus points where $3z - i$ is negative or zero are disallowed; a little thought shows that these points lie on the horizontal ray $x \le 0$, $y = \frac{1}{3}$ (see Fig. 3.4). In this slit plane, then, from Eq. (8):

$$f'(z) = \frac{d}{dz} \text{Log}(3z - i) = \frac{1}{3z - i} \frac{d}{dz}(3z - i) = \frac{3}{3z - i}. \quad \blacksquare$$

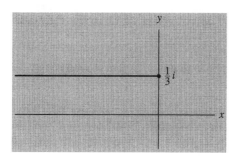

**Figure 3.4**  Analyticity domain for $\text{Log}(3z - i)$.

Other branches of $\log z$ can be employed if the location of the discontinuities on the negative axis is inconvenient. Clearly the specification

$$\mathscr{L}_\tau(z) := \text{Log}\,|z| + i\,\text{arg}_\tau z \tag{12}$$

results in a single-valued function whose imaginary part lies in the interval $(\tau, \tau + 2\pi]$. Moreover, the same reasoning used in the proof of Theorem 2 shows that this function is analytic in the complex plane excluding the ray $\theta = \tau$ and the origin, and in this domain,

$$\frac{d}{dz} \mathscr{L}_\tau(z) = \frac{1}{z}.$$

Figure 3.5 on page 86 depicts the domain of analyticity for $\mathscr{L}_{-\pi/4}(z)$. Of course, no branch of $\log z$ is analytic at the origin, which is called a *branch point* for $\log z$.

Thus far we have used the phrase "branch of $\log z$" in a somewhat informal manner to denote specific values for this multiple-valued function. To make matters more precise, we give the following definition.

**Definition 4.**  $F(z)$ is said to be a **branch** of a multiple-valued function $f(z)$ in a domain $D$ if $F(z)$ is single-valued and continuous in $D$ and has the property that, for each $z$ in $D$, the value $F(z)$ is one of the values of $f(z)$.

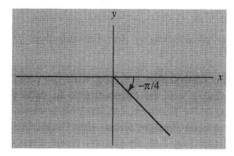

**Figure 3.5**  Domain of $\mathscr{L}_{-\pi/4}(z)$.

For example, Arg $z$ is a branch of arg $z$ and Log $z$ is a branch of log $z$ in the plane slit along the negative real axis. In this same domain, the function defined by $e^{(1/2)\text{Log }z}$ gives a branch of $z^{1/2}$ whose values all lie in the right half-plane.

**Example 2**

> Determine a branch of $f(z) = \log(z^3 - 2)$ that is analytic at $z = 0$, and find $f(0)$ and $f'(0)$.
>
> **Solution.**  $f(z)$ is the composition of the logarithm with the analytic function $g(z) = z^3 - 2$. Thus, by the chain rule, it suffices to choose any branch of the logarithm that is analytic at $g(0) = -2$. In particular, $f(z) = \mathscr{L}_{-\pi/4}(g(z))$ solves the problem. For this choice,
>
> $$f(0) = \mathscr{L}_{-\pi/4}(0^3 - 2) = \text{Log } 2 + i\pi,$$
>
> $$f'(0) = \mathscr{L}'_{-\pi/4}(g(0))g'(0) = \frac{g'(0)}{g(0)} = 0. \quad \blacksquare$$

We conclude this section with a word of warning. When complex arithmetic is incorporated into computer packages, all functions must of necessity be programmed as single-valued. The complex logarithm, for instance, is usually programmed as our "principal value," Log $z$. This invalidates some identities, such as Eq. (5), since it is not true in general that Log $z_1 z_2 = $ Log $z_1 + $ Log $z_2$. (See Prob. 3.)

## EXERCISES 3.2

1. Evaluate each of the following.
   - (a) log $i$
   - (b) log$(1 - i)$
   - (c) Log$(-i)$
   - (d) Log$(\sqrt{3} + i)$
2. Verify formulas (5) and (6).
3. Show that if $z_1 = i$ and $z_2 = i - 1$, then

$$\text{Log } z_1 z_2 \neq \text{Log } z_1 + \text{Log } z_2.$$

**4.** Prove that $\text{Log } e^z = z$ if and only if $-\pi < \text{Im } z \le \pi$.

**5.** Solve the following equations.

(a) $e^z = 2i$        (b) $\text{Log}(z^2 - 1) = \dfrac{i\pi}{2}$        (c) $e^{2z} + e^z + 1 = 0$

**6.** Find the error in the following "proof" that $z = -z$:
Since $z^2 = (-z)^2$, $2\text{Log } z = 2\text{Log}(-z)$, and hence $\text{Log } z = \text{Log}(-z)$, which implies that $z = e^{\text{Log } z} = e^{\text{Log}(-z)} = -z$.

**7.** Use the polar form of the Cauchy-Riemann equations (Prob. 6 in Exercises 2.4, page 62) to give another proof of Theorem 2.

**8.** Without directly verifying Laplace's equation, explain why the function $\text{Log }|z|$ is harmonic in every domain that does not contain the origin.

**9.** Determine the domain of analyticity for the function $f(z) = \text{Log}(4 + i - z)$. Compute $f'(z)$.

**10.** Show that the function $\text{Log}(-z) + i\pi$ is a branch of $\log z$ analytic in the domain $D_0$ consisting of all points in the plane except those on the nonnegative real axis.

**11.** Determine a branch of $\log(z^2 + 2z + 3)$ that is analytic at $z = -1$, and find its derivative there.

**12.** Find a branch of $\log(z^2 + 1)$ that is analytic at $z = 0$ and takes the value $2\pi i$ there.

**13.** Find a branch of $\log(2z - 1)$ that is analytic at all points in the plane except those on the following rays.

(a) $\{x + iy \,|\, x \le \tfrac{1}{2}, y = 0\}$

(b) $\{x + iy \,|\, x \ge \tfrac{1}{2}, y = 0\}$

(c) $\{x + iy \,|\, x = \tfrac{1}{2}, y \ge 0\}$

**14.** Prove that there exists no function $F(z)$ analytic in the annulus $D: 1 < |z| < 2$ such that $F'(z) = 1/z$ for all $z$ in $D$. [HINT: Assume that $F$ exists and show that for $z$ in $D$, $z$ not a negative real number, $F(z) = \text{Log } z + c$, where $c$ is a constant.]

**15.** Find a one-to-one analytic mapping of the upper half-plane $\text{Im } z > 0$ onto the infinite horizontal strip

$$\mathscr{H} := \{u + iv \,|\, -\infty < u < \infty, 0 < v < 1\}.$$

[HINT: Start by considering $w = \text{Log } z$.]

**16.** Sketch the level curves for the real and imaginary parts of $\text{Log } z$ and verify the orthogonality property discussed in Sec. 2.5.

**17.** Prove that any branch of $\log z$ (cf. Definition 4) is analytic in its domain and has derivative $1/z$.

**18.** Prove that if $F$ is a branch of $\log z$ analytic in a domain $D$, then the totality of branches of $\log z$ analytic in $D$ are the functions $F + 2k\pi i$, $k = 0, \pm 1, \pm 2, \ldots$. [HINT: Use the result of Prob. 17.]

**19.** How would you construct a branch of $\log z$ that is analytic in the domain $D$ consisting of all points in the plane except those lying on the half-parabola $\{x + iy : x \ge 0, y = \sqrt{x}\}$?

**20.** Using only real arithmetic operations, write a computer program whose input $(x, y)$ is the real and imaginary parts of $z = x + iy$ and whose output is the real and imaginary parts of

(a) $\text{Log } z$      (b) $\mathscr{L}_{-\pi/4}(z)$      (c) $\mathscr{L}_0(z)$      (d) $\mathscr{L}_{4\pi}(z)$

**21.** Find a counterexample to the rule $\log(z_1 z_2) = \log z_1 + \log z_2$ for the software system your computer uses.

## *3.3 COMPLEX POWERS AND INVERSE TRIGONOMETRIC FUNCTIONS*

One important use of the logarithmic function is to define complex powers of $z$. The definition is motivated by the identity

$$z^n = (e^{\log z})^n = e^{n \log z},$$

which holds for any integer $n$.

---

**Definition 5.**   If $\alpha$ is a complex constant and $z \neq 0$, then we define $z^\alpha$ by

$$z^\alpha := e^{\alpha \log z}.$$

---

This means that each value of $\log z$ leads to a particular value of $z^\alpha$.

**Example 1**

Find all the values of $(-2)^i$.

**Solution.**   Since $\log(-2) = \text{Log } 2 + (\pi + 2k\pi)i$, we have

$$(-2)^i = e^{i \log(-2)} = e^{i \, \text{Log } 2} e^{-\pi - 2k\pi} \qquad (k = 0, \pm 1, \pm 2, \ldots).$$

Thus $(-2)^i$ has infinitely many different values.   ∎

Using the representations of Sec. 3.2, we can write

$$z^\alpha = e^{\alpha(\text{Log } |z| + i \, \text{Arg } z + 2k\pi i)} = e^{\alpha(\text{Log } |z| + i \, \text{Arg } z)} e^{\alpha 2k\pi i}, \tag{1}$$

where $k = 0, \pm 1, \pm 2, \ldots$. The values of $z^\alpha$ obtained by taking $k = k_1$ and $k = k_2 \ (\neq k_1)$ in Eq. (1) will therefore be the same when

$$e^{\alpha 2k_1 \pi i} = e^{\alpha 2k_2 \pi i}.$$

But by Theorem 1 this occurs only if

$$\alpha 2k_1 \pi i = \alpha 2k_2 \pi i + 2m\pi i,$$

where $m$ is an integer. Solving the equation we get $\alpha = m/(k_1 - k_2)$; i.e., *formula* (1) *yields some identical values of* $z^\alpha$ *only when* $\alpha$ *is a real rational number.* Consequently, if $\alpha$ is not a real rational number, we obtain infinitely many different values for $z^\alpha$, one for each choice of the integer $k$ in Eq. (1). On the other hand, if $\alpha = m/n$, where $m$ and $n > 0$ are integers having no common factor, then one can verify that there are exactly $n$ distinct values of $z^{m/n}$, namely,

$$z^{m/n} = \exp\left(\frac{m}{n} \text{Log } |z|\right) \exp\left(i \frac{m}{n} (\text{Arg } z + 2k\pi)\right) \qquad (k = 0, 1, \ldots, n - 1). \tag{2}$$

This is entirely consistent with the theory of roots discussed in Sec. 1.5. In summary,

$z^\alpha$ is single-valued when $\alpha$ is a real integer;

$z^\alpha$ takes finitely many values when $\alpha$ is a real rational number;

$z^\alpha$ takes infinitely many values in all other cases.

It is clear from Definitions 4 and 5 that each branch of $\log z$ yields a branch of $z^\alpha$. For example, using the principal branch of $\log z$ we obtain the *principal branch of* $z^\alpha$, namely, $e^{\alpha \, \text{Log} \, z}$. Since $e^z$ is entire and $\text{Log} \, z$ is analytic in the slit domain $D^* = \mathbf{C} \backslash (-\infty, 0]$ of Theorem 2, the chain rule implies that *the principal branch of* $z^\alpha$ *is also analytic in* $D^*$. Furthermore, for $z$ in $D^*$, we have

$$\frac{d}{dz}(e^{\alpha \, \text{Log} \, z}) = e^{\alpha \, \text{Log} \, z} \frac{d}{dz}(\alpha \, \text{Log} \, z) = e^{\alpha \, \text{Log} \, z} \frac{\alpha}{z}.$$

Other branches of $z^\alpha$ can be constructed by using other branches of $\log z$, and since each branch of the latter has derivative $1/z$, the formula[†]

$$\frac{d}{dz}(z^\alpha) = \alpha z^\alpha \frac{1}{z} \tag{3}$$

is valid for each corresponding branch of $z^\alpha$ (provided the same branch is used on both sides of the equation). Observe that if $z_0$ is any given nonzero point, then by selecting a branch cut for $\log z$ that avoids $z_0$, we get a branch of $z^\alpha$ that is analytic in a neighborhood of $z_0$.

"Branch chasing" with complicated functions is often a tedious task; fortunately, it is seldom necessary for elementary applications. Some of the subtleties are demonstrated in the following example.

**Example 2**

Define a branch of $(z^2 - 1)^{1/2}$ that is analytic in the exterior of the unit circle, $|z| > 1$.

**Solution.** Our task, restated, is to find a function $w = f(z)$ that is analytic outside the unit circle and satisfies

$$w^2 = z^2 - 1. \tag{4}$$

Note that the principal branch of $(z^2 - 1)^{1/2}$, namely,

$$e^{(1/2) \, \text{Log}(z^2 - 1)},$$

will not work; it has branch cuts wherever $z^2 - 1$ is negative real, and this constitutes the whole $y$-axis as well as a portion of the $x$-axis (see Fig. 3.6). But if we experiment with some alternative expressions for $w$, we are led to consider

---

[†] We remark that Eq. (3) can be written in the more familiar form

$$\frac{d}{dz}(z^\alpha) = \alpha z^{\alpha - 1}$$

with the proviso that the branch of the logarithm used in defining $z^\alpha$ is the same as the branch of the logarithm used in defining $z^{\alpha - 1}$.

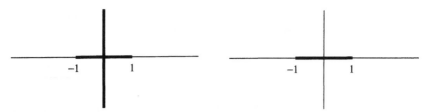

**Figure 3.6**  Branch cut for $e^{(1/2)\text{Log}(z^2-1)}$.     **Figure 3.7**  Branch cut for $e^{(1/2)\text{Log}(1-1/z^2)}$.

$z(1 - 1/z^2)^{1/2}$ as a solution to (4). The principal branch of $(1 - 1/z^2)^{1/2}$, i.e.,

$$e^{(1/2)\text{Log}\,(1-1/z^2)},$$

has cuts where $1 - 1/z^2$ is negative real, and this occurs only when $1/z^2$ is real and greater than one—i.e., the cut is the segment $[-1, 1]$, as shown in Fig. 3.7. Thus

$$w = f(z) = ze^{(1/2)\,\text{Log}(1-1/z^2)}$$

satisfies the required condition of analyticity outside the unit circle.  ■

Now that we have exponentials expressed in terms of trig functions (Sec. 1.4), trig functions expressed as exponentials (Sec. 3.1), and logs interpreted as inverses of exponentials, the following example should come as no surprise. It demonstrates that the arcsine is, in fact, a logarithm.

**Example 3**

The inverse sine function $w = \sin^{-1} z$ is defined by the equation $z = \sin w$. Show that $\sin^{-1} z$ is a multiple-valued function given by

$$\sin^{-1} z = -i \log[iz + (1 - z^2)^{1/2}]. \tag{5}$$

**Solution.**  From the equation

$$z = \sin w = \frac{e^{iw} - e^{-iw}}{2i},$$

we deduce that

$$e^{2iw} - 2ize^{iw} - 1 = 0. \tag{6}$$

Using the quadratic formula we can solve Eq. (6) for $e^{iw}$:

$$e^{iw} = iz + (1 - z^2)^{1/2},$$

where, of course, the square root is two-valued. Formula (5) now follows by taking logarithms.[†]  ■

---

[†] To ensure that our method has not introduced extraneous solutions, one should verify that every value $w$ given by Eq. (5) satisfies $z = \sin w$.

We can obtain a branch of the multiple-valued function $\sin^{-1} z$ by first choosing a branch of the square root and then selecting a suitable branch of the logarithm. Using the chain rule and formula (5) one can show that any such branch of $\sin^{-1} z$ satisfies

$$\frac{d}{dz} (\sin^{-1} z) = \frac{1}{(1 - z^2)^{1/2}} \qquad (z \neq \pm 1), \tag{7}$$

where the choice of the square root on the right must be the same as that used in the branch of $\sin^{-1} z$.

**Example 4**

Suppose $z$ is real and lies in the interval $(-1, 1)$. If principal values are used in formula (5), what is the range of $\sin^{-1} z$?

**Solution.**   With principal values, Eq. (5) is realized as

$$\text{Sin}^{-1} z = -i \, \text{Log}[iz + e^{(1/2) \, \text{Log}(1 - z^2)}]. \tag{8}$$

For $|z| = |x| < 1$, clearly $1 - z^2$ lies in the interval $(0, 1]$, and its Log is real; hence the exponential in (8), which represents $(1 - z^2)^{1/2}$, is positive real. The term $iz$, on the other hand, is pure imaginary. Consequently, the bracketed expression in (8) lies in the right half-plane. As a matter of fact, it also lies on the unit circle, since

$$\left| iz + (1 - z^2)^{1/2} \right| = \sqrt{x^2 + (1 - x^2)} = 1$$

(see Fig. 3.8). Taking the Log, then, results in values $i\theta$, where $-\pi/2 < \theta < \pi/2$, and the leading factor $(-i)$ in (8) produces

$$-\frac{\pi}{2} < \text{Sin}^{-1} x < \frac{\pi}{2}$$

(in keeping with the usual interpretation).  ■

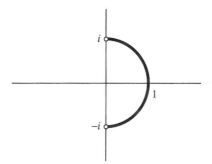

Figure 3.8   Points $ix + (1 - x^2)^{1/2}$ for $|x| < 1$.

For the inverse cosine and inverse tangent functions, calculations similar to those in Example 3 lead to the expressions

$$\cos^{-1} z = -i \log[z + (z^2 - 1)^{1/2}], \tag{9}$$

$$\tan^{-1} z = \frac{i}{2} \log \frac{i + z}{i - z} = \frac{i}{2} \log \frac{1 - iz}{1 + iz} \qquad (z \neq \pm i) \tag{10}$$

and the formulas

$$\frac{d}{dz}(\cos^{-1} z) = \frac{-1}{(1 - z^2)^{1/2}} \qquad (z \neq \pm 1), \tag{11}$$

$$\frac{d}{dz}(\tan^{-1} z) = \frac{1}{1 + z^2} \qquad (z \neq \pm i). \tag{12}$$

Notice that the derivative in Eq. (12) is independent of the branch chosen for $\tan^{-1} z$, whereas the derivative in Eq. (11) depends on the choice of the square root used in the branch of $\cos^{-1} z$.

The same methods can be applied to the inverse hyperbolic functions. The results are

$$\sinh^{-1} z = \log[z + (z^2 + 1)^{1/2}], \tag{13}$$

$$\cosh^{-1} z = \log[z + (z^2 - 1)^{1/2}], \tag{14}$$

$$\tanh^{-1} z = \frac{1}{2} \log \frac{1 + z}{1 - z} \qquad (z \neq \pm 1). \tag{15}$$

These identities shed some light on a curiosity that puzzles most students in elementary calculus. In many published integral tables the following entry appears:

$$\int \frac{dx}{a^2 - b^2 \sin^2 cx} = \begin{cases} \dfrac{1}{ac\sqrt{a^2 - b^2}} \arctan \dfrac{\sqrt{a^2 - b^2} \tan cx}{a} & \text{if } a^2 > b^2, \\[4mm] \dfrac{1}{2ac\sqrt{b^2 - a^2}} \log \dfrac{\sqrt{b^2 - a^2} \tan cx + a}{\sqrt{b^2 - a^2} \tan cx - a} & \text{if } a^2 < b^2. \end{cases}$$

Would the two forms be equivalent if we used complex numbers to obtain the square roots? To see what happens, consider the log form but now assume $a^2 > b^2$:

$$\frac{1}{2ac\sqrt{b^2 - a^2}} \log \frac{\sqrt{b^2 - a^2} \tan cx + a}{\sqrt{b^2 - a^2} \tan cx - a} = \frac{1}{2i} \frac{1}{ac\sqrt{a^2 - b^2}} \log \frac{i\sqrt{a^2 - b^2} \tan cx + a}{i\sqrt{a^2 - b^2} \tan cx - a}$$

$$= \frac{-i}{2} \frac{1}{ac\sqrt{a^2 - b^2}} \log \frac{iz + 1}{iz - 1},$$

where $z = (\sqrt{a^2 - b^2} \tan cx)/a$. Continuing, we derive from properties of the log

$$\frac{-i}{2} \frac{1}{ac\sqrt{a^2 - b^2}} \log \frac{iz + 1}{iz - 1} = \frac{1}{ac\sqrt{a^2 - b^2}} \left(\frac{+i}{2}\right) \log \frac{iz - 1}{iz + 1}$$

$$= \frac{1}{ac\sqrt{a^2 - b^2}} \left(\frac{+i}{2}\right) \left[\log \frac{1 - iz}{1 + iz} + \log(-1)\right]$$

$$= \frac{1}{ac\sqrt{a^2 - b^2}} \arctan z - \frac{1}{ac\sqrt{a^2 - b^2}} \left(\frac{\pi}{2} + k\pi\right)$$

$(k = 0, \pm 1, \pm 2, \ldots)$ by Eq. (10). Thus, except for the constant term, which is immaterial in an indefinite integral, the two entries in the integral table are consistent.

## EXERCISE♪ 3.3

1. Find all the values of the following.
   (a) $i^i$                  (b) $(-1)^{2/3}$                  (c) $2^{\pi i}$
   (d) $(1 + i)^{1-i}$        (e) $(1 + i)^3$

2. Show that if $z \neq 0$, then $z^0 = 1$.

3. Find the principal value (i.e., the value given by the principal branch) of each of the following.
   (a) $4^{1/2}$                  (b) $i^{2i}$                  (c) $(1 + i)^{1+i}$

4. Is 1 raised to any power always equal to 1?

5. Give an example to show that the principal value of $(z_1 z_2)^\alpha$ need not be equal to the product of principal values $z_1^\alpha z_2^\alpha$.

6. Let $\alpha$ and $\beta$ be complex constants and let $z \neq 0$. Show that the following identities hold when each power function is given by its principal branch.

   (a) $z^{-\alpha} = \frac{1}{z^\alpha}$            (b) $z^\alpha z^\beta = z^{\alpha + \beta}$            (c) $\frac{z^\alpha}{z^\beta} = z^{\alpha - \beta}$

7. Find the derivative of the principal branch of $z^{1+i}$ at $z = i$.

8. Show that all solutions of the equation $\sin z = 2$ are given by $\pi/2 + 2k\pi \pm i \operatorname{Log}(2 + \sqrt{3})$, where $k = 0, \pm 1, \pm 2, \ldots$. [REMARK: This solution set can also be represented as $\pi/2 + 2k\pi - i \operatorname{Log}(2 \pm \sqrt{3}), k = 0, \pm 1, \pm 2, \ldots$.]

9. Derive formulas (9) and (11) concerning $\cos^{-1} z$.

10. Show that all the solutions of the equation $\cos z = 2i$ are given by
    $\pi/2 + 2k\pi - i \operatorname{Log}(\sqrt{5} + 2), -\pi/2 + 2k\pi + i \operatorname{Log}(\sqrt{5} + 2)$, for $k = 0, \pm 1, \pm 2, \ldots$.
    [REMARK: This solution set can also be represented as $\pi/2 + 2k\pi - i \operatorname{Log}(\sqrt{5} + 2)$, $-\pi/2 + 2k\pi - i \operatorname{Log}(\sqrt{5} - 2), k = 0, \pm 1, \pm 2, \ldots$.]

11. Find all solutions of the equation $\sin z = \cos z$.

12. Derive formulas (10) and (12) concerning $\tan^{-1} z$.

13. Derive formulas (13) and (14) for $\sinh^{-1} z$ and $\cosh^{-1} z$.

14. Derive the formula $d(\sinh^{-1} z)/dz = 1/(1 + z^2)^{1/2}$ and explain the conditions under which it is valid.

15. Find a branch of each of the following multiple-valued functions that is analytic in the given domain:
    (a) $(z^2 - 1)^{1/2}$ in the unit disk, $|z| < 1$.
    (b) $(4 + z^2)^{1/2}$ in the complex plane slit along the imaginary axis from $-2i$ to $2i$.

(c) $(z^4 - 1)^{1/2}$ in the exterior of the unit circle, $|z| > 1$.

(d) $(z^3 - 1)^{1/3}$ in the exterior of the unit circle, $|z| > 1$.

16. According to Definition 5 the multiple-valued function $c^z$, where $c$ is a nonzero constant, is given by $c^z = e^{z \log c}$. Show that by selecting a particular value of $\log c$ we obtain a branch of $c^z$ that is entire. Find the derivative of such a branch.

17. Derive the identity

$$\sec^{-1} z = -i \log\left[\frac{1}{z} + \left(\frac{1}{z^2} - 1\right)^{1/2}\right].$$

Using principal values determine the range of $\operatorname{Sec}^{-1} x$ when $x > 1$, and when $x < -1$. Compare this with the ranges listed in standard mathematical handbooks.

18. Show that the following are consistent with the identities listed in this section:

(a) $\displaystyle\int \frac{dx}{\sqrt{a + bx + cx^2}} = \begin{cases} \dfrac{1}{\sqrt{c}} \log\{2\sqrt{c(a + bx + cx^2)} + 2cx + b\} & \text{if } c > 0 \\[2ex] \dfrac{1}{\sqrt{c}} \sinh^{-1} \dfrac{2cx + b}{\sqrt{4ac - b^2}} & \text{if } c > 0 \text{ and } b^2 < 4ac \\[2ex] \dfrac{-1}{\sqrt{-c}} \sin^{-1} \dfrac{2cx + b}{\sqrt{b^2 - 4ac}} & \text{if } c < 0 \text{ and } b^2 > 4ac \end{cases}$

(b) $\displaystyle\int \frac{dx}{a + bx + cx^2} = \begin{cases} \dfrac{2}{\sqrt{4ac - b^2}} \tan^{-1} \dfrac{2cx + b}{\sqrt{4ac - b^2}} & \text{if } b^2 < 4ac \\[2ex] \dfrac{-2}{\sqrt{b^2 - 4ac}} \tanh^{-1} \dfrac{2cx + b}{\sqrt{b^2 - 4ac}} \\[1ex] \qquad\qquad\text{or} \\[1ex] \dfrac{1}{\sqrt{b^2 - 4ac}} \log \dfrac{2cx + b - \sqrt{b^2 - 4ac}}{2cx + b + \sqrt{b^2 - 4ac}} & \text{if } b^2 > 4ac \end{cases}$

(c) $\displaystyle\int \frac{dx}{\sqrt{a^2 - x^2}} = \sin^{-1} \frac{x}{|a|}$    if $x^2 < a^2$,

$\displaystyle\int \frac{dx}{\sqrt{x^2 - a^2}} = \log[x + \sqrt{x^2 - a^2}]$    if $x^2 > a^2$

19. Using only real arithmetic operations, write a computer program whose input $(x, y)$ is the real and imaginary parts of $z = x + iy$ and whose output is the real and imaginary parts of

(a) the principal branch of $z^\alpha$

(b) $\operatorname{Sin}^{-1} z$

(c) $\operatorname{Sec}^{-1} z$

(d) the branch of $(z^2 - 1)^{1/2}$ discussed in Example 2

## *3.4 APPLICATION TO OSCILLATING SYSTEMS

Many engineering problems are ultimately based upon the response of a system to a sinusoidal input. Naturally, all the parameters in such a situation are real, and the models can be analyzed using the techniques of real variables. However, the utilization of complex variables can greatly simplify the computations and lend some insight into the roles played by the various parameters. In this section we shall illustrate the technique for the analysis of a simple *RLC* (resistor-inductor-capacitor) circuit.

The electric circuit is shown in Fig. 3.9. We suppose that the power supply is driving the system with a sinusoidal voltage $V_s$ oscillating at a frequency of $v$ cycles per second. To be precise, let us say that at time $t$

$$V_s = A \sin \omega t, \tag{1}$$

where $A$ is the amplitude of the signal and $\omega = 2\pi v$. Our goal is to find the current output $I_s$ of the power supply.

Across each element of the circuit there is a voltage drop $V$ which is related to the current $I$ flowing through the element. The relationships between these two quantities are as follows: For the resistor

$$V_r = I_r R, \tag{2}$$

where $R$ is a constant known as the *resistance*; for the capacitor

$$C \frac{dV_c}{dt} = I_c, \tag{3}$$

where $C$ is the *capacitance*; and for the inductor

$$V_l = L \frac{dI_l}{dt}, \tag{4}$$

in terms of the *inductance L*.

One can incorporate formulas (1)–(4) together with Kirchhoff's laws,[†] which express conservation of charge and energy, to produce a system of differential equations involving all the currents and voltages. The solution of this system is, however, a laborious process, and a simpler technique is clearly desirable.

Before we demonstrate the utilization of complex variables in this problem, we shall make some observations based upon physical considerations.

First, notice that if the capacitor and inductor were replaced by resistors, as

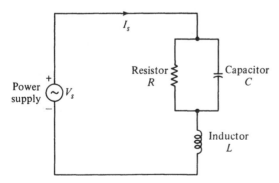

**Figure 3.9** *RLC circuit.*

---

[†] See the references at the end of this chapter.

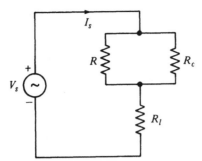

**Figure 3.10** Resistor circuit.

we illustrate in Fig. 3.10, the solution would be simple. The pair of resistances $R$ and $R_c$ are wired in parallel, so they can be replaced by an equivalent resistance $R_1$ given by the familiar law

$$\frac{1}{R_1} = \frac{1}{R} + \frac{1}{R_c},$$

or

$$R_1 = \frac{RR_c}{R + R_c}. \tag{5}$$

This resistance then appears in series with $R_l$, yielding an effective total resistance $R_{\text{eff}}$ given by

$$R_{\text{eff}} = R_1 + R_l, \tag{6}$$

and the circuit is equivalently (from the point of view of the power supply) represented by Fig. 3.11.

Equations (1) and (2) now yield the current output $I_s$:

$$I_s = \frac{A \sin \omega t}{R_{\text{eff}}} = \frac{A \sin \omega t}{\dfrac{RR_c}{R + R_c} + R_l}.$$

Since this model is so easy to solve, it would clearly be advantageous to replace, if possible, the capacitor and inductor by equivalent resistors.

The second point we wish to make involves the nature of the solution $I_s$ for the original circuit. If the power supply is turned on at time $t = 0$, there will be a fairly complicated initial current response; this so-called "transient," however, eventually dies out, and the system enters a steady state in which all the currents and voltages oscillate sinusoidally at the same frequency as the driving voltage. This behavior is common to all damped linear systems, and it will be familiar to anyone

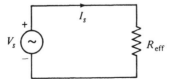

**Figure 3.11** Equivalent resistor circuit.

who has ridden a bicycle over railroad ties or mastered the art of dribbling a basket-ball. For many applications it is this steady-state response, which is independent of the initial state of the system, that is of interest. The complex-variable technique that we shall describe is ideally suited for finding this response.[†]

The technique is based upon the equation

$$e^{i\omega t} = \cos \omega t + i \sin \omega t \tag{7}$$

for real $t$. Observe first that the simple sinusoids can be written, from Eq. (7), as

$$\cos \omega t = \text{Re } e^{i\omega t}, \qquad \sin \omega t = \text{Im } e^{i\omega t}. \tag{8}$$

Second, the most general sinusoid with this frequency, $a \cos \omega t + b \sin \omega t$, can be written in the form

$$a \cos \omega t + b \sin \omega t = \text{Re } \alpha e^{i\omega t}$$
$$= \text{Im } \alpha' e^{i\omega t}, \tag{9}$$

where $\alpha = a - ib$ and $\alpha' = i\alpha$. The advantage of the exponential representation for these functions lies in its compactness and in the simplicity of the expression for the derivative; differentiating $e^{i\omega t}$ (with respect to $t$) merely amounts to multiplying by $i\omega$. Consequently, the derivative of the general sinusoid in Eq. (9) becomes

$$\frac{d}{dt}(a \cos \omega t + b \sin \omega t) = \frac{d}{dt} \text{Re } \alpha e^{i\omega t} = \text{Re } \frac{d}{dt}(\alpha e^{i\omega t})$$
$$= \text{Re } i\omega\alpha e^{i\omega t}$$
$$= -a\omega \sin \omega t + b\omega \cos \omega t;$$

alternatively,

$$\frac{d}{dt}(a \cos \omega t + b \sin \omega t) = \text{Im } i\omega\alpha' e^{i\omega t}.$$

To take advantage of these notions, we must make one more observation about the circuit in Fig. 3.9. Each of the elements is a *linear* device in the sense that the superposition principle holds; that is, if an element responds to the excitation voltages $V_1(t)$ and $V_2(t)$ by producing the currents $I_1(t)$ and $I_2(t)$, respectively, then the response to the voltage $V_1(t) + \beta V_2(t)$ will be the current $I_1(t) + \beta I_2(t)$, for any constant $\beta$. This is a mathematical consequence of the linearity of Eqs. (2)–(4), and it will hold true even if the functions $V_1(t)$ and $V_2(t)$ take complex values, although this corresponds to no physically meaningful situation. In particular, the response to the voltage $V_1(t) + iV_2(t)$ will be the current $I_1(t) + iI_2(t)$. And since we know that the circuit responds to real voltages with real currents, it follows that if the response to the complex voltage $V(t)$ is $I(t)$, then the responses to the (real) voltages $\text{Re } V(t)$ and $\text{Im } V(t)$ will be $\text{Re } I(t)$ and $\text{Im } I(t)$, respectively.

With all these mathematical tools at hand, let us return to the solution of the

---

[†] The widespread adoption of this technique in electrical engineering is traditionally attributed to Charles Proteus Steinmetz.

problem depicted in Fig. 3.9. The supply voltage, given in Eq. (1), can be represented by

$$V_s(t) = \text{Im } Ae^{i\omega t}. \tag{10}$$

Our strategy will be to find the steady-state response of the circuit to the (fictional) voltage $Ae^{i\omega t}$, and to take the imaginary part of the answer as our solution, in accordance with the previous paragraph.

From the earlier observations about the nature of the steady-state response and Eq. (9), we are led to postulate that the current or voltage in any part of the circuit can be written as a (possibly complex) constant times $e^{i\omega t}$; furthermore, differentiation of such a function is equivalent to multiplication by the factor $i\omega$. Thus, in the situation at hand, Eq. (3) for the behavior of the capacitor becomes

$$i\omega CV_c = I_c, \tag{11}$$

and Eq. (4) for the inductor becomes

$$V_l = i\omega LI_l. \tag{12}$$

But the voltage-current relationships (11) and (12) now have the same form as Eq. (2) for a resistor; in other words, operated at the frequency $v = \omega/2\pi$, a capacitor behaves mathematically like a resistor with resistance

$$R_c = \frac{1}{i\omega C}, \tag{13}$$

and an inductor behaves like a resistor with resistance

$$R_l = i\omega L. \tag{14}$$

These are pure imaginary numbers, but we have already abandoned physical reality in assuming a complex power-supply voltage. As a concession to realists, however, engineers have adopted the term "impedance" to describe a complex resistance.

Having replaced, formally, the capacitor and inductor by resistors, we are back in the situation of the simple circuit of Fig. 3.10. The effective impedance of the series-parallel arrangement is displayed in Eqs. (5) and (6), and substitution of the relations (13) and (14) yields the expression

$$R_{\text{eff}} = \frac{\dfrac{R}{i\omega C}}{R + \dfrac{1}{i\omega C}} + i\omega L.$$

The current output of the power supply is therefore

$$I_s = \text{Im } \frac{Ae^{i\omega t}}{R_{\text{eff}}},$$

which after some manipulation can be written as

$$I_s = \text{Im} \frac{Ae^{i\omega t}\left\{\dfrac{1}{R} + i\left[\omega C(1 - \omega^2 LC) - \dfrac{\omega L}{R^2}\right]\right\}}{(1 - \omega^2 LC)^2 + \dfrac{\omega^2 L^2}{R^2}}. \tag{15}$$

The expression for $I_s$ as a linear combination of sin $\omega t$ and cos $\omega t$ may be found by expanding Eq. (15), but complex variables can once more be used in providing a more meaningful interpretation of the answer. If we define $\phi_0$ and $R_0$ by

$$\phi_0 := \text{Arg} \, R_{\text{eff}}, \qquad R_0 := |R_{\text{eff}}|,$$

then $I_s$ can be expressed as

$$I_s = \text{Im} \frac{Ae^{i\omega t}}{R_0 e^{i\phi_0}} = \text{Im} \frac{A}{R_0} e^{i(\omega t - \phi_0)} = \frac{A}{R_0} \sin(\omega t - \phi_0). \tag{16}$$

This displays some easily visualized properties of the output current in relation to the input voltage (1). The current is, as we indicated earlier, a sinusoid with the same frequency, but its amplitude differs from the voltage amplitude by the factor $1/R_0$. Furthermore, the two sinusoids are out of phase, with $\phi_0$ measuring the phase difference in radians. The numbers $R_0$ and $\phi_0$ can be computed in terms of the circuit parameters and the frequency, and the circuit problem is solved.

In conclusion we wish to reiterate the two advantages of using complex notation in solving linear sinusoidal problems. First, the representation of general sinusoids (9) is compact and leads naturally to a reinterpretation of the sinusoid in terms of amplitudes and phases, as in Eqs. (15) and (16). And second, the process of differentiation is replaced by simple multiplication. It is important to keep in mind, however, that only the real (or imaginary) part of the solution corresponds to physical reality and that the condition of linearity is of the utmost importance.

# EXERCISES 3.4

1. Using the techniques of this section, find the steady-state current output $I_s$ of the circuits in Fig. 3.12 on page 100.
2. Verify the expression in Eq. (15).
3. In the limit of very low frequencies, a capacitor behaves like an open circuit (infinite resistance), while an inductor behaves like a short circuit (zero resistance). Draw and analyze the low-frequency limit of the circuit in Fig. 3.9, and verify expression (15) in this limit.
4. Repeat Prob. 3 for the high-frequency limit. (What are the capacitor's and inductor's behaviors in this case?)

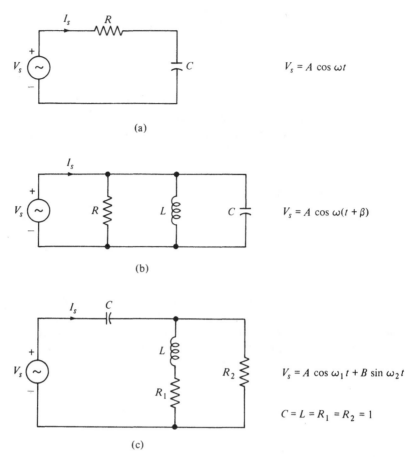

$V_s = A \cos \omega t$

(a)

$V_s = A \cos \omega(t + \beta)$

(b)

$V_s = A \cos \omega_1 t + B \sin \omega_2 t$

$C = L = R_1 = R_2 = 1$

(c)

**Figure 3.12**   Electrical circuits for Prob. 1.

## SUMMARY

The complex sine and cosine functions are defined in terms of the exponential function: $\sin z = (e^{iz} - e^{-iz})/2i$, $\cos z = (e^{iz} + e^{-iz})/2$. When $z$ is real these definitions agree with those given in calculus. Furthermore, the usual differentiation formulas and trigonometric identities remain valid for these functions. Complex hyperbolic functions are also defined in terms of the exponential function, and, again, they retain many of their familiar properties.

The complex logarithmic function $\log z$ is the inverse of the exponential function and is given by $\log z = \text{Log}\,|z| + i \arg z$, where the capitalization signifies the natural logarithmic function of real variables. Since for each $z \neq 0$, $\arg z$ has infinitely many values, the same is true of $\log z$. Thus $\log z$ is an example of a multiple-valued function.

The concept of analyticity for a multiple-valued function $f$ is discussed in terms of branches of $f$. A single-valued function $F$ is said to be a branch of $f$ if it is continuous in some domain, at each point of which $F(z)$ coincides with one of the values of $f(z)$.

Branches of log $z$ can be obtained by restricting arg $z$ so that it is single-valued and continuous. For example, the function Log $z = $ Log $|z| + i$ Arg $z$, which is the principal branch of log $z$, is analytic in the domain $D^*$ consisting of all points in the plane except those on the nonpositive real axis. The formula $d(\log z)/dz = 1/z$ is valid in the sense that it holds for every branch of log $z$.

Complex powers of $z$ are defined by means of logarithms. Specifically, $z^{\alpha} = e^{\alpha \log z}$ for $z \neq 0$. Unless $\alpha$ is an integer, $w = z^{\alpha}$ is a multiple-valued function whose branches can be obtained by selecting branches of log $z$. The inverse trigonometric and hyperbolic functions can also be expressed in terms of logarithms, and they too are multiple-valued.

The analysis of sinusoidally oscillating systems can be greatly simplified with the use of complex variables. In particular, the differentiation operation, applied to terms containing the factor $e^{i\omega t}$, reduces to simple multiplication (by $i\omega$).

## SUGGESTED READING

The "ambiguous" antiderivatives can be found in the following integral tables:

[1] *Standard Mathematical Tables*. The Chemical Rubber Company, Cleveland (continuing editions).

[2] *Handbook of Mathematical Tables and Formulas*. R. S. Burington. Handbook Publishers, Sandusky, Ohio (continuing editions).

The following texts may be helpful for further study of the concepts used in Sec. 3.4:

### *Electrical Circuits*

[3] Scott, D. E. *An Introduction to Circuit Analysis: A Systems Approach*, McGraw-Hill Book Company, New York, 1987.

[4] Smith, R. J., and Dorf, R. C. *Circuit Devices and Systems*, 5th ed. Wiley, New York, 1992.

### *Sinusoidal Analysis*

[5] Chen, W. H. *The Analysis of Linear Systems*. McGraw-Hill Book Company, New York, 1963.

[6] Guillemin, E. A. *Theory of Linear Physical Systems*. John Wiley & Sons, Inc., New York, 1963.

[7] Mikusinski, J. *Operational Calculus*, 2nd ed. Pergamon Press, New York, 1983.

# :4:

# Complex Integration

In Chapter 2 we saw that the *two*-dimensional nature of the complex plane required us to generalize our notion of a derivative because of the freedom of the variable to approach its limit along any of an infinite number of directions. This two-dimensional aspect will have an effect on the theory of integration as well, necessitating the consideration of integrals along general curves in the plane and not merely segments of the x-axis. Fortunately, such well-known techniques as using antiderivatives to evaluate integrals carry over to the complex case.

When the function under consideration is analytic the theory of integration becomes an instrument of profound significance in studying its behavior. The main result is the theorem of *Cauchy*, which roughly says that the integral of a function around a closed loop is zero if the function is analytic "inside and on" the loop. Using this result we shall derive the *Cauchy integral formula*, which explicitly displays many of the important properties of analytic functions.

## 4.1 CONTOURS

Let us turn to the problem of finding a mathematical explication of our intuitive concept of a *curve* in the xy-plane. Although most of the applications described in this book involve only two very simple types of curves—line segments and arcs of circles— it will be necessary for proving theorems to nail down the definitions of more general curves.

It is helpful in this regard to visualize an artist actually tracing the curve $\gamma$ on graph paper. At any particular instant of time $t$, a dot is drawn at, say, the point

$z = x + iy$; the locus of dots generated over an interval of time $a \leq t \leq b$ constitutes the curve. Clearly we can interpret the artist's actions as generating $z$ as a function of $t$, and then the curve $\gamma$ is the range of $z(t)$ as $t$ varies between $a$ and $b$. In such a case $z(t)$ is called a *parametrization* of $\gamma$.

To keep things simple, we shall set down certain ground rules for the artist to follow in drawing $\gamma$. First, we do not permit the pen to be lifted from the paper during the sketch; mathematically, we are requiring that $z(t)$ be continuous. Second, we insist that the curves be drawn with an even, steady stroke; specifically, the pen point must move with a well-defined (finite) velocity that, also, must vary continuously. Now the velocity of the point tracing out the trajectory $(x(t), y(t))$ is the vector $(dx/dt, dy/dt) = x'(t) + iy'(t)$. It makes sense to call this vector $z'(t)$, since $x(t) + iy(t) = z(t)$ and therefore

$$x'(t) + iy'(t) = \lim_{\Delta t \to 0} \frac{z(t + \Delta t) - z(t)}{\Delta t} .$$

Thus we insist that $z'(t)$ exist[†] and be continuous on $[a, b]$.

Finally, it is sometimes convenient to stipulate that no point be drawn twice; in other words, $z(t)$ must be one-to-one. Occasionally, however, we allow the possibility that the initial and terminal points coincide, as in the case of a circle.

Putting this all together, we now specify the class of *smooth curves*. These fall into two separate categories: *smooth arcs*, which have distinct endpoints, and *smooth closed curves*, whose endpoints coincide.

---

**Definition 1.** A point set $\gamma$ in the complex plane is said to be a **smooth[‡] arc** if it is the range of some continuous complex-valued function $z = z(t)$, $a \leq t \leq b$, that satisfies the following conditions:

  (i) $z(t)$ has a continuous derivative[§] on $[a, b]$,

  (ii) $z'(t)$ never vanishes on $[a, b]$,

  (iii) $z(t)$ is one-to-one on $[a, b]$.

A point set $\gamma$ is called a **smooth closed curve** if it is the range of some continuous function $z = z(t)$, $a \leq t \leq b$, satisfying conditions (i) and (ii) and the following:

  (iii)' $z(t)$ is one-to-one on the half-open interval $[a, b)$, but $z(b) = z(a)$ and $z'(b) = z'(a)$.

---

[†] Observe, however, that since $t$ is a real variable in this context, the existence of the derivative $z'(t)$ is not nearly so profound as it was in Chapter 2, where the independent variable was complex.

[‡] The term *regular* is sometimes used instead of smooth.

[§] At the endpoint $t = a$, $z'(t)$ denotes the right-hand derivative, while at $t = b$, $z'(t)$ is the left-hand derivative.

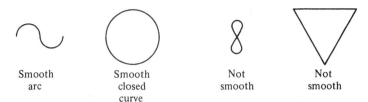

**Figure 4.1**   Examples of smooth and nonsmooth curves.

The phrase "$\gamma$ is a smooth curve" means that $\gamma$ is either a smooth arc or a smooth closed curve.

In elementary calculus it is shown that the vector $(x'(t), y'(t))$, if it exists and is nonzero, can be interpreted geometrically as being tangent to the curve at the point $(x(t), y(t))$. Hence the conditions of Definition 1 imply that a smooth curve possesses a unique tangent at every point and that the tangent direction varies continuously along the curve. Consequently a smooth curve has no corners or cusps; see Fig. 4.1.

The functional relationship $z = z(t)$, $a \leq t \leq b$, that describes the curve $\gamma$ is, of course, somewhat arbitrary. Another artist could draw the same curve faster or use a different starting point, resulting in a different description $z = z_1(t)$, $a_1 \leq t \leq b_1$. For example, the straight-line segment joining 0 to $1 + i$ is generated by each of the following functions:

$$z_1(t) = t + it \qquad\qquad (0 \leq t \leq 1),$$

$$z_2(t) = (1 - 2t) + i(1 - 2t) \qquad \left(0 \leq t \leq \frac{1}{2}\right),$$

$$z_3(t) = \tan t + i \tan t \qquad \left(0 \leq t \leq \frac{\pi}{4}\right).$$

*Any* function $z(t)$ satisfying the conditions of Definition 1 for a given smooth curve $\gamma$ is called an *admissible parametrization* of $\gamma$, and the variable $t$ is called the *parameter*; the corresponding interval $[a, b]$ is known as the *parametric interval*. So we can describe our situation in the following manner: A given smooth curve $\gamma$ will have many different admissible parametrizations, but we need produce only *one* admissible parametrization in order to demonstrate that a given curve is smooth.

**Example 1**

Find an admissible parametrization for each of the following smooth curves: (a) the straight-line segment joining $-2 - 3i$ and $5 + 6i$, (b) the circle of radius 2 centered at $1 - i$, and (c) the graph of the function $y = x^3$ for $0 \leq x \leq 1$.

**Solution.** (a) Given any two distinct points $z_1$ and $z_2$, every point on the line segment joining $z_1$ and $z_2$ is of the form $z_1 + t(z_2 - z_1)$, where $0 \leq t \leq 1$ (see Prob. 18 in Exercises 1.3). Therefore, the given curve constitutes the range of

$$z(t) = -2 - 3i + t(7 + 9i) \qquad (0 \leq t \leq 1).$$

The verification of conditions (i), (ii), and (iii) is immediate; thus this is an admissible parametrization.

**(b)** In Section 1.4 it was shown that any point on the unit circle centered at the origin can be written in the form $e^{it} = \cos t + i \sin t$ for $0 \le t < 2\pi$; therefore, an admissible parametrization for this smooth closed curve is $z_0(t) = e^{it}$, $0 \le t \le 2\pi$ (notice that the endpoints are joined properly). To parametrize the given circle (b) we simply shift the center and double the radius:

$$z(t) = 1 - i + 2e^{it} \qquad (0 \le t \le 2\pi).$$

**(c)** The parametrization of the graph of any function $y = f(x)$ is quite trivial; simply let $x = t$ and $y = f(t)$. Of course one must verify that this is an admissible parametrization. For the graph (c) we have

$$z(t) = t + it^3 \qquad (0 \le t \le 1). \quad \blacksquare$$

Let us carry our analysis of curve sketching a little further. Suppose that the artist is to draw a smooth *arc* like that in Fig. 4.2 and is to abide by our ground rules (in particular it is not permitted to retrace points). Then it is intuitively clear that the artist must either start at $z_I$ and work toward $z_{II}$, or start at $z_{II}$ and terminate at $z_I$. Either mode produces an ordering of the points along the curve (Fig. 4.3).

Thus we see that there are exactly two such "natural" orderings of the points of a smooth arc $\gamma$, and either one can be specified by declaring which endpoint of $\gamma$ is the initial point. A smooth arc, together with a specific ordering of its points, is called a *directed smooth arc*. The ordering can be indicated by an arrow, as in Fig. 4.3.

The ordering that the artist generates while drawing $\gamma$ is reflected in the parametrization function $z(t)$ that describes the pen's trajectory; specifically, the point $z(t_1)$ will precede the point $z(t_2)$ whenever $t_1 < t_2$. Since there are only two possible (natural) orderings, *any* admissible parametrization must fall into one of two categories, according to the particular ordering it respects. (The proof of this fact forms the

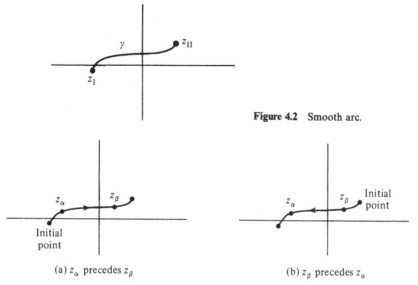

**Figure 4.2**  Smooth arc.

(a) $z_\alpha$ precedes $z_\beta$          (b) $z_\beta$ precedes $z_\alpha$

**Figure 4.3**  Directed smooth arcs.

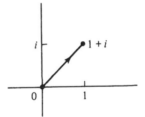

**Figure 4.4**  Directed line segment.

basis of the rigorous treatment of directed curves; see Ref. [1].) For example, consider the ordering indicated on the line segment in Fig. 4.4. The (admissible) parametrization functions $z_1(t) = (1 + i)t, 0 \le t \le 1$, and $z_2(t) = (1 + i)\tan t, 0 \le t \le \pi/4$, are both consistent with the given ordering, but $z_3(t) = -(1 + i)t, -1 \le t \le 0$, corresponds to the opposite ordering.

In general, if $z = z(t), a \le t \le b$, is an admissible parametrization consistent with one of the orderings, then $z = z(-t), -b \le t \le -a$, always corresponds to the opposite ordering.

The situation is slightly more complicated if the artist is to draw a smooth *closed* curve. First an initial point must be selected; then the artist must choose one of the two directions in which to trace the curve (see Fig. 4.5). Having made these decisions, the artist has established the ordering of the points of $\gamma$. Now, however, there is one anomaly; the initial point both precedes and is preceded by every other point, since it also serves as the terminal point. Ignoring this schizophrenic pest, we shall say that the points of a smooth closed curve have been ordered when (i) a designation of the initial point is made and (ii) one of the two "directions of transit" from this point is selected. A smooth closed curve whose points have been ordered is called a *directed smooth closed curve*.

As in the case of smooth arcs, the parametrization of the trajectory of the artist's pen reflects the ordering generated in sketching a smooth closed curve. If this parametrization is given by $z = z(t), a \le t \le b$, then (i) the initial point must be $w = z(a)$ and (ii) the point $z(t_1)$ precedes the point $z(t_2)$ whenever $a < t_1 < t_2 < b$. Any other admissible parametrization having the same initial point $w$ must reflect either the same or the opposite ordering. Thus in Fig. 4.6 the parametrizations $z = z_1(t) = e^{2\pi it}, 0 \le t \le 1$, and $z = z_2(t) = e^{it}, 0 \le t \le 2\pi$, are consistent with the ordering indicated. The parametrization functions $z_3(t) = e^{-2\pi it}, 0 \le t \le 1$, and

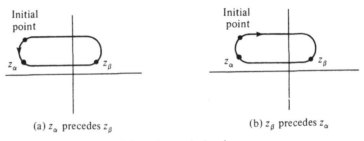

(a) $z_\alpha$ precedes $z_\beta$                    (b) $z_\beta$ precedes $z_\alpha$

**Figure 4.5**  Directed smooth closed curves.

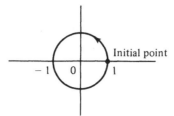

**Figure 4.6**   Ordered points on a circle.

$z_4(t) = -e^{2\pi it}$, $0 \le t \le 1$, are *not* consistent with this ordering; indeed $z_3(t)$ reflects the opposite ordering, and $z_4(t)$ has the wrong initial point.

The phrase *directed smooth curve* will be used to mean either a directed smooth arc or a directed smooth closed curve.

Now we are ready to specify the kinds of curves that will be used in the theory of integration. They are formed by joining directed smooth curves together end-to-end; this allows self-intersections, cusps, and corners. In addition, it will be convenient to include single isolated points as members of this class. Let us explore the possibilities uncovered by the following definition.

---

**Definition 2.**   A **contour** $\Gamma$ is either a single point $z_0$ or a finite sequence of directed smooth curves $(\gamma_1, \gamma_2, \ldots, \gamma_n)$ such that the terminal point of $\gamma_k$ coincides with the initial point of $\gamma_{k+1}$ for each $k = 1, 2, \ldots, n-1$.

---

Notice that a single directed smooth curve is a contour with $n = 1$.

Speaking loosely, we can say that the contour $\Gamma$ inherits a direction from its components $\gamma_k$: If $z_1$ and $z_2$ lie on the same directed smooth curve $\gamma_k$, they are ordered by the direction on $\gamma_k$, and if $z_1$ lies on $\gamma_i$ while $z_2$ lies on $\gamma_j$, we say that $z_1$ precedes $z_2$ if $i < j$. This is ambiguous because of the possibility that $z_1$, say, may lie on two different smooth curves, and therefore we must indicate which "occurrence" of $z_1$ is meant when we say $z_1$ precedes $z_2$.

Figure 4.7 illustrates four elementary examples of contours formed by joining directed smooth curves. In Fig. 4.7(d), $z_\alpha$ regarded as a point of $\gamma_1$ precedes $z_\beta$, but regarded as a point of $\gamma_3$, it is preceded by $z_\beta$.

In Fig. 4.8 we explore some of the more exotic possibilities allowed in Definition 2. The contour $\Gamma$ in Fig. 4.8(a) is traced by moving along the segment from left

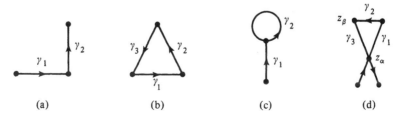

(a)                (b)                (c)                (d)

**Figure 4.7**   Examples of contours.

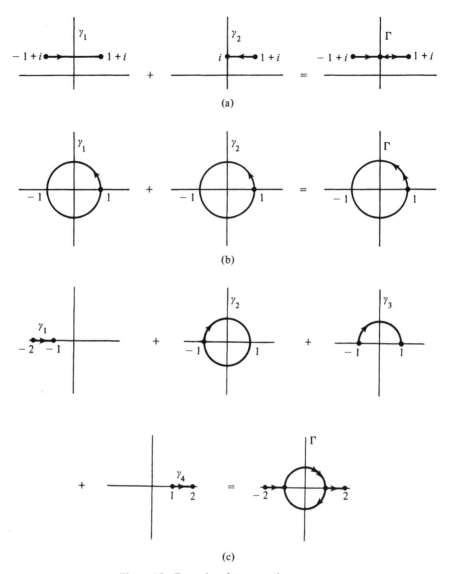

**Figure 4.8** Examples of more exotic contours.

to right and then halfway back from right to left. The circle in Fig. 4.8(b) is traversed twice in the counterclockwise direction. And in Fig. 4.8(c) the first segment of $\Gamma$ is traced left to right, then we make one and one-half clockwise revolutions around the circle, and finally trace out the second segment.

A parametrization of a contour is simply a "piecing together" of admissible parametrizations of its smooth-curve components. Although we rarely have need of a formal description of this process, we state it precisely as follows. One says that $z = z(t)$, $a \leq t \leq b$, is a parametrization of the contour $\Gamma = (\gamma_1, \gamma_2, \ldots, \gamma_n)$ if there is a subdivision of $[a, b]$ into $n$ subintervals $[\tau_0, \tau_1], [\tau_1, \tau_2], \ldots, [\tau_{n-1}, \tau_n]$, where

$$a = \tau_0 < \tau_1 < \cdots < \tau_{n-1} < \tau_n = b,$$

such that on each subinterval $[\tau_{k-1}, \tau_k]$ the function $z(t)$ is an admissible parametrization of the smooth curve $\gamma_k$, consistent with the direction on $\gamma_k$. Since the endpoints of consecutive $\gamma_k$'s are properly connected, $z(t)$ must be continuous on $[a, b]$. However, $z'(t)$ may have jump discontinuities at the points $\tau_k$. The contour parametrization of a point is simply a constant function.

When we have admissible parametrizations of the components $\gamma_k$ of a contour $\Gamma$, it is a simple matter to piece these together to get a contour parametrization for $\Gamma$; in fact, we can always use the "standard" parametric interval $[0, 1]$. We employ the techniques of *rescaling*, which are amply illustrated by the following example. (The general case is discussed in Prob. 6.)

**Example 2**

Parametrize the contour in Fig. 4.9, over the interval $[0, 1]$.

**Solution.**   We have already seen how to parametrize straight lines. The following functions are admissible parametrizations for $\gamma_1$, $\gamma_2$, and $\gamma_3$, consistent with their directions:

$$\gamma_1: \quad z_1(t) = t \qquad\qquad (0 \le t \le 1),$$

$$\gamma_2: \quad z_2(t) = 1 + t(i - 1) \quad (0 \le t \le 1),$$

$$\gamma_3: \quad z_3(t) = i - ti \qquad\quad (0 \le t \le 1).$$

Now we rescale so that $\gamma_1$ is traced as $t$ varies between 0 and $\frac{1}{3}$, $\gamma_2$ is traced for $\frac{1}{3} \le t \le \frac{2}{3}$, and $\gamma_3$ is traced for $\frac{2}{3} \le t \le 1$. This is simply a matter of shifting and stretching the variable $t$.

For $\gamma_1$, observe that the range of the function $z_1(t) = t$, $0 \le t \le 1$, is the same as the range of $z_I(t) = 3t$, $0 \le t \le \frac{1}{3}$, and that $z_I(t)$ is an admissible parametrization corresponding to the same ordering. The curve $\gamma_2$ is the range of $z_2(t) = 1 + t(i - 1)$, $0 \le t \le 1$, and this is the same as the range of $z_{II}(t) = 1 + 3(t - \frac{1}{3})(i - 1)$, $\frac{1}{3} \le t \le \frac{2}{3}$, again preserving admissibility and ordering. Handling $z_3(t)$ similarly, we find

$$z(t) = \begin{cases} 3t & (0 \le t \le \frac{1}{3}), \\ 1 + 3(t - \frac{1}{3})(i - 1) & (\frac{1}{3} \le t \le \frac{2}{3}), \\ i - 3(t - \frac{2}{3})i & (\frac{2}{3} \le t \le 1). \end{cases} \quad\blacksquare$$

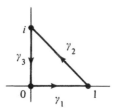

**Figure 4.9**   Contour for Example 2.

For obvious reasons, we call the (undirected) point set underlying a contour a *piecewise smooth curve*. We shall use the symbol $\Gamma$ ambiguously to refer to both the contour and its underlying curve, allowing the context to provide the proper interpretation.

Much of the terminology of directed smooth curves is readily applied to contours. The initial point of $\Gamma$ is the initial point of $\gamma_1$, and its terminal point is the terminal point of $\gamma_n$; therefore $\Gamma$ can be regarded as a path connecting these points. If the directions on all the components of $\Gamma$ are reversed and the components are taken in the opposite order, the resulting contour is called the *opposite contour* and is denoted by $-\Gamma$ (see Fig. 4.10). Notice that if $z = z(t)$, $a \le t \le b$, is a parametrization of $\Gamma$, then $z = z(-t)$, $-b \le t \le -a$, parametrizes $-\Gamma$.

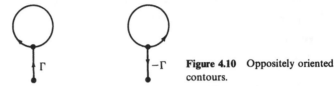

**Figure 4.10** Oppositely oriented contours.

$\Gamma$ is said to be a *closed contour* or a *loop* if its initial and terminal points coincide. A *simple closed contour* is a closed contour with no multiple points other than its initial-terminal point; in other words, if $z = z(t)$, $a \le t \le b$, is a parametrization of the closed contour, then $z(t)$ is one-to-one on the half-open interval $[a, b)$. Figure 4.9 illustrates a simple closed contour.

There is an alternative way of specifying the direction along a curve if the curve happens to be a simple closed contour. We employ the venerable *Jordan curve theorem* from topology, which guarantees the intuitively transparent observation that such a curve has an inside and an outside (see Fig. 4.11).

---

**Theorem 1.**   Any simple closed contour separates the plane into two domains, each having the curve as its boundary. One of these domains, called the *interior*, is bounded; the other, called the *exterior*, is unbounded.

---

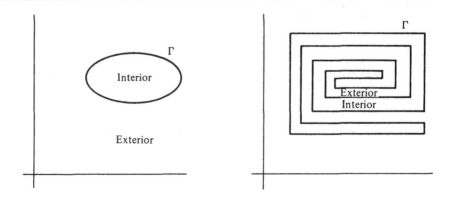

**Figure 4.11**   Jordan curve theorem.

Jordan's theorem actually holds for more general curves, but the proof is quite involved—even for contours.

Now given a simple closed contour $\Gamma$ we can imagine a child bicycling around the curve and tracing out its points in the order specified by its direction. If the bicycle has training wheels and if it is small enough, then one of the training wheels will always remain in the interior domain of the contour, while the other remains in the exterior (otherwise we would have a path connecting these domains without crossing $\Gamma$, in contradiction to the Jordan curve theorem). Consequently the direction along $\Gamma$ can be completely specified by declaring its initial-terminal point and stating which domain (interior or exterior) lies to the *left* of an observer tracing out the points in order. When the interior domain lies to the left, we say that $\Gamma$ is *positively* oriented. Otherwise $\Gamma$ is said to be oriented *negatively*. A positive orientation generalizes the concept of counterclockwise motion; see Fig. 4.12.

The final topic we want to discuss in this section is the length of a contour. We begin by considering a smooth curve $\gamma$, with any admissible parametrization $z = z(t)$, $a \leq t \leq b$. Let $s(t)$ denote the length of the arc of $\gamma$ traversed in going from the point $z(a)$ to the point $z(t)$. We wish to find an expression for $ds/dt$. Accordingly, we must consider $s(t + \Delta t) - s(t)$, which is the length of the curve between $z(t)$ and $z(t + \Delta t)$. See Fig. 4.13.

We appeal to the reader's geometric intuition and experience with differential calculus to establish the following claim: because the curve is smooth, the arc between $z(t)$ and $z(t + \Delta t)$ lies very close to the chord joining these points, and in fact the limit of the ratio $(s(t + \Delta t) - s(t))/|z(t + \Delta t) - z(t)|$ is 1 as $\Delta t$ decreases to zero.

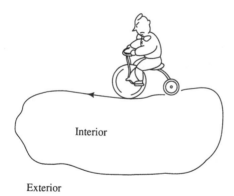

Interior

Exterior

**Figure 4.12**   Jordan, age four.

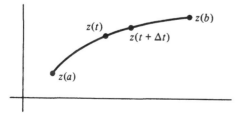

$z(t)$
$z(b)$
$z(t + \Delta t)$
$z(a)$

**Figure 4.13**   Computation of contour length.

Accepting this, we have

$$\frac{ds(t)}{dt} = \lim_{\Delta t \to 0+} \frac{s(t + \Delta t) - s(t)}{\Delta t}$$

$$= \lim_{\Delta t \to 0+} \frac{s(t + \Delta t) - s(t)}{|z(t + \Delta t) - z(t)|} \frac{|z(t + \Delta t) - z(t)|}{\Delta t}$$

$$= 1 \cdot \left| \frac{dz(t)}{dt} \right|,$$

i.e., $ds/dt = |dz/dt|$.

Now we can compute the length of the smooth curve by integrating $ds/dt$, and we have the important formula

$$l(\gamma) = \text{length of } \gamma = \int_a^b \frac{ds}{dt} \, dt = \int_a^b \left| \frac{dz}{dt} \right| \, dt;$$

i.e.,

$$l(\gamma) = \int_a^b \left| \frac{dz}{dt} \right| \, dt = \int_a^b \sqrt{\left( \frac{dx}{dt} \right)^2 + \left( \frac{dy}{dt} \right)^2} \, dt. \tag{1}$$

This formula is established rigorously in the references. Sometimes the shorthand $\int_\gamma |dz|$ is used to indicate $\int_a^b |dz/dt| \, dt$; it emphasizes the intuitively evident fact that $l(\gamma)$ is a geometric quantity that depends only on the point set $\gamma$ and is independent of the particular admissible parametrization used in the computation.

The *length of a contour* is simply defined to be the sum of the lengths of its component curves.

### Example 3

Compute the lengths of the contours in Fig. 4.8 on page 108.

**Solution.**  Simply sum the lengths of the components $\gamma_i$.

In Fig. 4.8(a), $l(\gamma_1) = 2$ and $l(\gamma_2) = 1$. Hence $l(\Gamma) = 2 + 1 = 3$.
In Fig. 4.8(b), $l(\Gamma) = 2\pi + 2\pi = 4\pi$.
In Fig. 4.8(c), $l(\Gamma) = 1 + 2\pi + \pi + 1 = 2 + 3\pi$.  ∎

The lengths of all the components in this example were determined by inspection (as will be true throughout our book). In Prob. 10 we invite the reader to verify formula (1) for these curves.

## EXERCISES 4.1

1. For each of the following smooth curves give an admissible parametrization that is consistent with the indicated direction.
   (a) the line segment from $z = 1 + i$ to $z = -2 - 3i$
   (b) the circle $|z - 2i| = 4$ traversed once in the clockwise direction starting from the point $z = 4 + 2i$
   (c) the arc of the circle $|z| = R$ lying in the second quadrant, from $z = Ri$ to $z = -R$
   (d) the segment of the parabola $y = x^2$ from the point $(1, 1)$ to the point $(3, 9)$

2. Show why the condition that $z'(t)$ never vanishes is necessary to ensure that smooth curves have no cusps. [HINT: Consider the curve traced by $z(t) = t^2 + it^3$, $-1 \le t \le 1$.]

3. Show that the ellipse $x^2/a^2 + y^2/b^2 = 1$ is a smooth curve by producing an admissible parametrization.

4. Show that the range of the function $z(t) = t^3 + it^6$, $-1 \le t \le 1$, is a smooth curve even though the given parametrization is not admissible.

5. Identify the interior of the simple closed contour $\Gamma$ in Fig. 4.14. Is $\Gamma$ positively oriented?

6. Let $\gamma$ be a directed smooth curve. Show that if $z = z(t)$, $a \le t \le b$, is an admissible parametrization of $\gamma$ consistent with the ordering on $\gamma$, then the same is true of

$$z_1(t) = z\left(\frac{b - a}{d - c}t + \frac{ad - bc}{d - c}\right) \qquad (c \le t \le d).$$

7. Parametrize the contour consisting of the perimeter of the square with vertices $-1 - i$, $1 - i$, $1 + i$, and $-1 + i$ traversed once in that order. What is the length of this contour?

8. Parametrize the contour $\Gamma$ indicated in Fig. 4.15. Also give a parametrization for the opposite contour $-\Gamma$.

9. Parametrize the barbell-shaped contour in Fig. 4.16; it has initial point $-1$ and terminal point $1$.

**Figure 4.14** Contour for Prob. 5.

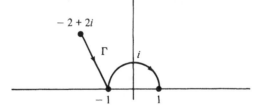

**Figure 4.15** Contour for Prob. 8.

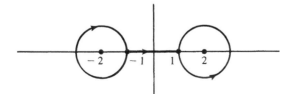

**Figure 4.16** Contour for Prob. 9.

10. Using an admissible parametrization verify from formula (1) that
    (a) the length of the line segment from $z_1$ to $z_2$ is $|z_2 - z_1|$;
    (b) the length of the circle $|z - z_0| = r$ is $2\pi r$.
11. Find the length of the contour $\Gamma$ parametrized by $z = z(t) = 5e^{3it}$, $0 \leq t \leq \pi$.
12. Does formula (1) remain valid when $\gamma$ is a *contour* parametrized by $z = z(t)$, $a \leq t \leq b$?
13. Interpreting $t$ as time and the admissible parametrization $z = z(t)$, $a \leq t \leq b$, as the position function of a moving particle, give the physical meaning of the following quantities.
    (a) $z'(t)$                      (b) $|z'(t)|$
    (c) $|z'(t)\, dt|$            (d) $\int_a^b |z'(t)|\, dt$
14. Let $z = z_1(t)$ be an admissible parametrization of the smooth curve $\gamma$. If $\phi(s)$, $c \leq s \leq d$, is a strictly increasing function such that (i) $\phi'(s)$ is positive and continuous on $[c, d]$ and (ii) $\phi(c) = a$, $\phi(d) = b$, then the function $z_2(s) = z_1(\phi(s))$, $c \leq s \leq d$, is also an admissible parametrization of $\gamma$. Verify that

$$\int_a^b |z_1'(t)|\, dt = \int_c^d |z_2'(s)|\, ds,$$

which demonstrates the invariance of formula (1).

## 4.2 CONTOUR INTEGRALS

In calculus the definite integral of a real-valued function $f$ over an interval $[a, b]$ is defined as the limit of certain sums $\sum_{k=1}^{n} f(c_k)\, \Delta x_k$ (called *Riemann sums*). However, the fundamental theorem of calculus lets us evaluate integrals more directly when an antiderivative is known. The aim of the present section is to use this notion of Riemann sums to define the definite integral of a *complex*-valued function $f$ along a contour $\Gamma$ in the plane. We will accomplish this by first defining the integral along a single directed smooth curve and then defining integrals along a contour in terms of the integrals along its smooth components. When we are finished, however, we will once again obtain simple rules for evaluating integrals in terms of antiderivatives.

Consider then a function $f$ defined over a directed smooth curve $\gamma$ with initial point $\alpha$ and terminal point $\beta$ (possibly coinciding with $\alpha$). As in the previous section, the points on $\gamma$ are ordered in accordance with the direction.

For any positive integer $n$, we define a *partition* $\mathscr{P}_n$ of $\gamma$ to be a finite number of points $z_0, z_1, \ldots, z_n$ on $\gamma$ such that $z_0 = \alpha$, $z_n = \beta$, and $z_{k-1}$ precedes $z_k$ on $\gamma$ for $k = 1, 2, \ldots, n$ (see Fig. 4.17). If we compute the arc length along $\gamma$ between every

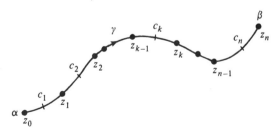

**Figure 4.17** Partitioned curve.

consecutive pair of points $(z_{k-1}, z_k)$, the largest of these lengths provides a measure of the "fineness" of the subdivision; this maximum length is called the *mesh* of the partition and is denoted by $\mu(\mathscr{P}_n)$. It follows that if a given partition $\mathscr{P}_n$ has a "small" mesh, then $n$ must be large and the successive points of the partition must be close to one another.

Now let $c_1, c_2, \ldots, c_n$ be any points on $\gamma$ such that $c_1$ lies on the arc from $z_0$ to $z_1$, the point $c_2$ lies on the arc from $z_1$ to $z_2$, etc. Under these circumstances the sum $S(\mathscr{P}_n)$ defined by

$$S(\mathscr{P}_n) := f(c_1)(z_1 - z_0) + f(c_2)(z_2 - z_1) + \cdots + f(c_n)(z_n - z_{n-1})$$

is called a *Riemann sum* for the function $f$ corresponding to the partition $\mathscr{P}_n$. On writing $z_k - z_{k-1} = \Delta z_k$, this becomes

$$S(\mathscr{P}_n) = \sum_{k=1}^{n} f(c_k)(z_k - z_{k-1}) = \sum_{k=1}^{n} f(c_k)\, \Delta z_k.$$

Now we can generalize the definition of definite integral given in calculus.

---

**Definition 3.** Let $f$ be a complex-valued function defined on the directed smooth curve $\gamma$. We say that $f$ **is integrable along** $\gamma$ if there exists a complex number $L$ that is the limit of *every* sequence of Riemann sums $S(\mathscr{P}_1), S(\mathscr{P}_2), \ldots,$ $S(\mathscr{P}_n), \ldots$ corresponding to *any* sequence of partitions of $\gamma$ satisfying $\lim_{n \to \infty} \mu(\mathscr{P}_n) = 0$; i.e.,

$$\lim_{n \to \infty} S(\mathscr{P}_n) = L \quad \text{whenever} \quad \lim_{n \to \infty} \mu(\mathscr{P}_n) = 0.$$

---

The constant $L$ is called the *integral of $f$ along $\gamma$*, and we write

$$L = \lim_{n \to \infty} \sum_{k=1}^{n} f(c_k)\, \Delta z_k = \int_{\gamma} f(z)\, dz \quad \text{or} \quad L = \int_{\gamma} f.$$

Because Definition 3 is analogous to the definition of the integral given in calculus, certain familiar properties of the latter integrals carry over to the complex case. For example, if $f$ and $g$ are integrable along $\gamma$, then

$$\int_{\gamma} [f(z) \pm g(z)]\, dz = \int_{\gamma} f(z)\, dz \pm \int_{\gamma} g(z)\, dz, \tag{1}$$

$$\int_{\gamma} cf(z)\, dz = c \int_{\gamma} f(z)\, dz \qquad (c \text{ any complex constant}), \tag{2}$$

and

$$\int_{-\gamma} f(z)\, dz = -\int_{\gamma} f(z)\, dz, \tag{3}$$

where $-\gamma$ denotes the curve directed opposite to $\gamma$.

As we know from calculus, not all functions $f$ are integrable. However, if we require that $f$ be continuous, then its integral must exist.

> **Theorem 2.** If $f$ is continuous[†] on the directed smooth curve $\gamma$, then $f$ is integrable along $\gamma$.

For a proof of this theorem see Ref. [2] at the end of the chapter.

While Theorem 2 is of great theoretical importance, it gives us no information on how to compute the integral $\int_\gamma f(z)\,dz$. However, since we are already skilled in evaluating the definite integrals of calculus, it would certainly be advantageous if we could express the complex integral in terms of real integrals.

For this purpose we first consider the special case when $\gamma$ is the real line segment $[a, b]$ directed from left to right. Notice that if $f$ happened to be a real-valued function defined on $[a, b]$, then Definition 3 would agree precisely with the definition of the integral $\int_a^b f(t)\,dt$ given in calculus. Hence, even when $f$ is complex-valued, we shall use the symbol

$$\int_a^b f(t)\,dt$$

to denote the integral of $f$ along the directed real line segment. In this case, when $f(t)$ is a complex-valued function continuous on $[a, b]$, we can write $f(t) = u(t) + iv(t)$, where $u$ and $v$ are each real-valued and continuous on $[a, b]$. Then from properties (1) and (2) we have

$$\int_a^b f(t)\,dt = \int_a^b [u(t) + iv(t)]\,dt$$

$$= \int_a^b u(t)\,dt + i \int_a^b v(t)\,dt; \tag{4}$$

this expresses the complex integral in terms of two real integrals.

If $f(t)$ has the antiderivative $F(t) = U(t) + iV(t)$, then $U' = u$, $V' = v$, and Eq. (4) leads immediately to the following generalization of the fundamental theorem of calculus.

> **Theorem 3.** If the complex-valued function $f$ is continuous on $[a, b]$ and $F'(t) = f(t)$ for all $t$ in $[a, b]$, then
>
> $$\int_a^b f(t)\,dt = F(b) - F(a).$$

This result is illustrated in the following example.

**Example 1**

Compute $\int_0^\pi e^{it}\,dt$.

---

[†] The meaning of continuity for a function $f$ having an arbitrary set $S$ as its domain of definition is as follows: $f$ is continuous on the set $S$ if for any point $z_0$ in $S$ and for every $\varepsilon > 0$, there exists a $\delta > 0$ such that $|f(z) - f(z_0)| < \varepsilon$ *whenever z belongs to S and* $|z - z_0| < \delta$.

**Solution.**  Since $F(t) = e^{it}/i$ is an antiderivative of $f(t) = e^{it}$, we have by Theorem 3

$$\int_0^\pi e^{it}\,dt = \frac{e^{it}}{i}\bigg|_0^\pi = \frac{e^{i\pi}}{i} - \frac{e^{i0}}{i} = \frac{-2}{i} = 2i. \quad \blacksquare$$

Now we move on to the general case where $\gamma$ is any directed smooth curve along which $f$ is continuous. We can obtain a formula for the integral $\int_\gamma f(z)\,dz$ by considering an admissible parametrization $z = z(t)$, $a \le t \le b$, for $\gamma$ (consistent with its direction). Indeed, if $\mathscr{P}_n = \{z_0, z_1, \ldots, z_n\}$ is a partition of $\gamma$, then we can write

$$z_0 = z(t_0), \quad z_1 = z(t_1), \quad \ldots, \quad z_n = z(t_n),$$

where

$$a = t_0 < t_1 < \cdots < t_n = b.$$

Furthermore, since the function $z(t)$ has a continuous derivative on $[a, b]$, the difference $\Delta z_k = z(t_k) - z(t_{k-1})$ is approximately equal to $z'(t_k)(t_k - t_{k-1}) = z'(t_k)\,\Delta t_k$, the error going to zero faster than $\Delta t_k$. Hence we see that the sum

$$\sum_{k=1}^n f(z_k)\,\Delta z_k = \sum_{k=1}^n f(z(t_k))\,\Delta z_k,$$

which is a Riemann sum for $f$ along $\gamma$, can be approximated by the sum

$$\sum_{k=1}^n f(z(t_k))z'(t_k)\,\Delta t_k,$$

which is a Riemann sum for the continuous function $f(z(t))z'(t)$ over the interval $[a, b]$. These considerations suggest the following theorem (and provide the essential ingredients for its justification):

---

**Theorem 4.**  Let $f$ be a function continuous on the directed smooth curve $\gamma$. Then if $z = z(t)$, $a \le t \le b$, is any admissible parametrization of $\gamma$ consistent with its direction, we have

$$\int_\gamma f(z)\,dz = \int_a^b f(z(t))z'(t)\,dt. \tag{5}$$

---

The precise details of the proof of Theorem 4, though not difficult, are quite laborious and not particularly illuminating for our subject matter. Hence we shall omit them. A rigorous treatment of this theorem can be found in Ref. [2].

Since Eq. (5) is valid for all suitable parametrizations of $\gamma$ and since the integral of $f$ along $\gamma$ was defined independently of any parametrization, we immediately deduce the following.

**Corollary 1.** If $f$ is continuous on the directed smooth curve $\gamma$ and if $z = z_1(t)$, $a \le t \le b$, and $z = z_2(t)$, $c \le t \le d$, are any two admissible parametrizations of $\gamma$ consistent with its direction, then

$$\int_a^b f(z_1(t)) z_1'(t)\, dt = \int_c^d f(z_2(t)) z_2'(t)\, dt.$$

## Example 2

Compute the integral $\int_{C_r} (z - z_0)^n\, dz$, where $n$ is an integer and $C_r$ is the circle $|z - z_0| = r$ traversed once in the counterclockwise direction,[†] as indicated in Fig. 4.18.

**Solution.** A suitable parametrization for $C_r$ is given by $z(t) = z_0 + re^{it}$, $0 \le t \le 2\pi$. Setting $f(z) = (z - z_0)^n$, we have

$$f(z(t)) = (z_0 + re^{it} - z_0)^n = r^n e^{int}$$

and

$$z'(t) = ire^{it}.$$

Hence, by formula (5),

$$\int_{C_r} (z - z_0)^n\, dz = \int_0^{2\pi} (r^n e^{int})(ire^{it})\, dt = ir^{n+1} \int_0^{2\pi} e^{i(n+1)t}\, dt.$$

The evaluation of the last integral requires two separate computations. If $n \ne -1$, we obtain

$$ir^{n+1} \int_0^{2\pi} e^{i(n+1)t}\, dt = ir^{n+1} \left. \frac{e^{i(n+1)t}}{i(n+1)} \right|_0^{2\pi} = ir^{n+1} \left[ \frac{1}{i(n+1)} - \frac{1}{i(n+1)} \right] = 0,$$

while if $n = -1$, then

$$ir^{n+1} \int_0^{2\pi} e^{i(n+1)t}\, dt = i \int_0^{2\pi} dt = 2\pi i.$$

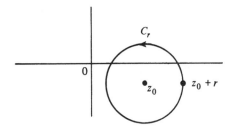

**Figure 4.18**  Directed smooth curve for Example 2.

---

[†] Occasionally we write

$$\oint_{C_r} f(z)\, dz$$

to emphasize the fact that the integration is taken in the positive direction.

Thus (regardless of the value of $r$)

$$\int_{C_r} (z - z_0)^n \, dz = \begin{cases} 0 & \text{for } n \neq -1, \\ 2\pi i & \text{for } n = -1. \end{cases} \quad \blacksquare \qquad (6)$$

Integrals along a contour are computed according to the following definition.

---

**Definition 4.** Suppose that $\Gamma$ is a contour consisting of the directed smooth curves $(\gamma_1, \gamma_2, \ldots, \gamma_n)$, and let $f$ be a function continuous on $\Gamma$. Then the **contour integral of $f$ along** $\Gamma$ is denoted by the symbol $\int_\Gamma f(z) \, dz$ and is defined by the equation

$$\int_\Gamma f(z) \, dz := \int_{\gamma_1} f(z) \, dz + \int_{\gamma_2} f(z) \, dz + \cdots + \int_{\gamma_n} f(z) \, dz. \text{†} \qquad (7)$$

---

If $\Gamma$ consists of a single point, then for obvious reasons we set

$$\int_\Gamma f(z) \, dz := 0.$$

## Example 3

Compute $\int_\Gamma 1/(z - z_0) \, dz$, where $\Gamma$ is the circle $|z - z_0| = r$ traversed twice in the counterclockwise direction starting from the point $z_0 + r$.

**Solution.**   Letting $C_r$ denote the circle traversed once in the counterclockwise direction, we have $\Gamma = (C_r, C_r)$. Hence, from formula (6) obtained in the solution of Example 2, there follows

$$\int_\Gamma \frac{dz}{z - z_0} = \int_{C_r} \frac{dz}{z - z_0} + \int_{C_r} \frac{dz}{z - z_0} = 2\pi i + 2\pi i = 4\pi i. \quad \blacksquare$$

## Example 4

Compute $\int_\Gamma \bar{z}^2 \, dz$ along the simple closed contour $\Gamma$ of Fig. 4.19 on page 120.

**Solution.**   According to Definition 4 we have

$$\int_\Gamma \bar{z}^2 \, dz = \int_{\gamma_1} \bar{z}^2 \, dz + \int_{\gamma_2} \bar{z}^2 \, dz + \int_{\gamma_3} \bar{z}^2 \, dz.$$

Suitable parametrizations for the line segments $\gamma_k$ are

$$\begin{array}{lll} \gamma_1: & z_1(t) = t & (0 \leq t \leq 2), \\ \gamma_2: & z_2(t) = 2 + ti & (0 \leq t \leq 2), \\ \gamma_3: & z_3(t) = -t(1 + i) & (-2 \leq t \leq 0), \end{array}$$

---

† For this reason we sometimes write $\Gamma = \gamma_1 + \gamma_2 + \cdots + \gamma_n$. More generally, if $\Gamma_1, \Gamma_2, \ldots, \Gamma_m$ are contours, we use the symbol $\int_{\Gamma_1 + \Gamma_2 + \cdots + \Gamma_m} f(z) \, dz$ to denote the sum $\sum_{k=1}^m \int_{\Gamma_k} f(z) \, dz$.

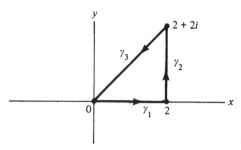

**Figure 4.19** Contour for Example 4.

and so by Theorem 4 we have

$$\int_{\gamma_1} \bar{z}^2\, dz = \int_0^2 \overline{z_1(t)}^2 z_1'(t)\, dt = \int_0^2 t^2\, dt = \frac{t^3}{3}\Big|_0^2 = \frac{8}{3},$$

$$\int_{\gamma_2} \bar{z}^2\, dz = \int_0^2 \overline{z_2(t)}^2 z_2'(t)\, dt = \int_0^2 (2 - ti)^2 i\, dt$$

$$= \frac{i(2 - ti)^3}{-3i}\Big|_0^2 = \frac{-(2 - 2i)^3}{3} + \frac{8}{3},$$

and

$$\int_{\gamma_3} \bar{z}^2\, dz = \int_{-2}^0 \overline{z_3(t)}^2 z_3'(t)\, dt = \int_{-2}^0 [-t(1 - i)]^2 [-(1 + i)]\, dt$$

$$= -(1 + i)(1 - i)^2 \int_{-2}^0 t^2\, dt = -(1 + i)(1 - i)^2 \frac{8}{3}.$$

Therefore,

$$\int_\Gamma \bar{z}^2\, dz = \frac{8}{3} + \left[ \frac{-(2 - 2i)^3}{3} + \frac{8}{3} \right] + \left[ -(1 + i)(1 - i)^2 \frac{8}{3} \right],$$

which after some computations turns out to equal $16/3 + 32i/3$. ∎

Using Definition 4 it is easy to see that the results discussed previously for integrals along a directed smooth curve carry over to integrals along a contour. In particular, we have

$$\int_\Gamma [f(z) \pm g(z)]\, dz = \int_\Gamma f(z)\, dz \pm \int_\Gamma g(z)\, dz, \qquad (8)$$

$$\int_\Gamma cf(z)\, dz = c \int_\Gamma f(z)\, dz \qquad (c \text{ any complex constant}), \qquad (9)$$

and

$$\int_{-\Gamma} f(z)\, dz = -\int_\Gamma f(z)\, dz, \qquad (10)$$

where $f$ and $g$ are both continuous on the contour $\Gamma$.

Furthermore, if we have a parametrization $z = z(t)$, $a \le t \le b$, for the whole contour $\Gamma = (\gamma_1, \gamma_2, \ldots, \gamma_n)$, then we know that there is a subdivision

$$a = \tau_0 < \tau_1 < \cdots < \tau_{n-1} < \tau_n = b$$

such that the function $z(t)$ restricted to the $k$th subinterval $[\tau_{k-1}, \tau_k]$ constitutes a suitable parametrization of $\gamma_k$. Hence by formula (5)

$$\int_{\gamma_k} f(z)\,dz = \int_{\tau_{k-1}}^{\tau_k} f(z(t))z'(t)\,dt \qquad (k = 1, 2, \ldots, n),$$

and so

$$\int_{\Gamma} f(z)\,dz = \sum_{k=1}^{n} \int_{\tau_{k-1}}^{\tau_k} f(z(t))z'(t)\,dt,$$

which we can write as

$$\int_{\Gamma} f(z)\,dz = \int_{a}^{b} f(z(t))z'(t)\,dt.$$

Using this formula it is not difficult to prove that integration around simple closed contours is independent of the choice of the initial-terminal point (see Prob. 18). Consequently, in problems dealing with integrals along such contours, we need only specify the direction of transit, not the starting point.

Many times in theory and in practice, it is not actually necessary to evaluate a contour integral. What may be required is simply a good upper bound on its magnitude. We therefore turn to the problem of estimating contour integrals.

Suppose that the function $f$ is continuous on the directed smooth curve $\gamma$ and that $f(z)$ is bounded by the constant $M$ on $\gamma$; i.e., $|f(z)| \leq M$ for all $z$ on $\gamma$. If we consider a Riemann sum $\sum_{k=1}^{n} f(c_k)\,\Delta z_k$ corresponding to a partition $\mathscr{P}_n$ of $\gamma$, then we have, by the generalized triangle inequality,

$$\left| \sum_{k=1}^{n} f(c_k)\,\Delta z_k \right| \leq \sum_{k=1}^{n} |f(c_k)|\,|\Delta z_k| \leq M \sum_{k=1}^{n} |\Delta z_k|.$$

Furthermore, notice that the sum of the chordal lengths $\sum_{k=1}^{n} |\Delta z_k|$ cannot be greater than the length of $\gamma$. Hence

$$\left| \sum_{k=1}^{n} f(c_k)\,\Delta z_k \right| \leq M\,l(\gamma). \tag{11}$$

Since inequality (11) is valid for all Riemann sums of $f(z)$, it follows by taking the limit [as $\mu(\mathscr{P}_n) \to 0$] that

$$\left| \int_{\gamma} f(z)\,dz \right| \leq M\,l(\gamma). \tag{12}$$

Applying this fact and the triangle inequality to the Eq. (7) defining a contour integral, we deduce

---

**Theorem 5.**  If $f$ is continuous on the contour $\Gamma$ and if $|f(z)| \leq M$ for all $z$ on $\Gamma$, then

$$\left| \int_{\Gamma} f(z)\,dz \right| \leq M\,l(\Gamma), \tag{13}$$

where $l(\Gamma)$ denotes the length of $\Gamma$. In particular, we have

$$\left| \int_{\Gamma} f(z)\,dz \right| \leq \max_{z\text{ on }\Gamma} |f(z)| \cdot l(\Gamma). \tag{14}$$

**Example 5**

Find an upper bound for $\left|\int_\Gamma e^z/(z^2 + 1)\, dz\right|$, where $\Gamma$ is the circle $|z| = 2$ traversed once in the counterclockwise direction.

**Solution.**  First observe that the path of integration has length $l = 4\pi$. Next we seek an upper bound $M$ for the function $e^z/(z^2 + 1)$ when $|z| = 2$. Writing $z = x + iy$ we have

$$|e^z| = |e^{x+iy}| = e^x \le e^2, \qquad \text{for } |z| = \sqrt{x^2 + y^2} = 2,$$

and by the triangle inequality

$$|z^2 + 1| \ge |z|^2 - 1 = 4 - 1 = 3, \qquad \text{for } |z| = 2.$$

Hence $|e^z/(z^2 + 1)| \le e^2/3$ for $|z| = 2$, and so by the theorem

$$\left|\int_\Gamma \frac{e^z}{z^2 + 1}\, dz\right| \le \frac{e^2}{3} \cdot 4\pi. \qquad \blacksquare$$

In concluding this section we remark that although the real definite integral can be interpreted, among other things, as an area, no corresponding geometric visualization is available for contour integrals. Nevertheless, the latter integrals are extremely useful in applied problems, as we shall see in subsequent chapters.

## EXERCISES 4.2

1. Let $\gamma$ be a directed smooth curve with initial point $\alpha$ and terminal point $\beta$. Show directly from Definition 3 that $\int_\gamma c\, dz = c(\beta - \alpha)$, where $c$ is any complex constant. Does the same formula hold for integration along an arbitrary contour joining $\alpha$ to $\beta$?

2. Using Definition 3, prove properties (1), (2), and (3).

3. Evaluate each of the following integrals.

   (a) $\displaystyle\int_0^1 (2t + it^2)\, dt$ 

   (b) $\displaystyle\int_{-2}^0 (1 + i)\cos(it)\, dt$

   (c) $\displaystyle\int_0^1 (1 + 2it)^5\, dt$ 

   (d) $\displaystyle\int_0^2 \frac{t}{(t^2 + i)^2}\, dt$

4. Furnish the details of the proof of Theorem 3.

5. Utilize Example 2 to evaluate

$$\int_C \left[\frac{6}{(z - i)^2} + \frac{2}{z - i} + 1 - 3(z - i)^2\right] dz,$$

   where $C$ is the circle $|z - i| = 4$ traversed once counterclockwise.

6. Compute $\int_\Gamma \bar{z}\, dz$, where
   (a) $\Gamma$ is the circle $|z| = 2$ traversed once counterclockwise.
   (b) $\Gamma$ is the circle $|z| = 2$ traversed once clockwise.
   (c) $\Gamma$ is the circle $|z| = 2$ traversed three times clockwise.

7. Compute $\int_\gamma \operatorname{Re} z\, dz$ along the directed line segment from $z = 0$ to $z = 1 + 2i$.

8. Let $C$ be the perimeter of the square with vertices at the points $z = 0$, $z = 1$, $z = 1 + i$, and $z = i$ traversed once in that order. Show that

$$\int_C e^z \, dz = 0.$$

9. Evaluate $\int_\Gamma (x - 2xyi) \, dz$ over the contour $\Gamma : z = t + it^2$, $0 \le t \le 1$, where $x = \operatorname{Re} z$, $y = \operatorname{Im} z$.

10. Compute $\int_C \bar{z}^2 \, dz$ along the perimeter of the square in Prob. 8.

11. Evaluate $\int_\Gamma (2z + 1) \, dz$, where $\Gamma$ is the following contour from $z = -i$ to $z = 1$:
    **(a)** the simple line segment.
    **(b)** two simple line segments, the first from $z = -i$ to $z = 0$ and the second from $z = 0$ to $z = 1$.
    **(c)** the circular arc $z = e^{it}$, $-\pi/2 \le t \le 0$.

12. True or false: $\oint_{|z|=1} \bar{z} \, dz = \oint_{|z|=1} \dfrac{1}{z} \, dz$.

13. Compute $\int_\Gamma (|z - 1 + i|^2 - z) \, dz$ along the semicircle $z = 1 - i + e^{it}$, $0 \le t \le \pi$.

14. For each of the following use Theorem 5 to establish the indicated estimate.
    **(a)** If $C$ is the circle $|z| = 3$ traversed once, then

$$\left| \int_C \frac{dz}{z^2 - i} \right| \le \frac{3\pi}{4}.$$

   **(b)** If $\gamma$ is the vertical line segment from $z = R$ $(>0)$ to $z = R + 2\pi i$, then

$$\left| \int_\gamma \frac{e^{3z}}{1 + e^z} \, dz \right| \le \frac{2\pi e^{3R}}{e^R - 1}.$$

   **(c)** If $\Gamma$ is the arc of the circle $|z| = 1$ that lies in the first quadrant, then

$$\left| \int_\Gamma \operatorname{Log} z \, dz \right| \le \frac{\pi^2}{4}.$$

   **(d)** If $\gamma$ is the line segment from $z = 0$ to $z = i$, then

$$\left| \int_\gamma e^{\sin z} \, dz \right| \le 1.$$

15. Let $f$ be a continuous complex-valued function on the real interval $[a, b]$. Prove that

$$\left| \int_a^b f(t) \, dt \right| \le \int_a^b |f(t)| \, dt.$$

[HINT: Consider the Riemann sums of $f$ over $[a, b]$.]

16. Let $\gamma$ be a directed smooth curve with initial point $\alpha$ and terminal point $\beta$. Use formula (5) and Theorem 3 to show that

$$\int_\gamma z \, dz = \frac{\beta^2 - \alpha^2}{2}.$$

17. Using the result of Prob. 16, prove that for any closed contour $\Gamma$

$$\int_\Gamma z \, dz = 0.$$

18. Let $\Gamma_1$ be a closed contour parametrized by $z = z_1(t)$, $a \le t \le b$. We can shift the initial-terminal point of $\Gamma_1$ by choosing a number $c$ in the interval $(a, b)$ and letting $\Gamma_2$ be the

contour parametrized by

$$z_2(t) = \begin{cases} z_1(t) & \text{if } c \le t \le b, \\ z_1(t - b + a) & \text{if } b \le t \le b - a + c. \end{cases}$$

Prove that

$$\int_{\Gamma_1} f(z) \, dz = \int_{\Gamma_2} f(z) \, dz$$

for any function $f$ continuous on the points of $\Gamma_1$.

## 4.3 INDEPENDENCE OF PATH

One of the important results in the theory of complex analysis is the extension of the fundamental theorem of calculus to *contour* integrals. It implies that in certain situations the integral of a function is independent of the particular path joining the initial and terminal points; in fact, it completely characterizes the conditions under which this property holds. In this section we shall explore this phenomenon in detail. We begin with the Fundamental Theorem, which enables us to evaluate integrals without introducing parametrizations, provided that an antiderivative of the integrand is known.

> **Theorem 6.**   Suppose that the function $f(z)$ is continuous in a domain $D$ and has an antiderivative $F(z)$ throughout $D$; i.e., $dF(z)/dz = f(z)$ at each $z$ in $D$. Then for any contour $\Gamma$ lying in $D$, with initial point $z_1$ and terminal point $z_T$, we have
>
> $$\int_{\Gamma} f(z) \, dz = F(z_T) - F(z_1). \tag{1}$$

[Notice that the conditions of the theorem imply that $F(z)$ is analytic and hence continuous in $D$. The function Log $z$, for example, is *not* an antiderivative for $1/z$ in any domain containing points of the negative real axis.]

Before proceeding with the proof we shall show how Theorem 6 can greatly facilitate the computation of certain contour integrals.

**Example 1**

Compute the integral $\int_{\Gamma} \cos z \, dz$, where $\Gamma$ is the contour shown in Fig. 4.20.

**Solution.**   There is no need to parametrize $\Gamma$ since the integrand has the antiderivative $F(z) = \sin z$ for all $z$. Hence by Theorem 6, the value of the integral can be computed using only the endpoints of $\Gamma$:

$$\int_{\Gamma} \cos z \, dz = \sin z \Big|_{-1}^{2+i} = \sin(2 + i) - \sin(-1). \quad \blacksquare$$

*Proof of Theorem 6.*   The demonstration is quite straightforward; once we write the integral in terms of a parametrization, the conclusion will follow as a result of the chain rule and Theorem 3 of the last section.

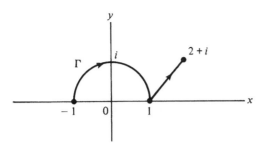

**Figure 4.20**   Contour for Example 1.

So suppose that $\Gamma$ is a contour in $D$ joining $z_1$ to $z_T$. We select a suitable parametrization $z = z(t)$, $a \le t \le b$, for $\Gamma$ and, as in the previous section, let $\{\tau_k\}_0^n$ denote the values of $t$ corresponding to the endpoints of the smooth components $\{\gamma_j\}_1^n$ of $\Gamma$ [in particular, $z(\tau_0) = z_1$ and $z(\tau_n) = z_T$]. Then we have

$$\int_\Gamma f(z)\, dz = \sum_{j=1}^n \int_{\gamma_j} f(z)\, dz = \sum_{j=1}^n \int_{\tau_{j-1}}^{\tau_j} f(z(t))z'(t)\, dt. \tag{2}$$

Using the fact that $F$ is an antiderivative of $f$, it is possible to rewrite the integrands appearing in Eq. (2). For this purpose we recall that on each separate interval $[\tau_{j-1}, \tau_j]$ the derivative $dz/dt$ exists and is continuous. Hence the chain rule implies that

$$\frac{d}{dt}\left[F(z(t))\right] = \frac{dF}{dz}\frac{dz}{dt} = f(z(t))z'(t) \qquad (\tau_{j-1} \le t \le \tau_j),$$

and so by Theorem 3

$$\int_{\tau_{j-1}}^{\tau_j} f(z(t))z'(t)\, dt = \int_{\tau_{j-1}}^{\tau_j} \frac{d}{dt}\left[F(z(t))\right] dt = F(z(\tau_j)) - F(z(\tau_{j-1})).$$

Therefore, we have

$$\int_\Gamma f(z)\, dz = \sum_{j=1}^n \left[F(z(\tau_j)) - F(z(\tau_{j-1}))\right]$$

$$= \left[F(z(\tau_1)) - F(z(\tau_0))\right] + \left[F(z(\tau_2)) - F(z(\tau_1))\right]$$

$$+ \cdots + \left[F(z(\tau_n)) - F(z(\tau_{n-1}))\right]. \tag{3}$$

But this sum telescopes, leaving

$$\int_\Gamma f(z)\, dz = F(z(\tau_n)) - F(z(\tau_0))$$

$$= F(z_T) - F(z_1). \quad \blacksquare$$

## Example 2

Compute $\int_\Gamma 1/z\, dz$, where **(a)** $\Gamma$ is the contour shown in Fig. 4.21 and **(b)** $\Gamma$ is the contour indicated in Fig. 4.22.

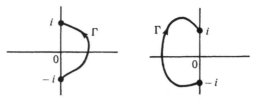

**Figure 4.21**  Contour for
Example 2(a).

**Figure 4.22**  Contour
for Example 2(b).

**Solution.** (a) At each point of the contour $\Gamma$ of Fig. 4.21 the function $1/z$ is
the derivative of the principal branch of $\log z$ (cf. Sec. 3.2). Hence

$$\int_\Gamma \frac{dz}{z} = \text{Log } z \Big|_{-i}^{i} = \frac{\pi}{2} i - \left(-\frac{\pi}{2} i\right) = \pi i.$$

(b) For the contour $\Gamma$ of Fig. 4.22 we cannot employ the function $\text{Log } z$,
since its branch cut intersects $\Gamma$. We use instead $\mathscr{L}_0(z) = \text{Log}|z| + i \arg z$,
$0 < \arg z < 2\pi$, which is a branch of the logarithm with cut along the
nonnegative $x$-axis. Thus

$$\int_\Gamma \frac{dz}{z} = \mathscr{L}_0(z) \Big|_{-i}^{i} = \frac{\pi}{2} i - \frac{3\pi}{2} i = -\pi i. \quad \blacksquare$$

Since the endpoints of a *loop*, i.e., a closed contour, are equal, we have the following immediate consequence of Theorem 6.

---

**Corollary 2.** If $f$ is continuous in a domain $D$ and has an antiderivative
throughout $D$, then

$$\int_\Gamma f(z)\, dz = 0$$

for all loops $\Gamma$ lying in $D$.

---

Corollary 2 provides an alternative solution to the problem of evaluating the
integral $\int_{C_r} (z - z_0)^n\, dz$ of Example 2 in Sec. 4.2 when $n \neq -1$. For if we set
$f(z) = (z - z_0)^n$, then $f(z)$ is the derivative of the function $F(z) = (z - z_0)^{n+1}/(n + 1)$,
which is analytic in the domain $D$ consisting of all points in the plane except $z = z_0$.
(Actually the point $z_0$ need be excluded only in the case when $n$ is negative. Why?)
Since $C_r$ is a closed contour which lies in $D$, we deduce from the corollary that
$\int_{C_r} (z - z_0)^n\, dz = 0$, $n \neq -1$.

Another important conclusion that can be drawn from Eq. (1) is that when a
function $f$ has an antiderivative throughout a domain $D$, its integral along a contour
in $D$ depends only on the endpoints $z_1$ and $z_T$; i.e., the integral is independent of the
path $\Gamma$ joining these two points! For instance, in Fig. 4.23 all the integrals $\int_{\Gamma_1} f(z)\, dz$,
$\int_{\Gamma_2} f(z)\, dz$, and $\int_{\Gamma_3} f(z)\, dz$ are equal under this condition. As a matter of fact, we shall
establish that the three properties we have discussed in this section amount to logically
equivalent statements when applied to a continuous function $f(z)$.

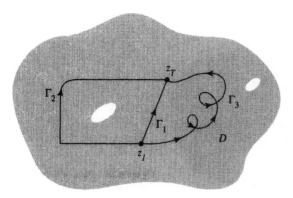

**Figure 4.23** Independence of path.

---

**Theorem 7.** Let $f$ be continuous in a domain $D$. Then the following are equivalent:

(i) $f$ has an antiderivative in $D$.

(ii) Every loop integral of $f$ in $D$ vanishes [i.e., if $\Gamma$ is any loop in $D$, then $\int_\Gamma f(z)\,dz = 0$].

(iii) The contour integrals of $f$ are independent of path in $D$ [i.e., if $\Gamma_1$ and $\Gamma_2$ are any two contours in $D$ sharing the same initial and terminal points, then $\int_{\Gamma_1} f(z)\,dz = \int_{\Gamma_2} f(z)\,dz$].

---

*Proof.* We have already seen from Theorem 6 that statement (i) implies (ii) [as well as (iii)]. Thus Theorem 7 will be proved if we can show that (ii) implies (iii) and that (iii) implies (i).

So assume that statement (ii) is true, and let $\Gamma_1$ and $\Gamma_2$ be any two contours in $D$ sharing the same initial point $z_I$ and terminal point $z_T$. Now define $\Gamma$ to be the contour generated by proceeding first along $\Gamma_1$ from $z_I$ to $z_T$ and then backwards from $z_T$ to $z_I$ along $-\Gamma_2$. Then by Eq. (10) of Sec. 4.2, we have

$$\int_\Gamma f(z)\,dz = \int_{\Gamma_1} f(z)\,dz + \int_{-\Gamma_2} f(z)\,dz = \int_{\Gamma_1} f(z)\,dz - \int_{\Gamma_2} f(z)\,dz.$$

On the other hand, since $\Gamma$ is closed, (ii) implies that

$$\int_\Gamma f(z)\,dz = 0.$$

Thus we deduce statement (iii).

We now show that whenever property (iii) holds, so does (i). To prove that $f$ has an antiderivative, we must define some function $F(z)$ and show that its derivative is $f(z)$. The clue as to where to look for $F(z)$ is provided by the earlier considerations; if $f$ *had* an antiderivative, Eq. (1) would hold. So we use Eq. (1) to *define* the function $F(z)$ and show that it is, indeed, an antiderivative.

Accordingly, we fix some point $z_0$ in $D$. Then for any point $z$ in $D$, let $F(z)$ be the integral of $f$ along some contour $\Gamma$ in $D$ joining $z_0$ to $z$. Since $D$ is connected,

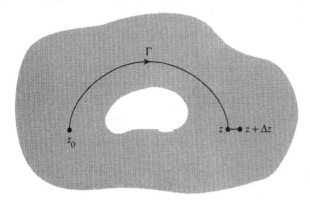

**Figure 4.24**   Path of integration for Theorem 7.

we know that there will be at least one such contour (a directed polygonal path), and by condition (iii) it does not matter which contour we choose; all the possible paths will yield the same value for $F(z)$. Hence $F(z)$ is a well-defined single-valued function in $D$. To prove (i) we compute $F(z + \Delta z) - F(z)$.

We prudently elect to evaluate $F(z + \Delta z)$ by first integrating $f$ along the contour $\Gamma$ from $z_0$ to $z$ and then along the straight-line segment from $z$ to $z + \Delta z$. This segment will lie in $D$ if $\Delta z$ is small enough, because $D$ is an open set; see Fig. 4.24. But now the difference $F(z + \Delta z) - F(z)$ is simply the integral of $f$ along this segment. Parametrizing the latter by $z(t) = z + t\,\Delta z$, $0 \le t \le 1$, we have

$$F(z + \Delta z) - F(z) = \int_0^1 f(z + t\,\Delta z)\,\Delta z\,dt,$$

and thus

$$\frac{F(z + \Delta z) - F(z)}{\Delta z} = \int_0^1 f(z + t\,\Delta z)\,dt.$$

Since $f$ is continuous, it is easy to see (Prob. 10) that as $\Delta z \to 0$ the last integral approaches $\int_0^1 f(z)\,dt = f(z)$. Thus $F'(z)$ exists and equals $f(z)$. This concludes the proof of the equivalence. ■

Theorem 7 probably appears useless at present; you may wonder how in the world one tests whether the integral of a function around *every* closed curve is zero. In the next section our efforts will be vindicated thanks to a surprising result known as Cauchy's theorem, which gives a simple condition for this property to hold. For now, we shall simply summarize by saying that a given continuous function has an antiderivative in $D$ *if and only if* its integral around every loop in $D$ is zero.

## EXERCISES 4.3

1. Calculate each of the following integrals along the indicated contours. (Observe that a standard table of integrals can be used. Explain why.)
    (a) $\int_\Gamma (3z^2 - 5z + i)\,dz$ along the line segment from $z = i$ to $z = 1$
    (b) $\int_\Gamma e^z\,dz$ along the upper half of the circle $|z| = 1$ from $z = 1$ to $z = -1$

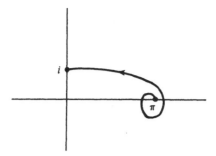

**Figure 4.25**   Contour for Prob. 1(e), (f), and (g).

(c) $\int_\Gamma 1/z \, dz$ for any contour in the right half-plane from $z = -3i$ to $z = 3i$

(d) $\int_\Gamma \csc^2 z \, dz$ for any closed contour that avoids the points $0, \pm\pi, \pm 2\pi, \ldots$

(e) $\int_\Gamma \sin^2 z \cos z \, dz$ along the contour in Fig. 4.25

(f) $\int_\Gamma e^z \cos z \, dz$ along the contour in Fig. 4.25

(g) $\int_\Gamma z^{1/2} \, dz$ for the principal branch of $z^{1/2}$ along the contour in Fig. 4.25

(h) $\int_\Gamma (\text{Log } z)^2 \, dz$ along the line segment from $z = 1$ to $z = i$

(i) $\int_\Gamma 1/(1 + z^2) \, dz$ along the line segment from $z = 1$ to $z = 1 + i$

2. If $P(z)$ is a polynomial and $\Gamma$ is any closed contour, explain why $\int_\Gamma P(z) \, dz = 0$.

3. In Chapter 5 we shall show that if $f$ is entire and $\Gamma$ is any contour, then for each $\varepsilon > 0$ there exists a polynomial $P(z)$ such that

$$|f(z) - P(z)| < \varepsilon \qquad \text{for all } z \text{ on } \Gamma.$$

Assuming this fact, prove that if $f$ is entire, then

(a) $\int_\Gamma f(z) \, dz = 0$ for all closed contours $\Gamma$. [HINT: Use the result of Prob. 2.]

(b) $f$ is the derivative of an entire function.

4. True or false: If $f$ is analytic at each point of a closed contour $\Gamma$, then $\int_\Gamma f(z) \, dz = 0$.

5. Explain why Example 2 shows that the function $f(z) = 1/z$ has no antiderivative in the punctured plane $\mathbf{C} \backslash \{0\}$.

6. Although Corollary 2 does not apply to the function $1/(z - z_0)$ in the plane punctured at $z_0$, Theorem 6 can be used as follows to show that

$$\int_C \frac{dz}{z - z_0} = 2\pi i$$

for any circle $C$ traversed once in the positive direction surrounding the point $z_0$. Introduce a horizontal branch cut from $z_0$ to $\infty$ as in Fig. 4.26 on page 130. In the resulting "slit plane" the function $1/(z - z_0)$ has the antiderivative $\text{Log}(z - z_0)$. Apply Theorem 6 to compute the integral along the portion of $C$ from $\alpha$ to $\beta$ as indicated in Fig. 4.26. Now let $\alpha$ and $\beta$ approach the point $\tau$ on the cut to evaluate the given integral over all of $C$.

7. Show that if $C$ is a positively oriented circle and $z_0$ lies outside $C$, then

$$\int_C \frac{dz}{z - z_0} = 0.$$

8. Show directly that property (iii) implies (ii) in Theorem 7.

9. As we know, an antiderivative is only specified up to a constant. How was this flexibility reflected in the proof that (iii) implies (i) in Theorem 7?

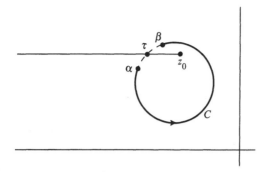

**Figure 4.26**   Contour for Prob. 6.

10. Verify the statement made in the text that if $f$ is continuous at the point $z$, then

$$\int_0^1 f(z + t\,\Delta z)\,dt \to f(z) \qquad \text{as} \ \ \Delta z \to 0.$$

[HINT: Estimate the difference

$$\int_0^1 f(z + t\,\Delta z)\,dt - f(z) = \int_0^1 \left[ f(z + t\,\Delta z) - f(z) \right] dt.]$$

11. Prove the *integration-by-parts formula*: If $f$ and $g$ have continuous first derivatives in a domain containing the contour $\Gamma$, then

$$\int_\Gamma f'(z)g(z)\,dz = f(z)g(z)\Big|_{z_I}^{z_T} - \int_\Gamma f(z)g'(z)\,dz,$$

where $z_I$ and $z_T$ are the initial and terminal points of $\Gamma$. [HINT: Use Theorem 6 on the function $d(fg)/dz$.]

12. Let $f$ be an analytic function with a continuous derivative satisfying $|f'(z)| \leq M$ for all $z$ in the disk $D: |z| < 1$. Show that

$$|f(z_2) - f(z_1)| \leq M|z_2 - z_1| \qquad (z_1, z_2 \text{ in } D).$$

[HINT: Observe that $f(z_2) - f(z_1) = \int f'(z)\,dz$, where the integration can be taken along the line segment from $z_1$ to $z_2$.]

## 4.4 CAUCHY'S INTEGRAL THEOREM

*The essential content of this section is the Cauchy integral theorem. We feel that a clear and intuitive approach to this subject is provided by the concept of continuous deformations of one contour into another. On the other hand, some instructors may feel that the theorem is better handled by appealing to vector analysis, in particular Green's theorem. Accordingly we have provided the reader with two alternative sections, 4.4a and 4.4b, and either one may be studied without affecting the subsequent development.*

*In order that each section may be self-contained, some duplication of text appears; for instance Theorem 12 in Sec. 4.4b restates Theorem 9 in Sec. 4.4a, and many of the same examples occur in both sections (though the methods of solution are different).*

*Exercises 4.4 are divided into three parts: problems appropriate to Sec. 4.4a, problems appropriate to Sec. 4.4b, and problems for all readers.*

### 4.4a Deformation of Contours Approach

In the last section we saw that if a continuous function $f$ possesses an (analytic) anti-derivative in a domain $D$, its integral around any loop in $D$ is zero and vice versa. Now we are going to show how this property ties in with the analyticity of $f$ itself. Our first task will be to develop the necessary geometry.

The critical notion in this regard is the *continuous deformation of one loop into another, in a given domain* of the plane. Deformations are quite easily visualized but somewhat harder to express in precise mathematical language. It is the visualization, however, that will suffice for most of our purposes. With this in mind, we first give an intuitive definition of deformations.

We say that the loop $\Gamma_0$ can be continuously deformed into the loop $\Gamma_1$ in the domain $D$ if $\Gamma_0$ (considered as an elastic string with indicated orientation) can be continuously moved about the plane *without leaving D* in such a manner that it ultimately coincides with $\Gamma_1$ (in position as well as direction).

The following examples serve to illustrate this notion:

**(a)** Let $D$ be the annulus and let $\Gamma_0$ and $\Gamma_1$ be the circular contours indicated in Fig. 4.27(a). Since both circles are positively oriented, the "elastic" circle $\Gamma_0$ can be continuously deformed to $\Gamma_1$ in $D$ by expanding the radius of $\Gamma_0$ from 1 to 2; i.e., we visualize a continuum of concentric circles varying in radii from 1 to 2. The dashed circles in Fig. 4.27(b) depict some of the intermediate loops; notice that all of them lie in $D$.

**(b)** Let $D$ be the annulus, $\Gamma_0$ the triangular contour, and $\Gamma_1$ the circular contour of Fig. 4.28. Then $\Gamma_0$ can be deformed to $\Gamma_1$ in $D$ by expanding $\Gamma_0$ and simultaneously making its sides more circular. Some intermediate loops are sketched in Fig. 4.28. Again they all remain in $D$.

**(c)** Let $D$ be the whole plane, $\Gamma_0$ the loop indicated in Fig. 4.29, and $\Gamma_1$ the point contour $z = 0$. Then $\Gamma_0$ can be continuously deformed to $\Gamma_1$ in $D$ by simply shrinking and shifting, as indicated in Fig. 4.29.

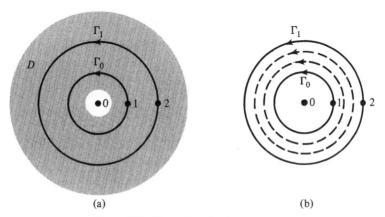

(a)                                             (b)

**Figure 4.27**   Expanding circular contours.

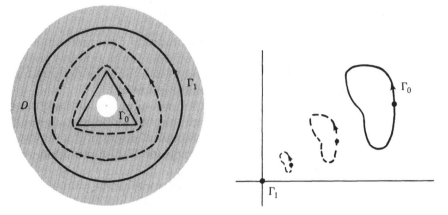

**Figure 4.28** Deformation of a triangle.     **Figure 4.29** Shrinking a contour.

**(d)** Let $D$ be the first quadrant and let $\Gamma_0$ and $\Gamma_1$ be the circular contours in Fig. 4.30(a). Notice that merely moving $\Gamma_0$ to the right will not yield the desired deformation, for while $\Gamma_0$ will eventually coincide in position with $\Gamma_1$, the orientations will be different. To circumvent this difficulty we first shrink $\Gamma_0$ to a point (which has no direction) and then expand the point to $\Gamma_1$, always remaining in $D$, as indicated in Fig. 4.30(b).

**(e)** Let $D$ be the plane minus the points $\pm i$, and let $\Gamma_0$ be the circular contour and $\Gamma_1$ be the "barbell" contour of Fig. 4.31. Then $\Gamma_0$ can be continuously deformed to $\Gamma_1$ in $D$, as illustrated by the dashed-line intermediate loops in Fig. 4.31.

Now let us be more precise. The preceding examples show that $\Gamma_0$ can be continuously deformed to $\Gamma_1$ in $D$ if $\Gamma_0$ and $\Gamma_1$ belong to a continuum of loops $\{\Gamma_s\}$, $0 \le s \le 1$, each lying in $D$, such that any pair $\Gamma_{s'}$ and $\Gamma_{s''}$ can be made "arbitrarily close" by taking $s'$ sufficiently near $s''$. Thus there must be parametrizations $\{z_s(t)\}$ for the contours $\{\Gamma_s\}$ which are continuous in the variable $s$. Using the standard $[0, 1]$ parametric interval for the loops and rewriting $z_s(t)$ as $z(s, t)$, we formalize these ideas in the following definition.

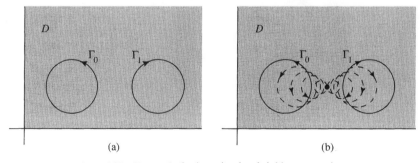

       (a)                                      (b)

**Figure 4.30** Reversal of orientation by shrinking to a point.

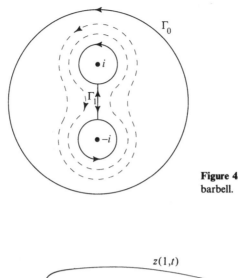

**Figure 4.31**    Circle deforming to a barbell.

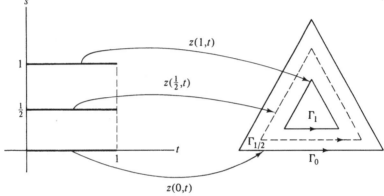

**Figure 4.32**    Parametrization of deformation.

---

**Definition 5.**    The loop $\Gamma_0$ is said to be **continuously deformable**[†] to the loop $\Gamma_1$ **in the domain** $D$ if there exists a function $z(s, t)$ continuous on the unit square $0 \le s \le 1, 0 \le t \le 1$, that satisfies the following conditions:

(i) For each fixed $s$ in $[0, 1]$, the function $z(s, t)$ parametrizes a loop lying in $D$.

(ii) The function $z(0, t)$ parametrizes the loop $\Gamma_0$.

(iii) The function $z(1, t)$ parametrizes the loop $\Gamma_1$. (See Fig. 4.32.)

---

**Example 1**

By exhibiting a deformation function $z(s, t)$, *prove* that the loop $\Gamma_0: z = e^{2\pi i t}$, $0 \le t \le 1$, can be continuously deformed to the loop $\Gamma_1: z = 2e^{2\pi i t}$, $0 \le t \le 1$, in the domain $D$ consisting of the annulus $\frac{1}{2} < |z| < 3$.

---

[†] The word *homotopic* is sometimes used.

**Solution.**  This is precisely the problem illustrated in Fig. 4.27; the intermediate loops $\Gamma_s$, $0 \le s \le 1$, are concentric circles with radii varying from 1 to 2. The function

$$z(s, t) = (1 + s)e^{2\pi it} \qquad (0 \le s \le 1, 0 \le t \le 1)$$

therefore effects the deformation.  ∎

**Example 2**

Exhibit a deformation function that shows that in the domain consisting of the whole plane any loop can be shrunk to the point contour $z = 0$.

**Solution.**   This is the situation of Fig. 4.29. If $\Gamma_0$ is parametrized by $z = z_0(t)$, $0 \le t \le 1$, then the shrinking can be accomplished by multiplying $z_0(t)$ by a scaling factor that varies from 1 to 0. The deformation function is therefore given by

$$z(s, t) = (1 - s)z_0(t) \qquad (0 \le s \le 1, 0 \le t \le 1). \quad ∎$$

A few elementary observations about continuous deformations are in order. First, notice that if $z(s, t)$ generates a deformation of loop $\Gamma_0$ into loop $\Gamma_1$, then $z(1 - s, t)$ deforms $\Gamma_1$ into $\Gamma_0$. Furthermore, if in a given domain $\Gamma_0$ can be deformed into a single point and $\Gamma_1$ can also be deformed into a point, then $\Gamma_0$ can be deformed into $\Gamma_1$ in the domain (see Prob. 2).

As we have observed in Example 2, in the domain $D$ consisting of the entire complex plane *any* loop can be deformed into the single point $z = 0$. Consequently, any two loops can be deformed one into the other in this domain. There are many other domains with this property, e.g., interiors of circles, interiors of regular polygons, half-planes, etc. We categorize such domains as follows.

---

**Definition 6.**   Any domain $D$ possessing the property that every loop in $D$ can be continuously deformed in $D$ to a point is called a **simply connected domain**.

---

Roughly speaking, we say that a simply connected domain cannot have any "holes," for if there were a hole in $D$, then a loop surrounding it could not be shrunk to a point without leaving $D$. It is shown in topology that if $\gamma$ is any simple closed contour, then its interior is a simply connected domain. Indeed, this fact is often regarded as part of the Jordan curve theorem. These considerations provide us with a quick method of identifying some simply connected domains (see Fig. 4.33).

Interesting situations arise when the domain is not simply connected (such a domain is called *multiply connected*). For example, let $D$ be the complex plane with the origin deleted, and let $\Gamma$ be the unit circle $|z| = 1$, traversed once in the counterclockwise direction starting from the point $z = 1$. We list some loops which are not deformable to $\Gamma$ in $D$:

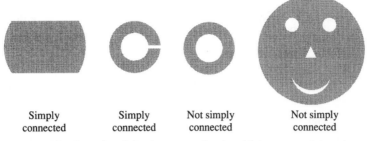

| Simply connected | Simply connected | Not simply connected | Not simply connected |

**Figure 4.33**   Examples of simply connected and multiply connected domains.

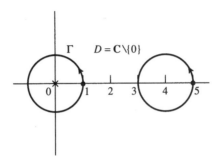

**Figure 4.34**   Nondeformable loops in punctured plane.

(a) The circle parametrized by $z(t) = 4 + e^{2\pi i t}$, $0 \leq t \leq 1$, cannot be deformed into $\Gamma$, because some intermediate loop would have to pass through $z = 0$ (see Fig. 4.34).

(b) $\Gamma$ cannot be shrunk to a point in $D$. [However, the circle in (a) can be so deformed.]

(c) $\Gamma$ cannot be continuously deformed into the opposite contour, $-\Gamma$. (The reader should mentally try to devise a family of loops linking these two, to see why the quarantining of the origin inhibits the deformation.)

(d) $\Gamma$ cannot be deformed into the same unit circle circumscribed *twice* in the positive direction.

At this point we would like to insert a word of comfort to the reader, who is probably developing some anxiety concerning his or her ability to construct the deformation function $z(s, t)$. Theorem 8 will show that, in practice, only the *existence* of the continuous deformation is important (at least for the theory of analytic functions). Consequently, we shall be content merely to visualize the deformation in most cases, without officially verifying its existence.

We are now ready to state the main theorem of this section.

---

**Theorem 8.** (*Deformation Invariance Theorem*) Let $f$ be a function analytic in a domain $D$ containing the loops $\Gamma_0$ and $\Gamma_1$. If these loops can be continuously deformed into one another in $D$, then

$$\int_{\Gamma_0} f(z)\, dz = \int_{\Gamma_1} f(z)\, dz.$$

---

A rigorous proof of Theorem 8 involves procedures that take us far afield from the basic techniques of complex analysis. Here we shall only prove a weaker version of this theorem for the special case when $\Gamma_0$ and $\Gamma_1$ are linked by a deformation function $z(s, t)$ whose second-order partial derivatives are continuous. We shall also assume that $f'(z)$ is continuous (recall that analyticity merely requires that $f'$ *exist*).

The fact that one need *not* assume continuity for $f'$ was first demonstrated by the mathematician Edouard Goursat; see Ref. [2]. The subsequent extension of the restricted theorem to the general statement of Theorem 8 can be effected by techniques of approximation theory (Ref. [5]).

*Proof of Weak Version of Theorem 8.*   As we mentioned before, we shall assume that the deformation function $z(s, t)$ has continuous partial derivatives up to order 2 for $0 \le s \le 1$, $0 \le t \le 1$, and that $f'(z)$ is continuous. Now, for each fixed $s$ the equation $z = z(s, t)$, $0 \le t \le 1$, defines a loop $\Gamma_s$ in $D$. Let $I(s)$ be the integral of $f$ along this loop, so that

$$I(s) := \int_{\Gamma_s} f(z)\, dz = \int_0^1 f(z(s, t)) \frac{\partial z(s, t)}{\partial t}\, dt. \tag{1}$$

We wish to take the derivative of $I(s)$ with respect to $s$. The assumptions guarantee that the integrand in Eq. (1) is continuously differentiable in $s$, so *Leibniz's rule for integrals* (Ref. [6]) sanctions differentiation under the integral sign. Using the chain rule we obtain

$$\frac{dI(s)}{ds} = \int_0^1 \left[ f'(z(s, t)) \frac{\partial z}{\partial s} \cdot \frac{\partial z}{\partial t} + f(z(s, t)) \frac{\partial^2 z}{\partial s\, \partial t} \right] dt. \tag{2}$$

On the other hand, observe that

$$\frac{\partial}{\partial t}\left[ f(z(s, t)) \frac{\partial z}{\partial s} \right] = f'(z(s, t)) \frac{\partial z}{\partial t} \cdot \frac{\partial z}{\partial s} + f(z(s, t)) \frac{\partial^2 z}{\partial t\, \partial s}.$$

Because of the continuity conditions the mixed partials of $z(s, t)$ are equal, so the last expression is the same as the integrand in Eq. (2). Thus

$$\frac{dI(s)}{ds} = \int_0^1 \frac{\partial}{\partial t}\left[ f(z(s, t)) \frac{\partial z}{\partial s} \right] dt$$

$$= f(z(s, 1)) \frac{\partial z}{\partial s}(s, 1) - f(z(s, 0)) \frac{\partial z}{\partial s}(s, 0).$$

But since each $\Gamma_s$ is closed, we have $z(s, 1) = z(s, 0)$ for all $s$, so $dI/ds$ is zero and, consequently, $I(s)$ is constant. In particular $I(0) = I(1)$, which is merely a disguised form of the conclusion

$$\int_{\Gamma_0} f(z)\, dz = \int_{\Gamma_1} f(z)\, dz. \quad\blacksquare$$

An easy consequence of Theorem 8 is the following, familiarly known as *Cauchy's integral theorem*.

---

**Theorem 9.**   If $f$ is analytic in a simply connected domain $D$ and $\Gamma$ is any loop (closed contour) in $D$, then

$$\int_{\Gamma} f(z)\, dz = 0. \tag{3}$$

---

*Proof.*   The proof is immediate; in a simply connected domain any loop can be shrunk to a point. The integral of a continuous function over a shrinking loop converges, of course, to zero.   $\blacksquare$

It can be shown by topological methods that if $\Gamma$ is a simple closed contour and $f$ is analytic at each point on and inside $\Gamma$, then $f$ must be analytic in some simply connected domain containing $\Gamma$. Thus, by Theorem 9, the integral along $\Gamma$ must vanish whenever the integrand is analytic "inside and on $\Gamma$."

Cauchy's theorem links the considerations of the last section with the property of analyticity. We can conclude the following.

---

**Theorem 10.**   In a simply connected domain, an analytic function has an anti-derivative, its contour integrals are independent of path, and its loop integrals vanish.

---

In an earlier section we showed that if $C$ is any circle centered at $z_0$ and $n$ is an integer, then

$$\oint_C (z - z_0)^n\, dz = \begin{cases} 0 & \text{for } n \neq -1, \\ 2\pi i & \text{for } n = -1 \end{cases} \tag{4}$$

(see Example 2, Sec. 4.2). Equation (4) with $z_0 = 0$ neatly exemplifies the theory we have been discussing. If $n$ is a positive integer or zero, $z^n$ is analytic in the whole plane, which is simply connected; thus Theorem 10 applies, $z^n$ has an antiderivative [the function $z^{n+1}/(n + 1)$], and its loop integrals are zero.

If $n$ is negative, $z^n$ is analytic only in the *punctured* plane, with the origin deleted. This domain is not simply connected, so Theorem 10 does not apply. In fact, for $n = -1$, the function $z^n$ does not even have an antiderivative in the punctured plane (since any branch of $\log z$ will be discontinuous on the branch cut), and sure enough the loop integral (4) fails to vanish. On the other hand, if $n \leq -2$, then $z^n$ regains its antiderivative $z^{n+1}/(n + 1)$ in the punctured plane, and the loop integrals (4) are zero. Thus either case can occur in multiply connected domains.

The main value of Theorem 8 is that it allows us to replace complicated contours with more familiar ones, for the purpose of integration. We shall illustrate this point with several examples.

### Example 3

Evaluate $\int_\Gamma 1/z \, dz$, where $\Gamma$ is the ellipse defined by $x^2 + 4y^2 = 1$, traversed once in the positive sense, as indicated in Fig. 4.35.

**Solution.** The integrand $1/z$ is analytic in the plane with the origin deleted. Furthermore, it is obvious that without passing through the origin $\Gamma$ can be continuously deformed into the unit circle $\Gamma_0$, oriented positively. Thus, from Eq. (4),

$$\int_\Gamma \frac{1}{z} \, dz = \int_{\Gamma_0} \frac{1}{z} \, dz = 2\pi i. \quad \blacksquare$$

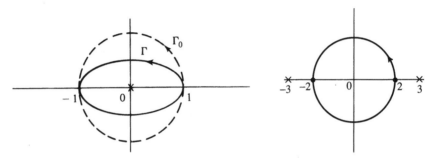

Figure 4.35  Contour for Example 3.        Figure 4.36  Contour for Example 4.

### Example 4

Evaluate

$$\oint_{|z|=2} \frac{e^z}{z^2 - 9} \, dz.$$

**Solution.** The notation employed signifies that the contour of integration is the circle $|z| = 2$ traversed once counterclockwise. The integrand $e^z/(z^2 - 9)$ is analytic everywhere except at $z = \pm 3$, where the denominator vanishes. From Fig. 4.36 we immediately see that the contour can be shrunk to a point in the domain of analyticity, and thus the integral is zero. (Alternatively, Cauchy's theorem can be applied to this example.) $\blacksquare$

### Example 5

Determine the possible values for $\int_\Gamma 1/(z - a) \, dz$, where $\Gamma$ is any circle not passing through $z = a$, traversed once in the counterclockwise direction.

**Solution.** The integrand is analytic in the domain $D$ consisting of the plane with the point $z = a$ deleted. If this point lies exterior to $\Gamma$, then $\Gamma$ can be continuously deformed to a point in $D$, and so the integral vanishes. If $a$ lies in the interior of $\Gamma$, the contour can be continuously deformed in $D$ to a positively oriented circle centered at $z = a$, and thus the integral is $2\pi i$ by Eq. (4). Summarizing, we have

$$\int_\Gamma \frac{dz}{z-a} = \begin{cases} 0 & \text{if } a \text{ lies outside } \Gamma, \\ 2\pi i & \text{if } a \text{ lies inside } \Gamma. \end{cases} \tag{5}$$

■

## Example 6

Find $\int_\Gamma (3z - 2)/(z^2 - z)\,dz$, where $\Gamma$ is the simple closed contour indicated in Fig. 4.37.

**Solution.** We don't need an exact description of $\Gamma$; since the integrand $f(z) = (3z - 2)/(z^2 - z)$ is analytic except at $z = 0$ and $z = 1$, the contour can be deformed to the "barbell"-shaped contour of Fig. 4.38 without affecting the value of the integral. This can be further simplified by observing that the

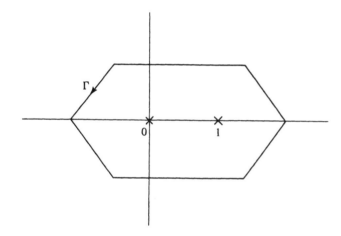

**Figure 4.37**   Contour for Example 6.

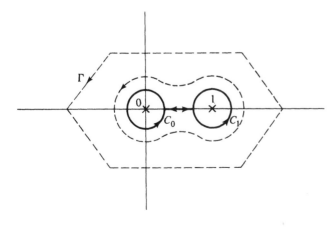

**Figure 4.38**   Deformation of contour in Fig. 4.37.

integration along the line segment proceeds forward and backward, the results canceling each other. Thus

$$\int_\Gamma f(z)\,dz = \int_{C_0} f(z)\,dz + \int_{C_1} f(z)\,dz,$$

where the circles $C_0$ and $C_1$ are as indicated in Fig. 4.38.

We shall derive a powerful method for evaluating these integrals in Sec. 6.1. For now we merely use the partial fraction expansion to rewrite the integrand as

$$\frac{3z - 2}{z^2 - z} = \frac{A}{z} + \frac{B}{z - 1}. \tag{6}$$

In Eq. (6) the constants $A$ and $B$ are determined by recombining the right-hand side:

$$\frac{A}{z} + \frac{B}{z - 1} = \frac{Az - A + Bz}{z(z - 1)} = \frac{3z - 2}{z(z - 1)}.$$

Thus $A = 2$, $B = 1$, and

$$\int_\Gamma \frac{3z - 2}{z(z - 1)}\,dz = \int_{C_0} \left(\frac{2}{z} + \frac{1}{z - 1}\right) dz + \int_{C_1} \left(\frac{2}{z} + \frac{1}{z - 1}\right) dz.$$

The right-hand side of the last equation can be viewed as the sum of four integrals, each of the form of Example 5. So by Eq. (5), the integral is

$$2(2\pi i) + 0 + 2 \cdot 0 + 2\pi i = 6\pi i. \quad \blacksquare$$

**Example 7**

Evaluate $\int_\Gamma 1/(z^2 - 1)\,dz$, where $\Gamma$ is depicted in Fig. 4.39.

**Solution.**    Observing that $1/(z^2 - 1)$ fails to be analytic only at $z = \pm 1$, we see that without passing through these points $\Gamma$ can be continuously deformed into a small positively oriented circle $C$ around $z = -1$. Again we use partial fractions to find

$$\frac{1}{z^2 - 1} = \frac{1}{2(z - 1)} - \frac{1}{2(z + 1)}.$$

Hence

$$\int_\Gamma \frac{1}{z^2 - 1}\,dz = \int_C \left[\frac{1}{2(z - 1)} - \frac{1}{2(z + 1)}\right] dz,$$

which, by Example 5, equals

$$\frac{1}{2} \cdot 0 - \frac{1}{2}(2\pi i) = -\pi i. \quad \blacksquare$$

*Note: Exercises appear at end of Sec. 4.4b.*

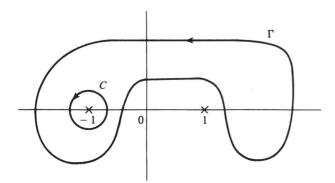

**Figure 4.39**  Contour deformation for Example 7.

### 4.4b Vector Analysis Approach

In Sec. 4.3 we deduced that a continuous function $f(z)$ possesses an (analytic) antiderivative in a domain $D$ if, and only if, its integral around every loop in $D$ is zero. Now we are going to show how this property ties in with the analyticity of $f(z)$ itself. To do this we shall employ some concepts and theorems from vector analysis (Ref. [8]). First we demonstrate that our definition of the integral of $f(z)$ over a contour $\Gamma$ can be related to the vector concept of a *line integral*.

Suppose that we have a two-dimensional vector $\mathbf{V} = (V_1, V_2)$ defined at every point $(x, y)$ in some domain $D$ in the plane; i.e., $\mathbf{V}$ is a *vector field*

$$\mathbf{V} = \mathbf{V}(x, y) = (V_1(x, y), V_2(x, y)).$$

For our purposes we require $V_1(x, y)$ and $V_2(x, y)$ to be continuous functions. Suppose furthermore that the (oriented) contour $\Gamma$, lying in $D$, has the parametrization

$$x = x(t), \qquad y = y(t) \qquad (a \le t \le b). \tag{1}$$

Then the *line integral of* $\mathbf{V}$ *along* $\Gamma$, denoted by

$$\int_\Gamma (V_1\, dx + V_2\, dy),$$

is given by

$$\int_\Gamma (V_1\, dx + V_2\, dy) := \int_a^b \left[ V_1(x(t), y(t)) \frac{dx}{dt} + V_2(x(t), y(t)) \frac{dy}{dt} \right] dt.$$

Students of physics can interpret this as the work done by a force $\mathbf{V}(x, y)$ exerted on a particle as it traverses the contour $\Gamma$.

To see how line integrals relate to complex integration we shall write out $\int_\Gamma f(z)\, dz$ in terms of its real and imaginary parts, utilizing the parametrization (1). With the usual notation $f(z) = u(x, y) + iv(x, y)$, we have

$$\int_\Gamma f(z)\, dz = \int_a^b f(z(t)) \frac{dz(t)}{dt}\, dt$$

$$= \int_a^b [u(x(t), y(t)) + iv(x(t), y(t))] \left( \frac{dx}{dt} + i \frac{dy}{dt} \right) dt$$

$$= \int_a^b \left[ u(x(t), y(t)) \frac{dx}{dt} - v(x(t), y(t)) \frac{dy}{dt} \right] dt$$

$$+ i \int_a^b \left[ v(x(t), y(t)) \frac{dx}{dt} + u(x(t), y(t)) \frac{dy}{dt} \right] dt;$$

that is,

$$\int_\Gamma f(z)\, dz = \int_\Gamma (u\, dx - v\, dy) + i \int_\Gamma (v\, dx + u\, dy). \tag{2}$$

From this equation we can see that *the real part of $\int_\Gamma f(z)\, dz$ equals the line integral over $\Gamma$ of the vector field $(u, -v)$, and that its imaginary part equals the line integral of the vector field $(v, u)$.*[†]

Now that we can express complex integrals in terms of line integrals, we can translate many of the theorems of vector analysis into theorems about complex analysis. In Probs. 6, 7, and 8 the reader is guided in rediscovering some of the earlier theorems by this route. Our immediate goal is to uncover the consequences of *Green's theorem* in the context of the complex integral.

To apply Green's theorem it is convenient to introduce one new geometric concept: the simply connected domain. Roughly speaking, a domain is said to be simply connected if it has no holes, e.g., the inside of a simple closed contour (recall the Jordan curve theorem). One way of characterizing such domains is given in Definition 7.

---

**Definition 7.**[‡]   A **simply connected domain** $D$ is a domain having the following property: If $\Gamma$ is any simple closed contour lying in $D$, then the domain interior to $\Gamma$ lies wholly in $D$. (See Fig. 4.40.)

---

In this context, one statement of *Green's theorem* is given in the next theorem (cf. Ref. [7]).

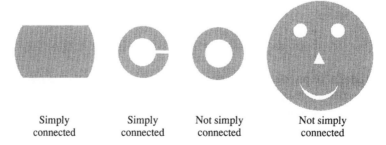

| Simply<br>connected | Simply<br>connected | Not simply<br>connected | Not simply<br>connected |

**Figure 4.40**   Examples of simply connected and multiply connected domains.

---

[†] Observe that the vector $(u, -v)$ corresponds to the complex number $u - iv = \bar{f}$ and that $(v, u)$ corresponds to $i\bar{f}$.

[‡] Definition 7 is equivalent to Definition 6.

**Theorem 11.**  Let $\mathbf{V} = (V_1,\ V_2)$ be a continuously differentiable[†] vector field defined on a simply connected domain $D$, and let $\Gamma$ be a positively oriented simple closed contour in $D$. Then the line integral of $\mathbf{V}$ around $\Gamma$ equals the integral of $(\partial V_2/\partial x - \partial V_1/\partial y)$, integrated with respect to area over the domain $D'$ interior to $\Gamma$; i.e.,

$$\int_\Gamma (V_1\, dx + V_2\, dy) = \iint_{D'} \left( \frac{\partial V_2}{\partial x} - \frac{\partial V_1}{\partial y} \right) dx\, dy.$$

Let's apply this to the line integrals that occur in $\int_\Gamma f(z)\, dz$. Using Eq. (2), we have

$$\int_\Gamma f(z)\, dz = \int_\Gamma (u\, dx - v\, dy) + i \int_\Gamma (v\, dx + u\, dy)$$

$$= \iint_{D'} \left( -\frac{\partial v}{\partial x} - \frac{\partial u}{\partial y} \right) dx\, dy + i \iint_{D'} \left( \frac{\partial u}{\partial x} - \frac{\partial v}{\partial y} \right) dx\, dy. \qquad (3)$$

Observe that we have assumed that $u$ and $v$ are continuously differentiable.

Now we take the big step. If $f(z)$ is *analytic* in $D$, the double integrals in Eq. (3) are zero because of the Cauchy-Riemann equations! In other words, we have shown that if a function is analytic in a simply connected domain and *if its derivative $f'(z)$ is continuous* (recall that analyticity only stipulates that $f'$ exist), then its integral around any simple closed contour in the domain is zero. This result, in a somewhat more general form, is known as *Cauchy's integral theorem.*

**Theorem 12.**[‡]  If $f$ is analytic in a simply connected domain $D$ and $\Gamma$ is any loop (closed contour) in $D$, then

$$\int_\Gamma f(z)\, dz = 0.$$

Observe that we have generalized in two directions. First, we require only that $\Gamma$ be a loop, not necessarily a simple closed curve. This is justified by the geometrically obvious fact that integration over a loop can always be decomposed into integrations over simple closed curves—see Fig. 4.41 on page 144 for an illustration.

The second generalization is that we have dropped the assumption that $f'(z)$ is continuous. The fact that this is possible was first demonstrated by the mathematician Edouard Goursat; see Ref. [2].

We remark that it can be shown by topological methods that if $\Gamma$ is a simple closed contour and $f$ is analytic at each point on and inside $\Gamma$, then $f$ must be

---

† Recall that this means the partials $\partial V_1/\partial x$, $\partial V_1/\partial y$, $\partial V_2/\partial x$, $\partial V_2/\partial y$ exist and are continuous.

‡ Theorem 12 is the same as Theorem 9.

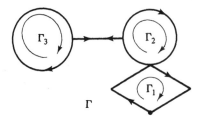

$$\int_\Gamma f(z)\,dz = \left(\int_{\Gamma_1} + \int_{\Gamma_2} + \int_{\Gamma_3}\right) f(z)\,dz$$     **Figure 4.41**   Decomposition of a loop integral.

analytic in some simply connected domain containing $\Gamma$. Thus, by Theorem 12, the integral along $\Gamma$ must vanish whenever the integrand is analytic "inside and on $\Gamma$."

Cauchy's theorem links the considerations of Sec. 4.3 with the property of analyticity. Combining Theorems 7 and 12 yields the following.

---

**Theorem 13.**[†]   In a simply connected domain, an analytic function has an antiderivative, its contour integrals are independent of path, and its loop integrals vanish.

---

In an earlier section we showed that if $C$ is any circle centered at $z_0$ and $n$ is an integer, then

$$\oint_C (z - z_0)^n\,dz = \begin{cases} 0 & \text{for } n \neq -1, \\ 2\pi i & \text{for } n = -1 \end{cases} \tag{4}$$

(see Example 2, Sec. 4.2). Equation (4) with $z_0 = 0$ neatly exemplifies the theory we have been discussing. If $n$ is a positive integer or zero, $z^n$ is analytic in the whole plane, which is simply connected; thus Theorem 13 applies, $z^n$ has an antiderivative [the function $z^{n+1}/(n+1)$], and its loop integrals are zero.

If $n$ is negative, $z^n$ is analytic only in the *punctured* plane, with the origin deleted. This domain is not simply connected, so Theorem 13 does not apply. In fact, for $n = -1$, the function $z^n$ does not even have an antiderivative in the punctured plane (since any branch of $\log z$ will be discontinuous on the branch cut), and sure enough the loop integral (4) fails to vanish. On the other hand, if $n \leq -2$, then $z^n$ regains its antiderivative $z^{n+1}/(n+1)$ in the punctured plane, and the loop integrals (4) are zero. Thus either case can occur in a domain that is not simply connected (such a domain is called *multiply connected*).

**Example 1**

Evaluate

$$\oint_{|z|=2} e^z/(z^2 - 9)\,dz.$$

---

[†] Theorem 13 is the same as Theorem 10.

**Solution.** The notation employed signifies that the contour of integration is the circle $|z| = 2$ traversed once counterclockwise. The integrand is analytic everywhere except at $z = \pm 3$, where the denominator vanishes. Since these points lie exterior to the contour, the integral is zero, by Cauchy's integral theorem. ■

Theorems 12 and 13 can often be used to change the contour of integration, as the following examples demonstrate:

## Example 2

Evaluate $\int_\Gamma 1/z \, dz$, where $\Gamma$ is the ellipse defined by $x^2 + 4y^2 = 1$, traversed once in the positive sense, as indicated in Fig. 4.42(a).

**Solution.** We shall show that one can change the contour from $\Gamma$ to the positively oriented unit circle without changing the integral. With reference to Fig. 4.42(b), observe that the complex plane, slit down the negative $y$-axis from the origin, constitutes a simply connected domain in which the function $1/z$ is analytic. Hence, by Theorem 13,

$$\int_{\gamma_2} \frac{1}{z} \, dz = \int_{\gamma_1} \frac{1}{z} \, dz.$$

Similarly, by considering the plane slit along the positive $y$-axis we have

$$\int_{\gamma_3} \frac{1}{z} \, dz = \int_{\gamma_4} \frac{1}{z} \, dz.$$

Hence

$$\int_\Gamma \frac{1}{z} \, dz = \int_{\gamma_2 + \gamma_3} \frac{1}{z} \, dz = \int_{\gamma_1 + \gamma_4} \frac{1}{z} \, dz = \oint_{|z|=1} \frac{1}{z} \, dz,$$

and by Eq. (4) the answer is $2\pi i$. ■

This technique is easily generalized in the next example.

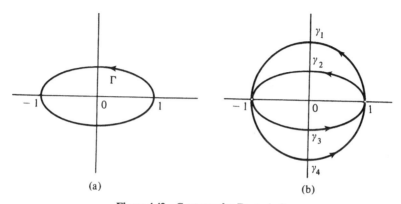

(a)                    (b)

**Figure 4.42**   Contours for Example 2.

**Example 3**

Determine the possible values for $\int_\Gamma 1/(z - a)\,dz$, where $\Gamma$ is any positively oriented simple closed contour not passing through $z = a$.

**Solution.** Observe that the integrand is analytic everywhere except at the point $z = a$. Thus if $a$ lies exterior to $\Gamma$, Cauchy's theorem yields the answer zero for the integral. If $a$ lies inside $\Gamma$, we choose a small circle $C_r$ centered at $a$ and lying within $\Gamma$, and we draw two segments from $\Gamma$ to $C_r$; see Fig. 4.43 for the construction. Then the endpoints $P_1$ and $P_2$ of the segments divide $\Gamma$ into two contours $\Gamma_1$ and $\Gamma_2$, and the endpoints $P_3$ and $P_4$ divide $C_r$ into $\gamma_1$ and $\gamma_2$. Now observe that both the contour $\Gamma_1$ and the composite contour consisting of the directed segment $P_1 P_3$, $\gamma_1$, and the segment $P_4 P_2$ can be enclosed in a simply connected domain that excludes the point $a$. Thus we deduce from Theorem 13 that the integral is the same along these contours:

$$\int_{\Gamma_1} \frac{dz}{z - a} = \left( \int_{P_1 P_3} + \int_{\gamma_1} + \int_{P_4 P_2} \right) \frac{dz}{z - a}.$$

Similarly,

$$\int_{\Gamma_2} \frac{dz}{z - a} = \left( \int_{P_2 P_4} + \int_{\gamma_2} + \int_{P_3 P_1} \right) \frac{dz}{z - a}.$$

Adding these and taking account of the cancellations along the line segments we find that

$$\int_{\Gamma} \frac{dz}{z - a} = \left( \int_{\Gamma_1} + \int_{\Gamma_2} \right) \frac{dz}{z - a} = \left( \int_{\gamma_1} + \int_{\gamma_2} \right) \frac{dz}{z - a} = \oint_{C_r} \frac{dz}{z - a} = 2\pi i$$

(recall Eq. (4)).

Summarizing, we have

$$\int_{\Gamma} \frac{dz}{z - a} = \begin{cases} 0 & \text{if } a \text{ lies outside } \Gamma, \\ 2\pi i & \text{if } a \text{ lies inside } \Gamma. \end{cases} \tag{5}$$

∎

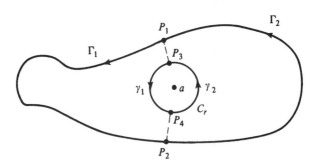

**Figure 4.43**   Contours for Example 3.

**Example 4**

Find $\int_\Gamma (3z - 2)/(z^2 - z) \, dz$, where $\Gamma$ is the simple closed contour indicated in Fig. 4.44(a).

**Solution.** The integrand $f(z) = (3z - 2)/(z^2 - z)$ is, of course, analytic everywhere except for the zeros of the denominator, $z = 0$ and $z = 1$. Referring to Fig. 4.44(b), we begin by enclosing these points in small circles $C_0$ and $C_1$, respectively, and observe by Theorem 13 that the integral over $\Gamma_1$, the upper portion of $\Gamma$, equals the integral over the contour indicated as $\Gamma_3$. Similarly, integration over $\Gamma_2$ can be replaced by integration over $\Gamma_4$. Combining these and taking into account the cancellations along the segments of the real axis we find

$$\int_\Gamma f(z) \, dz = \left( \int_{\Gamma_1} + \int_{\Gamma_2} \right) f(z) \, dz = \oint_{C_0} f(z) \, dz + \oint_{C_1} f(z) \, dz.$$

We shall derive a powerful method for evaluating these integrals in Sec. 6.1. For now we merely use the partial fraction expansion to write the integrand as

$$\frac{3z - 2}{z^2 - z} = \frac{A}{z} + \frac{B}{z - 1}. \tag{6}$$

In Eq. (6) the coefficients $A$ and $B$ are determined by recombining the right-hand side:

$$\frac{A}{z} + \frac{B}{z - 1} = \frac{Az - A + Bz}{z(z - 1)} = \frac{3z - 2}{z(z - 1)}.$$

Thus $A = 2$, $B = 1$, and

$$\int_\Gamma \frac{3z - 2}{z(z - 1)} \, dz = \oint_{C_0} \left( \frac{2}{z} + \frac{1}{z - 1} \right) dz + \oint_{C_1} \left( \frac{2}{z} + \frac{1}{z - 1} \right) dz.$$

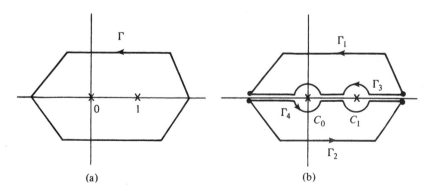

(a)                    (b)

**Figure 4.44**   Contours for Example 4.

The right-hand side of the last equation can be viewed as the sum of four integrals, each of the form of Example 3. So by Eq. (5), the integral is

$$2(2\pi i) + 0 + 2 \cdot 0 + 2\pi i = 6\pi i. \quad \blacksquare$$

### Example 5

Evaluate $\int_\Gamma 1/(z^2 - 1)\, dz$, where $\Gamma$ is depicted in Fig. 4.45(a).

**Solution.**    Observing that $1/(z^2 - 1)$ fails to be analytic only at $z = \pm 1$, the reader should be able by now to use the construction indicated in Fig. 4.45(b) to argue that the integral over $\Gamma$ is the same as the integral over the small circle $C$ enclosing $-1$. Using partial fractions again we find

$$\frac{1}{z^2 - 1} = \frac{1}{2(z - 1)} - \frac{1}{2(z + 1)}.$$

Hence

$$\int_\Gamma \frac{dz}{z^2 - 1} = \int_C \frac{dz}{z^2 - 1} = \int_C \left[ \frac{1}{2(z - 1)} - \frac{1}{2(z + 1)} \right] dz,$$

which, by Example 3, equals

$$\frac{1}{2} \cdot 0 - \frac{1}{2}(2\pi i) = -\pi i. \quad \blacksquare$$

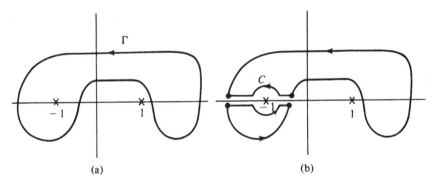

(a)                                        (b)

**Figure 4.45**    Contours for Example 5.

## EXERCISES 4.4

*Problems 1–5 refer to Sec. 4.4a.*

1. Let $D$ be the domain consisting of the complex plane with the three points $0$, $2i$, and $4$ deleted and let $\Gamma$ be the (solid-line) contour shown in Fig. 4.46. Decide which of the following contours are continuously deformable to $\Gamma$ in $D$.
   (a) the dashed line contour $\Gamma_0$ in Fig. 4.46

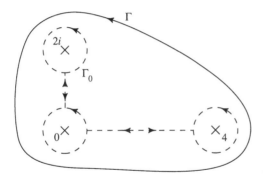

**Figure 4.46**  Contour for Prob. 1.

**(b)** the circle $|z| = 3$ traversed once in the positive direction starting from the point $z = 3$

**(c)** the circle $|z| = 10^4$ traversed once in the positive direction starting from the point $z = 10^4$

**(d)** the circle $|z - 2| = 1$ traversed once in the positive direction starting from the point $z = 3$

**2.** Prove the statement made in the text: If the contours $\Gamma_0$ and $\Gamma_1$ can each be shrunk to points in the domain $D$, then $\Gamma_0$ can be continuously deformed into $\Gamma_1$ in $D$. (Do not assume that $\Gamma_0$ and $\Gamma_1$ are deformable to the *same* point.)

**3.** Let $D$ be the annulus $1 < |z| < 5$, and let $\Gamma$ be the circle $|z - 3| = 1$ traversed once in the positive direction starting from the point $z = 4$. Decide which of the following contours are continuously deformable to $\Gamma$ in $D$.

**(a)** the circle $|z - 3| = 1$ traversed once in the positive direction starting from the point $z = 2$

**(b)** the point $z = 3i$

**(c)** the circle $|z| = 2$ traversed once in the positive direction starting from the point $z = 2$

**(d)** the circle $|z + 3| = 1$ traversed once in the positive direction starting from the point $z = -2$

**(e)** the circle $|z - 3| = 1$ traversed twice in the negative direction starting from the point $z = 4$

**4.** Let $\Gamma_0$ be the unit circle $|z| = 1$ traversed once counterclockwise and then once clockwise, starting from $z = 1$. Construct a function $z(s, t)$ which deforms $\Gamma_0$ to the single point $z = 1$ in *any* domain $D$ containing the unit circle. Verify directly that the conclusion of Theorem 8 is true for these two contours.

**5.** Write down a function $z(s, t)$ deforming $\Gamma_0$ into $\Gamma_1$ in the domain $D$, where $\Gamma_0$ is the ellipse $x^2/4 + y^2/9 = 1$ traversed once counterclockwise starting from $(2, 0)$, $\Gamma_1$ is the circle $|z| = 1$ traversed once counterclockwise starting from $(1, 0)$, and $D$ is the annulus $\frac{1}{2} < |z| < 4$. [HINT: Start with the parametrization $x(t) = 2 \cos 2\pi t$, $y(t) = 3 \sin 2\pi t$, $0 \le t \le 1$, for $\Gamma_0$.]

*Problems 6–8 refer to Sec. 4.4b.*

**6.** It is well known from potential theory that if the line integrals of a vector field $\mathbf{V}(x, y)$ are independent of path (i.e., if $\mathbf{V}$ is a "conservative" field), then there is a scalar function of position $\phi(x, y)$ such that $V_1 = \partial\phi/\partial x$ and $V_2 = \partial\phi/\partial y$ (under such conditions we say that $\phi$ is a *potential* for $\mathbf{V}$). Apply this result to the vector fields $\overline{f(z)}$ and $i\overline{f(z)}$ to prove that property (iii) implies (i) in Theorem 7, page 127. What is the relationship between the (analytic) antiderivative $F(z)$ and the potentials?

7. In vector analysis a vector field $\mathbf{V} = (V_1, V_2)$ is said to be *irrotational* if its components satisfy

$$\frac{\partial V_1}{\partial y} = \frac{\partial V_2}{\partial x};$$

it is called *solenoidal* if

$$\frac{\partial V_1}{\partial x} = -\frac{\partial V_2}{\partial y}.$$

(a) Show that, if $f(z)$ is an analytic function, the vector field corresponding to $\overline{f(z)}$ is both irrotational and solenoidal.

(b) Prove the converse to part (a), if the vector field is continuously differentiable.

8. An important result from potential theory says that if a vector field $\mathbf{V}$ is irrotational (see Prob. 7) in a simply connected domain $D$, then there is a potential function for $\mathbf{V}(x, y)$ in $D$. Applying this fact to $\overline{f(z)}$ and $i\overline{f(z)}$, prove the first assertion in Theorem 13.

*Problems 9–19 are for both Secs. 4.4a and 4.4b.*

9. Which of the following domains are simply connected?
   (a) the horizontal strip $|\operatorname{Im} z| < 1$
   (b) the annulus $1 < |z| < 2$
   (c) the set of all points in the plane except those on the nonpositive $x$-axis
   (d) the interior of the ellipse $4x^2 + y^2 = 1$
   (e) the exterior of the ellipse $4x^2 + y^2 = 1$
   (f) the domain $D$ in Fig. 4.47

**Figure 4.47**   Is $D$ simply connected?

10. Determine the domain of analyticity for each of the given functions $f$ and explain why

$$\oint_{|z|=2} f(z)\, dz = 0.$$

(a) $f(z) = \dfrac{z}{z^2 + 25}$

(b) $f(z) = e^{-z}(2z + 1)$

(c) $f(z) = \dfrac{\cos z}{z^2 - 6z + 10}$

(d) $f(z) = \operatorname{Log}(z + 3)$

(e) $f(z) = \sec\left(\dfrac{z}{2}\right)$

11. Explain why the function $e^{z^2}$ has an antiderivative in the whole plane.

12. Given that $D$ is a domain containing the closed contour $\Gamma$, that $z_0$ is a point not in $D$, and that $\int_\Gamma (z - z_0)^{-1}\, dz \neq 0$, explain why $D$ is not simply connected.

13. Evaluate $\int 1/(z^2 + 1)\, dz$ along the three closed contours $\Gamma_1, \Gamma_2, \Gamma_3$ in Fig. 4.48.

14. Consider the shaded domain $D$ in Fig. 4.49 bounded by the simple closed positively oriented contours $C, C_1, C_2,$ and $C_3$. If $f(z)$ is analytic on $D$ and on its boundary, explain why

$$\int_C f(z)\, dz = \int_{C_1} f(z)\, dz + \int_{C_2} f(z)\, dz + \int_{C_3} f(z)\, dz.$$

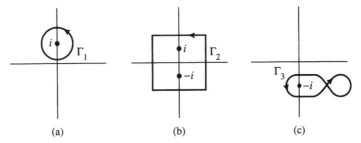

**Figure 4.48**  Contours for Prob. 13.

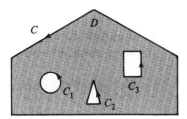

**Figure 4.49**  Domain for Prob. 14.

**15.** Evaluate

$$\int_\Gamma \frac{z}{(z+2)(z-1)}\, dz,$$

where $\Gamma$ is the circle $|z| = 4$ traversed twice in the clockwise direction.

**16.** Show that if $f$ is of the form

$$f(z) = \frac{A_k}{z^k} + \frac{A_{k-1}}{z^{k-1}} + \cdots + \frac{A_1}{z} + g(z) \qquad (k \geq 1),$$

where $g$ is analytic inside and on the circle $|z| = 1$, then

$$\oint_{|z|=1} f(z)\, dz = 2\pi i A_1.$$

**17.** Evaluate

$$\int_\Gamma \frac{2z^2 - z + 1}{(z-1)^2(z+1)}\, dz,$$

where $\Gamma$ is the figure-eight contour traversed once as shown in Fig. 4.50. [HINT: Use the partial fraction expansion $A/(z-1)^2 + B/(z-1) + C/(z+1)$.]

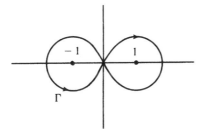

**Figure 4.50**  Contour for Prob. 17.

**18.** Let

$$I := \oint_{|z|=2} \frac{dz}{z^2(z-1)^3}.$$

Below is an outline of a proof that $I = 0$. Justify each step.
**(a)** For every $R > 2$, $I = I(R)$, where

$$I(R) := \oint_{|z|=R} 1/[z^2(z-1)^3]\, dz.$$

**(b)** $|I(R)| \leq \dfrac{2\pi}{R(R-1)^3}$ for $R > 2$.

**(c)** $\lim_{R \to +\infty} I(R) = 0$.

**(d)** $I = 0$.

**19.** Using the method of proof in Prob. 18, establish the following theorem. If $P$ is a polynomial of degree at least 2 and $P$ has all its zeros inside the circle $|z| = r$, then

$$\oint_{|z|=r} \frac{1}{P(z)}\, dz = 0.$$

## 4.5 CAUCHY'S INTEGRAL FORMULA AND ITS CONSEQUENCES

Given $f$ analytic inside and on the simple closed contour $\Gamma$, we know from Cauchy's theorem that $\int_\Gamma f(z)\, dz = 0$. However, if we consider the integral $\int_\Gamma f(z)/(z - z_0)\, dz$, where $z_0$ is a point in the interior of $\Gamma$, then there is no reason to expect that this integral is zero, because the integrand has a singularity inside the contour $\Gamma$. In fact, as the primary result of this section, we shall show that for all $z_0$ inside $\Gamma$ the value of this integral is proportional to $f(z_0)$.

---

**Theorem 14.** *(Cauchy's Integral Formula)* Let $\Gamma$ be a simple closed positively oriented contour. If $f$ is analytic in some simply connected domain $D$ containing $\Gamma$ and $z_0$ is any point inside $\Gamma$, then

$$f(z_0) = \frac{1}{2\pi i} \int_\Gamma \frac{f(z)}{z - z_0}\, dz. \tag{1}$$

---

*Proof.* The function $f(z)/(z - z_0)$ is analytic everywhere in $D$ except for the point $z_0$. Hence by the methods of Sec. 4.4 the integral over $\Gamma$ can be equated to the integral over some small positively oriented circle $C_r$: $|z - z_0| = r$; Fig. 4.51(a) illustrates the continuous deformation method of Sec. 4.4a, while Fig. 4.51(b) shows the construction appropriate to Sec. 4.4b. So we can write

$$\int_\Gamma \frac{f(z)}{z - z_0}\, dz = \int_{C_r} \frac{f(z)}{z - z_0}\, dz.$$

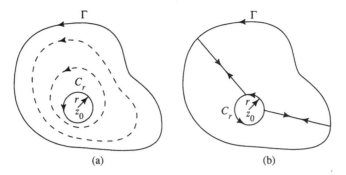

(a)                                    (b)

**Figure 4.51**  Contours for Cauchy's integral formula.

It is now convenient to express the right-hand side as the sum of two integrals:

$$\int_{C_r} \frac{f(z)}{z - z_0} \, dz = \int_{C_r} \frac{f(z_0)}{z - z_0} \, dz + \int_{C_r} \frac{f(z) - f(z_0)}{z - z_0} \, dz.$$

From our earlier deliberations we know that

$$\int_{C_r} \frac{f(z_0)}{z - z_0} \, dz = f(z_0) \int_{C_r} \frac{dz}{z - z_0} = f(z_0) 2\pi i;$$

consequently,

$$\int_{\Gamma} \frac{f(z)}{z - z_0} \, dz = f(z_0) 2\pi i + \int_{C_r} \frac{f(z) - f(z_0)}{z - z_0} \, dz. \tag{2}$$

Now observe that the first two terms in Eq. (2) are constants independent of $r$, and so the value of the last term does not change if we allow $r$ to decrease to zero; i.e.,

$$\int_{\Gamma} \frac{f(z)}{z - z_0} \, dz = f(z_0) 2\pi i + \lim_{r \to 0^+} \int_{C_r} \frac{f(z) - f(z_0)}{z - z_0} \, dz. \tag{3}$$

Therefore, Cauchy's formula will follow if we can prove that the last limit is zero.

For this purpose set $M_r := \max[|f(z) - f(z_0)|; z \text{ on } C_r]$. Then for $z$ on $C_r$ we have

$$\left| \frac{f(z) - f(z_0)}{z - z_0} \right| = \frac{|f(z) - f(z_0)|}{r} \le \frac{M_r}{r},$$

and hence by Theorem 5, page 121,

$$\left| \int_{C_r} \frac{f(z) - f(z_0)}{z - z_0} \, dz \right| \le \frac{M_r}{r} \, l(C_r) = \frac{M_r}{r} \, 2\pi r = 2\pi M_r.$$

But since $f$ is continuous at the point $z_0$, we know that $\lim_{r \to 0^+} M_r = 0$. Thus

$$\lim_{r \to 0^+} \int_{C_r} \frac{f(z) - f(z_0)}{z - z_0} \, dz = 0,$$

and so Eq. (3) reduces to formula (1).  ∎

One remarkable consequence of Cauchy's formula is that by merely knowing the values of the analytic function $f$ on $\Gamma$ we can compute the integral in Eq. (1) and hence all the values of $f$ inside $\Gamma$. In other words, the behavior of a function analytic in a region is completely determined by its behavior on the boundary.

We shall now present some examples that employ Cauchy's formula to evaluate certain integrals. The reader should keep in mind, however, that Chapter 6 will be devoted to more efficient and powerful techniques for computing integrals and that the present examples are intended primarily to illustrate the integral formula (1).

**Example 1**

Compute the integral

$$\int_{\Gamma} \frac{e^z + \sin z}{z} \, dz,$$

where $\Gamma$ is the circle $|z - 2| = 3$ traversed once in the counterclockwise direction.

**Solution.** Observe that the function $f(z) = e^z + \sin z$ is analytic inside and on $\Gamma$ and that the point $z_0 = 0$ lies inside this circle. Hence the integral has the format of formula (1) and the desired value is

$$2\pi i f(0) = 2\pi i [e^0 + \sin 0] = 2\pi i. \quad \blacksquare$$

**Example 2**

Evaluate the integral

$$\int_{\Gamma} \frac{\cos z}{z^2 - 4} \, dz$$

along the contour sketched in Fig. 4.52.

**Solution.** We first notice that the integrand fails to be analytic at the points $z = \pm 2$. However, only one of these, $z = 2$, occurs inside $\Gamma$. Thus if we write

$$\frac{\cos z}{z^2 - 4} = \frac{(\cos z)/(z + 2)}{z - 2}$$

we again have the format of Eq. (1). Thus the integral is

$$\int_{\Gamma} \frac{\cos z}{z^2 - 4} \, dz = 2\pi i \left. \frac{\cos z}{(z + 2)} \right|_{z=2} = \frac{2\pi i \cos 2}{4}. \quad \blacksquare$$

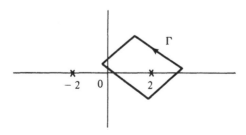

**Figure 4.52** Contour for Example 2.

**Example 3**

Compute

$$\int_C \frac{z^2 e^z}{2z + i} \, dz,$$

where $C$ is the unit circle $|z| = 1$ traversed once in the clockwise direction.

**Solution.** Two minor difficulties inhibit an immediate application of Cauchy's formula. First, the denominator is not of the form $z - z_0$, and second, the contour is *negatively* oriented. The former difficulty is easily resolved by writing

$$\frac{z^2 e^z}{2z + i} = \frac{\frac{1}{2} z^2 e^z}{z + i/2}$$

(notice that the singular point $z = -i/2$ lies *inside* $C$). And to compensate for the negative orientation of $C$ we have to introduce a minus sign in formula (1):

$$\int_C \frac{z^2 e^z}{2z + i} \, dz = \int_C \frac{\frac{1}{2} z^2 e^z}{z + i/2} \, dz = -2\pi i \cdot \frac{1}{2} z^2 e^z \Big|_{z = -i/2} = \frac{\pi i}{4} e^{-i/2}. \quad \blacksquare$$

If in Cauchy's formula (1) we replace $z$ by $\zeta$ and $z_0$ by $z$, then we obtain

$$f(z) = \frac{1}{2\pi i} \int_\Gamma \frac{f(\zeta)}{\zeta - z} \, d\zeta \qquad (z \text{ inside } \Gamma). \tag{4}$$

The advantage of this representation is that it suggests a formula for the derivative $f'(z)$, obtained by formally differentiating with respect to $z$ under the integral sign. Thus we are led to suspect that

$$f'(z) = \frac{1}{2\pi i} \int_\Gamma \frac{f(\zeta)}{(\zeta - z)^2} \, d\zeta \qquad (z \text{ inside } \Gamma). \tag{5}$$

In verifying this equation we shall actually establish a more general theorem.

---

**Theorem 15.** Let $g$ be continuous on the contour $\Gamma$, and for each $z$ not on $\Gamma$ set

$$G(z) := \int_\Gamma \frac{g(\zeta)}{\zeta - z} \, d\zeta. \tag{6}$$

Then the function $G$ is analytic at each point not on $\Gamma$, and its derivative is given by

$$G'(z) = \int_\Gamma \frac{g(\zeta)}{(\zeta - z)^2} \, d\zeta. \tag{7}$$

---

(Observe that we have generalized in two directions; we have not assumed that $\Gamma$ is closed or that $g$ is analytic. In fact, even if $\Gamma$ is a simple closed contour, the limiting values of $G$ need not agree with values of $g$ on $\Gamma$ in this generality; see Prob. 13.)

*Proof of Theorem 15.*   Let $z$ be any fixed point not on $\Gamma$. To prove the existence of $G'(z)$ and the formula (7) we need to show that

$$\lim_{\Delta z \to 0} \frac{G(z + \Delta z) - G(z)}{\Delta z} = \int_\Gamma \frac{g(\zeta)}{(\zeta - z)^2}\, d\zeta,$$

or, equivalently, that the difference

$$J := \frac{G(z + \Delta z) - G(z)}{\Delta z} - \int_\Gamma \frac{g(\zeta)}{(\zeta - z)^2}\, d\zeta \tag{8}$$

approaches zero as $\Delta z \to 0$. This is accomplished by first writing $J$ in a convenient form obtained as follows.

Using Eq. (6) we have

$$\frac{G(z + \Delta z) - G(z)}{\Delta z} = \frac{1}{\Delta z} \int_\Gamma \left[ \frac{1}{\zeta - (z + \Delta z)} - \frac{1}{\zeta - z} \right] g(\zeta)\, d\zeta$$

$$= \int_\Gamma \frac{g(\zeta)\, d\zeta}{(\zeta - z - \Delta z)(\zeta - z)},$$

where $\Delta z$ is chosen sufficiently small so that $z + \Delta z$ also lies off of $\Gamma$. Then from Eq. (8) and some elementary algebra we find

$$J = \int_\Gamma \frac{g(\zeta)\, d\zeta}{(\zeta - z - \Delta z)(\zeta - z)} - \int_\Gamma \frac{g(\zeta)}{(\zeta - z)^2}\, d\zeta$$

$$= \Delta z \int_\Gamma \frac{g(\zeta)\, d\zeta}{(\zeta - z - \Delta z)(\zeta - z)^2}. \tag{9}$$

To verify that $J \to 0$ as $\Delta z \to 0$, we estimate the last expression. In this regard, let $M$ equal the maximum value of $|g(\zeta)|$ on $\Gamma$, and set $d$ equal to the shortest distance from $z$ to $\Gamma$, so that $|\zeta - z| \ge d > 0$ for all $\zeta$ on $\Gamma$. Since we are letting $\Delta z$ approach zero, we may assume that $|\Delta z| < d/2$. Then, by the triangle inequality,

$$|\zeta - z - \Delta z| \ge |\zeta - z| - |\Delta z| \ge d - \frac{d}{2} = \frac{d}{2} \qquad (\zeta \text{ on } \Gamma)$$

(see Fig. 4.53) and so

$$\left| \frac{g(\zeta)}{(\zeta - z - \Delta z)(\zeta - z)^2} \right| \le \frac{M}{\dfrac{d}{2} \cdot d^2} = \frac{2M}{d^3}$$

for all $\zeta$ on $\Gamma$. Hence from Theorem 5, page 121, we see that

$$|J| = \left| \Delta z \int_\Gamma \frac{g(\zeta)\, d\zeta}{(\zeta - z - \Delta z)(\zeta - z)^2} \right| \le \frac{|\Delta z| 2M l(\Gamma)}{d^3},$$

where $l(\Gamma)$ denotes the length of $\Gamma$. Thus $J$ must approach zero as $\Delta z \to 0$. This implies that formula (7) is valid and completes the proof.  ∎

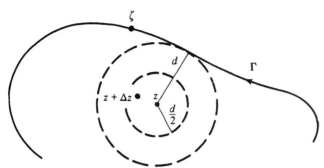

**Figure 4.53**   Contour for Theorem 15.

The preceding argument can be carried further; namely, starting with the function

$$H(z) := \int_\Gamma \frac{g(\zeta)}{(\zeta - z)^2} \, d\zeta \qquad (z \text{ not on } \Gamma), \tag{10}$$

it can be shown that $H$ is analytic off of $\Gamma$ and that $H'$ is given by the formula

$$H'(z) = 2 \int_\Gamma \frac{g(\zeta)}{(\zeta - z)^3} \, d\zeta \qquad (z \text{ not on } \Gamma), \tag{11}$$

obtained formally from (10) by differentiation under the integral sign. The proof of this fact parallels the proof of Theorem 15 and is left as an exercise.

One important consequence of these results is that the derivative of an analytic function is again analytic. For suppose that $f$ is analytic at the point $z_0$. We wish to argue that $f'$ is also analytic at $z_0$; i.e., $f'$ itself has a derivative in some neighborhood of $z_0$. To this end, we choose a positively oriented circle $C : |\zeta - z_0| = r$ so small that $f$ is analytic inside and on $C$. Since $f$ has the Cauchy integral representation

$$f(z) = \frac{1}{2\pi i} \int_C \frac{f(\zeta)}{\zeta - z} \, d\zeta \qquad (z \text{ inside } C),$$

it follows from Theorem 15 that

$$f'(z) = \frac{1}{2\pi i} \int_C \frac{f(\zeta)}{(\zeta - z)^2} \, d\zeta \qquad (z \text{ inside } C).$$

But the right-hand side is a function of the form (10) and hence has a derivative at each point inside $C$. Thus $f'$ must be analytic at $z_0$.

The same reasoning can be applied to the function $f'$ to deduce that *its* derivative, $f''$, is also analytic at $z_0$. More generally, the analyticity of $f^{(n)}$ implies that of $f^{(n+1)}$, and so by induction we obtain the following result.

---

**Theorem 16.**   If $f$ is analytic in a domain $D$, then all its derivatives $f', f'', \ldots,$ $f^{(n)}, \ldots$ exist and are analytic in $D$.

---

Theorem 16 is particularly surprising in light of the fact that its analogue in calculus fails to be true; for example, the function $f(x) = x^{5/3}$, $-\infty < x < \infty$, is differentiable for all real $x$, but $f'(x) = 5x^{2/3}/3$ has no derivative at $x = 0$.

Recall that if the analytic function $f$ is written in the form $f(z) = u(x, y) + iv(x, y)$, then as explained in Sec. 2.4 we have the alternative expressions

$$f'(z) = \frac{\partial u}{\partial x} + i\frac{\partial v}{\partial x} \quad \text{and} \quad f'(z) = \frac{\partial v}{\partial y} - i\frac{\partial u}{\partial y}. \tag{12}$$

We now know that $f'$ is analytic and hence continuous. Therefore, from (12), all the first-order partial derivatives of $u$ and $v$ must be continuous. Similarly, since $f''$ exists, the formulas (12) together with the Cauchy-Riemann equations for $f'$ lead to the expressions

$$f''(z) = \frac{\partial^2 u}{\partial x^2} + i\frac{\partial^2 v}{\partial x^2} = \frac{\partial^2 v}{\partial y\, \partial x} - i\frac{\partial^2 u}{\partial y\, \partial x},$$

$$f''(z) = \frac{\partial^2 v}{\partial x\, \partial y} - i\frac{\partial^2 u}{\partial x\, \partial y} = -\frac{\partial^2 u}{\partial y^2} - i\frac{\partial^2 v}{\partial y^2},$$

and so the continuity of $f''$ implies that all second-order partial derivatives of $u$ and $v$ are continuous at the points where $f$ is analytic. Continuing with this process we obtain the following theorem.

---

**Theorem 17.**   If $f = u + iv$ is analytic in a domain $D$, then all partial derivatives of $u$ and $v$ exist and are continuous in $D$.

---

(Recall that we presupposed this result in Sec. 2.5, where we argued that the real and imaginary parts of analytic functions are harmonic.)

Another way to phrase the results on the analyticity of derivatives is to say that whenever a given function $f$ has an antiderivative in a domain $D$, then $f$ must itself be analytic in $D$[†]. Now by Theorem 7 of Sec. 4.3 the existence of an antiderivative for a continuous function is equivalent to the property that all loop integrals vanish. Hence we deduce the following result, known as *Morera's theorem.*

---

**Theorem 18.**   If $f$ is continuous in a domain $D$ and if

$$\int_\Gamma f(z)\, dz = 0$$

for every closed contour $\Gamma$ in $D$, then $f$ is analytic in $D$.

---

Observe that in establishing Eqs. (7) and (11) we actually verify that for certain types of integrands the process of differentiation with respect to $z$ can be interchanged with the process of integration with respect to $\zeta$. In fact, starting with Cauchy's inte-

---

[†] Of course the antiderivative is analytic, since it has a derivative!

gral formula, it can be shown inductively that repeated differentiation with respect to $z$ under the integral sign yields valid formulas for the successive derivatives of $f$. Keeping track of the exponents, we have the following.

---

**Theorem 19.** If $f$ is analytic inside and on the simple closed positively oriented contour $\Gamma$ and if $z$ is any point inside $\Gamma$, then

$$f^{(n)}(z) = \frac{n!}{2\pi i} \int_\Gamma \frac{f(\zeta)}{(\zeta - z)^{n+1}} \, d\zeta \qquad (n = 1, 2, 3, \ldots).^\dagger \qquad (13)$$

---

For purposes of application it is convenient to write Eq. (13) in the equivalent form

$$\int_\Gamma \frac{f(z)}{(z - z_0)^m} \, dz = \frac{2\pi i f^{(m-1)}(z_0)}{(m-1)!} \qquad (z_0 \text{ inside } \Gamma, \; m = 1, 2, \ldots). \qquad (14)$$

**Example 4**

Compute $\int_\Gamma e^{5z}/z^3 \, dz$, where $\Gamma$ is the circle $|z| = 1$ traversed once counterclockwise.

**Solution.** Observe that $f(z) = e^{5z}$ is analytic inside and on $\Gamma$. Therefore, from Eq. (14) with $z_0 = 0$ and $m = 3$ we have

$$\int_\Gamma \frac{e^{5z}}{z^3} \, dz = \frac{2\pi i f''(0)}{2!} = 25\pi i. \quad \blacksquare$$

**Example 5**

Compute

$$\int_C \frac{2z + 1}{z(z - 1)^2} \, dz$$

along the figure-eight contour $C$ sketched in Fig. 4.54 on the next page.

**Solution.** Notice that integration along $C$ is equivalent to integrating once around the positively oriented right lobe $\Gamma_1$ and then integrating once around the negatively oriented left lobe $\Gamma_2$; i.e.,

$$\int_C \frac{2z + 1}{z(z - 1)^2} \, dz = \int_{\Gamma_1} \frac{(2z + 1)/z}{(z - 1)^2} \, dz + \int_{\Gamma_2} \frac{(2z + 1)/(z - 1)^2}{z} \, dz,$$

where we have written the integrand in each term so as to display the relevant singularity. These integrals along $\Gamma_1$ and $\Gamma_2$ can be evaluated by using Eq. (14) with $m = 2$ and $m = 1$; the desired value is

$$\frac{2\pi i}{1!} \frac{d}{dz} \left( \frac{2z + 1}{z} \right) \Bigg|_{z=1} - 2\pi i \frac{2z + 1}{(z - 1)^2} \Bigg|_{z=0} = -2\pi i - 2\pi i = -4\pi i. \quad \blacksquare$$

---

$\dagger$ When $n = 0$, this reduces to Cauchy's formula. In fact, (13) is sometimes called the *generalized Cauchy integral formula*.

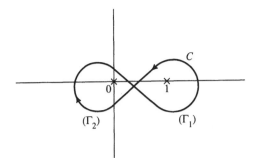

**Figure 4.54** Contour for Example 5.

## EXERCISES 4.5

1. Let $f$ be analytic inside and on the simple closed contour $\Gamma$. What is the value of

$$\frac{1}{2\pi i} \int_\Gamma \frac{f(z)}{z - z_0} \, dz$$

when $z_0$ lies outside $\Gamma$?

2. Let $f$ and $g$ be analytic inside and on the simple loop $\Gamma$. Prove that if $f(z) = g(z)$ for all $z$ on $\Gamma$, then $f(z) = g(z)$ for all $z$ inside $\Gamma$.

3. Let $C$ be the circle $|z| = 2$ traversed once in the positive sense. Compute each of the following integrals.

   **(a)** $\displaystyle \int_C \frac{\sin 3z}{z - \dfrac{\pi}{2}} \, dz$
   **(b)** $\displaystyle \int_C \frac{ze^z}{2z - 3} \, dz$
   **(c)** $\displaystyle \int_C \frac{\cos z}{z^3 + 9z} \, dz$

   **(d)** $\displaystyle \int_C \frac{5z^2 + 2z + 1}{(z - i)^3} \, dz$
   **(e)** $\displaystyle \int_C \frac{e^{-z}}{(z + 1)^2} \, dz$
   **(f)** $\displaystyle \int_C \frac{\sin z}{z^2(z - 4)} \, dz$

4. Compute

$$\int_C \frac{z + i}{z^3 + 2z^2} \, dz,$$

   where $C$ is
   **(a)** the circle $|z| = 1$ traversed once counterclockwise.
   **(b)** the circle $|z + 2 - i| = 2$ traversed once counterclockwise.
   **(c)** the circle $|z - 2i| = 1$ traversed once counterclockwise.

5. Let $C$ be the ellipse $x^2/4 + y^2/9 = 1$ traversed once in the positive direction, and define

$$G(z) := \int_C \frac{\zeta^2 - \zeta + 2}{\zeta - z} \, d\zeta \qquad (z \text{ inside } C).$$

   Find $G(1)$, $G'(i)$, and $G''(-i)$.

6. Evaluate

$$\int_\Gamma \frac{e^{iz}}{(z^2 + 1)^2} \, dz,$$

where $\Gamma$ is the circle $|z| = 3$ traversed once counterclockwise. [HINT: Show that the integral can be written as the sum of two integrals around small circles centered at the singularities.]

**7.** Compute

$$\int_\Gamma \frac{\cos z}{z^2(z-3)}\, dz$$

along the contour indicated in Fig. 4.55.

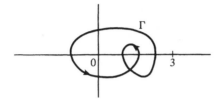

**Figure 4.55**   Contour for Prob. 7.

**8.** Use Cauchy's formula to show that if $f$ is analytic inside and on the circle $|z - z_0| = r$, then

$$f(z_0) = \frac{1}{2\pi} \int_0^{2\pi} f(z_0 + re^{i\theta})\, d\theta.$$

Prove more generally that

$$f^{(n)}(z_0) = \frac{n!}{2\pi r^n} \int_0^{2\pi} f(z_0 + re^{i\theta})e^{-in\theta}\, d\theta.$$

**9.** Suppose that $f$ is analytic inside and on the unit circle $|z| = 1$. Prove that if $|f(z)| \le M$ for $|z| = 1$, then $|f(0)| \le M$ and $|f'(0)| \le M$. What estimate can you give for $|f^{(n)}(0)|$?

**10.** Let $f$ be analytic inside and on the simple closed contour $\Gamma$. Verify from the theorems in this section that

$$\int_\Gamma \frac{f'(z)}{z - z_0}\, dz = \int_\Gamma \frac{f(z)}{(z - z_0)^2}\, dz$$

for all $z_0$ not on $\Gamma$.

**11.** Let $f = u + iv$ be analytic in a domain $D$. Explain why $\partial^2 u / \partial x^2$ is harmonic in $D$.

**12.** Prove that the function $H$ defined in Eq. (10) is analytic inside $\Gamma$ and that its derivative is given by formula (11).

**13.** According to Theorem 15, when $\Gamma$ is a simple closed contour the function $G$ defined by

$$G(z) := \frac{1}{2\pi i} \int_\Gamma \frac{g(\zeta)}{\zeta - z}\, d\zeta$$

is analytic in the domain enclosed by $\Gamma$ (assuming only that $g$ is continuous on $\Gamma$). Show that the limiting values of $G(z)$ as $z$ approaches $\Gamma$ need not coincide with the values of $g$, by considering the situation where $\Gamma$ is the positively oriented circle $|z| = 1$ and $g(z) = 1/z$. Why doesn't this violate Cauchy's formula? [HINT: Use partial fractions to evaluate $G$.]

**14.** Let $\Gamma$ be a simple closed positively oriented contour that passes through the point $2 + 3i$. Set

$$G(z) := \frac{1}{2\pi i} \int_\Gamma \frac{\cos \zeta}{\zeta - z}\, d\zeta.$$

Find the following limits:

**(a)** $\lim_{z \to 2+3i} G(z)$, where $z$ approaches $2 + 3i$ from inside $\Gamma$.

**(b)** $\lim_{z \to 2+3i} G(z)$, where $z$ approaches $2 + 3i$ from outside $\Gamma$.

[The curious reader may speculate on the interpretation of the integral for points $z$, such as $2 + 3i$, which actually lie on $\Gamma$. The theory of Sokhotskyi and Plemelj, which states that in such a case $G(z)$ equals the *average* of the interior and exterior limiting values, is developed in Sec. 8.5.]

15. Suppose that $f(z)$ is analytic at each point of the closed disk $|z| \leq 1$ and that $f(0) = 0$. Prove that the function

$$F(z) := \begin{cases} \dfrac{f(z)}{z} & z \neq 0, \\ f'(0) & z = 0, \end{cases}$$

is analytic on $|z| \leq 1$. [HINT: To show that $F$ is analytic at $z = 0$ note that by Theorem 15 the function

$$G(z) := \frac{1}{2\pi i} \oint_{|\zeta|=1} \frac{f(\zeta)/\zeta}{\zeta - z} \, d\zeta$$

is analytic at this point. Using partial fractions deduce that $G(z) = F(z)$ for $|z| < 1$.]

16. Below is an outline of a proof of the fact that *for any analytic function $f(z)$ that is never zero in a simply connected domain $D$ there exists a single-valued branch of $\log f(z)$ analytic in $D$.* Justify each step in the proof.

**(a)** $f'(z)/f(z)$ is analytic in $D$.

**(b)** $f'(z)/f(z)$ has an (analytic) antiderivative in $D$, say $H(z)$.

**(c)** The function $f(z)e^{-H(z)}$ is constant in $D$, so that $f(z) = ce^{H(z)}$.

**(d)** Letting $\alpha$ be a value of $\log c$, the function $H(z) + \alpha$ is a branch of $\log f(z)$ analytic in $D$.

17. Use the result of Prob. 16 to prove that there exists a single-valued analytic branch of $(z^3 - 1)^{1/2}$ in the unit disk $|z| < 1$.


## 4.6 BOUNDS FOR ANALYTIC FUNCTIONS

Many interesting facts about analytic functions are uncovered when one considers upper bounds on their moduli. We already have one result in this direction, namely, the integral estimate Theorem 5 of Sec. 4.2. When this is judiciously applied to the Cauchy integral formulas we obtain the *Cauchy estimates* for the derivatives of an analytic function.

---

**Theorem 20.** Let $f$ be analytic inside and on a circle $C_R$ of radius $R$ centered about $z_0$. If $|f(z)| \leq M$ for all $z$ on $C_R$, then the derivatives of $f$ at $z_0$ satisfy

$$|f^{(n)}(z_0)| \leq \frac{n! \, M}{R^n} \qquad (n = 1, 2, 3, \ldots). \tag{1}$$

---

*Proof.* Giving $C_R$ a positive orientation in Fig. 4.56, we have, by the generalized Cauchy formula (Theorem 19),

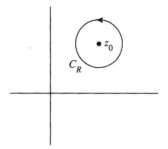

**Figure 4.56**   Circle for Cauchy estimates.

$$f^{(n)}(z_0) = \frac{n!}{2\pi i} \int_{C_R} \frac{f(\zeta)}{(\zeta - z_0)^{n+1}} \, d\zeta.$$

For $\zeta$ on $C_R$, the integrand is bounded by $M/R^{n+1}$; the length of $C_R$ is $2\pi R$. Thus from Theorem 5, page 121, there follows

$$|f^{(n)}(z_0)| \le \frac{n!}{2\pi} \frac{M}{R^{n+1}} 2\pi R,$$

which reduces to (1). ■

This innocuous-looking theorem actually places rather severe restrictions on the behavior of analytic functions. Suppose, for instance, that $f$ is analytic and bounded by some number $M$ over the whole plane $\mathbf{C}$. Then the conditions of the theorem hold for any $z_0$ and for any $R$. Taking $n = 1$ in (1) and letting $R \to \infty$, we conclude that $f'$ vanishes everywhere; i.e., $f$ must be constant. This startling result is known as *Liouville's theorem*.

---

**Theorem 21.**   The only bounded entire functions are the constant functions.

---

Of course, nonconstant polynomials are unbounded over the whole plane. Loosely speaking, we expect a polynomial of degree $n$ to behave like $z^n$ for large $|z|$, because the leading term will dominate the lower powers. To be precise, we establish the following technical lemma.

---

**Lemma 1.**   Let $P(z) = z^n + a_{n-1}z^{n-1} + \cdots + a_1 z + a_0$, where $n \ge 1$, and set $A := \max\{1, |a_{n-1}|, \ldots, |a_1|, |a_0|\}$. Then for $|z| \ge 2nA$, we have $|P(z)| \ge |z|^n/2$.

---

*Proof.*   Write $P(z) = z^n(1 + a_{n-1}/z + \cdots + a_1/z^{n-1} + a_0/z^n)$. By the triangle inequality, we have

$$\left| 1 + \frac{a_{n-1}}{z} + \cdots + \frac{a_1}{z^{n-1}} + \frac{a_0}{z^n} \right| \ge 1 - \left| \frac{a_{n-1}}{z} + \cdots + \frac{a_1}{z^{n-1}} + \frac{a_0}{z^n} \right|. \tag{2}$$

But for $|z| \geq 2nA \ (\geq 1)$,

$$\left| \frac{a_{n-k}}{z^k} \right| \leq \frac{A}{|z|^k} \leq \frac{A}{|z|} \leq \frac{1}{2n},$$

so

$$\left| \frac{a_{n-1}}{z} + \cdots + \frac{a_1}{z^{n-1}} + \frac{a_0}{z^n} \right| \leq \frac{1}{2n} + \frac{1}{2n} + \cdots + \frac{1}{2n} = \frac{1}{2}.$$

Using this estimate in (2), we obtain

$$|P(z)| = |z|^n \left| 1 + \frac{a_{n-1}}{z} + \cdots + \frac{a_1}{z^{n-1}} + \frac{a_0}{z^n} \right|$$

$$\geq |z|^n \left( 1 - \frac{1}{2} \right)$$

$$= \frac{|z|^n}{2}. \quad \blacksquare$$

Probably the reader has heard about the *Fundamental Theorem of Algebra*, which states that every equation of the form $P(z) = 0$, where $P$ is a nonconstant polynomial, has a root. We can now prove this result by applying Liouville's theorem to $1/P(z)$, which would be entire if $P(z)$ had no zeros.

---

**Theorem 22.**  Every nonconstant polynomial with complex coefficients has at least one zero.

---

*Proof.*  Suppose, to the contrary, that $P(z) = a_n z^n + a_{n-1} z^{n-1} + \cdots + a_1 z + a_0$ is a nonconstant polynomial ($n \geq 1$) having no zeros. There is no loss of generality in assuming $a_n = 1$. (Why?) As we just remarked, $1/P(z)$ must be entire. We shall show that it is bounded over the whole plane.

(i) For $|z| \geq 2nA$ as in the lemma, we have

$$\left| \frac{1}{P(z)} \right| \leq \frac{1}{|z|^n / 2} \leq \frac{2}{(2nA)^n}.$$

(ii) For $|z| \leq 2nA$, the situation becomes that of a continuous function, $|1/P(z)|$, on a closed disk. Under such circumstances it is known from calculus that the function is bounded there (indeed, it achieves its maximum!).

But then $1/P(z)$ is bounded and entire so it must be constant. Thus $P(z)$, itself, is constant, violating our assumptions.  $\blacksquare$

Once we grant that every nonconstant polynomial has a zero, it is easy to use a "deflation" argument to prove that any polynomial of degree $n$ must possess *exactly* $n$ zeros, counting multiplicities. This refinement is discussed in the exercises.

Let us now return to the Cauchy formula for the function $f$, analytic inside and on the circle $C_R$ of radius $R$ around $z_0$. We have

$$f(z_0) = \frac{1}{2\pi i} \oint_{C_R} \frac{f(z)}{z - z_0} \, dz. \tag{3}$$

Parametrizing $C_R$ by $z = z_0 + Re^{it}$, $0 \le t \le 2\pi$, we write Eq. (3) as

$$f(z_0) = \frac{1}{2\pi i} \int_0^{2\pi} \frac{f(z_0 + Re^{it})}{Re^{it}} iRe^{it} \, dt,$$

or

$$f(z_0) = \frac{1}{2\pi} \int_0^{2\pi} f(z_0 + Re^{it}) \, dt. \tag{4}$$

Equation (4), which is known as the *mean-value property*, displays $f(z_0)$ as the average of its values around the circle $C_R$. Clearly, if $|f(z)| \le M$ on $C_R$, then $|f(z_0)|$ is bounded by $M$ also (this verifies the case $n = 0$ of Theorem 20, with $0! = 1$). But more importantly, we can utilize Eq. (4) to establish the following.

**Lemma 2.**  Suppose that $f$ is analytic in a disk centered at $z_0$ and that the maximum value of $|f(z)|$ over this disk is $|f(z_0)|$. Then $|f(z)|$ is constant in the disk.

*Proof.*  Assume to the contrary that $|f(z)|$ is not constant. Then there must exist a point $z_1$ in the disk such that $|f(z_0)| > |f(z_1)|$. Let $C_R$ denote the circle centered at $z_0$ which passes through $z_1$. Then by hypothesis $|f(z_0)| \ge |f(z)|$ for all $z$ on $C_R$. Moreover, by the continuity of $f$, the strict inequality $|f(z_0)| > |f(z)|$ must hold for $z$ on a portion of $C_R$ containing $z_1$. This leads to a contradiction of Eq. (4), because the portion containing $z_1$ would contribute "less than its share" to the average in Eq. (4), and the "deficit" cannot be made up anywhere else on $C_R$. (We invite the reader to provide a rigorous version of this argument in Prob. 9.)  ■

Observe that the lemma says that the modulus of an analytic function cannot achieve its maximum at the center of the disk unless $|f|$ is constant. We shall use the lemma to extend this idea in the following version of the *maximum modulus principle*.

**Theorem 23.**  If $f$ is analytic in a domain $D$ and $|f(z)|$ achieves its maximum value at a point $z_0$ in $D$, then $f$ is constant in $D$.

*Proof.*  We shall prove that $|f|$ is constant in $D$; by Prob. 12 in Exercises 2.4, we can then conclude that $f$, itself, is constant.

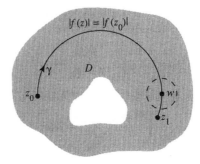

**Figure 4.57**  Geometry for proof of Theorem 23.

So for the moment let us suppose that $|f(z)|$ is not constant. Then there must be a point $z_1$ in $D$ such that $|f(z_1)| < |f(z_0)|$. Let $\gamma$ be a path in $D$ running from $z_0$ to $z_1$, as in Fig. 4.57. Now we consider the values of $|f(z)|$ for $z$ on $\gamma$, starting at $z_0$. Intuitively, we expect to encounter a point $w$ where $|f(z)|$ first starts to decrease on $\gamma$. That is, there should be a point $w$ on $\gamma$ with the following properties:

(i) $|f(z)| = |f(z_0)|$ for all $z$ preceding $w$ on $\gamma$.
(ii) There are points $z$ on $\gamma$, arbitrarily close to $w$, where $|f(z)| < |f(z_0)|$.

(It is possible that $w$ may coincide with $z_0$.) The existence of $w$ is a fact that can be rigorously demonstrated from the axioms of the real number system, but we shall omit the details here. Naturally, from property (i) and the continuity of $f$, we have $|f(w)| = |f(z_0)|$.

Now since every point of a domain is an interior point, there must be a disk centered at $w$ that lies in $D$. But Lemma 2 applies and says that $|f|$ is constant in this disk, contradicting property (ii) above. We are forced to conclude, therefore, that our initial supposition about the existence of $z_1$ is erroneous. Consequently $|f|$, and hence $f$ itself, is constant in $D$. ∎

Next, consider a function $f(z)$ analytic in a *bounded* domain $D$ and continuous on $D$ and its boundary. From calculus we know that the continuous function $|f(z)|$ must achieve its maximum on this closed bounded region; on the other hand, Theorem 23 says that the maximum cannot occur at an interior point unless $f$ is constant. In any case, we deduce the following modification of the maximum modulus principle.

---

**Theorem 24.**  A function analytic in a bounded domain and continuous up to and including its boundary attains its maximum modulus on the boundary.

---

**Example 1**

Find the maximum value of $|z^2 + 3z - 1|$ in the disk $|z| \leq 1$.

**Solution.**  The triangle inequality immediately gives us

$$|z^2 + 3z - 1| \leq |z^2| + 3|z| + 1 \leq 5 \qquad \text{(for } |z| \leq 1\text{).} \tag{5}$$

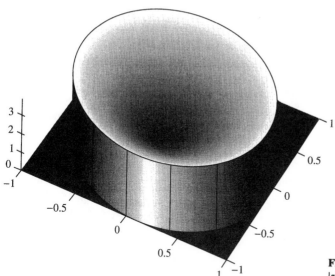

**Figure 4.58** Graph of $|z^2 + 3z - 1|$ for $|z| < 1$.

However, the maximum is actually smaller than this, as the following analysis shows.

The maximum of $|z^2 + 3z - 1|$ must occur on the boundary of the disk ($|z| = 1$). The latter can be parametrized $z = e^{it}$, $0 \le t \le 2\pi$; whence

$$|z^2 + 3z - 1|^2 = (e^{i2t} + 3e^{it} - 1)(e^{-i2t} + 3e^{-it} - 1).$$

Expanding and gathering terms reduces this to $(11 - 2\cos 2t)$. Thus the maximum of $|z^2 + 3z - 1|$ is $\sqrt{13}$, which occurs at $z = \pm i$. A sketch of the graph of the function $|z^2 + 3z - 1|$ appears in Fig. 4.58.  ∎

## EXERCISES 4.6

**1.** Let $f(z) = 1/(1 - z)^2$, and let $0 < R < 1$. Verify that $\max_{|z| = R} |f(z)| = 1/(1 - R)^2$, and also show $f^{(n)}(0) = (n + 1)!$, so that by the Cauchy estimates

$$(n + 1)! \le \frac{n!}{R^n(1 - R)^2}.$$

**2.** Suppose that $f$ is analytic in $|z| < 1$ and that $|f(z)| < 1/(1 - |z|)$. Prove that

$$|f^{(n)}(0)| \le \frac{n!}{R^n(1 - R)} \qquad (0 < R < 1),$$

and show that the upper bound is smallest when $R = n/(n + 1)$.

**3.** Let $f$ be analytic and bounded by $M$ in $|z| \le r$. Prove that

$$|f^{(n)}(z)| \le \frac{n!M}{(r - |z|)^n} \qquad (|z| < r).$$

4. If $p(z) = a_0 + a_1 z + \cdots + a_n z^n$ is a polynomial and $\max |p(z)| = M$ for $|z| = 1$, show that each coefficient $a_k$ is bounded by $M$.

5. Let $f$ be entire and suppose that $\mathrm{Re}\, f(z) \leq M$ for all $z$. Prove that $f$ must be a constant function. [HINT: Apply Liouville's theorem to the function $e^f$.]

6. Let $f$ be entire and suppose that $f^{(5)}$ is bounded in the whole plane. Prove that $f$ must be a polynomial of degree at most 5.

7. Suppose that $f$ is entire and that $|f(z)| \leq |z|^2$ for all sufficiently large values of $|z|$, say $|z| > r_0$. Prove that $f$ must be a polynomial of degree at most 2. [HINT: Show by the Cauchy estimates that for $R$ sufficiently large $|f^{(3)}(z_0)| \leq 3! \, (R + |z_0|)^2/R^3$, and thereby conclude that $f^{(3)}$ vanishes everywhere.] Generalize this result.

8. If $f$ is analytic in the annulus $1 \leq |z| \leq 2$ and $|f(z)| \leq 3$ on $|z| = 1$ and $|f(z)| \leq 12$ on $|z| = 2$, prove that $|f(z)| \leq 3|z|^2$ for $1 \leq |z| \leq 2$. [HINT: Consider $f(z)/3z^2$.]

9. Using formula (4) prove that if the analytic function $f$ satisfies $|f(z_0)| \geq |f(z)|$ for all $z$ on $C_R: z_0 + Re^{it}, 0 \leq t \leq 2\pi$, then there is no point $z_1$ on $C_R$ for which $|f(z_0)| > |f(z_1)|$. [HINT: If $z_1$ exists, then by the continuity of $f$ there is an $\varepsilon > 0$ such that $|f(z_0 + Re^{it})| \leq |f(z_0)| - \varepsilon$ over some interval of $t$. Using this interval divide up the integration in Eq. (4) to reach the contradiction $|f(z_0)| < |f(z_0)|$.]

10. Find all functions $f$ analytic in $D: |z| < R$ that satisfy $f(0) = i$ and $|f(z)| \leq 1$ for all $z$ in $D$. [HINT: Where does the maximum modulus occur?]

11. Suppose that $f$ is analytic inside and on the simple closed curve $C$ and that $|f(z) - 1| < 1$ for all $z$ on $C$. Prove that $f$ has no zeros inside $C$. [HINT: Suppose $f(z_0) = 0$ for some $z_0$ inside $C$ and consider the function $g(z) := f(z) - 1$.]

12. It is proved in Chapter 6 that every analytic function which is nonconstant on domains maps open sets onto open sets. Using this fact give another proof of the maximum modulus principle (Theorem 23).

13. Let $f$ and $g$ be analytic in the bounded domain $D$ and continuous up to and including its boundary $B$. Suppose that $g$ never vanishes. Prove that if the inequality $|f(z)| \leq |g(z)|$ holds for all $z$ on $B$, then it must hold for all $z$ in $D$. [HINT: Consider the function $f(z)/g(z)$.]

14. Prove the *minimum modulus principle*: Let $f$ be analytic in a bounded domain $D$ and continuous up to and including its boundary. Then *if $f$ is nonzero in $D$*, the modulus $|f(z)|$ attains its minimum value on the boundary of $D$. [HINT: Consider the function $1/f(z)$.] Give an example to show why the italicized condition is essential.

15. Let the nonconstant function $f$ be analytic in the bounded domain $D$ and continuous up to and including its boundary $B$. Prove that if $|f(z)|$ is constant on $B$, then $f$ must have at least one zero in $D$.

### Polynomial Problems

16. Modify the proof of Lemma 1 to show that, with the same hypotheses, $|P(z)| \leq 3 \, |z|^n/2$ for $|z| \geq 2nA$.

17. Find $\max_{|z| \leq 1} |(z - 1)(z + \frac{1}{2})|$. [HINT: Use calculus.]

18. Show that $\max_{|z| \leq 1} |az^n + b| = |a| + |b|$.

19. Prove that if $P(z)$ is a polynomial of degree $n$ and $P(\alpha) = 0$, then $P(z) = (z - \alpha)Q(z)$, where $Q(z)$ is a polynomial of degree $n - 1$. The process of factoring out a zero of a polynomial is known as *deflation*. [HINT: Observe that $P(z) = P(z) - P(\alpha)$ and that for each integer $k \geq 1$,

$$z^k - \alpha^k = (z - \alpha)(z^{k-1} + z^{k-2}\alpha + \cdots + z\alpha^{k-2} + \alpha^{k-1}).]$$

**20.** Prove that every polynomial of degree $n$ ($\geq 1$) has exactly $n$ zeros (counting multiplicity). [HINT: Proceed by induction using the result of Prob. 19.]

**21.** Prove that if $P(z)$ is a polynomial of degree $\leq n$ that vanishes on the $n$ distinct points $z_1, z_2, \ldots, z_n$, then

$$P(z) = c(z - z_1)(z - z_2) \cdots (z - z_n),$$

where $c$ is a constant.

**22.** Show that if the polynomial $P(z)$ has real coefficients, it can be expressed as a product of linear and quadratic factors, each having real coefficients.

**23.** Prove that if $p(z)$ and $q(z)$ are polynomials of degree $\leq n$ that agree on $n + 1$ distinct points, then $q(z) = p(z)$ for all $z$.

**24.** Let $P$ be a polynomial which has no zeros on the simple closed positively oriented contour $\Gamma$. Prove that the number of zeros of $P$ (counting multiplicity) that lie *inside* $\Gamma$ is given by the integral

$$\frac{1}{2\pi i} \int_\Gamma \frac{P'(z)}{P(z)} \, dz.$$

[HINT: First show that

$$\frac{P'(z)}{P(z)} = \sum_{k=1}^n \frac{1}{z - z_k},$$

where $z_1, z_2, \ldots, z_n$ are all the zeros of $P$ listed according to multiplicity.]

**25.** Prove that for any polynomial $P$ of the form $P(z) = z^n + a_{n-1}z^{n-1} + \cdots + a_1 z + a_0$, we have $\max_{|z|=1} |P(z)| \geq 1$. HINT: Consider the polynomial $Q(z) = z^n P(1/z)$, and note that $Q(0) = 1$ and that

$$\max_{|z|=1} |Q(z)| = \max_{|z|=1} |P(z)|.$$

## *4.7 APPLICATIONS TO HARMONIC FUNCTIONS

In the last part of Chapter 2 we discussed the harmonic functions, which are twice-continuously differentiable solutions of Laplace's equation

$$\frac{\partial^2 \phi}{\partial x^2} + \frac{\partial^2 \phi}{\partial y^2} = 0.$$

In particular, we showed that the real and imaginary parts of analytic functions are harmonic, subject to an assumption about their differentiability. This assumption has now been vindicated by Theorem 17, page 158. Conversely, we showed how a given harmonic function can be regarded as the real (or imaginary) part of an analytic function by giving a method that constructs the harmonic conjugate, at least in certain elementary domains such as disks. We shall now exploit this interpretation to derive some more facts about harmonic functions.

Our first step will be to extend the harmonic-analytic dualism to simply connected domains, via the next theorem.

---

**Theorem 25.**   Let $\phi$ be a function harmonic on a simply connected domain $D$. Then there is an analytic function $f$ such that $\phi = \mathrm{Re}\, f$ on $D$.

---

*Proof.*   For motivation, suppose that we *had* such an analytic function, say, $f = \phi + i\psi$. Then one expression for $f'(z)$ would be given by $f' = \partial\phi/\partial x - i\,\partial\phi/\partial y$ (using the Cauchy-Riemann equations), and $f$ would be an antiderivative of this analytic function.

Accordingly, we begin the proof by defining $g(z) := \partial\phi/\partial x - i\,\partial\phi/\partial y$. We now claim that $g$ satisfies the Cauchy-Riemann equations in $D$:

$$\frac{\partial}{\partial x}\left(\frac{\partial\phi}{\partial x}\right) = \frac{\partial}{\partial y}\left(-\frac{\partial\phi}{\partial y}\right)$$

because $\phi$ is harmonic, and

$$\frac{\partial}{\partial y}\left(\frac{\partial\phi}{\partial x}\right) = -\frac{\partial}{\partial x}\left(-\frac{\partial\phi}{\partial y}\right)$$

because of the equality of mixed second partial derivatives. Of course, these partials are continuous since $\phi$ is harmonic. Hence $g(z)$ is analytic, and by Theorem 10 (or 13) it has an analytic antiderivative $G = u + iv$ in the simply connected domain $D$. Since $G' = g$, we can write

$$\frac{\partial u}{\partial x} - i\frac{\partial u}{\partial y} = \frac{\partial\phi}{\partial x} - i\frac{\partial\phi}{\partial y},$$

showing that $u$ and $\phi$ have identical first partial derivatives in $D$. Thus by Theorem 1 of Sec. 1.6, we conclude that $\phi - u$ is constant in $D$; i.e., $\phi = u + c$. It follows that $f(z) := G(z) + c$ is an analytic function of the kind predicted by the theorem. ∎

With Theorem 25 in hand we are fully equipped to study harmonic functions in simply connected domains, using the theory of analytic functions. Let $\phi(x, y)$ be a harmonic function and $f = \phi + i\psi$ be an "analytic completion" for $\phi$ in the simply connected domain $D$. Now observe the following about the function $e^{f(z)}$:

$$|e^f| = |e^{\phi + i\psi}| = |e^\phi||e^{i\psi}| = e^\phi. \tag{1}$$

Because the exponential is a monotonically increasing function of a real variable, Eq. (1) implies that the maximum points of $\phi$ coincide with the maximum points of the modulus of the analytic function $e^f$. Thus we immediately have a maximum principle for harmonic functions! Furthermore, since the minimum points of $\phi$ are the same as the maximum points of the harmonic function $-\phi$, we can state the following versions of the *maximum-minimum principle for harmonic functions*.

---

**Theorem 26.**   If $\phi$ is harmonic in a simply connected domain $D$ and $\phi(z)$ achieves its maximum or minimum value at some point $z_0$ in $D$, then $\phi$ is constant in $D$.

**Theorem 27.** A function harmonic in a bounded simply connected domain and continuous up to and including the boundary attains its maximum and minimum on the boundary.

Actually, these principles can easily be extended to multiply connected domains by appropriately modifying the proof of Theorem 23; see Prob. 3. We shall utilize this extended form hereafter.

An important problem that arises in electromagnetism, fluid mechanics, and heat transfer is the following.

*Dirichlet Problem*    Find a function $\phi(x, y)$ continuous on a domain $D$ and its boundary, harmonic in $D$, and taking specified values on the boundary of $D$.

The function $\phi$ can be interpreted as electric potential, velocity potential, or steady-state temperature. In studying the Dirichlet problem we are concerned with two main questions: Does a solution exist, and, if so, is it uniquely determined by the given boundary values? For the case of bounded domains the question of uniqueness is answered by the next theorem.

**Theorem 28.** Let $\phi_1(x, y)$ and $\phi_2(x, y)$ each be harmonic in a bounded domain $D$ and continuous on $D$ and its boundary. Furthermore, suppose that $\phi_1 = \phi_2$ on the boundary of $D$. Then $\phi_1 = \phi_2$ throughout $D$.

*Proof.* Consider the harmonic function $\phi := \phi_1 - \phi_2$. It must attain its maximum and minimum on the boundary, but it vanishes there! Hence $\phi = 0$ throughout $D$. ∎

A solution to the Dirichlet problem could be expressed by a formula giving $\phi$ inside $D$ in terms of its (specified) values on the boundary. Surely this suggests experimenting with the Cauchy integral formula. We are, in fact, able to solve the Dirichlet problems for the disk and the half-plane by this technique. Leaving the latter as an exercise for the reader, we proceed with the former.

For simplicity we consider the disk that is bounded by the positively oriented circle $C_R: |z| = R$. The Cauchy integral formula gives the values of an analytic function $f$ inside the disk in terms of its values on the circle:

$$f(z) = \frac{1}{2\pi i} \int_{C_R} \frac{f(\zeta)}{\zeta - z} \, d\zeta \qquad (|z| < R) \tag{2}$$

(assuming the domain of analyticity includes the circle $C_R$ as well as its interior). We wish to transform Eq. (2) into a formula that involves only the real part of $f$. To this end, we observe that for fixed $z$, with $|z| < R$, the function

$$\frac{f(\zeta)\bar{z}}{R^2 - \zeta\bar{z}}$$

is an *analytic function of* $\zeta$ inside and on $C_R$ (think about this; the denominator does not vanish). Hence by the Cauchy theorem

$$\frac{1}{2\pi i}\int_{C_R}\frac{f(\zeta)\bar{z}}{R^2 - \zeta\bar{z}}\,d\zeta = 0.$$

The utility of this relationship will become apparent when we add it to Eq. (2):

$$f(z) = \frac{1}{2\pi i}\int_{C_R}\left(\frac{1}{\zeta - z} + \frac{\bar{z}}{R^2 - \zeta\bar{z}}\right)f(\zeta)\,d\zeta$$

$$= \frac{1}{2\pi i}\int_{C_R}\frac{R^2 - |z|^2}{(\zeta - z)(R^2 - \zeta\bar{z})}f(\zeta)\,d\zeta. \tag{3}$$

If we parametrize $C_R$ by $\zeta = Re^{it}$, $0 \le t \le 2\pi$, Eq. (3) becomes

$$f(z) = \frac{1}{2\pi i}\int_0^{2\pi}\frac{R^2 - |z|^2}{(Re^{it} - z)(R^2 - Re^{it}\bar{z})}f(Re^{it})Rie^{it}\,dt$$

$$= \frac{R^2 - |z|^2}{2\pi}\int_0^{2\pi}\frac{f(Re^{it})}{(Re^{it} - z)(Re^{-it} - \bar{z})}\,dt$$

$$= \frac{R^2 - |z|^2}{2\pi}\int_0^{2\pi}\frac{f(Re^{it})}{|Re^{it} - z|^2}\,dt.$$

Finally, by taking the real part of this equation, identifying Re $f$ as the harmonic function $\phi$, and writing $z$ in the polar form $z = re^{i\theta}$, we arrive at *Poisson's integral formula*, which we state as follows.

---

**Theorem 29.**  Let $\phi$ be harmonic in a domain containing the disk $|z| \le R$. Then for $0 \le r < R$, we have

$$\phi(re^{i\theta}) = \frac{R^2 - r^2}{2\pi}\int_0^{2\pi}\frac{\phi(Re^{it})}{R^2 + r^2 - 2rR\cos(t - \theta)}\,dt. \tag{4}$$

---

Actually, Poisson's formula is more general than is indicated in this statement. We direct the interested reader to Ref. [9] or [10] for a proof of the next theorem.

**Theorem 30.** Let $U$ be a real-valued function defined on the circle $C_R: |z| = R$ and continuous there except for a finite number of jump discontinuities. Then the function

$$u(re^{i\theta}) := \frac{R^2 - r^2}{2\pi} \int_0^{2\pi} \frac{U(Re^{it})}{R^2 + r^2 - 2rR\cos(t - \theta)} \, dt \tag{5}$$

is harmonic inside $C_R$, and as $re^{i\theta}$ approaches any point on $C_R$ where $U$ is continuous, $u(re^{i\theta})$ approaches the value of $U$ at that point.

Naturally at the points of discontinuity the behavior is more complicated. As an example, consider the harmonic function $\text{Arg}(z + 1)$, which is the imaginary part of $\text{Log}(z + 1)$, in the domain depicted in Fig. 4.59. Clearly, $\text{Arg}(z + 1)$ approaches its boundary values in a reasonable manner except at $z = -1$, where it is erratic (the boundary value jumps from $\pi/2$ to $-\pi/2$ there).

Thus Poisson's formula solves the Dirichlet problem for the disk under very general circumstances.

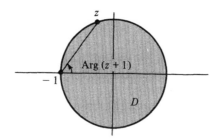

Figure 4.59   Arg($z$ + 1).

**Example 1**

Find the steady-state temperature $T$ at each point inside the unit disk if the temperature is prescribed to be $+1$ on the part of the rim in the first quadrant, and 0 elsewhere (Fig. 4.60).

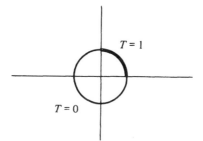

Figure 4.60   Find the temperature inside the disk.

**Solution.** The temperature is given by the harmonic function taking the pre-scribed boundary values. Here formula (5) becomes

$$T(re^{i\theta}) = \frac{1 - r^2}{2\pi} \int_0^{\pi/2} \frac{1}{1 + r^2 - 2r \cos(t - \theta)} \, dt.$$

For example, at the center,

$$T(0) = \frac{1}{2\pi} \int_0^{\pi/2} 1 \, dt = \frac{1}{4},$$

which is the average of its values on the rim. ∎

## EXERCISES 4.7

1. Find all functions $\phi$ harmonic in the unit disk $D: |z| < 1$ that satisfy $\phi(i/2) = -5$ and $\phi(z) \geq -5$ for all $z$ in $D$.

2. Show by example that a harmonic function need not have an analytic completion in a multiply connected domain. [HINT: Consider Log $|z|$.]

3. Prove the maximum-minimum principle for harmonic functions in an arbitrary domain. [HINT: Theorem 26 can be used to establish a lemma analogous to Lemma 2 in Sec. 4.6. Then argue as in Theorem 23.]

4. What is the physical interpretation of the maximum-minimum principle in steady-state heat flow?

5. Show by example that the solution to the Dirichlet problem need not be unique for unbounded domains. [HINT: Construct two functions that are harmonic in the upper half-plane, each vanishing on the x-axis.]

6. Prove the *circumferential mean-value theorem* for harmonic functions: If $\phi$ is harmonic in a domain containing the disk $|z| \leq \rho$, then

$$\phi(0) = \frac{1}{2\pi} \int_0^{2\pi} \phi(\rho e^{it}) \, dt.$$

7. Prove the following version of the *solid mean-value theorem* for harmonic functions: If $\phi$ is harmonic in a domain containing the closed disk $D: |z| \leq R$, then

$$\phi(0) = \frac{1}{\pi R^2} \iint_D \phi \, dx \, dy.$$

[HINT: Multiply the equation in Prob. 6 by $\rho \, d\rho$ and integrate.]

8. Without doing any computations, explain why

$$\frac{R^2 - r^2}{2\pi} \int_0^{2\pi} \frac{1}{R^2 + r^2 - 2rR \cos(t - \theta)} \, dt = 1 \qquad \text{for } 0 \leq r < R.$$

9. Prove *Harnack's inequality*: If $\phi(z)$ is harmonic and nonnegative in a domain containing the disk $|z| \leq R$, then for $0 \leq r < R$

$$\phi(0) \frac{R - r}{R + r} \leq \phi(re^{i\theta}) \leq \phi(0) \frac{R + r}{R - r}.$$

[HINT: Use Poisson's formula and the mean-value property of Prob. 6, observing also that

$$(R - r)^2 \le R^2 + r^2 - 2rR\cos(t - \theta) \le (R + r)^2.]$$

**10.** Prove *Liouville's theorem* for harmonic functions: If $\phi$ is harmonic in the whole plane and bounded from above or below there, then $\phi$ is constant. [HINT: Modify $\phi$ so that Harnack's inequality (Prob. 9) can be applied.]

**11.** The temperature of the rim of the unit disk is maintained at the levels indicated in Fig. 4.61. What is the temperature at the center?

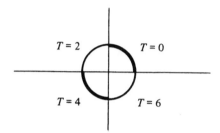

*Figure 4.61*    Find the temperature at the center.

**12.** (*Schwarz Integral Formula*) Let $f = u + iv$ be analytic on the disk $|z| \le R$. Then Poisson's integral formula expresses the values of the real part $u$ inside the disk in terms of the values of $u$ on the boundary of the disk. In this problem you will derive an expression for the imaginary part $v$ in terms of the boundary values of $u$.

**(a)** Show that the Poisson integral formula (4) can be written

$$u(z) = \frac{1}{2\pi}\int_0^{2\pi} P(Re^{it}, z)u(Re^{it})\, dt,$$

where the *Poisson kernel P* is given by

$$P(\zeta, z) := \frac{|\zeta|^2 - |z|^2}{|\zeta - z|^2}.$$

**(b)** Show that

$$\frac{\zeta + z}{\zeta - z} = P(\zeta, z) + 2i\frac{\operatorname{Im} z\bar{\zeta}}{|\zeta - z|^2};$$

in other words, $P(\zeta, z)$ is the real part of $(\zeta + z)/(\zeta - z)$.

**(c)** Utilizing Theorem 15, page 155, argue that

$$H(z) := \frac{1}{2\pi i}\oint_{|\zeta| = R}\frac{\zeta + z}{\zeta - z}\frac{u(\zeta)}{\zeta}\, d\zeta$$

defines an analytic function of $z$ for $|z| < R$.

**(d)** Insert the parametrization $\zeta = Re^{it}$ into the integral for $H(z)$ and derive

$$\operatorname{Re} H(z) = \frac{1}{2\pi}\int_0^{2\pi} P(Re^{it}, z)u(Re^{it})\, dt$$

which, by the Poisson integral formula, equals $u(z)$.

**(e)** Since $H(z)$ and $f(z)$ are two analytic functions whose real parts coincide, use the theory developed in Sec. 2.4 to argue that they can differ only by an imaginary constant,

$$f(z) = H(z) + iC.$$

Insert the value $z = 0$ into this identity and use the circumferential mean-value theorem (Prob. 6) to demonstrate that $C$ must be $v(0)$.

Assemble the results to obtain the *Schwarz integral formula*

$$f(z) = \frac{1}{2\pi i} \oint_{|\zeta| = R} \frac{\zeta + z}{\zeta - z} \frac{u(\zeta)}{\zeta} \, d\zeta + iv(0), \qquad \text{for } |z| < R.$$

**(f)** Equate the imaginary parts in the Schwarz integral formula to derive the representation for $v$ in terms of the boundary values of $u$:

$$v(z) = \frac{1}{2\pi} \int_0^{2\pi} Q(Re^{it}, z)u(Re^{it}) \, dt + v(0), \qquad \text{for } |z| < R,$$

where

$$Q(\zeta, z) := 2 \frac{\operatorname{Im} z\overline{\zeta}}{|\zeta - z|^2}.$$

13. Let $f$ be an entire function whose real part satisfies $|\operatorname{Re} f(z)| \le M|z|^2$ for all sufficiently large values of $|z|$, where $M$ is a constant. Show that $f$ must be a polynomial by arguing as follows.
    **(a)** Use the Schwarz formula [Prob. 12(e)] with $R = 2|z|$ to show that $|f(z)|$ is bounded for large $|z|$ by some multiple of $|z|^2$.
    **(b)** Use the result of Prob. 7 of Exercises 4.6 to conclude that $f$ is a polynomial of degree at most 2.

14. (*Poisson Integral Formula for the Half-Plane*) If $f = \phi + i\psi$ is analytic in a domain containing the x-axis and the upper half-plane and $|f(z)| \le K$ in this domain, then the values of the harmonic function $\phi$ in the upper half-plane are given in terms of its values on the x-axis by

$$\phi(x, y) = \frac{y}{\pi} \int_{-\infty}^{\infty} \frac{\phi(\xi, 0) \, d\xi}{(\xi - x)^2 + y^2} \qquad (y > 0).$$

Here is an outline of the derivation; justify the steps.
    **(a)** For the situation depicted in Fig. 4.62,

$$f(z) = \frac{1}{2\pi i} \int_{\Gamma_R} \frac{f(\zeta)}{\zeta - z} \, d\zeta.$$

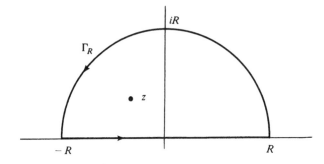

**Figure 4.62**

**(b)** For the same situation,

$$0 = \frac{1}{2\pi i} \int_{\Gamma_R} \frac{f(\zeta)}{\zeta - \bar{z}} \, d\zeta.$$

**(c)** Subtract these two equations to conclude that

$$f(z) = \frac{1}{2\pi i} \int_{-R}^{R} f(\xi) \frac{2i \, \text{Im} \, z}{|\xi - z|^2} \, d\xi$$

$$+ \frac{1}{2\pi i} \int_{C_R^+} f(\zeta) \frac{2i \, \text{Im} \, z}{(\zeta - z)(\zeta - \bar{z})} \, d\zeta,$$

where $C_R^+$ is the semicircular portion of $\Gamma_R$.

**(d)** Show that the integral along $C_R^+$ is bounded by

$$\frac{K}{\pi} \frac{\text{Im} \, z}{(R - |z|)^2} \, \pi R.$$

**(e)** Let $R \to \infty$ in the last equation and take the real part.

15. The Poisson integral formula in Prob. 14 admits a generalization; for suitable functions $U(\xi)$ the integral

$$\phi(x, y) := \frac{y}{\pi} \int_{-\infty}^{\infty} \frac{U(\xi)}{(\xi - x)^2 + y^2} \, d\xi$$

will define a harmonic function in either half-plane taking the limiting value $U(x)$ as $z$ approaches a point of continuity $x$ of $U$. Use the formula to find the temperature in the upper half-plane if the temperature on the $x$-axis is maintained as in Fig. 4.63.

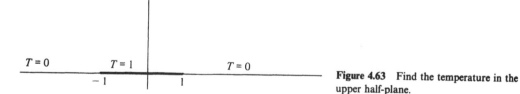

*T* = 0          *T* = 1            *T* = 0

          −1         1

**Figure 4.63** Find the temperature in the upper half-plane.

16. (*Schwarz Integral Formula for the Half-Plane*) If $f = u + iv$ is an analytic function in the closed upper half-plane and $|f|$ is bounded there, then the Poisson integral formula expresses the values of $u(x, y)$ for $y > 0$ in terms of the values of $u$ on the real axis (Prob. 14). If $f$ satisfies a slightly stronger condition, then the derivation in Prob. 14 can be modified to express also the values of the imaginary part $v(x, y)$ for $y > 0$ in terms of these boundary values of $u$.

**(a)** Retrace parts (a) and (b) of Prob. 14, but in part (c) *add* the two integrals to obtain the identity

$$f(z) = \frac{1}{2\pi i} \int_{-R}^{R} f(\xi) \frac{2(\xi - x)}{|\xi - z|^2} \, d\xi + \frac{1}{2\pi i} \int_{C_R^+} f(\zeta) \frac{2(\zeta - x)}{(\zeta - z)(\zeta - \bar{z})} \, d\zeta.$$

**(b)** Show that the integral along $C_R^+$ goes to zero as $R \to \infty$ if $f$ satisfies an inequality of the form $|f(z)| \le K/|z|^\alpha$ with $\alpha > 0$, for sufficiently large $|z|$ in the closed upper half-plane. Thus, for such an $f$,

$$f(z) = \frac{1}{\pi i} \int_{-\infty}^{\infty} f(\xi) \frac{(\xi - x)}{|\xi - z|^2} \, d\xi, \qquad \text{for Im } z > 0.$$

**(c)** Equate imaginary parts in the last equation to derive

$$v(z) = -\frac{1}{\pi} \int_{-\infty}^{\infty} u(\xi, 0) \frac{(\xi - x)}{|\xi - z|^2} \, d\xi, \qquad \text{Im } z > 0.$$

**(d)** Combine this formula with the Poisson integral formula to obtain the *Schwarz integral formula* for the upper half-plane:

$$\boxed{f(z) = \frac{1}{\pi i} \int_{-\infty}^{\infty} \frac{u(\xi, 0)}{\xi - z} \, d\xi, \qquad \text{Im } z > 0.}$$

17. With the assumptions of Prob. 16, the Schwarz integral formula also defines a function in the *lower* half-plane; call this function $\tilde{f}(z)$.
    **(a)** Show that $\tilde{f}(z) = -\overline{f(\bar{z})}$, for Im $z < 0$, and conclude from this that $\tilde{f}(z)$ is analytic in the lower half-plane.
    **(b)** Use the equation in part (a) to show that as $z$ crosses the real axis from above, the change in

$$\frac{1}{\pi i} \int_{-\infty}^{\infty} \frac{u(\xi, 0)}{\xi - z} \, d\xi$$

is the real number $-2u(x, 0)$.

18. Use the theory developed in Problems 16 and 17 to argue that

$$\frac{1}{z + i} = \frac{1}{\pi i} \int_{-\infty}^{\infty} \frac{\xi}{(\xi^2 + 1)(\xi - z)} \, d\xi \qquad \text{for Im } z > 0,$$

and that

$$\frac{-1}{z - i} = \frac{1}{\pi i} \int_{-\infty}^{\infty} \frac{\xi}{(\xi^2 + 1)(\xi - z)} \, d\xi \qquad \text{for Im } z < 0.$$

(These values will be confirmed directly in Prob. 12, Exercises 6.3.) Verify that the jump in the value of the integral as $z$ crosses the real axis equals $-2x/(x^2 + 1)$.

## SUMMARY

Integration in the complex plane takes place along contours, which are continuous chains of directed smooth curves. The definite integral over each smooth part is defined by imitating the Riemann sum definition used in calculus, and the contour integral $\int_\Gamma f(z) \, dz$ is the sum of the integrals over the smooth components of $\Gamma$.

A "brute-force" technique for computing an integral along a contour $\Gamma$ involves finding a parametrization $z = z(t)$, $a \le t \le b$, for $\Gamma$; then the contour integral can be obtained by performing an integration with respect to the real variable $t$, in accordance with the formula

$$\int_\Gamma f(z) \, dz = \int_a^b f(z(t)) z'(t) \, dt.$$

When the integrand $f(z)$ is analytic in a domain containing $\Gamma$, the following considerations may be useful in evaluating $\int_\Gamma f(z) \, dz$:

1. (*Fundamental Theorem of Calculus*) If $f$ has an antiderivative $F$ in a domain containing $\Gamma$, then

$$\int_\Gamma f(z)\,dz = F(z_T) - F(z_I),$$

where $z_I$ and $z_T$ are the initial and terminal points of $\Gamma$.

2. (*Cauchy's Theorem*) If $\Gamma$ is a simple closed contour (i.e., $z_I = z_T$ but no other self-intersections occur) and $f$ is analytic inside and on $\Gamma$, then $\int_\Gamma f(z)\,dz = 0$.

3. (*Cauchy's Integral Formula*) If $\Gamma$ is a simple closed positively oriented contour and $f$ has the form $f(z) = g(z)/(z - z_0)$, with $g$ analytic inside and on $\Gamma$ and $z_0$ lying inside $\Gamma$, then

$$\int_\Gamma f(z)\,dz = \int_\Gamma \frac{g(z)}{z - z_0}\,dz = 2\pi i g(z_0).$$

More generally,

$$\int_\Gamma \frac{g(z)\,dz}{(z - z_0)^m} = \frac{2\pi i g^{(m-1)}(z_0)}{(m-1)!}.$$

4. If $f$ is of the form $f(z) = g(z)/(z - z_1)^{m_1}(z - z_2)^{m_2}$, with $g$ as in (3) and the points $z_1$ and $z_2$ lying inside the simple closed contour $\Gamma$, then $\int_\Gamma f(z)\,dz$ can be reduced to case (3) by using partial fractions.

5. (*Deformation Invariance Theorem*) If the closed contour $\Gamma$ can be continuously deformed to another closed contour $\Gamma'$ without passing through any singularities of $f$, then

$$\int_\Gamma f(z)\,dz = \int_{\Gamma'} f(z)\,dz.$$

When the domain of analyticity of $f$ is simply connected (i.e., has no "holes"), then $f$ has an antiderivative and integrals of $f$ are independent of path.

The Cauchy integral formula has many consequences for analytic functions. Among these are the infinite differentiability of analytic functions (see 3), Liouville's theorem (bounded entire functions are constant), the Fundamental Theorem of Algebra (every nonconstant polynomial has a zero), and the maximum modulus theorem (the maximum of $|f|$ is attained on the boundary of a bounded domain).

In simply connected domains harmonic functions can be identified as real parts of analytic functions. Hence these results have analogues for harmonic functions, such as infinite differentiability, maximum principles, and Poisson's formula (the analogue of Cauchy's integral formula).

## SUGGESTED READING

The following references will be helpful to the reader interested in seeing a more detailed treatment of the topics of this chapter:

### Theory of Curves, Arc Length

[1] Apostol, T. M. *Mathematical Analysis*, 2nd ed. Addison-Wesley Publishing Company, Inc., Reading, Mass., 1974.

[2] Nevanlinna, R., and Paatero, V. *Introduction to Complex Analysis*. Addison-Wesley Publishing Company, Inc., Reading, Mass., 1969.

### Jordan Curve Theorem

[3] Pederson, R. N. "The Jordan Curve Theorem for Piecewise Smooth Curves." *American Mathematical Monthly*, 76 (1969), 605–610.

### Riemann Sums, Integration

Reference 2.

### Continuous Deformation of Curves, Cauchy's Theorem

[4] Conway, J. B. *Functions of One Complex Variable*, 2nd ed. Springer-Verlag, New York, 1978.

[5] Flatto, L., and Shisha, O. "A Proof of Cauchy's Integral Theorem." *Journal of Approximation Theory*, 7 (1973), 386–90.

### Leibniz's Rule for Integrals

[6] Taylor, A. E., and Mann, W. R. *Advanced Calculus*, 3rd ed. Wiley, New York, 1983.

### Green's Theorem

[7] Borisenko, A. I., and Tarapov, I. E. *Vector and Tensor Analysis with Applications* (trans. from Russian by R. A. Silverman). Dover, New York, 1979.

[8] Davis, H., and Snider, A. D. *Introduction to Vector Analysis*, 6th ed. Allyn and Bacon, Inc., Boston, 1991.

### Poisson's Formula

[9] Fisher, S. D. *Function Theory on Planar Domains*. Wiley-Interscience, New York, 1983.

[10] Hille, E. *Analytic Function Theory*, Vol. 2, 2nd ed. Chelsea, New York, 1973.

### Additional Problems

[11] Spiegel, M. R. *Theory and Problems of Complex Variables*. Schaum's Outline Series, McGraw-Hill Book Company, New York, 1964.

# :5:

# Series Representations for Analytic Functions

## 5.1 SEQUENCES AND SERIES

In Chapter 2 we defined what is meant by convergence of a sequence of complex numbers; recall that the sequence $\{A_n\}_{n=1}^{\infty}$ has $A$ as a limit if $|A - A_n|$ can be made arbitrarily small by taking $n$ large enough. For computational convenience it is often advantageous to use an element $A_n$ of the sequence as an approximation to $A$. Indeed, when we calculate the area of a circle we usually use an element of the sequence 3.14, 3.141, 3.1415, 3.14159, ... as an approximation to $\pi$. The use of sequences, and in particular the kind of sequences associated with *series*, is an important tool in both the theory and applications of analytic functions, and the present chapter is devoted to the development of this subject.

The possibility of summing an infinite string of numbers must have occurred to anyone who has toyed with adding

$$\frac{1}{2} + \frac{1}{4} = \frac{3}{4}, \quad \frac{1}{2} + \frac{1}{4} + \frac{1}{8} = \frac{7}{8}, \quad \frac{1}{2} + \frac{1}{4} + \frac{1}{8} + \frac{1}{16} = \frac{15}{16}, \quad \text{etc.}$$

The sequence of sums thus derived obviously has 1 as a limit, and it seems sensible to say $\frac{1}{2} + \frac{1}{4} + \frac{1}{8} + \cdots = 1$. We are motivated to generalize this by saying that an *infinite series* of the form $c_0 + c_1 + c_2 + \cdots$ has the sum $S$ if the sums of the first $n$ terms approach $S$ as a limit as $n$ goes to infinity. The customary nomenclature is summarized in the following definition.

> **Definition 1.**   A **series** is a formal expression of the form $c_0 + c_1 + c_2 + \cdots$, or equivalently $\sum_{j=0}^{\infty} c_j$, where the **terms** $c_j$ are complex numbers. The **nth partial sum** of the series, usually denoted $S_n$, is the sum of the first $n + 1$ terms, i.e., $S_n := \sum_{j=0}^{n} c_j$. If the sequence of partial sums $\{S_n\}_{n=0}^{\infty}$ has a limit $S$, the series is said to **converge**, or **sum**, to $S$, and we write $S = \sum_{j=0}^{\infty} c_j$. A series that does not converge is said to **diverge**.

Notice that the notion of convergence for a series has been defined in terms of convergence for a sequence. As an illustration, observe that $\pi$ is the sum of the series $3 + .1 + .04 + .001 + .0005 + .00009 + \cdots$.

Clearly one way to demonstrate that a series converges to $S$ is to show that the *remainder* after summing the first $n + 1$ terms, $S - \sum_{j=0}^{n} c_j$, goes to zero as $n \to \infty$. We use this technique in describing the convergence of the simple, but extremely useful, *geometric series* $\sum_{j=0}^{\infty} c^j$.

> **Lemma 1.**   The series $\sum_{j=0}^{\infty} c^j$ converges to $1/(1 - c)$ if $|c| < 1$, that is,
>
> $$1 + c + c^2 + c^3 + \cdots = \frac{1}{1 - c} \qquad \text{if } |c| < 1. \tag{1}$$
>
> (In Prob. 6 we shall see that such a series diverges if $|c| \geq 1$.)

*Proof.*   Observe that

$$(1 - c)(1 + c + c^2 + \cdots + c^{n-1} + c^n)$$

$$= 1 + c + c^2 + \cdots + c^{n-1} + c^n - c - c^2 - \cdots - c^{n-1} - c^n - c^{n+1}$$

$$= 1 - c^{n+1}.$$

Rearranging this yields

$$\frac{1}{1 - c} - (1 + c + c^2 + \cdots + c^{n-1} + c^n) = \frac{c^{n+1}}{1 - c}. \tag{2}$$

Since $|c| < 1$, the lemma follows immediately; Eq. (2) displays the remainder as $c^{n+1}/(1 - c)$, which certainly goes to zero as $n \to \infty$ (recall Prob. 4 in Exercises 2.2).   ∎

Another important way to establish the convergence of a series involves comparing it with another series whose convergence is known. The following theorem, which generalizes a result from calculus, seems so transparent that we shall spare our trusting readers the proof and refer the skeptics to Sec. 5.4.

**Theorem 1.** (*Comparison Test*) Suppose that the terms $c_j$ satisfy the inequality

$$|c_j| \le M_j$$

for all integers $j$ larger than some number $J$. Then, if the series $\sum_{j=0}^{\infty} M_j$ converges, so does $\sum_{j=0}^{\infty} c_j$.

## Example 1

Show that the series $\sum_{j=0}^{\infty} (3 + 2i)/(j + 1)^j$ converges.

**Solution.**   We compare the series

$$\sum_{j=0}^{\infty} \frac{3 + 2i}{(j + 1)^j} = (3 + 2i) + \frac{(3 + 2i)}{2} + \frac{(3 + 2i)}{9} + \frac{(3 + 2i)}{64} + \cdots \qquad (3)$$

with the *convergent* geometric series

$$\sum_{j=0}^{\infty} \frac{1}{2^j} = 1 + \frac{1}{2} + \frac{1}{4} + \frac{1}{8} + \cdots . \qquad (4)$$

Since $|3 + 2i| = \sqrt{13} < 4$, the reader can easily verify that for $j \ge 3$,

$$\left| \frac{3 + 2i}{(j + 1)^j} \right| < \frac{4}{(j + 1)^j} \le \frac{1}{2^j};$$

that is, the terms of (4) dominate those of (3) and hence (3) converges. ∎

Sometimes the *ratio test* can be applied to a series to establish convergence.

**Theorem 2.** (*Ratio Test*) Suppose that the terms of the series $\sum_{j=0}^{\infty} c_j$ have the property that the ratios $|c_{j+1}/c_j|$ approach a limit $L$ as $j \to \infty$. Then the series converges if $L < 1$ and diverges if $L > 1$.

The proof of this theorem involves comparing the given series with a series obtained by judiciously modifying the geometric series $\sum_{j=0}^{\infty} L^j$. See Prob. 15.

## Example 2

Show that the series $\sum_{j=0}^{\infty} 4^j/j!$ converges.

**Solution.**   We have

$$\left| \frac{c_{j+1}}{c_j} \right| = \frac{4^{j+1}}{(j + 1)!} \frac{j!}{4^j} = \frac{4}{j + 1}.$$

This ratio approaches zero as $j \to \infty$; thus the series converges. ∎

We remark that a series $\sum_{j=0}^{\infty} c_j$ is said to be *absolutely convergent* if the series $\sum_{j=0}^{\infty} |c_j|$ converges. Any absolutely convergent series is convergent, by a trivial application of the comparison test..

The kinds of sequences and series that often arise in complex analysis are those where the terms are functions of a complex variable $z$. Thus if we have a sequence of functions $F_1(z), F_2(z), F_3(z), \ldots$, we must consider the possibility that for some values of $z$ the sequence converges, while for others it diverges. As an example, the sequence $(z/2i)^n$, $n = 1, 2, 3, \ldots$, approaches zero for $|z| < |2i| = 2$, approaches 1 for $z = 2i$ (obviously!), and has no limit otherwise (see Prob. 4). Similarly, a *series* of complex functions $\sum_{j=0}^{\infty} f_j(z)$ may converge for some values of $z$ and diverge for others.

**Example 3**

If $z_0$ ($\neq 0$) is fixed, show that the series $\sum_{j=0}^{\infty} (z/z_0)^j$ [which is quite distinct from the *sequence* $(z/z_0)^j$] converges for $|z| < |z_0|$.

**Solution.** This is merely a thinly disguised resurrection of Lemma 1. In fact, setting $c = z/z_0$ in Eq. (2) yields

$$\frac{1}{1 - \dfrac{z}{z_0}} - \left[ 1 + \frac{z}{z_0} + \left(\frac{z}{z_0}\right)^2 + \cdots + \left(\frac{z}{z_0}\right)^n \right] = \frac{\left(\dfrac{z}{z_0}\right)^{n+1}}{1 - \dfrac{z}{z_0}}. \tag{5}$$

We conclude, as before, that for $|z| < |z_0|$ the series sums to the function $1/(1 - z/z_0)$. ∎

In applying this theory to analytic functions we need a somewhat stronger notion of convergence. By way of illustration, consider the sequence of real functions $F_n(x) = x^n$, depicted over the half-open interval $0 \leq x < 1$ in Fig. 5.1. Clearly, on this set the sequence $\{F_n(x)\}_{n=1}^{\infty}$ converges to the function $F(x) \equiv 0$; that is, for any *given* $x$ the powers $x^n$ become minuscule, for sufficiently large $n$. But none of the *curves*

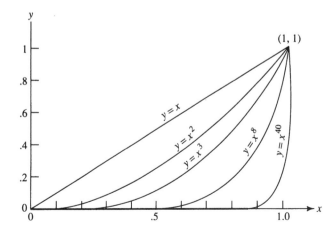

**Figure 5.1** The functions $F_n(x) = x^n$ converge to zero pointwise but not uniformly on [0, 1).

$y = x^n$ $(0 \le x < 1)$ would be regarded as good approximations to the *curve* $y = 0$, since each of the former has points near its right edge that generate (relatively) large values of $x^n$. We say that the convergence is *pointwise*, but not *uniform*. Thus we formulate the following.

---

**Definition 2.**  The sequence $\{F_n(z)\}_{n=1}^\infty$ is said to **converge uniformly to $F(z)$ on the set** $\mathscr{S}$ if for any $\varepsilon > 0$ there exists an integer $N$ such that when $n > N$,

$$|F(z) - F_n(z)| < \varepsilon \qquad \text{for } all \ z \text{ in } \mathscr{S}.$$

Accordingly, the series $\sum_{j=0}^\infty f_j(z)$ converges uniformly to $f(z)$ on $\mathscr{S}$ if the sequence of its partial sums converges uniformly to $f(z)$ there.

---

The essential feature of uniform convergence is that for a given $\varepsilon > 0$, one must be able to find an integer $N$ *that is independent of $z$ in $\mathscr{S}$* such that the error $|F(z) - F_n(z)|$ is less than $\varepsilon$ for $n > N$. In contrast, for pointwise convergence, $N$ can depend upon $z$. (See Prob. 17.) Of course uniform convergence on $\mathscr{S}$ implies pointwise convergence on $\mathscr{S}$.

### Example 4

Show that the series $\sum_{j=0}^\infty (z/z_0)^j$ of Example 3 is uniformly convergent in every closed disk $|z| \le r < |z_0|$.

**Solution.**  Given $\varepsilon > 0$, we have to show that the remainder after $n + 1$ terms will be less than $\varepsilon$ for all $z$ in the disk, when $n$ is large enough. This is easy; from Eq. (5) we can find an upper bound, independent of $z$, for the remainder:

$$\left| \frac{\left(\dfrac{z}{z_0}\right)^{n+1}}{1 - \dfrac{z}{z_0}} \right| \le \frac{\left(\dfrac{r}{|z_0|}\right)^{n+1}}{1 - \dfrac{r}{|z_0|}}, \qquad \text{for } |z| \le r.$$

This can be made arbitrarily small since $r < |z_0|$.  ∎

Combining Examples 3 and 4 we see that the series $\sum_{j=0}^\infty (z/z_0)^j$ converges pointwise in the open disk $|z| < |z_0|$ and uniformly on any closed subdisk $|z| \le r < |z_0|$.

## EXERCISES 5.1

1. Find the sum of the following convergent series.

(a) $\displaystyle\sum_{j=0}^\infty \left(\frac{i}{3}\right)^j$

(b) $\displaystyle\sum_{k=0}^\infty \frac{3}{(1+i)^k}$

(c) $\displaystyle\sum_{j=0}^\infty (-1)^j \left(\frac{2}{3}\right)^j$

(d) $\displaystyle\sum_{k=14}^\infty \left(\frac{1}{2i}\right)^k$

(e) $\displaystyle\sum_{j=0}^\infty \left(\frac{1}{3}\right)^{2j}$

(f) $\displaystyle\sum_{j=0}^\infty \left[\frac{1}{j+2} - \frac{1}{j+1}\right]$

**2.** Using the ratio test, show that the following series converge.

(a) $\sum_{j=1}^{\infty} \dfrac{1}{j!}$　　　(b) $\sum_{k=1}^{\infty} \dfrac{(3+i)^k}{k!}$　　　(c) $\sum_{j=0}^{\infty} \dfrac{j^2}{4^j}$　　　(d) $\sum_{k=1}^{\infty} \dfrac{k!}{k^k}$

**3.** Prove that if the sequence $\{z_n\}_{n=1}^{\infty}$ converges, then $(z_n - z_{n-1}) \to 0$ as $n \to \infty$.

**4.** Let $z_0 \neq 0$. Prove that the sequence $(z/z_0)^n$, $n = 1, 2, \dots$, diverges if $|z| \geq |z_0|$, $z \neq z_0$. [HINT: For $|z| = |z_0|$, observe that

$$\left| \left(\frac{z}{z_0}\right)^n - \left(\frac{z}{z_0}\right)^{n-1} \right| = \left| \frac{z}{z_0} - 1 \right| > 0$$

and use the result of Prob. 3.]

**5.** Prove that if the series $\sum_{j=0}^{\infty} c_j$ converges, then $c_j \to 0$ as $j \to \infty$. [HINT: Consider the difference $S_n - S_{n-1}$ of consecutive partial sums.]

**6.** Prove that the series $\sum_{j=0}^{\infty} c^j$ diverges if $|c| \geq 1$. [HINT: See Prob. 5.]

**7.** For each of the following determine if the given series converges or diverges.

(a) $\sum_{k=0}^{\infty} \left(\dfrac{1+2i}{1-i}\right)^k$　　　(b) $\sum_{j=1}^{\infty} \dfrac{1}{j^2 3^j}$　　　(c) $\sum_{n=1}^{\infty} \dfrac{n\,i^n}{2n+1}$

(d) $\sum_{j=1}^{\infty} \dfrac{j!}{5^j}$　　　(e) $\sum_{k=1}^{\infty} \dfrac{(-1)^k k^3}{(1+i)^k}$　　　(f) $\sum_{k=1}^{\infty} \left(i^k - \dfrac{1}{k^2}\right)$

**8.** Prove the following statements.

(a) If $\sum_{j=0}^{\infty} c_j$ sums to $S$, then $\sum_{j=0}^{\infty} \bar{c}_j$ sums to $\bar{S}$.

(b) If $\sum_{j=0}^{\infty} c_j$ sums to $S$ and $\lambda$ is any complex number, then $\sum_{j=0}^{\infty} \lambda c_j$ sums to $\lambda S$.

(c) If $\sum_{j=0}^{\infty} c_j$ sums to $S$ and $\sum_{j=0}^{\infty} d_j$ sums to $T$, then $\sum_{j=0}^{\infty} (c_j + d_j)$ sums to $S + T$.

**9.** Prove that the series $\sum_{j=0}^{\infty} z_j$ converges if and only if both of the series $\sum_{j=0}^{\infty} \text{Re}(z_j)$ and $\sum_{j=0}^{\infty} \text{Im}(z_j)$ converge.

**10.** Show that the sequence of functions $F_n(z) = z^n/(z^n - 3^n)$, $n = 1, 2, \dots$, converges to zero for $|z| < 3$ and to 1 for $|z| > 3$.

**11.** Using the ratio test, find a domain in which convergence holds for each of the following series of functions.

(a) $\sum_{j=1}^{\infty} jz^j$　　　　　　　(b) $\sum_{k=0}^{\infty} \dfrac{(z-i)^k}{2^k}$

(c) $\sum_{j=0}^{\infty} \dfrac{z^j}{j!}$　　　　　　　(d) $\sum_{k=0}^{\infty} (z + 5i)^{2k}(k+1)^2$

**12.** Let $F_n(z) = [nz/(n+1)] + (3/n)$, $n = 1, 2, \dots$. Prove that the sequence $\{F_n(z)\}_1^{\infty}$ converges uniformly to $F(z) = z$ on every closed disk $|z| \leq R$.

**13.** Prove that $\sum_{j=1}^{\infty} 1/j^p$ converges if $p > 1$. [HINT: Interpret the integral $\int_1^N (1/x^p)\,dx$ as an area; then interpret $\sum_{j=2}^{N} 1/j^p$ as an area, and compare.]

**14.** Using the comparison test and the result of Prob. 13, show that the following series converge.

(a) $\sum_{j=1}^{\infty} \dfrac{1}{j(j+i)}$　　　　　　　(b) $\sum_{k=1}^{\infty} \dfrac{\sin(k^2)}{k^{3/2}}$

(c) $\sum_{k=1}^{\infty} \dfrac{k^2 i^k}{k^4 + 1}$　　　　　　　(d) $\sum_{k=2}^{\infty} (-1)^k \left(\dfrac{5k+8}{k^3 - 1}\right)$

**15.** Prove the ratio test (Theorem 2). [HINT: If $L < 1$, choose $\varepsilon > 0$ and $J$ so that $|c_{j+1}/c_j| < L + \varepsilon < 1$ for $j \geq J$. Then show that $|c_k| \leq |c_J|(L + \varepsilon)^{k-J}$ for $k > J$ and use the comparison test.]

**16.** Prove that the sequence $\{z_n\}_1^\infty$ converges if and only if the series $\sum_{k=1}^\infty (z_{k+1} - z_k)$ converges.

**17.** Consider the sequence of functions $F_n(x) = x^n$ on the real interval $\mathcal{S} = [0, 1)$, which converges pointwise to $F(x) = 0$ on $\mathcal{S}$. Show that, for $0 < x < 1$,

$$|F_n(x) - F(x)| < \tfrac{1}{2}$$

when and only when $n > N_x$, where

$$N_x := \frac{\mathrm{Log}\, 2}{\mathrm{Log}(x^{-1})}.$$

(Observe that $N_x \to \infty$ as $x \to 1^-$, so that it is not possible to fulfill Definition 2 with an $N$ *independent* of $x$ in $\mathcal{S}$ when $\varepsilon = \tfrac{1}{2}$. This proves that this sequence does not converge uniformly on $\mathcal{S}$.)

**18.** Assume that the sequence of functions $\{F_n(z)\}_1^\infty$ converges uniformly to $F(z)$ on a set $\mathcal{S}$. Prove that if $|F(z)| \geq \rho > 0$ for all $z$ in $\mathcal{S}$, then there exists an integer $N$ such that for $n > N$ the inequality $|F_n(z)| > \rho/2$ holds for all $z$ in $\mathcal{S}$. [HINT: Take $\varepsilon = \rho/2$ in Definition 2 and apply the triangle inequality.]

**19.** It will be shown in the next section that the series $\sum_{k=0}^\infty z^k/k!$ converges uniformly to $e^z$ on every disk $|z| \leq R$. Accepting this fact, prove that, for $n$ sufficiently large, none of the polynomials $S_n(z) = \sum_{k=0}^n z^k/k!$ has zeros on $|z| \leq 5$. [HINT: Use Prob. 18.] (There is nothing special about the disk $|z| \leq 5$; the same assertion holds on any bounded set.)

**20.** Prove that the series $\sum_{j=0}^\infty z^j$ does not converge *uniformly* to $1/(1 - z)$ on the open disk $|z| < 1$.

## 5.2 TAYLOR SERIES

Suppose that we wish to find a polynomial $p_n(z)$ of degree at most $n$ that approximates an analytic function $f(z)$ in a neighborhood of a point $z_0$. Naturally there are differing criteria as to how well the polynomial approximates the function. We shall construct a polynomial that "looks like" $f(z)$ at the point $z_0$ in the sense that its derivatives match those of $f$ at $z_0$, insofar as possible:

$$p_n(z_0) = f(z_0)$$
$$p_n'(z_0) = f'(z_0)$$
$$\vdots$$
$$p_n^{(n)}(z_0) = f^{(n)}(z_0).$$

[Of course, the $(n + 1)$st derivative of any polynomial of degree $n$ must equal *zero*.] The constant polynomial $p_0(z)$ that matches $f$ at $z_0$ is simply $f(z_0)$:

$$p_0(z) \equiv f(z_0).$$

The polynomial of degree 1 that matches $f$ and $f'$ at $z_0$ is

$$p_1(z) = f(z_0) + f'(z_0)(z - z_0).$$

The second-degree polynomial matching $f$, $f'$, and $f''$ at $z_0$ is

$$p_2(z) = f(z_0) + f'(z_0)(z - z_0) + \frac{f''(z_0)}{2!}(z - z_0)^2.$$

From this we can see a pattern emerging; the third-degree polynomial is

$$p_3(z) = f(z_0) + f'(z_0)(z - z_0) + \frac{f''(z_0)}{2!}(z - z_0)^2 + \frac{f'''(z_0)}{3!}(z - z_0)^3,$$

and *the nth-degree polynomial that matches* $f, f', f'', \ldots, f^{(n)}$ *at* $z_0$ *is*

$$p_n(z) = f(z_0) + f'(z_0)(z - z_0) + \frac{f''(z_0)}{2!}(z - z_0)^2 + \cdots + \frac{f^{(n)}(z_0)}{n!}(z - z_0)^n.\dagger \qquad (1)$$

Naturally we conjecture that as $n$ tends to infinity, $p_n(z)$ becomes a better and better approximation to $f(z)$ near $z_0$. In fact, the astute reader may have noticed that $p_n(z)$ looks like a partial sum of a series, and so might anticipate that this series converges to $f(z)$. The precise state of affairs is given in the following definition and theorem.

---

**Definition 3.**  If $f$ is analytic at $z_0$, then the series

$$f(z_0) + f'(z_0)(z - z_0) + \frac{f''(z_0)}{2!}(z - z_0)^2 + \cdots = \sum_{j=0}^{\infty} \frac{f^{(j)}(z_0)}{j!}(z - z_0)^j \qquad (2)$$

is called the **Taylor series** for $f$ around $z_0$. When $z_0 = 0$, it is also known as the **Maclaurin series** for $f$.

---

**Theorem 3.**  If $f$ is analytic in the disk $|z - z_0| < R$, then the Taylor series (2) converges to $f(z)$ for all $z$ in this disk. Furthermore, the convergence of the series is uniform in any closed subdisk $|z - z_0| \leq R' < R$.

---

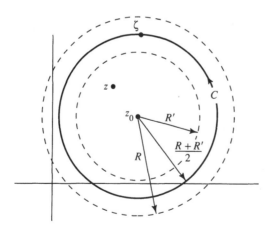

**Figure 5.2**  Uniform convergence in closed subdisks.

---

$\dagger$ Actually $p_n(z)$ will have degree less than $n$ if $f^{(n)}(z_0) = 0$.

*Proof.* Notice that if we prove uniform convergence in *every* closed subdisk $|z - z_0| \leq R' < R$, we will have pointwise convergence for each $z$ in the open disk $|z - z_0| < R$ (why?); thus we deal only with the closed subdisk statement. Let $C$ be the circle $|z - z_0| = (R + R')/2$, positively oriented (see Fig. 5.2). Then by Cauchy's integral formula we have, for any $z$ within the closed subdisk,

$$f(z) = \frac{1}{2\pi i} \int_C \frac{f(\zeta)}{\zeta - z} \, d\zeta.$$

We modify the integrand by writing

$$\frac{1}{\zeta - z} = \frac{1}{(\zeta - z_0) - (z - z_0)} = \frac{1}{\zeta - z_0} \cdot \frac{1}{1 - \dfrac{z - z_0}{\zeta - z_0}}$$

$$= \frac{1}{\zeta - z_0} \left[ 1 + \frac{z - z_0}{\zeta - z_0} + \frac{(z - z_0)^2}{(\zeta - z_0)^2} + \cdots + \frac{(z - z_0)^n}{(\zeta - z_0)^n} + \frac{\dfrac{(z - z_0)^{n+1}}{(\zeta - z_0)^{n+1}}}{1 - \dfrac{z - z_0}{\zeta - z_0}} \right], \tag{3}$$

using identity (2) of Sec. 5.1 with $c = (z - z_0)/(\zeta - z_0)$. Putting this into the Cauchy integral formula we find

$$f(z) = \frac{1}{2\pi i} \int_C \frac{f(\zeta)}{\zeta - z_0} \, d\zeta + \frac{z - z_0}{2\pi i} \int_C \frac{f(\zeta)}{(\zeta - z_0)^2} \, d\zeta$$

$$+ \frac{(z - z_0)^2}{2\pi i} \int_C \frac{f(\zeta)}{(\zeta - z_0)^3} \, d\zeta + \cdots + \frac{(z - z_0)^n}{2\pi i} \int_C \frac{f(\zeta)}{(\zeta - z_0)^{n+1}} \, d\zeta + T_n(z), \tag{4}$$

where $T_n(z)$ can be expressed (with a little algebra) as

$$T_n(z) = \frac{1}{2\pi i} \int_C \frac{f(\zeta)}{(\zeta - z)} \frac{(z - z_0)^{n+1}}{(\zeta - z_0)^{n+1}} \, d\zeta. \tag{5}$$

Now by Cauchy's integral formula for derivatives

$$\frac{1}{2\pi i} \int_C \frac{f(\zeta)}{\zeta - z_0} \, d\zeta = f(z_0), \qquad \frac{1}{2\pi i} \int_C \frac{f(\zeta)}{(\zeta - z_0)^2} \, d\zeta = f'(z_0),$$

$$\frac{1}{2\pi i} \int_C \frac{f(\zeta)}{(\zeta - z_0)^3} \, d\zeta = \frac{f''(z_0)}{2!}, \quad \text{etc.,}$$

and so the first $n + 1$ terms on the right-hand side of Eq. (4) yield the $n$th partial sum of the series in (2). Thus $T_n(z)$ is the remainder, and we must show that it can be made "uniformly small" for all $z$ in the subdisk, by taking $n$ sufficiently large.

This is simply a matter of applying the integral inequality of Chapter 4; for the terms in the integrand of Eq. (5) we have

$$|z - z_0| \leq R', \qquad |\zeta - z_0| = \frac{R + R'}{2},$$

and, from Fig. 5.2,

$$|\zeta - z| \geq \frac{R + R'}{2} - R' = \frac{R - R'}{2}.$$

The length of $C$ is $2\pi(R + R')/2$. Thus for $z$ in the subdisk

$$|T_n(z)| \leq \frac{1}{2\pi} \cdot \max_{\zeta \text{ on } C} |f(\zeta)| \frac{2}{R - R'} \left(\frac{2R'}{R + R'}\right)^{n+1} 2\pi \left(\frac{R + R'}{2}\right).$$

The right-hand side is independent of $z$, and since $2R' < R + R'$, it can be made less than any positive $\varepsilon$ by taking $n$ large enough. ■

Notice that the theorem implies that *the Taylor series will converge to $f(z)$ everywhere inside the largest open disk, centered at $z_0$, over which $f$ is analytic.*

**Example 1**

Compute and state the convergence properties of the Taylor series for **(a)** Log $z$ around $z_0 = 1$, **(b)** $1/(1 - z)$ around $z_0 = 0$, and **(c)** $e^z$ around $z_0 = 0$.

**Solution.** **(a)** The consecutive derivatives of Log $z$ are Log $z$, $z^{-1}$, $-z^{-2}$, $2z^{-3}$, $-3 \cdot 2z^{-4}$, etc.; in general,

$$\frac{d^j \operatorname{Log} z}{dz^j} = (-1)^{j+1}(j - 1)! \, z^{-j} \qquad (j = 1, 2, \ldots).$$

Evaluating these at $z = 1$ we find

$$\operatorname{Log} z = 0 + (z - 1) - \frac{(z - 1)^2}{2!} + 2! \frac{(z - 1)^3}{3!} - 3! \frac{(z - 1)^4}{4!} + \cdots$$

$$= \sum_{j=1}^{\infty} \frac{(-1)^{j+1}(z - 1)^j}{j}. \tag{6}$$

This is valid for $|z - 1| < 1$, the largest open disk centered at $+1$ over which Log $z$ is analytic.

**(b)** The consecutive derivatives of $(1 - z)^{-1}$ are $(1 - z)^{-1}$, $(1 - z)^{-2}$, $2(1 - z)^{-3}$, $3 \cdot 2(1 - z)^{-4}$, and, in general,

$$\frac{d^j}{dz^j}(1 - z)^{-1} = j!(1 - z)^{-j-1}.$$

Evaluating these at $z = 0$ gives the Taylor series

$$\frac{1}{1 - z} = 1 + z + \frac{2!z^2}{2!} + \frac{3!z^3}{3!} + \cdots = \sum_{j=0}^{\infty} z^j, \tag{7}$$

valid for $|z| < 1$ (why?). In fact, this is just the geometric series considered in Lemma 1.

**(c)** Since

$$\frac{d^j e^z}{dz^j} = e^z$$

for all $j = 0, 1, \ldots$, the common value of these derivatives at $z = 0$ is 1. Thus

$$e^z = 1 + z + \frac{z^2}{2!} + \frac{z^3}{3!} + \cdots = \sum_{j=0}^{\infty} \frac{z^j}{j!},$$

which is valid for all $z$ because $e^z$ is entire.  ■

Let's experiment with Eq. (6), the expansion for Log $z$. If we differentiate the *series* term by term, we get

$$1 - (z - 1) + (z - 1)^2 - (z - 1)^3 + \cdots = \sum_{j=0}^{\infty} (1 - z)^j, \qquad (8)$$

properly incorporating the negative signs. But the series (8) has the same form as the series in Eq. (7), with $(1 - z)$ in place of $z$. Thus (8) actually converges to the function $1/[1 - (1 - z)] = 1/z$; in other words, by differentiating the *series* for Log $z$ we obtained the *series* for $1/z$, which is, in fact, the derivative of Log $z$. We are led to conjecture the following.

---

**Theorem 4.**  If $f$ is analytic at $z_0$, the Taylor series for $f'$ around $z_0$ can be obtained by termwise differentiation of the Taylor series for $f$ around $z_0$ and converges in the same disk as the series for $f$.

---

*Proof.*  The $j$th derivative of $f'$ is, of course, the $(j + 1)$st derivative of $f$. Thus the Taylor series for $f'$ is given by

$$f'(z_0) + f''(z_0)(z - z_0) + \frac{f'''(z_0)}{2!} (z - z_0)^2 + \cdots. \qquad (9)$$

On the other hand, termwise differentiation of the Taylor series (2) yields

$$0 + 1 \cdot f'(z_0) + \frac{2}{2!} \cdot f''(z_0)(z - z_0) + \frac{3}{3!} \cdot f'''(z_0)(z - z_0)^2 + \cdots,$$

which is the same as (9). Furthermore, application of Theorem 3 to the function $f'(z)$ establishes that (9) converges in the largest open disk around $z_0$ over which $f'$ is analytic. But according to the theory of Chapter 4, $f'$ is analytic wherever $f$ is analytic. This completes the proof.  ■

## Example 2

Find the Maclaurin series for sin $z$ and cos $z$.

**Solution.**  We expand sin $z$ as usual. The sequence of derivatives is sin $z$, cos $z$, $-\sin z$, $-\cos z$, sin $z$, .... Evaluating at the origin yields

$$\sin z = z - \frac{z^3}{3!} + \frac{z^5}{5!} - \frac{z^7}{7!} + \cdots, \qquad (10)$$

which holds for all $z$ since $\sin z$ is entire. To get $\cos z$, we differentiate Eq. (10):

$$\cos z = 1 - \frac{z^2}{2!} + \frac{z^4}{4!} - \frac{z^6}{6!} + \cdots. \tag{11}$$

The reader should by now be able to predict what would result if we differentiate Eq. (11). ∎

The next two theorems may sometimes simplify the computation of a Taylor series.

---

**Theorem 5.** Let $f$ and $g$ be analytic functions with Taylor series
$f(z) = \sum_{j=0}^{\infty} a_j(z - z_0)^j$ and $g(z) = \sum_{j=0}^{\infty} b_j(z - z_0)^j$ around the point $z_0$
[i.e., $a_j = f^{(j)}(z_0)/j!$ and $b_j = g^{(j)}(z_0)/j!$]. Then

(i) the Taylor series for $cf(z)$, $c$ a constant, is $\sum_{j=0}^{\infty} ca_j(z - z_0)^j$;

(ii) the Taylor series for $f(z) \pm g(z)$ is $\sum_{j=0}^{\infty} (a_j \pm b_j)(z - z_0)^j$.

---

The proof is left as an easy exercise. The disk of convergence for $f \pm g$ is, of course, at least as big as the smaller of the convergence disks for $f$ and $g$.

**Example 3**

Find the Maclaurin series for $\cos z + i \sin z$.

**Solution.**    Using the expansions (10) and (11) we find

$$\cos z + i \sin z = 1 + iz - \frac{z^2}{2!} - \frac{iz^3}{3!} + \frac{z^4}{4!} + \frac{iz^5}{5!} - \cdots$$

$$= e^{iz}$$

for all $z$. (This validates a computation we made in Example 1, Sec. 1.4.) ∎

Theorem 5 naturally leads us to cogitate the corresponding statement for products. First we must find a sensible way of multiplying two Taylor series. The *Cauchy product* of two Taylor series around a point $z_0$ is defined in the manner suggested by applying the distributive law and then grouping the terms in powers of $(z - z_0)$. Thus, if $z_0 = 0$, we find for the Cauchy product

$$[a_0 + a_1 z + a_2 z^2 + a_3 z^3 + \cdots] \cdot [b_0 + b_1 z + b_2 z^2 + b_3 z^3 + \cdots]$$

$$= a_0 b_0 + (a_1 b_0 + a_0 b_1)z + (a_2 b_0 + a_1 b_1 + a_0 b_2)z^2$$

$$+ (a_3 b_0 + a_2 b_1 + a_1 b_2 + a_0 b_3)z^3 + \cdots. \tag{12}$$

The coefficient, $c_j$, of $z^j$ is therefore given by

$$c_j = a_j b_0 + a_{j-1} b_1 + a_{j-2} b_2 + \cdots + a_1 b_{j-1} + a_0 b_j = \sum_{l=0}^{j} a_{j-l} b_l. \tag{13}$$

**Definition 4.**  The **Cauchy product** of two Taylor series $\sum_{j=0}^{\infty} a_j(z - z_0)^j$ and $\sum_{j=0}^{\infty} b_j(z - z_0)^j$ is defined to be the (formal) series $\sum_{j=0}^{\infty} c_j(z - z_0)^j$, where $c_j$ is given by formula (13).

**Theorem 6.**  Let $f$ and $g$ be analytic functions with Taylor series $f(z) = \sum_{j=0}^{\infty} a_j(z - z_0)^j$ and $g(z) = \sum_{j=0}^{\infty} b_j(z - z_0)^j$ around the point $z_0$. Then the Taylor series for the product $fg$ around $z_0$ is given by the Cauchy product of these two series.

Actually, we anticipated this result in electing to write the Cauchy product in (12) as if it were an ordinary product. As in Theorem 5, the Taylor series for $fg$ converges at least in the smaller of the convergence disks for $f$ and $g$.

*Proof of Theorem 6.*   We compute the consecutive derivatives of the product $fg$:

$$(fg)' = f'g + fg',$$
$$(fg)'' = f''g + 2f'g' + fg'',$$
$$(fg)''' = f'''g + 3f''g' + 3f'g'' + fg''',$$

and, in general, we have *Leibniz's formula* for the $j$th derivative of $fg$:

$$(fg)^{(j)} = \sum_{l=0}^{j} j! \frac{f^{(j-l)}}{(j-l)!} \cdot \frac{g^{(l)}}{l!}, \tag{14}$$

which we invite the reader to prove in Prob. 10.

On the other hand, if we identify the constants $a_k$ and $b_k$ in Eq. (13) with their expressions in terms of derivatives of $f$ and $g$ [e.g., $a_k = f^{(k)}(z_0)/k!$], we see from Eq. (14) that $(fg)^{(j)}/j!$ evaluated at $z_0$ is precisely $c_j$. This completes the proof.  ■

**Example 4**

Use the Cauchy product to find the Maclaurin series for $\sin z \cdot \cos z$.

**Solution.**   We have

$$\left( z - \frac{z^3}{3!} + \frac{z^5}{5!} - \frac{z^7}{7!} + \cdots \right) \cdot \left( 1 - \frac{z^2}{2!} + \frac{z^4}{4!} - \frac{z^6}{6!} + \cdots \right)$$

$$= z - \left( \frac{1}{3!} + \frac{1}{2!} \right) z^3 + \left( \frac{1}{5!} + \frac{1}{3!} \frac{1}{2!} + \frac{1}{4!} \right) z^5$$

$$- \left( \frac{1}{7!} + \frac{1}{5!} \frac{1}{2!} + \frac{1}{3!} \frac{1}{4!} + \frac{1}{6!} \right) z^7 + \cdots.$$

It is amusing to try to simplify the coefficients; the reader can verify that

$$\sin z \cdot \cos z = z - \frac{4}{3!} z^3 + \frac{16}{5!} z^5 - \frac{64}{7!} z^7 + \cdots,$$

which, when rewritten as

$$\frac{1}{2}\left[ (2z) - \frac{(2z)^3}{3!} + \frac{(2z)^5}{5!} - \frac{(2z)^7}{7!} + \cdots \right],$$

will be recognized as the Taylor series for $\frac{1}{2}\sin w$, with $w = 2z$. We have reproduced a well-known trigonometric identity!   ∎

## Example 5

Find the first few terms of the Maclaurin series for $\tan z$.

**Solution.**   The expressions for the higher derivatives of $\tan z$ are cumbersome, so let's try to use the Cauchy product. First observe that $\cos z \cdot \tan z = \sin z$. Now set $\tan z = \sum_{j=0}^{\infty} a_j z^j$ for $|z| < \pi/2$. (Why $\pi/2$?) The product $\cos z \cdot \tan z$ then becomes

$$\left(1 - \frac{z^2}{2!} + \frac{z^4}{4!} - \cdots \right) \cdot (a_0 + a_1 z + a_2 z^2 + a_3 z^3 + a_4 z^4 + a_5 z^5 + \cdots)$$

$$= a_0 + a_1 z + \left(a_2 - \frac{a_0}{2!}\right)z^2 + \left(a_3 - \frac{a_1}{2!}\right)z^3 + \left(a_4 - \frac{a_2}{2!} + \frac{a_0}{4!}\right)z^4$$

$$+ \left(a_5 - \frac{a_3}{2!} + \frac{a_1}{4!}\right)z^5 + \cdots.$$

Identifying this with $\sin z = z - z^3/3! + z^5/5! - \cdots$, we solve recursively and find

$$a_0 = 0, \quad a_1 = 1, \quad a_2 = 0, \quad a_3 = \frac{1}{3}, \quad a_4 = 0, \quad a_5 = \frac{2}{15}, \quad \text{etc.}$$

Thus

$$\tan z = z + \frac{z^3}{3} + \frac{2z^5}{15} + \cdots.$$

The shrewd reader will observe that we have actually discovered an indirect method of dividing Taylor series!   ∎

In closing this section we would like to point out that the proof of the validity of the Taylor expansion substantiates the claim, made in Sec. 2.3, that any *analytic* function can be displayed with a formula involving $z$ alone, and not $\bar{z}$, $x$, or $y$.

## EXERCISES 5.2

1. Using Definition 3, verify each of the following Taylor expansions by finding a general formula for $f^{(j)}(z_0)$.

   **(a)** $e^{-z} = \sum_{j=0}^{\infty} \frac{(-z)^j}{j!} = 1 - z + \frac{z^2}{2!} - \frac{z^3}{3!} + \cdots,$    $z_0 = 0$

   **(b)** $\cosh z = \sum_{j=0}^{\infty} \frac{z^{2j}}{(2j)!} = 1 + \frac{z^2}{2!} + \frac{z^4}{4!} + \cdots,$    $z_0 = 0$

   **(c)** $\sinh z = \sum_{j=0}^{\infty} \frac{z^{2j+1}}{(2j+1)!} = z + \frac{z^3}{3!} + \frac{z^5}{5!} + \cdots,$    $z_0 = 0$

   **(d)** $\dfrac{1}{1-z} = \sum_{j=0}^{\infty} \dfrac{(z-i)^j}{(1-i)^{j+1}},$    $z_0 = i$

   **(e)** $\operatorname{Log}(1-z) = \sum_{j=1}^{\infty} \dfrac{-z^j}{j},$    $z_0 = 0$

   **(f)** $z^3 = 1 + 3(z-1) + 3(z-1)^2 + (z-1)^3,$    $z_0 = 1$

2. Determine the disks over which the Taylor expansions in Prob. 1 are valid.

3. Let $f(z) = \sum_{j=0}^{\infty} a_j z^j$ be the Maclaurin expansion of a function $f(z)$ analytic at the origin. Prove each of the following statements.

   **(a)** $\sum_{j=0}^{\infty} a_j z^{2j}$ is the Maclaurin expansion of $g(z) := f(z^2)$.

   **(b)** $\sum_{j=0}^{\infty} a_j c^j z^j$ is the Maclaurin expansion of $h(z) := f(cz)$.

   **(c)** $\sum_{j=0}^{\infty} a_j z^{m+j}$ is the Maclaurin expansion of $H(z) := z^m f(z)$.

   **(d)** $\sum_{j=0}^{\infty} a_j (z - z_0)^j$ is the Taylor expansion of $G(z) := f(z - z_0)$ around $z_0$.

4. Let $\alpha$ be a complex number. Show that if $(1 + z)^{\alpha}$ is taken as $e^{\alpha \operatorname{Log}(1+z)}$, then for $|z| < 1$

   $$(1 + z)^{\alpha} = 1 + \frac{\alpha}{1} z + \frac{\alpha(\alpha - 1)}{1 \cdot 2} z^2 + \frac{\alpha(\alpha - 1)(\alpha - 2)}{1 \cdot 2 \cdot 3} z^3 + \cdots.$$

   [REMARK: This generalizes the binomial theorem.]

5. Find, and state the convergence properties of, the Taylor series for the following.

   **(a)** $\dfrac{1}{1 + z}$ around $z_0 = 0$          **(b)** $e^{-z^2}$ around $z_0 = 0$

   **(c)** $z^3 \sin 3z$ around $z_0 = 0$          **(d)** $2 \cos z - i e^z$ around $z_0 = 0$

   **(e)** $\dfrac{1 + z}{1 - z}$ around $z_0 = i$          **(f)** $\cos z$ around $z_0 = \dfrac{\pi}{4}$

   **(g)** $\dfrac{z}{(1 - z)^2}$ around $z_0 = 0$

6. Prove that the Taylor expansion of $1/(\zeta - z)$ around $z_0$ ($\neq \zeta$) is given by

   $$\frac{1}{\zeta - z} = \sum_{j=0}^{\infty} \frac{(z - z_0)^j}{(\zeta - z_0)^{j+1}} \qquad \text{for } |z - z_0| < |\zeta - z_0|.$$

   [REMARK: The expansion lies at the heart of the proof of Theorem 3.]

7. Verify that the identity

   $$\operatorname{Log}\left(\frac{1 + z}{1 - z}\right) = \operatorname{Log}(1 + z) - \operatorname{Log}(1 - z)$$

holds when $|z| < 1$. Then, using the Maclaurin expansions of $\text{Log}(1 + z)$ and $\text{Log}(1 - z)$, find the Maclaurin expansion of $\text{Log}[(1 + z)/(1 - z)]$.

8. Use Taylor series to verify the following identities.

(a) $\sin(-z) = -\sin z$

(b) $\dfrac{de^z}{dz} = e^z$

(c) $e^{-iz} = \cos z - i \sin z$

(d) $e^{2z} = e^z \cdot e^z$

9. Prove Theorem 5.

10. Prove Leibniz's formula, Eq. (14). [HINT: Use mathematical induction.]

11. Using Theorem 6 for computing the product of Taylor series, find the first three nonzero terms in the Maclaurin expansion of the following.

(a) $e^z \cos z$

(b) $\dfrac{e^z}{z - 1}$

(c) $\sec z = \dfrac{1}{\cos z}$

(d) $\tanh z = \dfrac{\sinh z}{\cosh z}$

12. Prove that the polynomial $p_n(z)$ of Eq. (1) is the *only* polynomial of degree at most $n$ that matches $f, f', f'', \dots, f^{(n)}$ at $z_0$.

13. Find an explicit formula for the analytic function $f(z)$ that has the Maclaurin expansion $\sum_{k=0}^{\infty} k^2 z^k$. [HINT: Starting with the expression $(1 - z)^{-1} = \sum_{k=0}^{\infty} z^k$, differentiate, multiply by $z$, differentiate again, and finally multiply by $z$.]

14. Let $f(z)$ be analytic in the disk $D: |z - z_0| < R$. Prove that if $f^{(k)}(z_0) = 0$ for every $k = 0$, $1, 2, \dots$, then $f(z)$ is identically zero in $D$.

15. Let $f(z)$ be analytic in the unit disk $|z| < 1$. Prove that if $f'(0) = f^{(3)}(0) = f^{(5)}(0) = \cdots = 0$, then $f(-z) = f(z)$ for all $z$ in this disk. That is, show that $f$ is an even function.

16. Rewrite the polynomial $p(z) = a_0 + a_1 z + \cdots + a_n z^n$ in powers of $(z - 1)$; i.e., find the coefficients $c_i$ of the expansion $p_n(z) = c_0 + c_1(z - 1) + \cdots + c_n(z - 1)^n$ in terms of the $a_j$. [HINT: Do not rearrange; use Taylor series.]

17. Recall from Exercises 5.1, Prob. 20, page 187, that the Taylor series $\sum_{j=0}^{\infty} z^j$ does not converge *uniformly* to $(1 - z)^{-1}$ on the open disk $D: |z| < 1$. Why doesn't this contradict Theorem 3?

18. The Taylor series provides a workable method of numerically tabulating the functions of mathematical physics when the remainder term can be estimated. Establish each of the following error estimates.

(a) $\left| e^z - \sum_{k=0}^{n} \dfrac{z^k}{k!} \right| \le \dfrac{1}{(n + 1)!} \cdot \left( 1 + \dfrac{1}{n + 1} \right)$  for $|z| \le 1$

(b) $\left| \sin z - \sum_{k=0}^{n} \dfrac{(-1)^k z^{2k+1}}{(2k + 1)!} \right| \le \dfrac{1}{(2n + 3)!} \left( \dfrac{4n^2 + 18n + 20}{4n^2 + 18n + 19} \right)$  for $|z| \le 1$

[HINT: Write the error as an infinite series, factor out the first term, and then compare with a geometric series.]

19. According to the estimate in Prob. 18(a), how many terms of the expansion $\sum_{k=0}^{\infty} z^k/k!$ are needed to compute $e^z$ to within $\pm 10^{-5}$ for $|z| \le 1$?

20. (*Hermite Formula*) From the proof of Theorem 3 deduce the following formula for the remainder in the Maclaurin series for a function $f$ analytic on $|z| \le R$:

$$f(z) - \sum_{j=0}^{n} \frac{f^{(j)}(0)}{j!} z^j = \frac{1}{2\pi i} \oint_{|\zeta| = R} \frac{z^{n+1}}{\zeta^{n+1}} \frac{f(\zeta)}{(\zeta - z)} \, d\zeta \qquad \text{for } |z| < R.$$

## 5.3 POWER SERIES

A Taylor series for an analytic function appears to be a special instance of a certain general type of series of the form $\sum_{j=0}^{\infty} a_j(z - z_0)^j$. Such series have a name:

---

**Definition 5.**   A series of the form $\sum_{j=0}^{\infty} a_j(z - z_0)^j$ is called a **power series**. The constants $a_j$ are the **coefficients** of the power series.

---

Suppose that we are presented with an arbitrary power series, such as

$$\sum_{j=0}^{\infty} \frac{z^j}{(j + 1)^2} = 1 + \frac{z}{4} + \frac{z^2}{9} + \frac{z^3}{16} + \cdots. \tag{1}$$

Certain questions then arise naturally. For what values of $z$ does the series converge? Is the sum an analytic function? Is the power series representation of a function unique? In short, is every power series a Taylor series? This section is devoted to answering these questions.

The issue of convergence is settled by the following result, which smacks of the Taylor expansion.

---

**Theorem 7.**   For any power series $\sum_{j=0}^{\infty} a_j(z - z_0)^j$ there is a real number $R$ between 0 and $\infty$, inclusive, which depends only on the coefficients $\{a_j\}$, such that

(i) the series converges for $|z - z_0| < R$,
(ii) the series converges uniformly in any closed subdisk $|z - z_0| \le R' < R$,
(iii) the series diverges for $|z - z_0| > R$.

The number $R$ is called the **radius of convergence** of the power series.

---

In particular, when $R = 0$ the power series converges only at $z = z_0$, and when $R = \infty$ the series converges for all $z$. For $0 < R < \infty$, the circle $|z - z_0| = R$ is called the *circle of convergence*, but no general convergence statement can be made for $z$ lying on this circle (see Prob. 1). The situation is depicted in Fig. 5.3 on page 198.

Although a rigorous proof of Theorem 7 is deferred to (optional) Sec. 5.4, we can give an informal argument here which shows why the region of convergence has to be a disk. The essential ingredient is the following lemma, which we state for the special case $z_0 = 0$:

---

**Lemma 2.**   If the power series $\sum_{j=0}^{\infty} a_j z^j$ converges at a point having modulus $r$, then it converges at every point in the disk $|z| < r$.

---

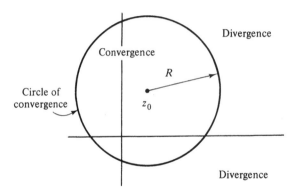

**Figure 5.3** Circle of convergence.

*Proof of Lemma 2.* By hypothesis, there exists a point $z_1$, with $|z_1| = r$, such that the series $\sum_{j=0}^{\infty} a_j z_1^j$ converges. This implies that the sequence of terms $a_j z_1^j$ is bounded; i.e., there exists a constant $M$ such that

$$|a_j z_1^j| = |a_j| r^j \le M \qquad \text{(for all } j\text{)}.$$

Now for $|z| < r$ we can write

$$|a_j z^j| = |a_j| r^j \cdot \left(\frac{|z|}{r}\right)^j \le M\left(\frac{|z|}{r}\right)^j.$$

Thus, as $|z|/r < 1$, the terms of the series $\sum_{j=0}^{\infty} a_j z^j$ are dominated by the terms of a convergent geometric series. By the comparison test, therefore, we conclude that the series converges. ∎

To see the existence of the number $R$ in Theorem 7 for the power series $\sum_{j=0}^{\infty} a_j z^j$ we reason informally as follows: Consider the set of all real numbers $r$ such that the series converges at some point having modulus $r$. Let $R$ be the "largest"[†] of these numbers $r$. Then, by Lemma 2, the series converges for $|z| < R$, and from the definition of $R$ the series diverges for all $z$ with $|z| > R$.

If $z$ is replaced by $(z - z_0)$ in the above argument, we deduce that the region of convergence of the general power series $\sum_{j=0}^{\infty} a_j(z - z_0)^j$ must be a disk with center $z_0$.

A formula for the radius of convergence $R$ can be given, but we shall postpone it also until Sec. 5.4. However, in Prob. 2, the ratio test is used to show that in the special case when $|a_{j+1}/a_j|$ has a limit as $j$ goes to infinity, $R$ is the reciprocal of this limit. For example, the coefficients of the power series (1) satisfy

$$\lim_{j \to \infty} \left|\frac{a_{j+1}}{a_j}\right| = \lim_{j \to \infty} \frac{(j+1)^2}{(j+2)^2} = 1,$$

so its circle of convergence is $|z| = R = \frac{1}{1} = 1$.

---

[†] More advanced students will recognize that $R$ is precisely defined as the *least upper bound* of the numbers $r$.

Uniform convergence [cf. (ii) of Theorem 7] is a powerful feature of a sequence, as the next three results show. The first says that the uniform limit of continuous functions is itself continuous.

---

**Lemma 3.**   Let $f_n$ be a sequence of functions continuous on a set $\mathscr{S} \subset \mathbf{C}$ and converging uniformly to $f$ on $\mathscr{S}$. Then $f$ is also continuous on $\mathscr{S}$.

---

*Proof.*   To prove that $f$ is continuous at a point $z_0$ of $\mathscr{S}$, we must show that for any $\varepsilon > 0$ there is a $\delta > 0$ such that if $z$ belongs to $\mathscr{S}$ and $|z_0 - z| < \delta$, then $|f(z_0) - f(z)| < \varepsilon$. We proceed by first choosing an integer $N$ so large that $|f(z) - f_N(z)| < \varepsilon/3$ for all $z$ in $\mathscr{S}$; this is possible thanks to uniform convergence. Now since $f_N$ is continuous, there is a number $\delta' > 0$ such that $|f_N(z_0) - f_N(z)| < \varepsilon/3$ for any $z$ in $\mathscr{S}$ satisfying $|z_0 - z| < \delta'$. But then $|f(z_0) - f(z)| < \varepsilon$ for such $z$, since

$$|f(z_0) - f(z)| = |f(z_0) - f_N(z_0) + f_N(z_0) - f_N(z) + f_N(z) - f(z)|$$

$$\leq |f(z_0) - f_N(z_0)| + |f_N(z_0) - f_N(z)| + |f_N(z) - f(z)|$$

$$< \frac{\varepsilon}{3} + \frac{\varepsilon}{3} + \frac{\varepsilon}{3} = \varepsilon.$$

Thus $f$ is continuous at every point of $\mathscr{S}$.   ∎

Knowing that the uniform limit of a sequence of continuous functions is continuous, we can integrate this limit. In fact the integral of the limit is the limit of integrals.

---

**Theorem 8.**   Let $f_n$ be a sequence of functions continuous on a set $\mathscr{S} \subset \mathbf{C}$ containing the contour $\Gamma$, and suppose that $f_n$ converges uniformly to $f$ on $\mathscr{S}$. Then the sequence $\int_\Gamma f_n(z)\,dz$ converges to $\int_\Gamma f(z)\,dz$.

---

*Proof.*   This is easy. Let $l$ be the length of $\Gamma$, and choose $N$ so that $|f(z) - f_n(z)| < \varepsilon/l$ for any $n > N$ and for all $z$ on $\Gamma$. Then

$$\left| \int_\Gamma f(z)\,dz - \int_\Gamma f_n(z)\,dz \right| = \left| \int_\Gamma [f(z) - f_n(z)]\,dz \right|$$

$$< \frac{\varepsilon}{l} \cdot l = \varepsilon.   ∎$$

Combining these results with Morera's theorem (Theorem 18, Chapter 4, page 158), we can prove the following.

> **Theorem 9.** Let $f_n$ be a sequence of functions analytic in a simply connected domain $D$ and converging uniformly to $f$ in $D$. Then $f$ is analytic in $D$.[†]

*Proof.* By Lemma 3, the function $f$ is continuous. Let $\Gamma$ be any loop contained in $D$. Then by Theorem 8, the integral $\int_\Gamma f(z)\,dz$ is the limit of $\int_\Gamma f_n(z)\,dz$; but the latter is zero for all $n$, since each $f_n$ is analytic inside and on $\Gamma$. Thus $\int_\Gamma f(z)\,dz = 0$, and Morera's result applies. ∎

Now we have everything we need to render an account of power series. Since the partial sums of a power series are analytic functions (indeed, polynomials) and since they converge uniformly in any closed subdisk interior to the circle of convergence, Theorem 9 tells us that the limit function is analytic inside every such subdisk. But *any* point within the circle of convergence lies inside such a subdisk, so we can state the following.

> **Theorem 10.** A power series sums to a function that is analytic at every point inside its circle of convergence.

For example, the power series (1) defines an analytic function for $|z| < 1$.

Notice that Theorems 7 and 8 justify integrating a power series termwise, as long as the contour lies inside the circle of convergence. Using this fact, we can identify every power series (with $R > 0$) as a Taylor series, in accordance with the next result.

> **Theorem 11.** If $\sum_{j=0}^{\infty} a_j(z - z_0)^j$ converges to $f(z)$ in some neighborhood of $z_0$ (i.e., the radius of its circle of convergence is nonzero), then
>
> $$a_j = \frac{f^{(j)}(z_0)}{j!} \qquad (j = 0, 1, 2, \ldots).$$
>
> Consequently, $\sum_{j=0}^{\infty} a_j(z - z_0)^j$ is the Taylor expansion of $f(z)$ around $z_0$.

*Proof.* Let $C$ be a positively oriented circle centered at $z_0$ and lying inside the circle of convergence. Since the limit $f(z)$ is analytic, we can write the generalized Cauchy integral formula

$$f^{(n)}(z_0) = \frac{n!}{2\pi i} \int_C \frac{f(\zeta)}{(\zeta - z_0)^{n+1}}\,d\zeta \qquad (n = 0, 1, 2, \ldots). \tag{2}$$

---

[†] In fact this result holds in *any* domain, since a domain is a union of open disks.

Now plug in the series $\sum_{j=0}^{\infty} a_j(\zeta - z_0)^j$ for $f(\zeta)$ and integrate termwise; since

$$\int_C \frac{(\zeta - z_0)^j}{(\zeta - z_0)^{n+1}} \, d\zeta = \begin{cases} 2\pi i & \text{if } j = n, \\ 0 & \text{otherwise,} \end{cases}$$

the only term that survives the integration in (2) is $n! \, a_n$.  ■

At this point we have specified two disks around $z_0$ in which a power series such as (1) converges. One is the interior of the "circle of convergence" of Theorem 7. The other is the largest disk over whose interior the limit function $f$ is analytic. Are these two disks the same? Observe that the "Taylor disk" cannot extend beyond the circle of convergence because the series is known to *diverge* outside the latter. On the other hand, the limit function is analytic inside the circle of convergence, so the Taylor disk must enclose the interior of this circle. Thus the disks actually do coincide.

Summarizing, we have shown that if a power series converges inside some circle, it is the Taylor series of its (analytic) limit function and can be integrated and differentiated term by term inside this circle; moreover, this limit function must fail to be analytic somewhere on the circle of convergence.

**Example 1**

Find a function $f$ that is analytic and satisfies the differential equation

$$\frac{df(z)}{dz} = 3if(z) \tag{3}$$

in a neighborhood of $z = 0$, taking the value 1 at $z = 0$.

**Solution.**  Since $f$ is analytic at the origin, it must have a Maclaurin series representation. We can find this series by using Eq. (3) to compute the derivatives.

We are given $f(0) = 1$, and from Eq. (3) we have $f'(0) = 3i \cdot 1 = 3i$. By differentiating Eq. (3), we see recursively that

$$f''(0) = 3if'(0) = (3i)^2,$$
$$f'''(0) = 3if''(0) = (3i)^3,$$

and, in general,

$$f^{(j)}(0) = 3if^{(j-1)}(0) = (3i)^2 f^{(j-2)}(0) = \cdots = (3i)^j.$$

Thus we can write the solution as

$$f(z) = 1 + 3iz + \frac{(3i)^2 z^2}{2!} + \cdots = \sum_{j=0}^{\infty} \frac{(3iz)^j}{j!}. \tag{4}$$

Recalling the representation

$$e^w = \sum_{j=0}^{\infty} \frac{w^j}{j!}, \tag{5}$$

we can identify our solution (4) as

$$f(z) = e^{3iz}.$$

Indeed, direct computation quickly verifies that $e^{3iz}$ solves the problem.  ■

The classic initial value problem for the nth-*order linear homogeneous differential equation*

$$\frac{d^n f}{dz^n} + p_{n-1}(z)\frac{d^{n-1}f}{dz^{n-1}} + \cdots + p_2(z)\frac{d^2 f}{dz^2} + p_1(z)\frac{df}{dz} + p_0(z)f = 0$$

is the task of finding a solution that satisfies the initial conditions

$$f(z_0) = a_0, \quad f'(z_0) = a_1, \quad f''(z_0) = a_2, \quad \ldots, \quad f^{(n-1)}(z_0) = a_{n-1}$$

at some point $z_0$. In advanced differential equation texts (cf. Refs. [3] and [4]), it is shown that if each of the coefficients $p_j(z)$ is analytic inside a disk centered at $z_0$, then for arbitrary constants $\{a_0, a_1, \ldots, a_{n-1}\}$, there is one and only one solution of the initial value problem, and it, too, is analytic inside this disk.

**Example 2**

Let $g$ be a continuous complex-valued function of a real variable on $[0, 2]$, and for each complex number $z$ define

$$F(z) := \int_0^2 e^{zt} g(t)\, dt.$$

Prove that $F$ is entire, and find its power series around the origin.

**Solution.** We first find a power series representation for $F$. Let $z$ be fixed and define

$$h(t) := e^{zt}, \qquad \text{for all complex numbers } t.$$

Then $h$ is an entire function of $t$, and so its Maclaurin expansion

$$h(t) = e^{zt} = \sum_{k=0}^{\infty} \frac{z^k}{k!} \cdot t^k = 1 + zt + \frac{z^2 t^2}{2!} + \cdots$$

converges uniformly on every disk $|t| \le r$; in particular, the convergence is uniform for $t$ in the interval $[0, 2]$. Furthermore, termwise multiplication by the bounded function $g$ preserves this convergence; i.e.,

$$e^{zt} g(t) = \sum_{k=0}^{\infty} \frac{z^k}{k!} \cdot t^k g(t)$$

uniformly for $0 \le t \le 2$ (and fixed $z$). Now, by Theorem 8, we can integrate term by term with respect to $t$ to obtain

$$F(z) = \int_0^2 e^{zt} g(t)\, dt = \sum_{k=0}^{\infty} \left[\frac{1}{k!} \int_0^2 t^k g(t)\, dt\right] z^k.$$

Since the bracketed quantity in the series is a constant dependent only on $k$, the above expansion is a power series in $z$; moreover, it converges to $F(z)$ for all $z$. Hence $F$ is entire by Theorem 10. ∎

## EXERCISES 5.3

1. (a) Prove that the power series $\sum_{j=0}^{\infty} z^j$ converges at no point on its circle of convergence $|z| = 1$.

   (b) Prove that the power series $\sum_{j=1}^{\infty} z^j/j^2$ converges at every point on its circle of convergence $|z| = 1$. [HINT: See Prob. 13 in Exercises 5.1.]

2. Assume that for the power series $\sum_{j=0}^{\infty} a_j(z - z_0)^j$ we have $\lim_{j \to \infty} |a_{j+1}/a_j| = L$. Prove, by the ratio test, that the radius of convergence of the power series is given by $R = 1/L$.

3. Using the result of Prob. 2, find the circle of convergence of each of the following power series.

   (a) $\sum_{j=0}^{\infty} j^3 z^j$  (b) $\sum_{k=0}^{\infty} 2^k(z - 1)^k$  (c) $\sum_{j=0}^{\infty} j! z^j$

   (d) $\sum_{k=0}^{\infty} \frac{(-1)^k k}{3^k}(z - i)^k$  (e) $\sum_{k=1}^{\infty} \frac{(3 - i)^k}{k^2}(z + 2)^k$  (f) $\sum_{j=0}^{\infty} \frac{z^{2j}}{4^j}$

4. Does there exist a power series $\sum_{j=0}^{\infty} a_j z^j$ that converges at $z = 2 + 3i$ and diverges at $z = 3 - i$?

5. Let $f(z) = \sum_{k=0}^{\infty} (k^3/3^k) z^k$. Compute each of the following.

   (a) $f^{(6)}(0)$  (b) $\oint_{|z|=1} \frac{f(z)}{z^4} dz$

   (c) $\oint_{|z|=1} e^z f(z) dz$  (d) $\oint_{|z|=1} \frac{f(z) \sin z}{z^2} dz$

6. Define

   $$f(z) := \begin{cases} \dfrac{\sin z}{z} & \text{for } z \neq 0, \\ 1 & \text{for } z = 0. \end{cases}$$

   (a) Using the Maclaurin expansion for $\sin z$, show that for *all* $z$

   $$f(z) = 1 - \frac{z^2}{3!} + \frac{z^4}{5!} - \frac{z^6}{7!} + \cdots.$$

   (b) Explain why $f(z)$ is analytic at the origin.

   (c) Find $f^{(3)}(0)$ and $f^{(4)}(0)$.

7. Find the first three nonzero terms in the Maclaurin expansion of $f(z) := \int_0^z e^{\zeta^2} d\zeta$. [HINT: First expand $e^{\zeta^2}$.]

8. Assume that $f(z)$ is analytic at the origin and that $f(0) = f'(0) = 0$. Prove that $f(z)$ can be written in the form $f(z) = z^2 g(z)$, where $g(z)$ is analytic at $z = 0$.

9. Suppose that $g$ is continuous on the circle $C : |z| = 1$, and that there exists a sequence of polynomials which converge uniformly to $g$ on $C$. Prove that

   $$\oint_C g(z) dz = 0.$$

10. Explain why the two power series $\sum_{k=0}^{\infty} a_k z^k$ and $\sum_{k=1}^{\infty} k a_k z^{k-1}$ have the same radius of convergence.

11. Let $\sum_{k=0}^{\infty} a_k z^k$ and $\sum_{k=0}^{\infty} b_k z^k$ be two power series having a positive radius of convergence.

    (a) Show that if $\sum_{k=0}^{\infty} a_k z^k = \sum_{k=0}^{\infty} b_k z^k$ in some neighborhood of the origin, then $a_k = b_k$ for all $k$.

    (b) Show, more generally, that if $\sum_{k=0}^{\infty} a_k x^k = \sum_{k=0}^{\infty} b_k x^k$ for all real $x$ in some open interval containing the origin, then $a_k = b_k$ for all $k$.

12. Prove by means of power series that the only solution of the initial-value problem

$$\begin{cases} \dfrac{df}{dz} = f \\ f(0) = 1 \end{cases}$$

that is analytic at $z = 0$ is $f(z) = e^z$.

13. Each of the following initial-value problems has a unique solution that is analytic at the origin. Find the power series expansion $\sum_{j=0}^{\infty} a_j z^j$ of the solution by determining a recurrence relation for the coefficients $a_j$.

(a) $\begin{cases} \dfrac{d^2 f}{dz^2} - z \dfrac{df}{dz} - f = 0 \\ f(0) = 1, \qquad f'(0) = 0 \end{cases}$      (b) $\begin{cases} \dfrac{d^2 f}{dz^2} + 4f = 0 \\ f(0) = 1, \qquad f'(0) = 1 \end{cases}$

(c) $\begin{cases} (1 - z^2) \dfrac{d^2 f}{dz^2} - 6z \dfrac{df}{dz} - 4f = 0 \\ f(0) = 1, \qquad f'(0) = 0 \end{cases}$

[HINT: The technique demonstrated in Example 1 becomes laborious for more complicated equations such as (c). It is more efficient to substitute the power series expression for $f(z)$ into the equation and collect like powers of $z$.]

14. Prove by means of power series that the only solution of the initial-value problem

$$\begin{cases} \dfrac{d^2 f}{dz^2} + f = 0 \\ f(0) = 0, \qquad f'(0) = 1 \end{cases}$$

that is analytic at $z = 0$ is $f(z) = \sin z$.

15. Let $g$ be continuous on the real interval $[-1, 2]$, and define

$$F(z) := \int_{-1}^{2} g(t) \sin(zt) \, dt.$$

(a) Prove that $F$ is entire and find its power series expansion around the origin.
(b) Prove that for all $z$

$$F'(z) = \int_{-1}^{2} tg(t) \cos(zt) \, dt.$$

16. Let $g$ be continuous on the real interval $[0, 1]$ and define

$$H(z) := \int_{0}^{1} \frac{g(t)}{1 - zt^2} \, dt \qquad (|z| < 1).$$

Prove that $H$ is analytic in the open disk $|z| < 1$.

17. Using the notation

$$(a)_j := a(a + 1) \cdots (a + j - 1), \qquad j \geq 1, \qquad (a)_0 := 1,$$

for any complex number $a$, the *Gaussian hypergeometric series* $_2F_1(b, c; d; z)$ is defined by

$$_2F_1(b, c; d; z) := \sum_{j=0}^{\infty} \frac{(b)_j (c)_j}{(d)_j} \cdot \frac{z^j}{j!},$$

and the *confluent hypergeometric series* $_1F_1(c; d; z)$ is given by

$$_1F_1(c; d; z) := \sum_{j=0}^{\infty} \frac{(c)_j}{(d)_j} \cdot \frac{z^j}{j!}.$$

**(a)** Verify that

$$\sum_{j=0}^{\infty} \frac{z^j}{j+1} = {}_2F_1(1, 1; 2; z),$$

$$e^z - \sum_{k=0}^{n-1} \frac{z^k}{k!} = \frac{z^n}{n!} {}_1F_1(1; n + 1; z) \qquad (n = 1, 2, \ldots).$$

**(b)** Prove that if $d \neq 0, -1, -2, \ldots$, then the series $_2F_1(b, c; d; z)$ converges for $|z| < 1$ and satisfies the differential equation

$$z(1 - z)\frac{d^2f}{dz^2} + [d - (b + c + 1)z]\frac{df}{dz} - bcf = 0.$$

**(c)** Prove that if $d \neq 0, -1, -2, \ldots$, then the series $_1F_1(c; d; z)$ converges for all $z$ and satisfies the differential equation

$$z\frac{d^2f}{dz^2} + (d - z)\frac{df}{dz} - cf = 0.$$

**18.** (*Generalized L'Hospital's Rule*). Use power series to prove that if $f$, $g$ are both analytic at $z_0$ and

$$f(z_0) = g(z_0) = f'(z_0) = g'(z_0) = \cdots = f^{(m-1)}(z_0) = g^{(m-1)}(z_0) = 0$$

but $g^{(m)}(z_0) \neq 0$, then

$$\lim_{z \to z_0} \frac{f(z)}{g(z)} = \frac{f^{(m)}(z_0)}{g^{(m)}(z_0)}.$$

## *5.4 MATHEMATICAL THEORY OF CONVERGENCE

In this section we shall backtrack somewhat and provide the mathematical details of the unproved theorems of this chapter. Applications-oriented students may wish to skip to Sec. 5.5.

So far all the conditions we have seen for convergence of a sequence involve the limit explicitly. However, there is a way of testing whether or not a sequence is convergent without mentioning a limit at all. It is known as the *Cauchy criterion* for convergence.

---

**Theorem 12.** A necessary and sufficient condition for the sequence of complex numbers $\{A_n\}_{n=1}^{\infty}$ to converge is the following: For any $\varepsilon > 0$ there exists an integer $N$ such that $|A_n - A_m| < \varepsilon$ for every pair of integers $m$ and $n$ satisfying $m > N, n > N$.

---

*Proof (necessity).* If the sequence does converge, say, to $A$, we choose $N$ so that each $A_l$ is within $\varepsilon/2$ of $A$ for $l > N$. Then any two such $A_l$ must lie within $\varepsilon$ of each other.

The proof that the Cauchy criterion is sufficient for convergence requires a rigorous axiomatization for the real number system; indeed, the criterion can be used to define the concept of an irrational real number. We shall not explore this here.  ∎

A sequence that satisfies the Cauchy criterion is often called a *Cauchy sequence*. By Theorem 12, every convergent sequence is a Cauchy sequence and vice versa.

---

**Corollary 1.**  If $\{A_n\}_{n=1}^{\infty}$ is a Cauchy sequence and $N$ is chosen so that $|A_n - A_m| < \varepsilon$ for every $m$ and $n$ greater than $N$, then each $A_n$ with $n > N$ is within $\varepsilon$ of the limit.

---

*Proof.*   Let $m \to \infty$ in the inequality $|A_n - A_m| < \varepsilon$. The result is

$$|A_n - A| \leq \varepsilon. \quad \blacksquare$$

The Cauchy criterion, applied to the sequence of partial sums of a series, reads as follows:

---

**Corollary 2.**  A necessary and sufficient condition for the series $\sum_{j=0}^{\infty} c_j$ to converge is the following: For any $\varepsilon > 0$ there exists an $N$ such that $\left|\sum_{j=n+1}^{m} c_j\right| < \varepsilon$ for every pair of integers $m$ and $n$ satisfying $m > n > N$.

---

The proof is immediate. Such a series is (naturally) called a *Cauchy series*. Corollary 2 justifies the following, almost obvious, result: If $\sum_{j=0}^{\infty} c_j$ converges, then $c_j \to 0$ as $j \to \infty$.

With the Cauchy criterion in hand we can give a proof of the comparison test, Theorem 1 of this chapter. However, it takes only a little more effort to prove the following, more general theorem, known as the *Weierstrass M-test*:

---

**Theorem 13.**  (*M-test*) Suppose $\sum_{j=0}^{\infty} M_j$ is a convergent series with real nonnegative terms and suppose, for all $z$ in some set $\mathscr{S}$ and for all $j$ greater than some number $J$, that $|f_j(z)| \leq M_j$. Then the series $\sum_{j=0}^{\infty} f_j(z)$ converges uniformly on $\mathscr{S}$.

---

*Proof.*   Since $\sum_{j=0}^{\infty} M_j$ is a Cauchy series, we can choose $N > J$ so that for any $m$ and $n$ satisfying $m > n > N$ we have $\sum_{j=n+1}^{m} M_j < \varepsilon$. But then for $z$ in $\mathscr{S}$, $\sum_{j=0}^{\infty} f_j(z)$ is a Cauchy series also, because

$$\left| \sum_{j=n+1}^{m} f_j(z) \right| \leq \sum_{j=n+1}^{m} |f_j(z)| \leq \sum_{j=n+1}^{m} M_j < \varepsilon. \tag{1}$$

Hence $\sum_{j=0}^{\infty} f_j(z)$ converges for each $z$ in $\mathscr{S}$, say to the function $F(z)$. It is easy to see that the convergence is uniform; observe that inequality (1) can be rewritten in terms of the partial sums as

$$\left| \sum_{j=0}^{m} f_j(z) - \sum_{j=0}^{n} f_j(z) \right| < \varepsilon, \qquad \text{for all } z \text{ in } \mathscr{S}, \text{ and } m > n > N.$$

Therefore, by Corollary 1,

$$\left| F(z) - \sum_{j=0}^{n} f_j(z) \right| \le \varepsilon, \qquad \text{for all } z \text{ in } \mathscr{S}, \text{ and } n > N.$$

This proves uniform convergence. ■

The comparison test can be regarded as a special case of the M-test wherein each $f_j(z)$ is a constant function.

Now we are ready to analyze Theorem 7 of the previous section, specifying the convergence properties of power series. For convenience, we restate the theorem here.

---

**Theorem 7.**   For any power series $\sum_{j=0}^{\infty} a_j(z - z_0)^j$ there is a real number $R$ between 0 and $\infty$, inclusive, which depends only on the coefficients $\{a_j\}$ such that

   (i)  the series converges for $|z - z_0| < R$,
  (ii)  the series converges uniformly in any closed subdisk $|z - z_0| \le R' < R$,
 (iii)  the series diverges for $|z - z_0| > R$.

---

To specify this number $R$, we must introduce the concept of the *limit superior* of an infinite sequence of real numbers; it generalizes the notion of limit. For motivation, let us first consider a *convergent* sequence of real numbers $\{x_n\}$, with limit $x$. Then for any $\varepsilon > 0$ there is an $N$ such that all the elements $x_n$ for $n > N$ will lie within $\varepsilon$ of $x$. So, in particular, $x$ has the following property: Given any $\varepsilon > 0$, for only a finite number of values of $n$ does $x_n$ exceed $x + \varepsilon$. Moreover, no number less than $x$ has this property. We now extend this notion to arbitrary sequences.

---

**Definition 6.**   The **limit superior** of a sequence of real numbers $\{x_n\}_{n=1}^{\infty}$, abbreviated lim sup $x_n$, is defined to be the smallest real number $l$ with the property that for any $\varepsilon > 0$ there are only a finite number of values of $n$ such that $x_n$ exceeds $l + \varepsilon$; if there are no such numbers with this property, we set lim sup $x_n := \infty$; if all real numbers have this property, we set lim sup $x_n := -\infty$.

---

As we indicated before, if $\{x_n\}$ converges to $x$, then lim sup $x_n = x$. Other examples are $\limsup(-1)^n = 1$, $\limsup(n) = \infty$, and $\limsup(-n) = -\infty$. A less trivial example is given in Lemma 4.

---

**Lemma 4.**   Lim sup $\sqrt[n]{n!} = \infty$.

---

*Proof of Lemma 4.* Let $v$ be any fixed positive integer. Then if $n$ is greater than $2v$, say $n = 2v + \lambda$ with $\lambda > 0$, we have

$$\frac{n!}{v^n} = \frac{(2v)!}{v^{2v}} \frac{(2v+1)}{v} \frac{(2v+2)}{v} \cdots \frac{(2v+\lambda)}{v} > \frac{(2v)!}{v^{2v}} 2^\lambda.$$

Thus if we choose $\lambda$, and hence $n$, large enough, the ratio $n!/v^n$ will exceed 1; that is $n!$ will exceed $v^n$. Therefore, $\sqrt[n]{n!}$ will eventually exceed *any number* $v$. ∎

The number $R$ in Theorem 7 is now specified as follows. From the set of coefficients $\{a_j\}$ we form the sequence $\sqrt[n]{|a_n|}$. Then $R$ is equal to the reciprocal of the limit superior of this sequence,

$$\boxed{R = \frac{1}{\limsup \sqrt[n]{|a_n|}},} \tag{2}$$

with the usual conventions $1/0 = \infty$, $1/\infty = 0$. Equation (2) is known as the *Cauchy-Hadamard formula*. Starting with this formula, let's address Theorem 7.

*Proof of Theorem 7.* First we consider the convergence statements (i) and (ii). If $R = 0$, there is nothing to prove. When $R > 0$, the convergence for $|z - z_0| < R$ follows from the uniform convergence in all closed subdisks $|z - z_0| \leq R' < R$, so we attack the latter problem.

Choose a number $k$ in the interval

$$\frac{1}{R} < k < \frac{1}{R'}.$$

Then because of Eq. (2), all but a finite number of the $a_j$ will satisfy $\sqrt[j]{|a_j|} < k$. Consequently, if $z$ lies in the closed subdisk $|z - z_0| \leq R'$, we have the inequality

$$|a_j(z - z_0)^j| = (\sqrt[j]{|a_j|}\, |z - z_0|)^j < (kR')^j, \tag{3}$$

valid for $j$ sufficiently large. But inequality (3) tells us that the M-test (Theorem 13) is satisfied when we compare $\sum_{j=0}^\infty a_j(z - z_0)^j$ to the series $\sum_{j=0}^\infty (kR')^j$, which converges since $kR' < 1$. Accordingly, $\sum_{j=0}^\infty a_j(z - z_0)^j$ is uniformly convergent in the closed subdisk and statement (ii) is proved.

To prove divergence when $|z - z_0| > R$, we choose $k$ in the interval

$$\frac{1}{|z - z_0|} < k < \frac{1}{R}.$$

Then it follows from the definition of lim sup and Eq. (2) that there must be an infinite number of $a_j$ satisfying $\sqrt[j]{|a_j|} > k$ (remember that the lim sup is the *smallest* number such that so-and-so). For such $a_j$

$$|a_j(z - z_0)^j| = (\sqrt[j]{|a_j|}\, |z - z_0|)^j > (k|z - z_0|)^j > 1;$$

that is, an infinite number of the terms of $\sum_{j=0}^\infty a_j(z - z_0)^j$ exceed 1 in modulus. This is clearly incompatible with the Cauchy criterion, so the series must diverge. ∎

**Example 1**

Show that the function

$$f(z) = \sum_{j=0}^{\infty} \frac{(-1)^j z^{2j}}{2^{2j}(j!)^2} \qquad (4)$$

is entire.

**Solution.**  Our goal is to prove that $R = \infty$. Keeping in mind that $a_j$ is the coefficient of $z^j$, we have

$$a_j = \begin{cases} 0 & \text{if } j \text{ is odd,} \\ \dfrac{(-1)^{j/2}}{2^j[(j/2)!]^2} & \text{if } j \text{ is even.} \end{cases}$$

Obviously $\sqrt[j]{|a_j|} = 0$ for odd $j$. For even $j$,

$$\sqrt[j]{|a_j|} = \frac{1}{2[(j/2)!]^{2/j}},$$

and the analysis of Lemma 4 shows that this goes to zero. Hence $\limsup \sqrt[j]{|a_j|} = 0$, and $R = \infty$. Consequently $f(z)$ is an entire function.  ∎

The function in (4) is customarily denoted $J_0(z)$, the *Bessel function of order zero*. The reader should verify (Prob. 7) that $J_0(z)$ satisfies *Bessel's equation of order zero*:

$$\frac{d^2f}{dz^2} + \frac{1}{z}\frac{df}{dz} + f = 0.$$

# EXERCISES 5.4

1. Find the limit superior of each of the following sequences $\{x_n\}_{n=1}^{\infty}$.

   (a) $x_n = (-1)^n\left(\dfrac{2n}{n+1}\right)$                    (b) $x_n = (-1)^n n$

   (c) $x_n = \dfrac{1}{n^2}$                    (d) $x_n = n \sin\left(\dfrac{n\pi}{2}\right)$

2. Prove that for any sequence $\{x_n\}$ of positive real numbers

$$\limsup \sqrt[n]{x_n} \le \limsup \frac{x_{n+1}}{x_n}.$$

3. Find the radius of convergence of each of the following power series.

   (a) $\displaystyle\sum_{j=1}^{\infty} \frac{2^j}{3^j + 4^j} z^j$                    (b) $\displaystyle\sum_{j=0}^{\infty} 2^j z^{j^2}$                    (c) $\displaystyle\sum_{j=0}^{\infty} [2 + (-1)^j]^j z^j$

   (d) $\displaystyle\sum_{j=1}^{\infty} \frac{j!}{j^j} z^j$                    (e) $\displaystyle\sum_{j=1}^{\infty} \frac{2}{3j} z^{2j}$                    (f) $\displaystyle\sum_{j=0}^{\infty} z^{j!}$

4. By considering the series $\sum_{j=1}^{\infty} z^j/j^2$, $\sum_{j=1}^{\infty} z^j/j$, and $\sum_{j=0}^{\infty} z^j$, show that a power series may converge on all, some, or none of the points on its circle of convergence.

5. If the radius of convergence for the series $\sum_{j=0}^{\infty} a_j z^j$ is $R$, find the radius of convergence for the following.

(a) $\sum_{j=0}^{\infty} j^3 a_j z^j$       (b) $\sum_{j=0}^{\infty} a_j^4 z^j$       (c) $\sum_{j=0}^{\infty} a_j z^{2j}$

(d) $\sum_{j=0}^{\infty} a_j z^{j+7}$       (e) $\sum_{j=1}^{\infty} j^{-j} a_j z^j$

6. Prove that if the radius of convergence for the series $\sum_{j=0}^{\infty} a_j z^j$ is $R$, then the radius of convergence for the series $\sum_{j=0}^{\infty} \text{Re}(a_j) z^j$ is greater than or equal to $R$.

7. Show that the Bessel function $J_0(z)$ satisfies Bessel's equation of order zero (as claimed after Example 1).

8. Bessel's equation of order $n$ is

$$\frac{d^2 f(z)}{dz^2} + \frac{1}{z} \frac{df(z)}{dz} + \left(1 - \frac{n^2}{z^2}\right) f(z) = 0.$$

Show that, for integers $n > 0$, the *Bessel function of order n*

$$J_n(z) = \sum_{j=0}^{\infty} \frac{(-1)^j}{j!(n+j)!} \cdot \left(\frac{z}{2}\right)^{2j+n}$$

is entire and satisfies the Bessel equation. (Bessel functions arise in the study of two-dimensional wave propagation in regions with cylindrical symmetry.)

9. Use power series to solve the *functional equation*

$$f(z) = z + f(z^2)$$

for the analytic function $f$.

10. The *Fibonacci sequence* 1, 1, 2, 3, 5, 8, 13, ... arises with surprising frequency in natural phenomena. The defining relations for the terms are

$$a_0 = a_1 = 1,$$
$$a_n = a_{n-1} + a_{n-2} \quad \text{(for } n \geq 2\text{)}.$$

Show that

$$f(z) := a_0 + a_1 z + a_2 z^2 + \cdots$$

defines an analytic function satisfying the equation

$$f(z) = 1 + zf(z) + z^2 f(z).$$

Solve for $f(z)$ and compute the Maclaurin series to derive the expression

$$a_j = \frac{1}{\sqrt{5}} \left[ \left(\frac{1+\sqrt{5}}{2}\right)^{j+1} - \left(\frac{1-\sqrt{5}}{2}\right)^{j+1} \right].$$

11. The *Legendre polynomials* $P_j(\zeta)$ are the coefficients of $z^j$ in the Maclaurin series for

$$(1 - 2\zeta z + z^2)^{-1/2} = \sum_{j=0}^{\infty} P_j(\zeta) z^j$$

(regarding $\zeta$ as a parameter). Show that $P_j(\zeta)$ is a polynomial in $\zeta$ of degree $j$, and compute $P_0$, $P_1$, $P_2$, and $P_3$. (These polynomials arise in three-dimensional potential theory.)

**12.** The *Riemann zeta function* has important applications to number theory. It is defined by

$$\zeta(z) := \sum_{j=1}^{\infty} \frac{1}{j^z} \qquad (\mathrm{Re}\, z > 1)$$

where $j^z := \exp(z \,\mathrm{Log}\, j)$. Prove that $\zeta(z)$ is analytic for $\mathrm{Re}\, z > 1$.

[HINT: Let $\mathrm{Re}\, z \geq \lambda > 1$. and show that $|1/j^z| \leq j^{-\lambda}$. Then use the Weierstrass M-test.]
(One of the most famous problems whose solution still eludes mathematicians is the *Riemann hypothesis*. It asserts that all the nonreal zeros of the analytic continuation (cf. Sec. 5.8) of $\zeta(z)$ lie on the vertical line $\mathrm{Re}\, z = \frac{1}{2}$.)

**13.** (*Abel's Limit Theorem*) Let $\sum_{j=0}^{\infty} a_j z^j$ be the power series expansion for a function $f$ analytic in $|z| < 1$ (so that $f(z) = \sum_{j=0}^{\infty} a_j z^j$ for $|z| < 1$). Suppose that $\lim_{r \to 1^-} f(r)$ (with $r$ real) exists and equals $A$. Prove that if $\sum_{j=0}^{\infty} a_j$ converges, then

$$A = \sum_{j=0}^{\infty} a_j.$$

[HINT: Set $M_n := \max_{j \geq n} \left| \sum_{k=j}^{\infty} a_k \right|$. Then $M_n \to 0$ as $n \to \infty$ and for $0 \leq r < 1$

$$\left| f(r) - \sum_{j=0}^{n} a_j r^j \right| = \left| \sum_{j=n+1}^{\infty} a_j r^j \right| = \left| \sum_{j=n+1}^{\infty} \left( \sum_{k=j}^{\infty} a_k - \sum_{k=j+1}^{\infty} a_k \right) r^j \right|$$

$$= \left| \left( \sum_{k=n+1}^{\infty} a_k \right) r^{n+1} + \sum_{j=n+2}^{\infty} \left( \sum_{k=j}^{\infty} a_k \right) r^{j-1}(r-1) \right|$$

$$\leq M_{n+1} + M_{n+2}.$$

Now let $r \to 1^-$ and then $n \to \infty$.]

**14.** Use Abel's limit theorem (Prob. 13) to prove that

$$\mathrm{Log}\, 2 = 1 - \frac{1}{2} + \frac{1}{3} - \frac{1}{4} + \cdots.$$

## 5.5 LAURENT SERIES

We now wish to investigate the possibility of a series representation of a function $f$ near a *singularity*, i.e., a point $z_0$ where $f$ is not analytic but which is the limit of points where $f$ is analytic. After all, if (for example) the occurrence of a singularity is merely due to a vanishing denominator, might it not be possible to express the function as something like $A/(z - z_0)^p + g(z)$, where $g$ is analytic and has a Taylor series around $z_0$? To be sure, not all singularities are of this type (recall $\mathrm{Log}\, z$ at $z_0 = 0$). However, if the function is analytic in an annulus surrounding one or more of its singularities (note that $\mathrm{Log}\, z$ does not have this property), we can display its "singular part" according to the following theorem.

**Theorem 14.** Let $f$ be analytic in an open annulus $r < |z - z_0| < R$. Then $f$ can be expressed there as the sum of two series

$$f(z) = \sum_{j=0}^{\infty} a_j(z - z_0)^j + \sum_{j=1}^{\infty} a_{-j}(z - z_0)^{-j},$$

both series converging in the annulus, and converging uniformly in any closed subannulus $r < \rho_1 \le |z - z_0| \le \rho_2 < R$. The coefficients $a_j$ are given by

$$a_j = \frac{1}{2\pi i} \oint_C \frac{f(\zeta)}{(\zeta - z_0)^{j+1}} \, d\zeta \qquad (j = 0, \pm 1, \pm 2, \dots), \tag{1}$$

where $C$ is any positively oriented simple closed contour lying in the annulus and containing $z_0$ in its interior.

Such an expansion, containing negative as well as positive powers of $(z - z_0)$, is called the *Laurent series* for $f$ in this annulus. It is usually abbreviated

$$\sum_{j=-\infty}^{\infty} a_j(z - z_0)^j.$$

Notice that if $f$ is analytic throughout the *disk* $|z - z_0| < R$, the coefficient in (1) with negative subscripts are zero, and the others reproduce the Taylor series for $f$.

*Proof of Theorem 14.*    It suffices to prove uniform convergence in every closed subannulus, for this implies (pointwise) convergence in the open annulus.

First we show that for any $z$ satisfying $r < \rho_1 \le |z - z_0| \le \rho_2 < R$ we have the representation

$$f(z) = \frac{1}{2\pi i} \oint_{C_1} \frac{f(\zeta)}{\zeta - z} \, d\zeta + \frac{1}{2\pi i} \oint_{C_2} \frac{f(\zeta)}{\zeta - z} \, d\zeta, \tag{2}$$

where $C_1$ is the negatively oriented circle around $z_0$ of radius $R_1 = (r + \rho_1)/2$, and $C_2$ is the positively oriented circle around $z_0$ of radius $R_2 = (R + \rho_2)/2$; see Fig. 5.4. Indeed, Eq. (2) is just a slight variation of the Cauchy integral formula in this case, as the following argument shows. Consider the contour $\Gamma$ of Fig 5.5(a); it is simple, closed, positively oriented, and contains $z$ in its interior. Therefore,

$$f(z) = \frac{1}{2\pi i} \int_\Gamma \frac{f(\zeta)}{\zeta - z} \, d\zeta. \tag{3}$$

Let's think of $\Gamma$ as a doughnut with a bite taken out of it, and let $\Gamma'$ denote the "bite," as in Fig. 5.5(b). Observe that

$$\frac{1}{2\pi i} \int_{\Gamma'} \frac{f(\zeta)}{\zeta - z} \, d\zeta = 0$$

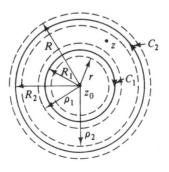

**Figure 5.4**    Circles of integration for Eq. (2).

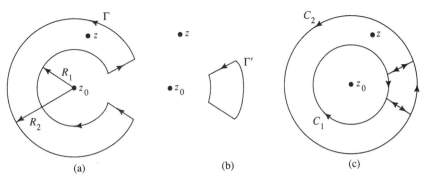

**Figure 5.5**  The hole equals the sum of its parts.

because the integrand is analytic inside and on $\Gamma'$. Consequently we can put the "bite" back into the doughnut and modify Eq. (3) to read

$$f(z) = \frac{1}{2\pi i} \int_{\Gamma+\Gamma'} \frac{f(\zeta)}{\zeta - z} \, d\zeta,$$

where $\Gamma + \Gamma'$ is the path indicated in Fig. 5.5(c). But the integrals along the line segments in Fig. 5.5(c) cancel and, keeping track of the orientation, we arrive at Eq. (2).[†] Now we are ready to proceed with the derivation of the Laurent expansion.

Since $z$ lies inside $C_2$, the integral over $C_2$ appearing in Eq. (2) is exactly like the integral that arose in the Taylor series theorem (Theorem 3); we treat it the same way, and find

$$\frac{1}{2\pi i} \oint_{C_2} \frac{f(\zeta)}{\zeta - z} \, d\zeta = \sum_{j=0}^{n} a_j (z - z_0)^j + T_n(z),$$

where $T_n(z) \to 0$ uniformly as $n \to \infty$ for $|z - z_0| \leq \rho_2$ and $a_j$ is given by

$$a_j = \frac{1}{2\pi i} \oint_{C_2} \frac{f(\zeta)}{(\zeta - z_0)^{j+1}} \, d\zeta \qquad (j = 0, 1, 2, \ldots). \tag{4}$$

Hence

$$\frac{1}{2\pi i} \oint_{C_2} \frac{f(\zeta)}{\zeta - z} \, d\zeta = \sum_{j=0}^{\infty} a_j (z - z_0)^j \qquad (|z - z_0| \leq \rho_2).$$

We now turn to the integral around $C_1$ in Eq. (2). Since $z$ lies outside $C_1$, we seek an expression for $1/(\zeta - z)$ in powers of $(\zeta - z_0)/(z - z_0)$; accordingly, we write

$$\frac{1}{\zeta - z} = \frac{1}{(\zeta - z_0) - (z - z_0)} = -\frac{1}{(z - z_0)} \frac{1}{1 - \dfrac{\zeta - z_0}{z - z_0}}$$

$$= -\frac{1}{z - z_0} \left[ 1 + \frac{\zeta - z_0}{z - z_0} + \frac{(\zeta - z_0)^2}{(z - z_0)^2} + \cdots + \frac{(\zeta - z_0)^m}{(z - z_0)^m} + \frac{\dfrac{(\zeta - z_0)^{m+1}}{(z - z_0)^{m+1}}}{1 - \dfrac{\zeta - z_0}{z - z_0}} \right.$$

---

[†] Some readers may be able to use a deformation-of-contour argument to derive Eq. (2), but the doughnut analogy is probably easier to digest.

Inserting this into the integral, we find

$$\frac{1}{2\pi i} \oint_{C_1} \frac{f(\zeta)}{\zeta - z} \, d\zeta = \sum_{j=1}^{m+1} a_{-j}(z - z_0)^{-j} + \mathcal{T}_m(z),$$

where

$$a_{-j} = -\frac{1}{2\pi i} \oint_{C_1} \frac{f(\zeta)}{(\zeta - z_0)^{-j+1}} \, d\zeta \qquad (j = 1, 2, 3, \dots) \tag{5}$$

(observe the exponent with care) and

$$\mathcal{T}_m(z) = \frac{1}{2\pi i} \oint_{C_1} \frac{f(\zeta)}{(\zeta - z)} \frac{(\zeta - z_0)^{m+1}}{(z - z_0)^{m+1}} \, d\zeta.$$

Now for $\zeta$ on $C_1$ we have $|\zeta - z| \geq \rho_1 - R_1$, $|\zeta - z_0| = R_1$, and $|z - z_0| \geq \rho_1$ (see Fig. 5.4). Thus

$$|\mathcal{T}_m(z)| \leq \frac{1}{2\pi} \cdot \max_{\zeta \, \text{on} \, C_1} |f(\zeta)| \frac{1}{\rho_1 - R_1} \left( \frac{R_1}{\rho_1} \right)^{m+1} 2\pi R_1.$$

Since $R_1/\rho_1 < 1$, $\mathcal{T}_m(z) \to 0$ uniformly for $|z - z_0| \geq \rho_1$, and so

$$\frac{1}{2\pi i} \oint_{C_1} \frac{f(\zeta)}{\zeta - z} \, d\zeta = \sum_{j=1}^{\infty} a_{-j}(z - z_0)^{-j} \qquad (|z - z_0| \geq \rho_1).$$

We have thus expressed both integrals in Eq. (2) as uniformly convergent series of the form mentioned in the theorem with the common region of convergence $\rho_1 \leq |z - z_0| \leq \rho_2$; we still have to verify formula (1) for the coefficients. If $j$ is non-negative, formula (4) applies, but the analysis of Chapter 4 justifies replacing the integral over $C_2$ by the integral over the contour $C$ mentioned in the theorem, since the intervening region contains no singularities of $f(\zeta)/(\zeta - z_0)^{j+1}$; hence Eq. (1) holds for $j \geq 0$. Similarly, the integral over $C_1$ in formula (5) can be changed into an integral over $C$, incorporating the minus sign to account for the change in orientation. Hence (1) is verified for every $j$. ∎

Replacing $(z - z_0)$ with $1/(z - z_0)$ in Theorem 7 (page 197), one easily sees that any formal series of the form $\sum_{j=1}^{\infty} c_{-j}(z - z_0)^{-j}$ will converge *outside* some "circle of convergence" $|z - z_0| = r$ whose radius depends on the coefficients, with uniform convergence holding in each region $|z - z_0| \geq r' > r$. Thus termwise integration is justified by Theorem 8, and proceeding in a manner analogous to that of Sec. 5.3 we can prove the following.

**Theorem 15.** Let $\sum_{j=0}^{\infty} c_j(z - z_0)^j$ and $\sum_{j=1}^{\infty} c_{-j}(z - z_0)^{-j}$ be any two series with the following properties:

(i) $\sum_{j=0}^{\infty} c_j(z - z_0)^j$ converges for $|z - z_0| < R$,

(ii) $\sum_{j=1}^{\infty} c_{-j}(z - z_0)^{-j}$ converges for $|z - z_0| > r$, and

(iii) $r < R$.

Then there is a function $f(z)$, analytic for $r < |z - z_0| < R$, whose Laurent series in this annulus is given by $\sum_{j=-\infty}^{\infty} c_j(z - z_0)^j$.

The proof is left to the exercises (Prob. 8).

This theorem, like Theorem 7, justifies the use of other methods of finding Laurent series since it implies that *any* convergent series of the form $\sum_{j=-\infty}^{\infty} c_j(z - z_0)^j$, however obtained, must be the Laurent series of its sum function. In Examples 1 and 2 we shall derive Laurent expansions by making judicious use of the fact that the geometric series $\sum_{j=0}^{\infty} w^j$ converges to $(1 - w)^{-1}$ when $|w| < 1$.

**Example 1**

Find the Laurent series for the function $(z^2 - 2z + 3)/(z - 2)$ in the region $|z - 1| > 1$.

**Solution.** Notice that this region is centered at $z_0 = 1$ and excludes the singularity at $z = 2$. First we manipulate $1/(z - 2)$ so that we can apply the geometric series result in the specified region:

$$\frac{1}{z - 2} = \frac{1}{(z - 1) - 1} = \frac{1}{z - 1} \cdot \frac{1}{1 - \dfrac{1}{z - 1}}.$$

Thus for $|1/(z - 1)| < 1$,

$$\frac{1}{z - 2} = \frac{1}{z - 1} \cdot \sum_{j=0}^{\infty} \frac{1}{(z - 1)^j}$$

$$= \frac{1}{z - 1} + \frac{1}{(z - 1)^2} + \frac{1}{(z - 1)^3} + \cdots.$$

Expressing the numerator $z^2 - 2z + 3$ in powers of $z - 1$, we find

$$z^2 - 2z + 3 = (z - 1)^2 + 0 \cdot (z - 1) + 2 = (z - 1)^2 + 2.$$

Therefore,

$$\frac{z^2 - 2z + 3}{z - 2} = [(z - 1)^2 + 2] \cdot \left[ \frac{1}{z - 1} + \frac{1}{(z - 1)^2} + \frac{1}{(z - 1)^3} + \cdots \right]$$

$$= \left[ (z - 1) + 1 + \frac{1}{(z - 1)} + \frac{1}{(z - 1)^2} + \cdots \right]$$

$$+ \left[ \frac{2}{(z - 1)} + \frac{2}{(z - 1)^2} + \cdots \right]$$

$$= (z - 1) + 1 + \sum_{j=1}^{\infty} \frac{3}{(z - 1)^j}. \quad \blacksquare$$

**Example 2**

For the function

$$\frac{1}{(z - 1)(z - 2)},$$

find the Laurent series expansion in

(a) the region $|z| < 1$,
(b) the region $1 < |z| < 2$,
(c) the region $|z| > 2$.

**Solution.**  Using partial fractions, write

$$\frac{1}{(z - 1)(z - 2)} = \frac{1}{z - 2} - \frac{1}{z - 1}.$$

Now we proceed differently in each region in order to derive convergent series.

(a) For $|z| < 1$,

$$\frac{1}{z - 2} = -\frac{1}{2}\frac{1}{1 - \dfrac{z}{2}} = -\frac{1}{2}\sum_{j=0}^{\infty}\left(\frac{z}{2}\right)^j = -\sum_{j=0}^{\infty}\frac{z^j}{2^{j+1}}, \tag{6}$$

and

$$\frac{1}{z - 1} = -\frac{1}{1 - z} = -\sum_{j=0}^{\infty}z^j. \tag{7}$$

Subtracting Eq. (7) from Eq. (6) gives

$$\frac{1}{(z - 1)(z - 2)} = \sum_{j=0}^{\infty}\left(-\frac{1}{2^{j+1}} + 1\right)z^j = \frac{1}{2} + \frac{3}{4}z + \frac{7}{8}z^2 + \cdots.$$

(b) For $1 < |z| < 2$, Eq. (6) is still valid, but we have

$$\frac{1}{z - 1} = \frac{1}{z}\frac{1}{1 - \dfrac{1}{z}} = \frac{1}{z}\sum_{j=0}^{\infty}\frac{1}{z^j} = \sum_{j=0}^{\infty}\frac{1}{z^{j+1}}. \tag{8}$$

Thus

$$\frac{1}{(z - 1)(z - 2)} = -\sum_{j=0}^{\infty}\frac{z^j}{2^{j+1}} - \sum_{j=0}^{\infty}\frac{1}{z^{j+1}} = \cdots - \frac{1}{z^2} - \frac{1}{z} - \frac{1}{2} - \frac{z}{4} - \cdots.$$

(c) For $|z| > 2$, Eq. (8) is valid and

$$\frac{1}{z - 2} = \frac{1}{z}\frac{1}{1 - \dfrac{2}{z}} = \frac{1}{z}\sum_{j=0}^{\infty}\left(\frac{2}{z}\right)^j = \sum_{j=0}^{\infty}\frac{2^j}{z^{j+1}}.$$

Hence

$$\frac{1}{(z-1)(z-2)} = \sum_{j=0}^{\infty} \frac{2^j - 1}{z^{j+1}} = \frac{1}{z^2} + \frac{3}{z^3} + \frac{7}{z^4} + \cdots . \quad \blacksquare$$

## Example 3

Expand $e^{1/z}$ in a Laurent series around $z = 0$.

**Solution.**  As we already know,

$$e^w = 1 + w + \frac{w^2}{2!} + \frac{w^3}{3!} + \cdots$$

for all (finite) $w$. Thus if $z \neq 0$, we let $w = 1/z$ and find

$$e^{1/z} = 1 + \frac{1}{z} + \frac{1}{2!z^2} + \frac{1}{3!z^3} + \cdots . \quad \blacksquare \tag{9}$$

## Example 4

What is the Laurent expansion around $z = 0$ for the function

$$f(z) = \begin{cases} \sin z & \text{if } z \neq 0, \\ 5 & \text{if } z = 0? \end{cases}$$

**Solution.**  The alert reader, upon seeing this example, will undoubtedly accuse the authors of sophistry—and with good reason! The function $f$ is simply defined "incorrectly" at $z = 0$ in order to make a point. We ask, however, that the audience bear with us, because this example will be useful in the next section.

Observe that $f$ satisfies the hypothesis of Theorem 14, so it does have a Laurent series, valid for $|z| > 0$. But for such $z$ we have [recall Eq. (10), Sec. 5.2]

$$f(z) = \sin z = z - \frac{z^3}{3!} + \frac{z^5}{5!} - \cdots \qquad (|z| > 0). \tag{10}$$

Therefore, Eq. (10) must be the Laurent expansion for $f$. $\blacksquare$

## EXERCISES 5.5

1. Find the Laurent series for the function $1/(z + z^2)$ in each of the following domains.
   (a) $0 < |z| < 1$                          (b) $1 < |z|$
   (c) $0 < |z + 1| < 1$                     (d) $1 < |z + 1|$

2. Does the principal branch $\sqrt{z}$ have a Laurent series expansion in the domain $\mathbb{C}\backslash\{0\}$?

3. Find the Laurent series for the function $\dfrac{z}{(z + 1)(z - 2)}$ in each of the following domains.

   (a) $|z| < 1$                (b) $1 < |z| < 2$                (c) $2 < |z|$

4. Find the Laurent series for $(\sin 2z)/z^3$ in $|z| > 0$.

5. Find the Laurent series for $\dfrac{(z + 1)}{z(z - 4)^3}$ in $0 < |z - 4| < 4$.

6. Find the Laurent series for $z^2 \cos\left(\dfrac{1}{3z}\right)$ in $|z| > 0$.

7. Obtain the first few terms of the Laurent series for each of the following functions in the specified domains.

(a) $\dfrac{e^{1/z}}{z^2 - 1}$    for $|z| > 1$        (b) $\dfrac{1}{e^z - 1}$    for $0 < |z| < 2\pi$

(c) $\csc z$    for $0 < |z| < \pi$        (d) $1/e^{(1 - z)}$    for $1 < |z|$

8. Give a proof of Theorem 15.

9. Determine the annulus of convergence of the Laurent series

$$\sum_{j = -\infty}^{\infty} \frac{z^j}{2^{|j|}}.$$

10. Prove that the Laurent series expansion of the function

$$f(z) = \exp\left[\frac{\lambda}{2}\left(z - \frac{1}{z}\right)\right]$$

in $|z| > 0$ is given by

$$\sum_{k = -\infty}^{\infty} J_k(\lambda) z^k,$$

where

$$J_k(\lambda) = (-1)^k J_{-k}(\lambda) := \frac{1}{2\pi}\int_0^{2\pi} \cos(k\theta - \lambda \sin \theta)\, d\theta.$$

The $J_k(\lambda)$ are known as *Bessel functions* of the first kind. [HINT: Use the integral formula (1) with $C: |z| = 1$.]

11. Obtain a general formula for the Laurent expansion of

$$f_n(z) = \frac{1}{(z - \alpha)^n} \qquad (n = 1, 2, \ldots)$$

that is valid for $|z| > |\alpha|$.

12. Prove that if $f(z)$ has a Laurent series expansion of the form $\sum_{j=0}^{\infty} a_j z^j$ in $0 < |z| < \rho$, then $\lim_{z \to 0} f(z)$ exists.

13. Let $f(z)$ be analytic in the annulus $r < |z - z_0| < R$ and bounded by $M$ there. Prove that the coefficients $a_j$ of the Laurent expansion of $f(z)$ in the annulus satisfy

$$|a_j| \le \frac{M}{R^j}, \qquad |a_{-j}| \le Mr^j \qquad (\text{for } j = 0, 1, 2, \ldots).$$

## 5.6 ZEROS AND SINGULARITIES

In this section we shall use the Laurent expansion to classify, in general terms, the behavior of an analytic function near its zeros and isolated singularities. A *zero* of a function $f$ is a point $z_0$ where $f$ is analytic and $f(z_0) = 0$. An *isolated singularity* of $f$ is a point $z_0$ such that $f$ is analytic in some punctured disk $0 < |z - z_0| < R$ but not analytic at $z_0$ itself. For example, $\tan(\pi z/2)$ has a zero at each even integer and an isolated singularity at each odd integer.

We shall begin by examining the zeros of $f$.

---

**Definition 7.**    A point $z_0$ is called a **zero of order** $m$ for the function $f$ if $f$ is analytic at $z_0$ and $f$ and its first $m - 1$ derivatives vanish at $z_0$, but $f^{(m)}(z_0) \neq 0$.

---

In other words, we have

$$f(z_0) = f'(z_0) = f''(z_0) = \cdots = f^{(m-1)}(z_0) = 0 \neq f^{(m)}(z_0).$$

In this case the Taylor series for $f$ around $z_0$ takes the form

$$f(z) = a_m(z - z_0)^m + a_{m+1}(z - z_0)^{m+1} + a_{m+2}(z - z_0)^{m+2} + \cdots$$

or

$$f(z) = (z - z_0)^m[a_m + a_{m+1}(z - z_0) + a_{m+2}(z - z_0)^2 + \cdots], \tag{1}$$

where $a_m = f^{(m)}(z_0)/m! \neq 0$. The bracketed series in Eq. (1) clearly converges wherever the series for $f$ does (at any particular point one is just a multiple of the other); hence it defines a function $g(z)$ analytic in a neighborhood of $z_0$, with $g(z_0) \neq 0$. Conversely, any function with a representation like Eq. (1) must have a zero of order $m$, so we deduce the following.

---

**Theorem 16.**    Let $f$ be analytic at $z_0$. Then $f$ has a zero of order $m$ at $z_0$ if and only if $f$ can be written as

$$f(z) = (z - z_0)^m g(z),$$

where $g$ is analytic at $z_0$ and $g(z_0) \neq 0$.

---

A zero of order 1 is sometimes called a *simple zero*. For instance, the zeros of the function $\sin z$, which occur (as we saw in Chapter 3) at integer multiples of $\pi$, are all simple (at such points the first derivative, $\cos z$, is nonzero).

An easy consequence of Theorem 16 is the following result, which asserts that zeros of nonconstant analytic functions are isolated.

---

**Corollary 3.**    If $f$ is an analytic function such that $f(z_0) = 0$, then either $f$ is identically zero in a neighborhood of $z_0$ or there is a punctured disk about $z_0$ in which $f$ has no zeros.

---

*Proof.*    Let $\sum_{k=0}^{\infty} a_k(z - z_0)^k$ be the Taylor series for $f$ about $z_0$ (so that $a_k = f^{(k)}(z_0)/k!$). Then, as we know from Theorem 3, page 188, this series converges to $f(z)$ in some neighborhood of $z_0$. So if all the Taylor coefficients $a_k$ are zero, then $f(z)$ must be identically zero in this neighborhood. Otherwise, let $m (\geq 1)$ be the smallest subscript such that $a_m \neq 0$. Then, by Definition 7, $f$ has a zero of order $m$ at $z_0$, and so the representation $f(z) = (z - z_0)^m g(z)$ of Theorem 16 is valid. Since $g(z_0) \neq 0$ and $g$ is continuous at $z_0$ (indeed, it is analytic there), there exists a disk $|z - z_0| < \delta$ throughout which $g$ is nonzero. Consequently, $f(z) \neq 0$ for $0 < |z - z_0| < \delta$. ∎

Notice that if $f$ is nonconstant, analytic, and zero at $z_0$, the order of the zero must be a whole number; the condition of analyticity at $z_0$ in the analysis of Theorem 16 dictates that $m$ be an integer. The function $z^{1/2}$ could be said to have a zero of order $\frac{1}{2}$ at $z = 0$, but of course it is not analytic there.

We now turn to the isolated singularities of $f$. We know that $f$ has a Laurent expansion around any isolated singularity $z_0$;

$$f(z) = \sum_{j=-\infty}^{\infty} a_j (z - z_0)^j, \tag{2}$$

for, say, $0 < |z - z_0| < R$. (The "$r$" of Theorem 14 is zero for an isolated singularity.) We can classify $z_0$ into one of the following three categories.

---

**Definition 8.**  Let $f$ have an isolated singularity at $z_0$, and let (2) be the Laurent expansion of $f$ in $0 < |z - z_0| < R$. Then

(i) If $a_j = 0$ for all $j < 0$, we say that $z_0$ is a **removable singularity** of $f$;

(ii) If $a_{-m} \neq 0$ for some positive integer $m$ but $a_j = 0$ for all $j < -m$, we say that $z_0$ is a **pole of order** $m$ for $f$;

(iii) If $a_j \neq 0$ for an infinite number of negative values of $j$, we say that $z_0$ is an **essential singularity** of $f$.

---

By examining separately each of these three types of isolated singularities we shall show that they can be distinguished by the qualitative behavior of $f(z)$ near the singularity (i.e., without working out the Laurent expansion). The resulting characterizations are summarized in the final theorem of this section.

When $f$ has a removable singularity at $z_0$, its Laurent series takes the form

$$f(z) = a_0 + a_1(z - z_0) + a_2(z - z_0)^2 + \cdots \qquad (0 < |z - z_0| < R). \tag{3}$$

Example 4 of the previous section provides an illustration of this. Other examples of functions having removable singularities are

$$\frac{\sin z}{z} = \frac{1}{z}\left(z - \frac{z^3}{3!} + \frac{z^5}{5!} - \cdots\right) = 1 - \frac{z^2}{3!} + \frac{z^4}{5!} - \cdots \qquad (z_0 = 0),$$

$$\frac{\cos z - 1}{z} = \frac{1}{z}\left[\left(1 - \frac{z^2}{2!} + \frac{z^4}{4!} - \cdots\right) - 1\right] = -\frac{z}{2!} + \frac{z^3}{4!} - \cdots \qquad (z_0 = 0),$$

$$\frac{z^2 - 1}{z - 1} = z + 1 = 2 + (z - 1) + 0 + 0 + \cdots \qquad (z_0 = 1).$$

From (3) we can see that, except for the point $z_0$ itself, $f(z)$ is equal to a function $h(z)$, which is analytic at $z_0$. In other words, the only reason for the singularity is that $f(z)$ is undefined or defined "peculiarly" at $z_0$. Since the function $h(z)$ is analytic at $z_0$, it is obviously bounded[†] in some neighborhood of $z_0$, and so we have established the following lemma.

---

**Lemma 5.**    If $f$ has a removable singularity at $z_0$, then

   (i) $f(z)$ is bounded in some punctured neighborhood of $z_0$,
   (ii) $f(z)$ has a (finite) limit as $z$ approaches $z_0$, and
   (iii) $f(z)$ can be redefined at $z_0$ so that the new function is analytic at $z_0$.

---

Conversely, if a function is bounded in some punctured neighborhood of an isolated singularity, that singularity is removable; see Prob. 13 for a direct proof.

Clearly, removable singularities are not too important in the theory of analytic functions. But as we shall see in Lemmas 6 and 8, the concept is occasionally helpful in providing compact descriptions of the other kinds of singularities.

The Laurent series for a function with a pole of order $m$ looks like

$$f(z) = \frac{a_{-m}}{(z-z_0)^m} + \frac{a_{-(m-1)}}{(z-z_0)^{m-1}} + \cdots + \frac{a_{-1}}{z-z_0}$$

$$+ a_0 + a_1(z-z_0) + a_2(z-z_0)^2 + \cdots \qquad (a_{-m} \neq 0), \qquad \textbf{(4)}$$

valid in some punctured neighborhood of $z_0$. For example,

$$\frac{e^z}{z^2} = \frac{1}{z^2}\left(1 + z + \frac{z^2}{2!} + \cdots\right) = \frac{1}{z^2} + \frac{1}{z} + \frac{1}{2!} + \frac{z}{3!} + \cdots$$

has a pole of order 2, and

$$\frac{\sin z}{z^5} = \frac{1}{z^5}\left(z - \frac{z^3}{3!} + \frac{z^5}{5!} - \cdots\right) = \frac{1}{z^4} - \frac{1}{3!z^2} + \frac{1}{5!} - \frac{z^2}{7!} + \cdots$$

has a pole of order 4, at $z = 0$.

A pole of order 1 is called a *simple pole*. Obviously $z = 0$ is a simple pole of the function $(\sin z)/z^2$.

From Eq. (4), we can deduce the following characterization of a pole:

---

**Lemma 6.**    If the function $f$ has a pole of order $m$ at $z_0$, then $|(z-z_0)^l f(z)| \to \infty$ as $z \to z_0$ for all integers $l < m$, while $(z-z_0)^m f(z)$ has a removable singularity at $z_0$. In particular, $|f(z)| \to \infty$ as $z$ approaches a pole.[‡]

---

[†] That is, there exists a neighborhood of $z_0$ and a constant $M$ such that $|h(z)| \leq M$ for all $z$ in this neighborhood.

[‡] We remind the reader that the notation "$|h(z)| \to \infty$ as $z \to z_0$" means that $|h(z)|$ exceeds any given number for all $z$ sufficiently near $z_0$.

*Proof of Lemma 6.* Equation (4) implies that in some punctured neighborhood of $z_0$ we have

$$(z - z_0)^m f(z) = a_{-m} + a_{-m+1}(z - z_0) + \cdots, \tag{5}$$

and since there are no negative powers, the singularity of $(z - z_0)^m f(z)$ at $z_0$ is removable. Furthermore, $(z - z_0)^m f(z) \to a_{-m} \neq 0$ as $z \to z_0$. Thus for any integer $l < m$,

$$\left| (z - z_0)^l f(z) \right| = \left| \frac{1}{(z - z_0)^{m-l}} (z - z_0)^m f(z) \right| \to \infty \quad \text{as} \quad z \to z_0, \tag{6}$$

because $(z - z_0)^{m-l} \to 0$ and $a_{-m} \neq 0$. ∎

---

**Lemma 7.** A function $f$ has a pole of order $m$ at $z_0$ if and only if in some punctured neighborhood of $z_0$

$$f(z) = \frac{g(z)}{(z - z_0)^m}, \tag{7}$$

where $g$ is analytic at $z_0$ and $g(z_0) \neq 0$.

---

*Proof.* If $f$ has a pole of order $m$ at $z_0$, then it follows from the representation (5) that in some punctured neighborhood of $z_0$ we have $(z - z_0)^m f(z) = g(z)$, where

$$g(z) := a_{-m} + a_{-m+1}(z - z_0) + \cdots.$$

Setting $g(z_0) := a_{-m} \neq 0$, we see that $g$ is analytic and nonzero at $z_0$, so (7) follows.

Now suppose that the representation (7) holds, and write the Taylor series for $g(z)$:

$$g(z) = b_0 + b_1(z - z_0) + b_2(z - z_0)^2 + \cdots.$$

Then the Laurent series for $f$ near $z_0$ must be

$$f(z) = \frac{g(z)}{(z - z_0)^m} = \frac{b_0}{(z - z_0)^m} + \frac{b_1}{(z - z_0)^{m-1}} + \cdots.$$

Since $b_0 = g(z_0) \neq 0$, the expansion displays the predicted pole for $f$. ∎

**Example 1**

Classify the singularity at $z = 1$ of the function $(\sin z)/(z^2 - 1)^2$.

**Solution.** Since

$$\frac{\sin z}{(z^2 - 1)^2} = \frac{(\sin z)/(z + 1)^2}{(z - 1)^2}$$

and the numerator is analytic and nonzero at $z = 1$, Lemma 7 implies that the function has a pole of order 2. ∎

**Example 2**

Show that the only singularities of rational functions are removable singularities or poles.

**Solution.** Recall that a rational function is the ratio of two polynomials, $P(z)/Q(z)$, and is analytic everywhere except at the zeros of $Q(z)$. If $Q(z)$ has a zero, say of order $m$, at $z_0$, then $Q(z) = (z - z_0)^m q(z)$, where $q(z)$ is a polynomial and $q(z_0) \neq 0$.

If $P(z_0) \neq 0$, we apply Lemma 7 to the expression

$$\frac{P(z)}{Q(z)} = \frac{1}{(z - z_0)^m} \frac{P(z)}{q(z)}$$

to deduce that $P(z)/Q(z)$ has a pole of order $m$. If, on the other hand, $P(z_0) = 0$, we can write $P(z) = (z - z_0)^n p(z)$, where $n$ is the order of the zero at $z_0$ [we ignore the trivial case $P(z) \equiv 0$]; thus

$$\frac{P(z)}{Q(z)} = \frac{(z - z_0)^n}{(z - z_0)^m} \frac{p(z)}{q(z)},$$

and clearly $P(z)/Q(z)$ will have a pole if $n < m$ or a removable singularity if $n \geq m$. ■

The following lemma relating zeros and poles is easily derived using the above methods of analysis, so we simply state it here for reference purposes and assign the proof to the reader (Prob. 4).

---

**Lemma 8.** If $f$ has a zero of order $m$ at $z_0$, then $1/f$ has a pole of order $m$ at $z_0$. Conversely, if $f$ has a pole of order $m$ at $z_0$, then $1/f$ has a removable singularity at $z_0$, and if we define $1/f(z_0) = 0$, then $1/f$ has a zero of order $m$ at $z_0$.

---

Some students may have felt that it is obvious that $|f(z)| \to \infty$ as $z$ approaches a pole and that our painstaking analysis was unnecessary. They will probably be shocked to learn that such behavior does not occur as $z$ approaches an essential singularity; instead we have the following.

---

**Theorem 17.** (*Picard's Theorem*) A function with an essential singularity assumes every complex number, with possibly one exception, as a value in any neighborhood of this singularity.

---

The proof of this theorem is beyond our text, but we invite the student to prove a somewhat weaker result, the *Casorati-Weierstrass theorem*, in Prob. 14. We illustrate the Picard theorem in the next example.

**Example 3**

Verify Picard's result for $e^{1/z}$ near $z = 0$.

**Solution.** (Observe first of all that $z = 0$ is an essential singularity; see Example 3, Sec. 5.5.) Obviously $e^{1/z}$ is never zero. However, if $c \neq 0$, we can show that $e^{1/z}$ achieves the value $c$ for $|z|$ less than any positive $\varepsilon$. To this end, recall that

$$\log c = \text{Log } |c| + i \text{ Arg } c + 2n\pi i \qquad (n = 0, \pm 1, \pm 2, \ldots)$$

(Sec. 3.2). By picking $n$ sufficiently large, we can find a value $w$ of $\log c$ such that $|w| > 1/\varepsilon$. Then let $z = 1/w$. We will have $|z| < \varepsilon$, and

$$e^{1/z} = e^{w} = e^{\log c} = c.$$

(To gain further insight into the exotic behavior of $e^{1/z}$ near its essential singularity, the reader is invited (Prob. 16) to sketch the curves $|e^{1/z}| = s$, where $s = 1, \frac{1}{2}, 2, \frac{1}{3}, 3, \ldots$.) ∎

From the preceding results we observe that the three different kinds of isolated singularities produce qualitatively different behaviors near these points. *Thus boundedness indicates a removable singularity, approaching $\infty$ indicates a pole, and anything else must indicate an essential singularity.* These characterizations are often useful in determining the nature of a singularity when it is inconvenient to find the Laurent expansion, as illustrated in the next example.

**Example 4**

Classify the zeros and singularities of the function $\sin(1 - z^{-1})$.

**Solution.** Since the zeros of $\sin w$ occur only when $w$ is an integer multiple of $\pi$, the function $\sin(1 - z^{-1})$ has zeros when

$$1 - z^{-1} = n\pi,$$

i.e., at

$$z = \frac{1}{1 - n\pi} \qquad (n = 0, \pm 1, \pm 2, \ldots).$$

Furthermore, the zeros are simple because the derivative at these points is

$$\frac{d}{dz} \sin(1 - z^{-1}) \bigg|_{z = (1 - n\pi)^{-1}} = \frac{1}{z^2} \cos(1 - z^{-1}) \bigg|_{z = (1 - n\pi)^{-1}}$$

$$= (1 - n\pi)^2 \cos n\pi \neq 0.$$

The only singularity of $\sin(1 - z^{-1})$ appears at $z = 0$. If we let $z$ approach 0 through positive values, then $\sin(1 - z^{-1})$ oscillates between $\pm 1$. Such behavior can only characterize an essential singularity. ∎

**Example 5**

Classify the zeros and poles of the function $f(z) = (\tan z)/z$.

**Solution.** Since $(\tan z)/z = (\sin z)/(z \cos z)$, the only possible zeros are those of $\sin z$; i.e., $z = n\pi$ $(n = 0, \pm 1, \pm 2, \ldots)$. However, $z = 0$ is, in fact, a singularity. Furthermore, the points $z = (n + \frac{1}{2})\pi$, which are the zeros of $\cos z$, are also singularities. We shall investigate these in turn.

If $n$ is a nonzero integer, the reader should have no trouble convincing himself or herself that $z = n\pi$ is a *simple* zero for the given function.

Near the point $z = 0$ we can write

$$\frac{\tan z}{z} = \frac{\sin z}{z \cos z} = \frac{1}{z \cos z}\left(z - \frac{z^3}{3!} + \frac{z^5}{5!} - \cdots\right)$$

$$= \frac{1}{\cos z}\left(1 - \frac{z^2}{3!} + \frac{z^4}{5!} - \cdots\right),$$

and we see that $(\tan z)/z \to 1$ as $z \to 0$. Hence the origin is a *removable singularity*.

Finally, since $\cos z$ has simple zeros at $z = (n + \frac{1}{2})\pi$ $(n = 0, \pm 1, \pm 2, \ldots)$, it is easy to see that $f(z)$ has simple poles at these points.  ∎

We conclude this section by summarizing the various equivalent characterizations of the three types of isolated singularities. For economy of notation the statement of Theorem 18 utilizes the logician's symbol "$\Leftrightarrow$" to denote logical equivalence; it can be translated "if and only if."

---

**Theorem 18.** If $f$ has an isolated singularity at $z_0$, then the following equivalences hold:

(i) $z_0$ is a removable singularity $\Leftrightarrow |f|$ is bounded near $z_0 \Leftrightarrow f(z)$ has a limit as $z \to z_0 \Leftrightarrow f$ can be redefined at $z_0$ so that $f$ is analytic at $z_0$.

(ii) $z_0$ is a pole $\Leftrightarrow |f(z)| \to \infty$ as $z \to z_0 \Leftrightarrow f$ can be written $f(z) = g(z)/(z - z_0)^m$ for some integer $m > 0$ and some function $g$ analytic at $z_0$ with $g(z_0) \neq 0$.

(iii) $z_0$ is an essential singularity $\Leftrightarrow |f(z)|$ neither is bounded near $z_0$ nor goes to infinity as $z \to z_0 \Leftrightarrow f(z)$ assumes every complex number, with possibly one exception, as a value in every neighborhood of $z_0$.

---

# EXERCISES 5.6

1. Find and classify the isolated singularities of each of the following functions.

(a) $\dfrac{z^3 + 1}{z^2(z + 1)}$    (b) $z^3 e^{1/z}$    (c) $\dfrac{\cos z}{z^2 + 1} + 4z$    (d) $\dfrac{1}{e^z - 1}$

(e) $\tan z$      (f) $\cos\left(1 - \dfrac{1}{z}\right)$      (g) $\dfrac{\sin(3z)}{z^2} - \dfrac{3}{z}$      (h) $\cot\left(\dfrac{1}{z}\right)$

**2.** What is the order of the pole of

$$f(z) = \frac{1}{(2 \cos z - 2 + z^2)^2}$$

at $z = 0$? [HINT: Work with $1/f(z)$.]

**3.** For each of the following, construct a function $f$ analytic in the plane except for isolated singularities that satisfies the given conditions.

(a) $f$ has a zero of order 2 at $z = i$ and a pole of order 5 at $z = 2 - 3i$.

(b) $f$ has a simple zero at $z = 0$ and an essential singularity at $z = 1$.

(c) $f$ has a removable singularity at $z = 0$, a pole of order 6 at $z = 1$, and an essential singularity at $z = i$.

(d) $f$ has a pole of order 2 at $z = 1 + i$ and essential singularities at $z = 0$ and $z = 1$.

**4.** Give a proof of Lemma 8.

**5.** For each of the following, determine whether the statement made is always true or sometimes false.

(a) If $f$ and $g$ have a pole at $z_0$, then $f + g$ has a pole at $z_0$.

(b) If $f$ has an essential singularity at $z_0$ and $g$ has a pole at $z_0$, then $f + g$ has an essential singularity at $z_0$.

(c) If $f(z)$ has a pole of order $m$ at $z = 0$, then $f(z^2)$ has a pole of order $2m$ at $z = 0$.

(d) If $f$ has a pole at $z_0$ and $g$ has an essential singularity at $z_0$, then the product $f \cdot g$ has a pole at $z_0$.

(e) If $f$ has a zero of order $m$ at $z_0$ and $g$ has a pole of order $n$, $n \le m$, at $z_0$, then the product $f \cdot g$ has a removable singularity at $z_0$.

**6.** Prove that if $f(z)$ has a pole of order $m$ at $z_0$, then $f'(z)$ has a pole of order $m + 1$ at $z_0$.

**7.** If $f(z)$ is analytic in $D: 0 < |z| \le 1$, and $z^l \cdot f(z)$ is unbounded in $D$ for every integer $l$, then what kind of singularity does $f(z)$ have at $z = 0$?

**8.** Verify Picard's theorem for the function $\cos(1/z)$ at $z_0 = 0$.

**9.** Does there exist a function $f(z)$ having an essential singularity at $z_0$ which is bounded along some line segment emanating from $z_0$?

**10.** If the function $f(z)$ is analytic in a domain $D$ and has zeros at the distinct points $z_1, z_2, \ldots,$ $z_n$ of respective orders $m_1, m_2, \ldots, m_n$, then prove that there exists a function $g(z)$ analytic in $D$ such that

$$f(z) = (z - z_1)^{m_1}(z - z_2)^{m_2} \cdots (z - z_n)^{m_n}g(z).$$

**11.** If $f$ has a pole at $z_0$, show that $\operatorname{Re} f$ and $\operatorname{Im} f$ take on arbitrarily large positive as well as negative values in any punctured neighborhood of $z_0$.

**12.** Prove that if $f(z)$ has a pole of order $m$ at $z_0$, then $g(z) := f'(z)/f(z)$ has a simple pole at $z_0$. What is the coefficient of $(z - z_0)^{-1}$ in the Laurent expansion for $g(z)$?

**13.** Let $f(z)$ have an isolated singularity at $z_0$ and suppose that $f(z)$ is bounded in some punctured neighborhood of $z_0$. Prove directly from the integral formula for the Laurent coefficients that $a_{-j} = 0$ for all $j = 1, 2, \ldots$; i.e., $f(z)$ must have a removable singularity at $z_0$.

**14.** Without appealing to Picard's theorem, prove the *Casorati-Weierstrass theorem*: If $f(z)$ has an essential singularity at $z_0$, then in any punctured neighborhood of $z_0$ the function $f(z)$ comes arbitrarily close to any specified complex number. [HINT: Let the specified number be $c$ and assume to the contrary that $|f(z) - c| \ge \delta > 0$ in every small punctured neighborhood of $z_0$. Then, using Prob. 13, show that $f(z) - c$ [and hence $f(z)$ itself] must have either a pole or a removable singularity at $z_0$.]

15. Prove that if $f(z)$ has an essential singularity at $z_0$, then so does the function $e^{f(z)}$. [HINT: Argue that $e^{f(z)}$ is neither bounded nor tends (in modulus) to infinity as $z \to z_0$.]

16. Sketch the graph of the level curves $|e^{1/z}| = s$ for $s = 1, \frac{1}{2}, 2, \frac{1}{3}, 3, \ldots$, and observe that they all tend to the essential singularity $z = 0$ of $e^{1/z}$ [HINT: The level curves are circles.]

17. By completing each of the following steps, prove Schwarz's lemma.

> (*Schwarz's Lemma*) If $f$ is analytic in the unit disk $U: |z| < 1$ and satisfies the conditions
>
> $$f(0) = 0 \quad \text{and} \quad |f(z)| \le 1 \text{ for all } z \text{ in } U,$$
>
> then $|f(z)| \le |z|$ for all $z$ in $U$.

(a) Define $F(z) := f(z)/z$, for $z \ne 0$, and $F(0) = f'(0)$. Show that $F$ is analytic in $U$.

(b) Let $\zeta(\ne 0)$ be any fixed point in $U$, and $r$ be any real number that satisfies $|\zeta| < r < 1$. Show by means of the maximum-modulus principle that if $C_r$ denotes the circle $|z| = r$, then

$$|F(\zeta)| \le \max_{z \text{ on } C_r} \frac{|f(z)|}{r} \le \frac{1}{r}.$$

(c) Letting $r \to 1^-$ in part (b), deduce that $|f(\zeta)| \le |\zeta|$ for all $\zeta$ in $U$.

18. Let $f$ be a function satisfying the conditions of Schwarz's lemma (Prob. 17). Prove that if $|f(z_0)| = |z_0|$ for some nonzero $z_0$ in $U$, then $f$ must be a function of the form $f(z) = e^{i\theta}z$ for some real $\theta$. Show also that $f$ must be of this form if $|f'(0)| = 1$.

## 5.7 THE POINT AT INFINITY

From our discussion of singularities in Sec. 5.6 we know that if a mapping is given by an analytic function possessing a pole, it carries points near that pole to indefinitely distant points. It must have occurred to the reader that one might define the value of $f$ at the pole to be $\infty$. Before taking this plunge, however, we should be aware of all the ramifications. Let us look in detail at the behavior of $1/z$ near $z = 0$.

As $z \to 0$ along the positive real axis, $1/z$ goes to "plus infinity"; along the negative real axis, $1/z$ goes to "minus infinity"; and along the positive $y$-axis, $1/z$ goes to—what? "Minus $i$ times infinity?" If we are to assign the symbol $\infty$ to $1/0$ we must realize that we are identifying all these "limits" as a single number; geometrically, we are speaking of *the point at infinity*, which can be reached, in a manner of speaking, by proceeding infinitely far along any direction in the complex plane.

It is somewhat enlightening to try to visualize the situation in the following manner: Consider the rays emanating from the origin in the complex plane to all be joined at their "ends," deforming the complex plane into something like an upside-down parachute with its lines tied together. In fact, such an image motivates the *stereographic projection* depicted in Fig. 5.6 wherein the $xy$-plane is mapped onto the surface of the three-dimensional unit sphere which has the $xy$-plane as its equatorial

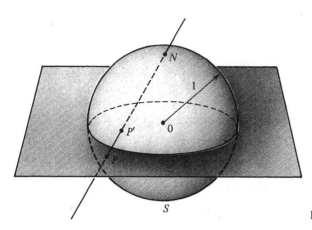

**Figure 5.6** Stereographic projection.

plane. The mapping is defined as follows: Starting from a point $P$ in the plane, we draw a line through the north pole of the sphere. Then the point $P$ is mapped to the point $P'$, where this line intersects the sphere, as illustrated in Fig. 5.6.

Thus the open disk $|z| < 1$ is mapped onto the southern hemisphere, the circle $|z| = 1$ onto the equator, and the exterior $|z| > 1$ onto the northern hemisphere excluding the north pole. Notice that all distant points get mapped onto "arctic zones" near the north pole; the latter is, in fact, the actual limit of such points, in the topology of the sphere. It corresponds to the "point at $\infty$."

With this model in mind, we officially append "$\infty$" to the set of complex numbers, calling the resulting collection the *extended complex plane*. A neighborhood of $\infty$ is any set of the form $|z| > M$, and a sequence of points $z_n$ ($n = 1, 2, 3, \ldots$) approaches $\infty$ if $|z_n|$ can be made arbitrarily large by taking $n$ large.

Consequently we shall write $f(z_0) = \infty$ when $|f(z)|$ increases without bound[†] as $z \to z_0$ and shall write $f(\infty) = w_0$ when $f(z) \to w_0$ as $z \to \infty$. For example, if

$$f(z) = \frac{2z + 1}{z - 1}, \tag{1}$$

then $f(1) = \infty$ and $f(\infty) = 2$.

Observe that for $h(z) = 2z + 1$, we have $h(\infty) = \infty$.

It is convenient in some applications to carry this notion still further and speak of functions that are "analytic at $\infty$." The analyticity properties of $f$ at $\infty$ are classified by first performing the mapping $w = 1/z$, which maps the point at infinity to the origin, and then examining the behavior of the composite function $g(w) := f(1/w)$ at the origin $w = 0$. Thus we say

1. $f(z)$ is analytic at $\infty$ if $f(1/w)$ is analytic (or has a removable singularity) at $w = 0$,

2. $f(z)$ has a pole of order $m$ at $\infty$ if $f(1/w)$ has a pole of order $m$ at $w = 0$, and

---

[†] Technically, $f(z_0) = \infty$ if for any $M > 0$ there is a $\delta > 0$ such that $0 < |z - z_0| < \delta$ implies that $|f(z)| > M$.

3. $f(z)$ has an essential singularity at $\infty$ if $f(1/w)$ has an essential singularity at $w = 0.$[†]

From Theorem 18, we can interpret these conditions for a function analytic outside some disk as follows:

1'. $f(z)$ is analytic at $\infty$ if $|f(z)|$ is bounded for sufficiently large $|z|$,
2'. $f(z)$ has a pole at $\infty$ if $f(z) \to \infty$ as $z \to \infty$, and
3'. $f(z)$ has an essential singularity at $\infty$ if $|f(z)|$ neither is bounded for large $|z|$ nor goes to infinity as $z \to \infty$.

## Example 1

Classify the behavior at $\infty$ of the functions $z^2 + 2$, $(iz + 1)/(z - 1)$, and $\sin z$.

**Solution.**  Obviously $f(z) = z^2 + 2$ has a pole at $\infty$. The pole is of order 2, because

$$f\left(\frac{1}{w}\right) = \frac{1}{w^2} + 2$$

has a pole of order 2 at $w = 0$.
    Since

$$\frac{iz + 1}{z - 1} \to i \quad \text{as} \quad z \to \infty,$$

this function is certainly analytic at $\infty$.
    Finally, $\sin z$ has no limit as $z \to \infty$, even for real $z$ (it oscillates). Hence $\infty$ must be an essential singularity.[‡]  ∎

## Example 2

Find all the functions $f$ that are analytic everywhere in the extended complex plane.

**Solution.**  Since $f$ is analytic at $\infty$, it is bounded for, say, $|z| > M$. By continuity, $f$ is also bounded for $|z| \leq M$. Consequently, $f$ is a bounded entire function. Hence $f$ is constant, by Liouville's theorem.  ∎

## Example 3

Classify all the functions that are everywhere analytic in the extended complex plane except for a pole at one point.

---

[†] Some authors also allow the possibility of a removable singularity at $\infty$, but we feel that nothing is gained by this generality. Of course a zero of order $m$ for $f(z)$ at $\infty$ corresponds to a zero of order $m$ for $f(1/w)$ at 0.

[‡] Alternatively, this can be seen directly from the Laurent expansion for $\sin(1/w)$ about $w = 0$.

**Solution.** If $f(z)$ has a pole, say, of order $m$ at some finite point $z_0$, then the Laurent series for $f$

$$f(z) = \frac{a_{-m}}{(z - z_0)^m} + \frac{a_{-m+1}}{(z - z_0)^{m-1}} + \cdots + \frac{a_{-1}}{z - z_0} + \sum_{n=0}^{\infty} a_n(z - z_0)^n \qquad (2)$$

converges for all $z \neq z_0$. Moreover, since we are assuming that $z_0 \neq \infty$, the function $f$ must be analytic, and hence bounded, at $\infty$. From Eq. (2), then, we see that the *entire* function defined by the series $\sum_{n=0}^{\infty} a_n(z - z_0)^n$ is also bounded at $\infty$; thus it must be constant, i.e., equal to $a_0$. Therefore, the most general form for such a function is

$$f(z) = \frac{a_{-m}}{(z - z_0)^m} + \frac{a_{-m+1}}{(z - z_0)^{m-1}} + \cdots + \frac{a_{-1}}{z - z_0} + a_0. \qquad (3)$$

If the pole occurs at $z = \infty$, then $f(1/w)$ has a pole at the origin and can be expressed in the form

$$f\left(\frac{1}{w}\right) = \frac{a_{-m}}{w^m} + \frac{a_{-m+1}}{w^{m-1}} + \cdots + \frac{a_{-1}}{w} + \sum_{n=0}^{\infty} a_n w^n. \qquad (4)$$

Since $f(z)$ is bounded near $z = 0$, it follows that $f(1/w)$ is bounded for large $|w|$, and, as before, we conclude that $a_n = 0$ for $n > 0$. Hence Eq. (4) becomes

$$f(z) = a_{-m}z^m + a_{-m+1}z^{m-1} + \cdots + a_{-1}z + a_0; \qquad (5)$$

i.e., $f(z)$ is a *polynomial* in $z$.

Equations (3) and (5) categorize the totality of all functions possessing one pole in the extended complex plane. ∎

We note in passing that the theory of *Fuchsian equations* is based upon considerations of singularities in the extended complex plane, and these have been extremely helpful in relating many of the so-called "special functions" which arise in mathematical physics; Ref. [3] discusses this application.

## EXERCISES 5.7

1. Classify the behavior at $\infty$ for each of the following functions (if a zero or pole, give its order):

(a) $e^z$

(b) $\cosh z$

(c) $\dfrac{z - 1}{z + 1}$

(d) $\dfrac{z}{z^3 + i}$

(e) $\dfrac{z^3 + i}{z}$

(f) $e^{\sinh z}$

(g) $\dfrac{\sin z}{z^2}$

(h) $\dfrac{1}{\sin z}$

(i) $e^{\tan 1/z}$

2. Prove that if $f(z)$ is analytic at $\infty$, then it has a series expansion of the form

$$f(z) = \sum_{n=0}^{\infty} \frac{a_n}{z^n}$$

converging uniformly outside some disk.

3. Construct the series mentioned in Prob. 2 for the following functions.

(a) $\dfrac{z-1}{z+1}$      (b) $\dfrac{z^2}{z^2+1}$      (c) $\dfrac{1}{z^3-i}$

4. State Picard's theorem (Sec. 5.6) for functions with an essential singularity at $\infty$. Verify for $e^z$.

5. What is the order of the zero at $\infty$ if $f(z)$ is a rational function of the form $P(z)/Q(z)$ with deg $P <$ deg $Q$?

6. Suppose that $f$ is analytic on and *outside* the simple closed *negatively* oriented contour $\Gamma$. Assume further that $f$ is analytic at $\infty$ and $f(\infty) = 0$. Prove that

$$f(z) = \frac{1}{2\pi i} \oint_\Gamma \frac{f(\zeta)}{\zeta - z}\, d\zeta$$

for all $z$ outside $\Gamma$. [HINT: Apply Cauchy's integral formula for $z$ in an annulus and let the outer radius tend to $\infty$.]

7. Prove that if $f$ is analytic on and outside the simple closed contour $\Gamma$ and has a zero of order 2 or more at $\infty$, then

$$\int_\Gamma f(z)\, dz = 0.$$

Does this integral vanish if we merely assume that $f$ has a simple zero at $\infty$?

8. Let the point $z = x + iy$ correspond to the point $(x_1, x_2, x_3)$ on the sphere $x_1^2 + x_2^2 + x_3^2 = 1$ under the stereographic projection. Show that

$$x_1 = \frac{2x}{|z|^2 + 1}, \qquad x_2 = \frac{2y}{|z|^2 + 1}, \qquad x_3 = \frac{|z|^2 - 1}{|z|^2 + 1}.$$

[HINT: The three points $(0, 0, 1)$, $(x_1, x_2, x_3)$, and $(x, y, 0)$ must be collinear.]

9. Prove that the image under the stereographic projection of a circle or line in the $z$-plane is a circle on the sphere $x_1^2 + x_2^2 + x_3^2 = 1$. [HINT: Show that the image is the intersection of the sphere with a plane $Ax_1 + Bx_2 + Cx_3 = D$.]

10. Show that if $z_1, z_2$ are (finite) points in the complex plane $\mathbf{C}$, then the distance between their stereographic projections is given by

$$\chi[z_1, z_2] = \frac{2|z_1 - z_2|}{\sqrt{1 + |z_1|^2}\,\sqrt{1 + |z_2|^2}},$$

which is called the *chordal distance* between $z_1$ and $z_2$. Show, also, that if $z_2 = \infty$, then the corresponding distance is given by

$$\chi[z_1, \infty] = \frac{2}{\sqrt{1 + |z_1|^2}}.$$

*Positive functions are discussed in Problems 11–15.*

11. (a) Argue that in some neighborhood of a zero, $z_0$, of order $m$ for the analytic function $f$ one can express $f(z)$ as $(z - z_0)^m[c + \varepsilon(z)]$, where $c$ is constant, $\varepsilon(z)$ is analytic, and $|\varepsilon(z)| < |c|/100$. (The fraction $\frac{1}{100}$ has no special significance.)

   (b) Argue that in some neighborhood of a pole, $z_0$, of order $m$ for the analytic function $f$ one can express $f(z)$ as $(z - z_0)^{-m}[c + \varepsilon(z)]$, where $c$ is constant, $\varepsilon(z)$ is analytic, and $|\varepsilon(z)| < |c|/100$.

12. A function $f(z)$ is said to be a *positive function* if $f(z)$ is rational (a ratio of polynomials) and if Re $f(z) > 0$ whenever Re $z > 0$. In 1931 Otto Brune proved that the complex *impedance* of any electric circuit must be a positive function when $z$ is interpreted as the "imaginary frequency" $i\omega$ (Sec. 3.4).

   (a) Show that the complex impedance for the electrical circuit studied in Sec. 3.4,

$$R_{\text{eff}} = \frac{R/i\omega C}{R + 1/i\omega C} + i\omega L,$$

   yields a positive function when $z$ is substituted for $i\omega$ (for positive $R$, $L$, and $C$).

   (b) Show that the complex *admittance* $1/R_{\text{eff}}$ is also a positive function.

   (c) Generalize: Show that the reciprocal of any positive function is also positive.

13. By considering the changes of sign in the factor $(z - z_0)^{\pm m}$, use the results of the previous problems to show that if $f$ is a positive function,

   (a) $f$ has no poles or zeros in the right half-plane;

   (b) the pure imaginary poles and zeros (if any) of $f$ are simple (order 1) and the corresponding constants $c$ in Prob. 11 must be real and positive.

14. Extend the reasoning of the previous problems to argue that a positive function is either analytic at the point at infinity or has a simple pole or a simple zero there. What does this say about the degrees of the numerator and denominator polynomials?

15. (a) Suppose that $f$ is a positive function that has no poles on the imaginary axis. Use the max-min theory of Sec. 4.7 to deduce that the minimal value of Re $f$ in the closed right half-plane occurs on the imaginary axis (including the point at infinity).

   (b) Using the characterization of the imaginary poles of $f$ as described in Prob. 13, remove the "no-poles" restriction in part (a). [HINT: Use an indented contour.]

   (c) Establish the following: If $f$ is a rational function that is analytic in the right half-plane, has only simple poles with positive "residues" (constants $c$ as in Prob. 11) on the imaginary axis (including the point at infinity), and satisfies Re $f \geq 0$ on the imaginary axis, then $f$ is a positive function.

16. (a) Prove that there does not exist a (single-valued) branch of $(z^2 - 1)^{1/3}$ that is analytic in $|z| > 1$. [HINT: Consider the point at infinity.]

   (b) Prove that there *does* exist a (single-valued) branch of $[(z - 1)(z - 2)(z - 3)]^{1/3}$ analytic in $|z| > 3$.

# *5.8 ANALYTIC CONTINUATION

When one is given a formula or algorithm for computing an analytic function $f$ in a domain $D$, it is often of interest to know if this "domain of analyticity" can be extended—that is, if there is a function $F$ analytic in a larger domain whose values agree with those of $f(z)$ for $z$ in $D$. In such a case we say that $F$ is an *analytic continuation* of $f$.

This terminology is also used in some situations when the original domain of definition of $f$ is not truly a "domain" in the sense of Chapter 1—and hence the original $f$ is not analytic. We have already encountered some trivial examples of analytic continuation in this sense; for instance, when we extend the real polynomial $x^2 + 1$ to complex numbers as $z^2 + 1$, we have analytically continued the function $f(x) = x^2 + 1$ from the $x$-axis to the entire plane $\mathbf{C}$. The functions $e^z$, $\sin z$, and $\log z$ can also be interpreted as analytic continuations.

Analytic continuation arises in the analysis of many engineering systems, wherein one is confronted with a function $f(\omega)$ describing the system response to an excitation at the frequency $\omega$ (recall Sec. 3.4). The analytic continuation of $f$ to "complex frequency values" can often be most instructive. In fact, we shall use this approach to relate the Fourier and Laplace transforms in Chapter 8.

Analytic continuation can be a subtle process. True, the continuation of $x^2 + 1$ to $z^2 + 1$ is only a matter of extending the formula; but the identity (recall Example 5, Sec. 4.4a, page 138)

$$\oint_C \frac{d\zeta}{\zeta - z} = \begin{cases} 2\pi i & \text{if } z \text{ lies inside } C, \\ 0 & \text{if } z \text{ lies outside } C, \end{cases}$$

demonstrates that one does *not* analytically continue the function $f(z) := 2\pi i$ by extending its "formula" $\int_C d\zeta/(\zeta - z)$ ($z$ inside $C$) to the exterior of the contour $C$.

The interesting questions about analytic continuation are these. Is analytic continuation across the boundary of $D$—or part of the boundary—always possible? Is the continued function $F$ unique? Are there any computational rules for obtaining $F$? We will try to address these issues at an elementary level in this section.

One tool for studying analytic continuations is the Taylor series. Let us postulate a situation where an analyst needs to investigate the function $f(z)$ defined by the series

$$f(z) := \sum_{j=1}^{\infty} \frac{(-1)^{j-1}(z-1)^j}{j}. \tag{1}$$

The series (1), for instance, could have been obtained as a result of using the power series method of Example 1, Sec. 5.3, page 201, on the differential equation

$$y'' + \frac{y'}{z} = 0 \tag{2}$$

with the initial conditions

$$y(1) = 0, \qquad y'(1) = 1. \tag{3}$$

[In fact, $f(z)$ is a well-known function, but for the purposes of exposition we choose not to identify it for the moment.]

Since the coefficients in the differential equation (2) fail to be analytic only at $z = 0$, the initial-value theorem quoted in Sec. 5.3 (page 202) guarantees that the circle of convergence for (1) extends at least to this point. Therefore, we can say that the analyst knows the values of this analytic function $f$ inside $C: |z - 1| = 1$ (or more realistically, she can compute the values of $f(z)$, and its derivatives, to arbitrary accuracy inside $C$). In particular, she "knows" the function and its derivatives at the point $z_1$ in Fig. 5.7 on the next page.

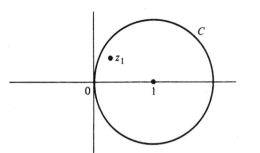

**Figure 5.7**    Circle of convergence for series (1).

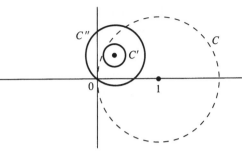

**Figure 5.8**    Proposed circles of convergence.

From these data she can construct the power series

$$\sum_{j=0}^{\infty} \frac{f^{(j)}(z_1)}{j!} (z - z_1)^j. \tag{4}$$

Of course, this is a Taylor series for the solution $f$ and is guaranteed to converge to $f$ at least inside the small circle $C'$ of Fig. 5.8, wherein $f$ is known to be analytic. The initial-value theorem, however, tells us more—namely, that (4) converges inside the larger circle $C''$, extending from $z_1$ to the origin (again, because the coefficients in the differential equation are analytic inside $C''$). Thus by employing (4) the analyst has analytically continued the function $f$, extending its domain of analyticity outside the circle $C$.

The function defined by the series (1) is, in fact, Log $z$. The reader can directly verify that (1) is a Taylor series for Log $z$ and that Log $z$ satisfies the differential equation (2) and the initial conditions (3). This solution does, indeed, fail to be analytic at $z = 0$, and the circles $C$ and $C''$ are true circles of convergence.

To continue our study we formalize some terminology covering the particular situation just analyzed.

---

**Definition 9.**   Suppose that $f$ is analytic in a domain $D_1$ and that $g$ is analytic in a domain $D_2$. Then we say that $g$ is a **direct analytic continuation** of $f$ to $D_2$ if $D_1 \cap D_2$ is nonempty and $f(z) = g(z)$ for all $z$ in $D_1 \cap D_2$.

---

In our example the first domain $D_1$ is the interior of $C$, the second domain $D_2$ is the interior of $C''$, and $g$ is the sum of the power series (4). We have not actually proved that $f = g$ over the whole lens-shaped domain $D_1 \cap D_2$; equality has been established only inside $C'$. So we shall use the following theorem to show that (4) is a bona fide direct analytic continuation of $f(z)$ to $D_2$.

---

**Theorem 19.**   If $F$ is analytic in a domain $D$ and vanishes on some open disk contained in $D$, then it vanishes throughout $D$.

---

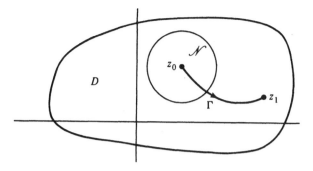

**Figure 5.9**  Geometry for Theorem 19.

*Proof.* By hypothesis, $D$ contains a disk $\mathscr{N}$, say with center $z_0$, such that $F(z) = 0$ for all $z$ in $\mathscr{N}$. Now let's assume, contrary to the conclusion of the theorem, that there is some point $z_1$ in $D$ such that $F(z_1) \neq 0$, and let $\Gamma$ be a path in $D$ joining $z_0$ to $z_1$. (See Fig. 5.9.) As we move along $\Gamma$ from $z_0$, at first we only observe points where $F(z) = 0$. Eventually, however, we must encounter a point $w$ with the properties

(i) For all $z$ on $\Gamma$ preceding $w$, we have $F(z) = 0$;

(ii) There are points $z$ on $\Gamma$ arbitrarily close to $w$ such that $F(z) \neq 0$. (Recall a similar situation in the proof of Theorem 23, Sec. 4.6, page 165.)

First we observe that condition (i) implies that the derivative of $F(z)$ is zero on the portion of $\Gamma$ *preceding* $w$, because at any point $z$ on this part of the curve we can evaluate

$$F'(z) = \lim_{\zeta \to z} \frac{F(\zeta) - F(z)}{\zeta - z}$$

by letting $\zeta$ approach $z$ along this portion of $\Gamma$, where $F(\zeta) = F(z) = 0$. Hence the limit is zero. Continuing in this fashion, we express

$$F''(z) = \lim_{\substack{\zeta \to z \\ \zeta \text{ on } \Gamma}} \frac{F'(\zeta) - F'(z)}{\zeta - z} = \lim_{\substack{\zeta \to z \\ \zeta \text{ on } \Gamma}} 0 = 0,$$

and so on, to conclude that *all derivatives of $F(z)$ vanish on the portion of $\Gamma$ preceding $w$.* By continuity, then, $F$ and all its derivatives also vanish at $w$, which implies that the Taylor coefficients for $F$ about $z = w$ are identically zero; i.e., $F$ must vanish in some *disk* around $w$. But this contradicts condition (ii), so our assumption that $F$ is not identically zero must be wrong.  ■

**Corollary 4.** If $f$ and $g$ are analytic in a domain $D$ and $f = g$ in some disk contained in $D$, then $f = g$ throughout $D$.

*Proof.* Simply apply Theorem 19 to the difference $F := f - g$.  ■

Returning to our discussion of the situation depicted in Fig. 5.8 we can now conclude from the fact that the series (1) and (4) agree inside the circle $C'$, that they must agree inside the lens-shaped domain formed by the intersection of the interiors of $C$ and $C''$. This, then, implies that (4) is a direct analytic continuation of $f(z)$ from the interior of $C$ to the interior of $C''$, in strict accordance with Definition 9.

It is worthwhile at this point to list some salient observations about direct analytic continuation. The first two are so trivial that we shall delete the proofs:

---

**Theorem 20.** If $f$ is analytic in a domain $D_1$ and $g$ is a direct analytic continuation of $f$ to the domain $D_2$, then the function

$$F(z) := \begin{cases} f(z) & \text{for } z \text{ in } D_1, \\ g(z) & \text{for } z \text{ in } D_2, \end{cases} \tag{5}$$

is single-valued and analytic on $D_1 \cup D_2$.

---

**Theorem 21.** If $f$ is analytic in a domain $D_1$, and $D_2$ is a domain such that $D_1 \cap D_2$ is nonempty, then the direct analytic continuation of $f$ to $D_2$, if it exists, is unique.

---

Theorem 19 and its corollary can be generalized as follows.

---

**Theorem 22.** Suppose that $f$ is analytic in a domain $D$ and that $\{z_n\}$ is an infinite sequence of distinct points converging to a point $z_0$ in $D$. Suppose, moreover, that $f(z_n) = 0$ for each $n = 1, 2, \dots$. Then $f(z) \equiv 0$ throughout $D$.

---

*Proof.* By continuity, $z_0$ is a zero of $f$. However, it is not an isolated zero because every punctured disk about $z_0$ contains points of the sequence $\{z_n\}$. Consequently, by Corollary 3, Sec. 5.6, $f$ must be identically zero in some neighborhood of $z_0$. Hence Theorem 19, page 219, implies $f(z) \equiv 0$ throughout $D$. ■

---

**Corollary 5.** If $f$ and $g$ are analytic functions in a domain $D$ and $f(z_n) = g(z_n)$ for an infinite sequence of distinct points $\{z_n\}$ converging to a point $z_0$ in $D$, then $f \equiv g$ throughout $D$.

---

*Proof.* Again, consider $f - g$. ■

Often this corollary is used to extend a known equality from a curve to a domain, as in the next example.

**Example 1**

> Prove, by direct analytic continuation, that
>
> $$\sin^2 z + \cos^2 z = 1 \qquad \text{for all } z.$$
>
> **Solution.** We know from elementary trigonometry that the equality is true for real $z$. In other words, the two entire functions $f(z) := \sin^2 z + \cos^2 z$ and $g(z) := 1$ agree on the real axis. Corollary 5 thus extends the equality to the whole plane. ■

Now let's turn to a related topic, the concept of *analytic continuation along a curve*. The situation is as follows (refer to Fig. 5.10): $f(z)$ is analytic in a domain $D$, $z_1$ is a point in $D$, and $\gamma$ is some path connecting $z_1$ to a point $z^*$.

We expand $f(z)$ in a Taylor series around $z_1$; the resulting power series converges to a function $f_1(z)$ inside, say, the circle $C_1$. Staying inside $C_1$, we proceed along $\gamma$ until we come to some point $z_2$, and then we expand $f_1(z)$ around $z_2$. This series converges to some analytic function $f_2(z)$ inside the circle $C_2$. Next we move further along $\gamma$, now staying inside $C_2$, to some point $z_3$, and expand around $z_3$. And so on. If this process eventually produces a circle of convergence, say $C_n$, enclosing the portion of $\gamma$ between $z_n$ and $z^*$, we say that the scheme derived from the sequence of points $\{z_1, z_2, \ldots, z_n, z^*\}$ and the corresponding functions $\{f, f_1, f_2, \ldots, f_n\}$ constitutes an *analytic continuation of $f(z)$ to $z^*$ along the curve $\gamma$*. The value of this analytic continuation at $z^*$ is, of course, $f_n(z^*)$.

(We remark that there can be situations in which $f_1$ is not a *direct* analytic continuation of $f$; Prob. 6 illustrates this possibility.)

Let's see what would happen if our aforementioned colleague investigating the series $\sum_{j=1}^{\infty} (-1)^{j-1}(z-1)^j/j$ sought to establish an analytic continuation along the curve $\gamma$ in Fig. 5.11. She expands around $z_1$ (as before), computing the function $f_1(z)$, which we know to equal Log $z$ inside $C_1$. Next she expands around $z_2$, deriving the function $f_2(z)$ inside the circle $C_2$; see Fig. 5.12 on the next page.

But now the domain of the analytic function $f_2$ extends beyond the negative real axis, so we cannot identify $f_2$ with Log $z$ over the whole interior of $C_2$. However, $f_2(z)$ is an analytic function, defined on the interior of $C_2$, whose value and derivatives coincide with those of Log $z$ at $z_2$; hence $f_2(z)$ must agree with some appropriately

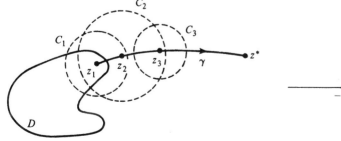

**Figure 5.10**   Analytic continuation along a curve.

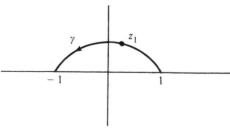

**Figure 5.11**   Curve for continuation.

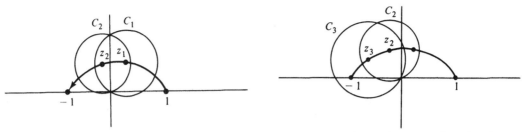

**Figure 5.12** Stages of continuation.    **Figure 5.13** Further stages of continuation.

chosen *branch* of log $z$ whose branch cut does not intersect $C_2$. For example, the branch given by $\text{Log}\,|z| + i\arg z$ with $0 < \arg z \le 2\pi$ matches $\text{Log}\,z$ at $z_2$; hence it agrees with $f_2(z)$ inside $C_2$.

Finally our tireless mathematician expands around the point $z_3$ and derives the function $f_3(z)$, analytic inside $C_3$ (Fig. 5.13). Once again we see that $f_3(z)$ is the branch $\text{Log}\,|z| + i\arg z$, $0 < \arg z \le 2\pi$, inside $C_3$; in particular, $f_3(-1) = \pi i$. In short, the mathematician has analytically continued the power series $f(z) = \sum_{j=1}^{\infty}(-1)^{j-1}(z-1)^j/j$ along the curve $\gamma$, and the value of the continuation at $-1$ is $\pi i$.

It is instructive to study the result of continuing this same function $f$ along the curve $\gamma'$ in Fig. 5.14. In this case, the scheme might consist of the points $\{+1, \xi_1, \xi_2, \xi_3, -1\}$ and the functions $\{f, g_1, g_2, g_3\}$. The power series computed around the point $\xi_2$ would sum to the function $g_2(z)$, whose derivatives would agree with those of $\text{Log}\,z$ at $\xi_2$ but whose domain of analyticity (the interior of $C_2'$) would enclose a portion of the negative real axis. Thus $g_2(z)$ and, in fact, $g_3(z)$ would agree with the branch $\mathscr{L}_{-2\pi}(z) = \text{Log}\,|z| + i\arg z$, $-2\pi < \arg z \le 0$. In particular, $g_3(-1) = -\pi i$.

Summarizing, we have seen that an analytic continuation of $f(z)$ along $\gamma$ gives the value $\pi i$ at $z = -1$, but an analytic continuation along $\gamma'$ gives the value $-\pi i$. We conclude that, in general, the value obtained by analytic continuation along a curve may depend on the curve itself, not merely on its terminal point.

The alert reader has probably surmised by now that the source of this anomaly is the singularity of all the branches of log $z$ at the origin. This is in fact the case, and the following result, known as the *Monodromy theorem*, can be considered a vindication of the procedure of analytic continuation along curves.

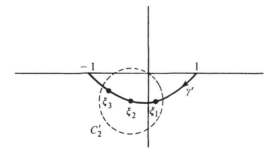

**Figure 5.14** Alternative route for continuation.

**Theorem 23.**   (*Monodromy Theorem*) Let $f(z)$ be analytic in a domain $D$, and suppose that $\gamma$ and $\gamma'$ are two directed smooth curves connecting the point $z_1$ in $D$ to some point $z^*$. Suppose further that there is some domain $D'$ with the following properties:

(i) The loop $\Gamma = \{\gamma, -\gamma'\}$ lies in $D'$ and can be continuously deformed[†] to a point in $D'$, and

(ii) $f(z)$ can be analytically continued along any smooth curve in $D'$.

Then the value at $z^*$ of the analytic continuation of $f$ along $\gamma$ agrees with the value of its continuation along $\gamma'$.

Thus, for instance, if we continue $\operatorname{Log} z$ from $z = +1$ to $z = -1$ along *any* curve lying in the upper half-plane, we shall arrive at the value $\pi i$, while the value of the continuation along any curve in the lower half-plane will be $-\pi i$.

The proof of the Monodromy theorem involves some topological constructions with which our readers may be unfamiliar, so we shall delete it (see Ref. [5]). It should be noted, however, that one consequence of the theorem is the fact (which we have tacitly assumed) that the analytic continuation of a function along a particular curve does not depend on the sequence of points $\{z_1, z_2, \ldots, z_n, z^*\}$ used in the scheme—again we direct the reader to the references for a rigorous proof.

### Example 2

Consider the function $f(z)$ defined for $\operatorname{Re} z > 0$ as the principal branch of $z^{1/2}$, i.e., that branch which takes the value $+1$ when $z = 1$. What is the value obtained by continuation of $f$ along a simple closed positively oriented curve $\gamma$ beginning at $z = 1$ and terminating at the same point, $z = 1$?

**Solution.**   We must consider three possibilities: Either the origin lies on $\gamma$, outside $\gamma$, or inside $\gamma$.

Case 1: The origin lies on $\gamma$. Then analytic continuation along $\gamma$ is not possible; there is no scheme of points and power series that can pass through the "barrier" $z = 0$, since this singularity will be excluded from every circle of convergence of a Taylor series for $z^{1/2}$. See Fig. 5.15.

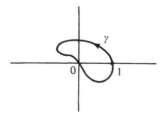

**Figure 5.15**   Origin lies on $\gamma$.

---

† Compare Sec. 4.4a.

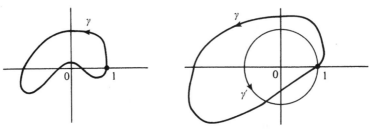

**Figure 5.16**  Origin lies outside $\gamma$.          **Figure 5.17**  Origin lies inside $\gamma$.

Case 2: The origin lies outside $\gamma$ (see Fig. 5.16). Then the conditions of the Monodromy theorem are satisfied with the curve $\gamma'$ consisting of the single point $z = 1$, and the domain $D'$ given by the entire plane with the origin deleted. The continuation along $\gamma'$ (and hence $\gamma$) results, of course, in the value $+1$.

Case 3: The origin lies inside $\gamma$. We can apply the Monodromy theorem, again identifying $D'$ as the punctured plane but now taking $\gamma'$ to be the unit circle $|z| = 1$, positively oriented (see Fig. 5.17). To get the continuation along $\gamma'$ we use De Moivre's formula for $z^{1/2}$. Then it is clear that if $\gamma'$ is parametrized by $z = e^{i\theta}$, $0 \le \theta \le 2\pi$, the value of the continuation of $z^{1/2}$ at $z = e^{i\theta}$ is $e^{i\theta/2}$; this gives $e^{i2\pi/2} = -1$ at the terminal point $z = 1$. Thus we learn that analytic continuation around a *closed* curve may yield functional values which are different from the original ones!  ■

We conclude this section by remarking that there exist functions analytic in a domain $D$ that cannot be analytically continued to *any* point outside $D$. In such a case the boundary of $D$ is called a *natural boundary*. This situation is illustrated in Prob. 10. *Schwarz reflection*, discussed in Probs. 13 through 15, can provide simple rules for implementing analytic continuations in some cases.

## EXERCISES 5.8

1. Given that $f(z)$ is analytic at $z = 0$ and that $f(1/n) = 1/n^2$, $n = 1, 2, \ldots$, find $f(z)$.
2. Prove that if $f(z)$ is analytic and agrees with a polynomial $\sum_{j=0}^{n} a_j x^j$ for $z = x$ on a segment of the real axis, then $f(z) = \sum_{j=0}^{n} a_j z^j$ everywhere.
3. Does there exist a function $f(z)$, not identically zero, which is analytic in the open disk $D: |z| < 1$ and vanishes at infinitely many points in $D$?
4. Prove that if $f$ is analytic in a deleted neighborhood of $z = 0$ and if $f(1/n) = 0$ for all $n = \pm 1, \pm 2, \ldots$, then either $f$ is identically zero or $f$ has an essential singularity at $z = 0$.
5. Let $f(z) = \sum_{j=0}^{\infty} z^j$ for $|z| < 1$. For what values of $\alpha$ ($|\alpha| < 1$) does the Taylor expansion of $f(z)$ about $z = \alpha$ yield a direct analytic continuation of $f(z)$ to a disk extending outside $|z| < 1$?
6. Show that when a function $f$ is analytically continued along a curve (as depicted in Fig. 5.10 on page 237), the first function $f_1(z)$ generated by the power series expansion around the initial point $z_1$ of $\gamma$ need not be a direct analytic continuation of $f$. [HINT: Take $f(z) = \text{Log}\, z$, with $D$ and $z_1$ as depicted in Fig. 5.18.]

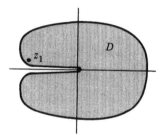

**Figure 5.18**  Example for Prob. 6.

7. By summing the series, show that $-\sum_{j=0}^{\infty}(2-z)^j$ is an analytic continuation, along some curve, of $\sum_{j=0}^{\infty} z^j$.

8. The *Gamma function* $\Gamma(z)$ is defined for Re $z > 0$ by the integral

$$\Gamma(z) := \int_0^{\infty} e^{-t} t^{z-1}\, dt.$$

(a) Show that $\Gamma(z+1) = z\Gamma(z)$ for Re $z > 0$.

(b) In most advanced texts it is shown that $\Gamma(z)$ is analytic in the right half-plane. Assuming this, argue that the functional equation in part (a) can be used to analytically continue $\Gamma(z)$ to the entire plane, except for the nonpositive integers $z = -n$, $n = 0, 1, 2, \dots$.

(c) Show that $\Gamma(n) = (n-1)!$ for positive integers $n$.

9. For each of the following functions, choose a branch that is analytic in the circle $|z-2| < 1$. Then analytically continue this branch along the curve $\gamma$ indicated in Fig. 5.19. Do the new functional values agree with the old?

(a) $3z^{2/3}$            (b) $\sin 5z$           (c) $(e^z)^{1/3}$

(d) $\sin(z^{1/2})$       (e) $(z^{1/2})^2$        (f) $(z^2)^{1/2}$

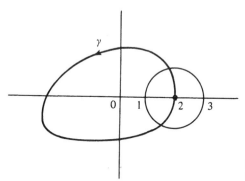

**Figure 5.19**  Continuation path for Prob. 8.

10. Show that the unit circle $|z| = 1$ is a natural boundary for the function $f(z) = \sum_{j=1}^{\infty} z^{j!}$, $|z| < 1$. [HINT: Argue that $|f(re^{ip\pi})| \to \infty$ as $r \to 1^-$ for any rational number $p$.]

11. Show that $|z| = 1$ is a natural boundary for the function $g(z) = \sum_{j=1}^{\infty} z^{j!}/j!$, although the series converges for $|z| = 1$. [HINT: Relate $g(z)$ to the function $f(z)$ of Prob. 10.]

12. Show that $|z| = 1$ is a natural boundary for $\sum_{j=0}^{\infty} z^{2^j}$.

13. The *Schwarz reflection principle* provides a formula for analytic continuation across a straight-line segment under certain circumstances. Its simplest form is stated as follows: Suppose that $f(z) = u(x, y) + iv(x, y)$ is analytic in a simply connected domain $D$ which

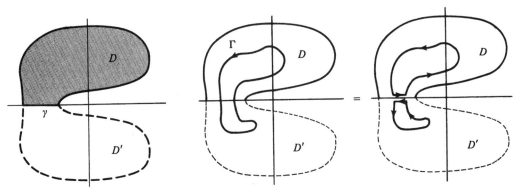

**Figure 5.20**   Domain for Prob. 13.          **Figure 5.21**   The integral of $F(z)$ along $\Gamma$ is zero.

lies in the upper half-plane and which has a segment $\gamma$ of the real axis as part of its boundary (see Fig. 5.20). Suppose furthermore that $v(x, y) \rightarrow 0$ as $(x, y)$ approaches $\gamma$ and that $u(x, y)$ also takes continuous limiting values on $\gamma$, denoted by $U(x)$ [so that $U(x)$ is continuous on $\gamma$]. Then the function $f$ can be analytically continued across $\gamma$ into the domain $D'$, which is the reflection of $D$ in the real axis. Specifically, the function $F(z)$ defined by

$$F(z) = \begin{cases} u(x, y) + iv(x, y) & \text{for } z = x + iy \text{ in } D, \\ U(x) & \text{for } z = x \text{ on } \gamma, \\ u(x, -y) - iv(x, -y) & \text{for } z = x + iy \text{ in } D' \end{cases}$$

is analytic in the domain $D \cup \gamma \cup D'$. Justify this principle based upon the following observations.

(a) $F(z)$ satisfies the Cauchy-Riemann equations in $D'$.
(b) $F(z)$ is continuous in $D \cup \gamma \cup D'$.
(c) Morera's theorem can be applied to $F(z)$, if the contour of integration $\Gamma$ is decomposed as illustrated in Fig. 5.21.

14. State and prove a generalization of the Schwarz reflection principle to the case where the boundary of $D$ contains an arbitrary line segment upon which the limiting values of $f(z)$ all lie on some straight line (as would be the case, for example, if these limiting values were all real).

15. State and prove two reflection principles for harmonic functions $\phi(x, y)$, based upon the Schwarz reflection principle. One should cover the case when $\phi(x, y) \rightarrow 0$ on the real axis, and the other should apply when $\partial\phi/\partial y \rightarrow 0$ on the real axis.

## SUMMARY

The principal achievement of this chapter was to establish an equivalence (roughly speaking) between analytic functions and convergent power series. The equivalence is as follows: Any function $f$ can be expressed as a power series around any point $z_0$ at which $f$ is analytic, and the series converges uniformly in every closed disk (centered at $z_0$) which excludes the singularities of $f$. This series is known as the Taylor series and has the form

$$\sum_{j=0}^{\infty} \frac{f^{(j)}(z_0)}{j!} (z - z_0)^j.$$

On the other hand, any power series $\sum_{j=0}^{\infty} a_j(z - z_0)^j$ converging in a disk $|z - z_0| < R$ sums to an analytic function and, in fact, is the Taylor series of this function.

Power series can be added, integrated, and differentiated termwise, as can any uniformly convergent sequence of analytic functions. Moreover, power series can be multiplied like polynomials.

If a function $f$ fails to be analytic at certain points but is analytic in an annulus surrounding or excluding these points, it can be expanded in a Laurent series,

$$f(z) = \sum_{j=-\infty}^{\infty} a_j(z - z_0)^j,$$

converging uniformly in any closed subannulus; the nonanalyticity is reflected in the appearance of negative exponents in the expansion. In fact, if $z_0$ is an isolated singularity of the function $f$, the Laurent series can be used to classify $z_0$ into one of three categories: a removable singularity ($f$ bounded near $z_0$), a pole ($|f| \to \infty$ at $z_0$), or an essential singularity (neither of the above). Similarly, the Taylor series allows one to classify the order of an isolated zero of $f$.

The proof of the Taylor and Laurent expansions, and the actual computation of many Taylor and Laurent series, is facilitated by the use of the geometric series $\sum_{j=0}^{\infty} w^j$, which converges uniformly to $(1 - w)^{-1}$ on any closed disk of the form $|w| \le \rho < 1$.

Finally, we have seen how the power series expansions lead one to the possibility of extending the domain of definition of $f$ as an analytic function. Analytic continuation, however, is a subtle process in that it may result in multiple-valuedness.

## SUGGESTED READING

More detailed treatments of some of the topics of this chapter can be found in the following references:

### *Theory of Series*

[1] Dienes, P. *The Taylor Series*. Dover Publications, Inc., New York, 1957.

[2] Knopp, K. *Infinite Sequences and Series*. Dover Publications, Inc., New York, 1956.

### *Differential Equations*

[3] Birkhoff, G., and Rota, G. C. *Ordinary Differential Equations*, 4th ed. John Wiley & Sons, New York, 1989.

[4] Rainville, E. D. *Intermediate Differential Equations*, 2nd ed. Chelsea, New York, 1972.

### *Analytic Continuation and the Reflection Principle*

[5] Hille, E. *Analytic Function Theory*, Vol. II, 2nd ed. Chelsea, New York, 1973.

[6] Nehari, Z. *Conformal Mapping*. Dover, New York, 1975.

### *Positive Functions and Circuit Analysis*

[7] Van Valkenburg, M. E. *Introduction to Modern Network Synthesis*. John Wiley & Sons, New York, 1960.

[8] Levinson, N., and Redheffer, R. M. *Complex Variables*. Holden-Day, Inc., San Francisco, 1970.

# :6:

# Residue Theory

We have already seen how the theory of contour integration lends great insight into the properties of analytic functions. In this chapter we shall explore another dividend of this theory, namely its usefulness in evaluating certain *real* integrals. We shall begin by presenting a technique for evaluating contour integrals which is known as *residue theory*.

## 6.1 THE RESIDUE THEOREM

Let us consider the problem of evaluating the integral

$$\int_\Gamma f(z)\, dz,$$

where $\Gamma$ is a simple closed positively oriented contour and $f(z)$ is analytic on and inside $\Gamma$ *except* for a single isolated singularity, $z_0$, lying interior to $\Gamma$. As we know, the function $f(z)$ has a Laurent series expansion

$$f(z) = \sum_{j=-\infty}^{\infty} a_j(z - z_0)^j, \tag{1}$$

converging in some punctured neighborhood of $z_0$; in particular Eq. (1) is valid for all $z$ on the small positively oriented circle $C$ indicated in Fig. 6.1. By the methods of Sec. 4.4, integration over $\Gamma$ can be converted to integration over $C$ without changing the integral:

$$\int_\Gamma f(z)\, dz = \int_C f(z)\, dz.$$

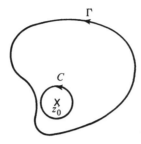

**Figure 6.1**    Contours for integration.

This last integral can be computed by termwise integration of the series (1) along $C$. For all $j \neq -1$ the integral is zero, and for $j = -1$ we obtain the value $2\pi i a_{-1}$. Consequently we have

$$\int_\Gamma f(z)\, dz = 2\pi i a_{-1}. \tag{2}$$

[Compare this with the formula for $a_{-1}$ given in Theorem 14 of Chapter 5 (page 211).]

Thus the constant $a_{-1}$ plays an important role in contour integration. Accordingly, we adopt the following terminology.

---

**Definition 1.**    If $f$ has an isolated singularity at the point $z_0$, then the coefficient $a_{-1}$ of $1/(z - z_0)$ in the Laurent expansion for $f$ around $z_0$ is called the **residue of $f$ at $z_0$** and is denoted by

$$\mathrm{Res}(f; z_0) \quad \text{or} \quad \mathrm{Res}(z_0).$$

---

**Example 1**

Find the residue at $z = 0$ of the function $f(z) = ze^{3/z}$ and compute

$$\oint_{|z|=4} ze^{3/z}\, dz.$$

**Solution.**    Since $e^w$ has the Taylor expansion

$$e^w = \sum_{j=0}^{\infty} \frac{w^j}{j!} \quad \text{(for all } w\text{)},$$

the Laurent expansion for $ze^{3/z}$ around $z = 0$ is given by

$$ze^{3/z} = z \sum_{j=0}^{\infty} \frac{1}{j!}\left(\frac{3}{z}\right)^j = z + 3 + \frac{3^2}{2!z} + \frac{3^3}{3!z^2} + \cdots.$$

Hence

$$\mathrm{Res}(0) = \frac{3^2}{2!} = \frac{9}{2},$$

and since $z = 0$ is the only singularity inside $|z| = 4$, we have, by formula (2),

$$\oint_{|z|=4} ze^{3/z}\, dz = 2\pi i \cdot \frac{9}{2} = 9\pi i. \quad \blacksquare$$

Now if $f$ has a *removable* singularity at $z_0$, all the coefficients of the negative powers of $(z - z_0)$ in its Laurent expansion are zero, and so, in particular, the residue at $z_0$ is zero. Furthermore, if $f$ has a *pole* at $z_0$, we shall see that its residue there can be computed from a formula. Suppose first that $z_0$ is a simple pole, i.e., a pole of order 1. Then for $z$ near $z_0$ we have

$$f(z) = \frac{a_{-1}}{z - z_0} + a_0 + a_1(z - z_0) + a_2(z - z_0)^2 + \cdots,$$

and so

$$(z - z_0)f(z) = a_{-1} + (z - z_0)[a_0 + a_1(z - z_0) + a_2(z - z_0)^2 + \cdots].$$

By taking the limit as $z \to z_0$ we deduce that

$$\lim_{z \to z_0} (z - z_0)f(z) = a_{-1} + 0.$$

Hence *at a simple pole*

$$\operatorname{Res}(f; z_0) = \lim_{z \to z_0} (z - z_0)f(z). \tag{3}$$

For example, the function $f(z) = \dfrac{e^z}{z(z + 1)}$ has simple poles at $z = 0$ and $z = -1$; therefore,

$$\operatorname{Res}(f; 0) = \lim_{z \to 0} z f(z) = \lim_{z \to 0} \frac{e^z}{z + 1} = 1,$$

and

$$\operatorname{Res}(f; -1) = \lim_{z \to -1} (z + 1)f(z) = \lim_{z \to -1} \frac{e^z}{z} = -e^{-1}.$$

Another consequence of formula (3) is illustrated in the next example.

## Example 2

Let $f(z) = P(z)/Q(z)$, where the functions $P(z)$ and $Q(z)$ are both analytic at $z_0$, and $Q$ has a simple zero at $z_0$, while $P(z_0) \neq 0$. Prove that

$$\operatorname{Res}(f; z_0) = \frac{P(z_0)}{Q'(z_0)}.$$

**Solution.**   Obviously $f$ has a simple pole at $z_0$ (see Sec. 5.6), so we can apply formula (3). Using the fact that $Q(z_0) = 0$ we see directly that

$$\operatorname{Res}(f; z_0) = \lim_{z \to z_0} (z - z_0) \frac{P(z)}{Q(z)} = \lim_{z \to z_0} \frac{P(z)}{\left[\dfrac{Q(z) - Q(z_0)}{z - z_0}\right]} = \frac{P(z_0)}{Q'(z_0)}. \quad \blacksquare$$

## Example 3

Compute the residue at each singularity of $f(z) = \cot z$.

**Solution.** Since $\cot z = \cos z/\sin z$, the singularities of this function are simple poles occurring at the points $z = n\pi$, $n = 0, \pm 1, \pm 2, \ldots$. Utilizing Example 2 with $P(z) = \cos z$, $Q(z) = \sin z$, the residues at these points are given by

$$\text{Res}(\cot z; n\pi) = \frac{\cos z}{(\sin z)'}\Big|_{z=n\pi} = \frac{\cos n\pi}{\cos n\pi} = 1. \quad \blacksquare$$

To obtain the general formula for the residue at a pole of order $m$ we need some method of picking out the coefficient $a_{-1}$ from the Laurent expansion. The reader should encounter no difficulty in following the derivation of the next formula.

---

**Theorem 1.**  If $f$ has a pole of order $m$ at $z_0$, then

$$\text{Res}(f; z_0) = \lim_{z \to z_0} \frac{1}{(m-1)!} \frac{d^{m-1}}{dz^{m-1}} [(z - z_0)^m f(z)]. \tag{4}$$

---

*Proof.*  Starting with the Laurent expansion for $f$ around $z_0$,

$$f(z) = \frac{a_{-m}}{(z-z_0)^m} + \cdots + \frac{a_{-2}}{(z-z_0)^2} + \frac{a_{-1}}{z-z_0} + a_0 + a_1(z-z_0) + \cdots,$$

we multiply by $(z - z_0)^m$,

$$(z-z_0)^m f(z) = a_{-m} + \cdots + a_{-2}(z-z_0)^{m-2} + a_{-1}(z-z_0)^{m-1}$$
$$+ a_0(z-z_0)^m + a_1(z-z_0)^{m+1} + \cdots,$$

and differentiate $m - 1$ times to derive

$$\frac{d^{m-1}}{dz^{m-1}} [(z-z_0)^m f(z)] = (m-1)! \, a_{-1} + m! \, a_0(z-z_0) + \frac{(m+1)!}{2} a_1(z-z_0)^2 + \cdots.$$

Hence

$$\lim_{z \to z_0} \frac{d^{m-1}}{dz^{m-1}} [(z-z_0)^m f(z)] = (m-1)! \, a_{-1} + 0,$$

which is equivalent to Eq. (4).  $\blacksquare$

**Example 4**

Compute the residues at the singularities of

$$f(z) = \frac{\cos z}{z^2(z-\pi)^3}.$$

**Solution.**  This function has a pole of order 2 at $z = 0$ and a pole of order 3 at $z = \pi$. Applying formula (4) we find

$$\text{Res}(0) = \lim_{z \to 0} \frac{1}{1!} \frac{d}{dz} [z^2 f(z)] = \lim_{z \to 0} \frac{d}{dz} \left[ \frac{\cos z}{(z-\pi)^3} \right]$$

$$= \lim_{z \to 0} \left[ \frac{-(z - \pi) \sin z - 3 \cos z}{(z - \pi)^4} \right] = \frac{-3}{\pi^4},$$

$$\text{Res}(\pi) = \lim_{z \to \pi} \frac{1}{2!} \frac{d^2}{dz^2} \left[ (z - \pi)^3 f(z) \right] = \lim_{z \to \pi} \frac{1}{2} \frac{d^2}{dz^2} \left[ \frac{\cos z}{z^2} \right]$$

$$= \lim_{z \to \pi} \frac{1}{2} \left[ \frac{(6 - z^2) \cos z + 4z \sin z}{z^4} \right] = \frac{-(6 - \pi^2)}{2\pi^4}. \quad \blacksquare$$

We have already seen how to compute the integral $\int_\Gamma f(z) \, dz$ when $f(z)$ has only one singularity inside $\Gamma$. Let's now turn to the more general case where $\Gamma$ is a simple closed positively oriented contour and $f(z)$ is analytic inside and on $\Gamma$ except for a finite number of isolated singularities at the points $z_1, z_2, \dots, z_n$ inside $\Gamma$ (see Fig. 6.2). Notice that by the methods of Sec. 4.4 we can express the integral along $\Gamma$ in terms of the integrals around the circles $C_j$ in Fig. 6.3:

$$\int_\Gamma f(z) \, dz = \sum_{j=1}^{n} \int_{C_j} f(z) \, dz.$$

However, because $z_j$ is the only singularity of $f$ inside $C_j$, we know that

$$\int_{C_j} f(z) \, dz = 2\pi i \, \text{Res}(z_j).$$

Hence we have established the following important result.

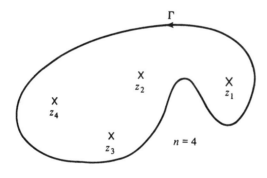

Figure 6.2  Isolated singularities inside contour.

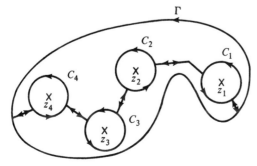

Figure 6.3  Equivalent contours for integration.

---

**Theorem 2.** (*Cauchy's Residue Theorem*) If $\Gamma$ is a simple closed positively oriented contour and $f$ is analytic inside and on $\Gamma$ except at the points $z_1, z_2, \ldots, z_n$ inside $\Gamma$, then

$$\int_\Gamma f(z)\, dz = 2\pi i \sum_{j=1}^{n} \text{Res}(z_j). \tag{5}$$

---

### Example 5

Evaluate

$$\oint_{|z|=2} \frac{1 - 2z}{z(z-1)(z-3)}\, dz.$$

**Solution.** The integrand $f(z) = (1 - 2z)/[z(z-1)(z-3)]$ has simple poles at $z = 0$, $z = 1$, and $z = 3$. However, only the first two of these points lie inside $\Gamma: |z| = 2$. Thus by the residue theorem

$$\oint_{|z|=2} f(z)\, dz = 2\pi i[\text{Res}(0) + \text{Res}(1)],$$

and since

$$\text{Res}(0) = \lim_{z \to 0} z f(z) = \lim_{z \to 0} \frac{1 - 2z}{(z-1)(z-3)} = \frac{1}{3},$$

$$\text{Res}(1) = \lim_{z \to 1} (z-1) f(z) = \lim_{z \to 1} \frac{1 - 2z}{z(z-3)} = \frac{1}{2},$$

we obtain

$$\oint_{|z|=2} f(z)\, dz = 2\pi i\left(\frac{1}{3} + \frac{1}{2}\right) = \frac{5\pi i}{3}. \quad \blacksquare$$

### Example 6

Compute

$$\oint_{|z|=5} \left[ z e^{3/z} + \frac{\cos z}{z^2(z-\pi)^3} \right] dz.$$

**Solution.** The given integral can obviously be expressed as the sum

$$\oint_{|z|=5} z e^{3/z}\, dz + \oint_{|z|=5} \frac{\cos z}{z^2(z-\pi)^3}\, dz,$$

which, by the residue theorem, equals

$$2\pi i \left[ \text{Res}(z e^{3/z}; 0) + \text{Res}\left(\frac{\cos z}{z^2(z-\pi)^3}; 0\right) + \text{Res}\left(\frac{\cos z}{z^2(z-\pi)^3}; \pi\right) \right].$$

These residues were computed in Examples 1 and 4; the desired answer is therefore

$$2\pi i \left[ \frac{9}{2} - \frac{3}{\pi^4} - \frac{(6 - \pi^2)}{2\pi^4} \right]. \quad \blacksquare$$

## EXERCISES 6.1

1. Determine all the isolated singularities of each of the following functions and compute the residue at each singularity.

   (a) $\dfrac{e^{3z}}{z - 2}$

   (b) $\dfrac{z + 1}{z^2 - 3z + 2}$

   (c) $\dfrac{\cos z}{z^2}$

   (d) $\left( \dfrac{z - 1}{z + 1} \right)^3$

   (e) $\dfrac{e^z}{z(z + 1)^3}$

   (f) $\sin\left( \dfrac{1}{3z} \right)$

   (g) $\tan z$

   (h) $\dfrac{z - 1}{\sin z}$

   (i) $z^2/(1 - \sqrt{z})$, where $\sqrt{z}$ denotes the principal branch.

2. Explain why Cauchy's integral formula can be regarded as a special case of the residue theorem.

3. Evaluate each of the following integrals by means of the Cauchy residue theorem.

   (a) $\displaystyle\oint_{|z| = 5} \frac{\sin z}{z^2 - 4} \, dz$

   (b) $\displaystyle\oint_{|z| = 3} \frac{e^z}{z(z - 2)^3} \, dz$

   (c) $\displaystyle\oint_{|z| = 2\pi} \tan z \, dz$

   (d) $\displaystyle\oint_{|z| = 3} \frac{e^{iz}}{z^2(z - 2)(z + 5i)} \, dz$

   (e) $\displaystyle\oint_{|z| = 1} \frac{1}{z^2 \sin z} \, dz$

   (f) $\displaystyle\oint_{|z| = 3} \frac{3z + 2}{z^4 + 1} \, dz$

   (g) $\displaystyle\oint_{|z| = 8} \frac{1}{z^2 + z + 1} \, dz$

4. Let $f$ have an isolated singularity at $z_0$ ($f$ analytic in a punctured neighborhood of $z_0$). Show that the residue of the derivative $f'$ at $z_0$ is equal to zero.

5. Does there exist a function $f$ having a simple pole at $z_0$ with $\text{Res}(f; z_0) = 0$? How about a function with a pole of order 2 at $z_0$ and $\text{Res}(f; z_0) = 0$?

6. Suppose that $f$ is analytic and has a zero of order $m$ at the point $z_0$. Show that the function $g(z) = f'(z)/f(z)$ has a simple pole at $z_0$ with $\text{Res}(g; z_0) = m$.

7. Evaluate

$$\oint_{|z| = 1} e^{1/z} \sin(1/z) \, dz.$$

## 6.2 TRIGONOMETRIC INTEGRALS OVER $[0, 2\pi]$

Our goal here is to apply the residue theorem to evaluate real integrals of the form

$$\int_0^{2\pi} U(\cos \theta, \sin \theta) \, d\theta, \tag{1}$$

where $U(\cos \theta, \sin \theta)$ is a rational function (with real coefficients) of $\cos \theta$ and $\sin \theta$ and is finite over $[0, 2\pi]$. An example of such an integral is

$$\int_0^{2\pi} \frac{\sin^2 \theta}{a + b \cos \theta} \, d\theta \qquad (0 < b < a).$$

We shall show that (1) can be identified as the parametrized form of a contour integral, $\int_C F(z) \, dz$, of some complex function $F$ around the positively oriented unit circle $C: |z| = 1$. To establish this identification we parametrize $C$ by

$$z = e^{i\theta} \qquad (0 \le \theta \le 2\pi).$$

For such $z$ we have[†]

$$\frac{1}{z} = \frac{1}{e^{i\theta}} = e^{-i\theta},$$

and since

$$\cos \theta = \frac{e^{i\theta} + e^{-i\theta}}{2}, \qquad \sin \theta = \frac{e^{i\theta} - e^{-i\theta}}{2i},$$

we have the identities

$$\cos \theta = \frac{1}{2}\left(z + \frac{1}{z}\right), \qquad \sin \theta = \frac{1}{2i}\left(z - \frac{1}{z}\right). \qquad (2)$$

Furthermore, when integrating along $C$,

$$dz = ie^{i\theta} \, d\theta = iz \, d\theta,$$

so that

$$d\theta = \frac{dz}{iz}. \qquad (3)$$

Making the substitutions (2) and (3) in the integral (1), we see that

$$\int_0^{2\pi} U(\cos \theta, \sin \theta) \, d\theta = \int_C F(z) \, dz, \qquad (4)$$

where the new integrand $F$ is

$$F(z) := U\left[\frac{1}{2}\left(z + \frac{1}{z}\right), \frac{1}{2i}\left(z - \frac{1}{z}\right)\right] \cdot \frac{1}{iz};$$

integration over $[0, 2\pi]$ has thus been replaced by integration around $C$.

Because of the form of $U$, the function $F$ must be a rational function of $z$. Hence it has only removable singularities (which can be ignored in evaluating integrals) or poles. Consequently, by the residue theorem, our trigonometric integral equals $2\pi i$ times the sum of the residues at those poles of $F$ which lie inside $C$.

The procedure is illustrated in the next example.

---

[†] Of course we could use $\bar{z}$ instead of $1/z$, but we would forfeit analyticity.

**Example 1**

Evaluate

$$I = \int_0^{2\pi} \frac{\sin^2 \theta}{5 + 4 \cos \theta} \, d\theta.$$

**Solution.**   First observe that the denominator, $5 + 4 \cos \theta$, is never zero, so the integrand is finite over $[0, 2\pi]$. Performing the substitutions (2) and (3) for $\cos \theta$, $\sin \theta$, and $d\theta$, we obtain

$$I = \int_C \frac{\left[\frac{1}{2i}\left(z - \frac{1}{z}\right)\right]^2}{5 + 4\left[\frac{1}{2}\left(z + \frac{1}{z}\right)\right]} \frac{dz}{iz},$$

which after some algebra reduces to

$$I = -\frac{1}{4i} \int_C \frac{(z^2 - 1)^2}{z^2(2z^2 + 5z + 2)} \, dz.$$

Clearly the integrand

$$g(z) := \frac{(z^2 - 1)^2}{z^2(2z^2 + 5z + 2)} = \frac{(z^2 - 1)^2}{2z^2(z + \frac{1}{2})(z + 2)}$$

has simple poles at $z = -\frac{1}{2}$ and $z = -2$, and has a pole of order 2 at the origin. However, only $-\frac{1}{2}$ and $0$ lie inside the unit circle $C$, so that

$$I = -\frac{1}{4i} \cdot 2\pi i \left[ \text{Res}\left(g; -\frac{1}{2}\right) + \text{Res}(g; 0) \right].$$

Utilizing the formulas of the preceding section we find

$$\text{Res}\left(g; -\frac{1}{2}\right) = \lim_{z \to -1/2} \left(z + \frac{1}{2}\right) g(z) = \lim_{z \to -1/2} \frac{(z^2 - 1)^2}{2z^2(z + 2)} = \frac{3}{4},$$

and

$$\text{Res}(g; 0) = \lim_{z \to 0} \frac{1}{1!} \frac{d}{dz} \left[z^2 g(z)\right] = \lim_{z \to 0} \frac{d}{dz} \left[\frac{(z^2 - 1)^2}{2z^2 + 5z + 2}\right]$$

$$= \frac{(2z^2 + 5z + 2) \cdot 2(z^2 - 1)2z - (z^2 - 1)^2(4z + 5)}{(2z^2 + 5z + 2)^2} \bigg|_{z=0}$$

$$= \frac{-5}{4}.$$

Hence

$$I = \frac{-1}{4i} 2\pi i \left[\frac{3}{4} - \frac{5}{4}\right] = \frac{\pi}{4}. \quad \blacksquare$$

As a simple check on our calculations we observe that the integrand of Example 1 is real and nonnegative, so $I$ must be a positive real number, which is consistent with our answer of $\pi/4$.

**Example 2**

Evaluate

$$I = \int_0^\pi \frac{d\theta}{2 - \cos \theta}.$$

**Solution.** The catch here is that the integral is taken over $[0, \pi]$ instead of $[0, 2\pi]$. However it is easy to see that, since $\cos \theta = \cos(2\pi - \theta)$,

$$\int_0^\pi \frac{d\theta}{2 - \cos \theta} = \int_\pi^{2\pi} \frac{d\theta}{2 - \cos \theta},$$

and, therefore,

$$\int_0^{2\pi} \frac{d\theta}{2 - \cos \theta} = 2I.$$

Substituting for $\cos \theta$ and $d\theta$ we have

$$2I = \int_C \frac{1}{2 - \frac{1}{2}\left(z + \frac{1}{z}\right)} \cdot \frac{dz}{iz} = -\frac{2}{i} \int_C \frac{dz}{z^2 - 4z + 1}. \tag{5}$$

By the quadratic formula the zeros of the denominator are

$$z_1 := 2 - \sqrt{3} \quad \text{and} \quad z_2 := 2 + \sqrt{3},$$

and so the integrand

$$g(z) := \frac{1}{z^2 - 4z + 1} = \frac{1}{(z - z_1)(z - z_2)}$$

has simple poles at these points. But only $z_1$ lies inside $C$, and the residue there is given by

$$\text{Res}(g; z_1) = \lim_{z \to z_1} (z - z_1)\, g(z) = \lim_{z \to z_1} \frac{1}{(z - z_2)}$$

$$= \frac{1}{z_1 - z_2} = -\frac{1}{2\sqrt{3}}.$$

Hence from Eq. (5)

$$2I = -\frac{2}{i} \cdot 2\pi i \left(-\frac{1}{2\sqrt{3}}\right) = \frac{2\pi}{\sqrt{3}},$$

or

$$I = \frac{\pi}{\sqrt{3}}. \quad \blacksquare$$

## EXERCISES 6.2

*Using the method of residues, verify each of the following.*

1. $\displaystyle\int_0^{2\pi} \frac{d\theta}{2 + \sin\theta} = \frac{2\pi}{\sqrt{3}}$

2. $\displaystyle\int_0^{\pi} \frac{8\,d\theta}{5 + 2\cos\theta} = \frac{8\pi}{\sqrt{21}}$

3. $\displaystyle\int_0^{\pi} \frac{d\theta}{(3 + 2\cos\theta)^2} = \frac{3\pi\sqrt{5}}{25}$

4. $\displaystyle\int_{-\pi}^{\pi} \frac{d\theta}{1 + \sin^2\theta} = \pi\sqrt{2}$

5. $\displaystyle\int_0^{2\pi} \frac{d\theta}{1 + a\cos\theta} = \frac{2\pi}{\sqrt{1 - a^2}}, \quad a^2 < 1$

6. $\displaystyle\int_0^{2\pi} \frac{\sin^2\theta}{a + b\cos\theta}\,d\theta = \frac{2\pi}{b^2}(a - \sqrt{a^2 - b^2}), \quad a > |b| > 0$

7. $\displaystyle\int_0^{\pi} \frac{d\theta}{(a + \sin^2\theta)^2} = \frac{\pi(2a + 1)}{2\sqrt{(a^2 + a)^3}}, \quad a > 0$

8. $\displaystyle\int_0^{2\pi} \frac{d\theta}{a^2 \sin^2\theta + b^2 \cos^2\theta} = \frac{2\pi}{ab}, \quad a, b > 0$

9. $\displaystyle\int_0^{2\pi} (\cos\theta)^{2n}\,d\theta = \frac{\pi \cdot (2n)!}{2^{2n-1}(n!)^2}, \quad n = 1, 2, \ldots$

10. $\displaystyle\int_0^{2\pi} e^{\cos\theta} \cos(n\theta - \sin\theta)\,d\theta = \frac{2\pi}{n!}, \quad n = 1, 2, \ldots$

11. $\displaystyle\int_0^{\pi} \tan(\theta + ia)\,d\theta = \pi i \cdot \text{sign } a, \quad a \text{ real and nonzero}$

## 6.3 IMPROPER INTEGRALS OF CERTAIN FUNCTIONS OVER $(-\infty, \infty)$

If $f(x)$ is a function continuous on the nonnegative real axis $0 \le x < \infty$, then the improper integral of $f$ over $[0, \infty)$ is defined by

$$\int_0^{\infty} f(x)\,dx := \lim_{b \to \infty} \int_0^b f(x)\,dx, \tag{1}$$

provided that this limit exists.[†] For example,

$$\int_0^{\infty} e^{-2x}\,dx = \lim_{b \to \infty} \int_0^b e^{-2x}\,dx = \lim_{b \to \infty} \frac{-e^{-2x}}{2}\Big|_0^b$$

$$= \lim_{b \to \infty} \left[\frac{-e^{-2b}}{2} + \frac{1}{2}\right] = \frac{1}{2}.$$

---

[†] More generally, $\int_a^{\infty} f(x)\,dx := \lim_{b \to \infty} \int_a^b f(x)\,dx$ if this limit exists.

Similarly, when $f(x)$ is continuous on $(-\infty, 0]$, we set

$$\int_{-\infty}^{0} f(x)\, dx := \lim_{c \to -\infty} \int_{c}^{0} f(x)\, dx. \tag{2}$$

If it turns out that both of the limits (1) and (2) exist for a function $f$ continuous on the whole real line, then $f$ is said to be *integrable over* $(-\infty, \infty)$, and we write

$$\int_{-\infty}^{\infty} f(x)\, dx := \lim_{c \to -\infty} \int_{c}^{0} f(x)\, dx + \lim_{b \to \infty} \int_{0}^{b} f(x)\, dx$$

$$= \int_{-\infty}^{0} f(x)\, dx + \int_{0}^{\infty} f(x)\, dx.$$

In such a case the value of the improper integral over $(-\infty, \infty)$ can be computed by taking a single limit, namely,

$$\int_{-\infty}^{\infty} f(x)\, dx = \lim_{\rho \to \infty} \int_{-\rho}^{\rho} f(x)\, dx.$$

However, we caution the reader that this last limit may exist even for certain *non-integrable* functions $f$. Indeed, consider $f(x) = x$. This function is not integrable over $(-\infty, \infty)$ because the limit

$$\lim_{b \to \infty} \int_{0}^{b} x\, dx = \lim_{b \to \infty} \frac{x^2}{2}\Big|_{0}^{b} = \lim_{b \to \infty} \frac{b^2}{2}$$

does not exist (as a finite number). However,

$$\lim_{\rho \to \infty} \int_{-\rho}^{\rho} x\, dx = \lim_{\rho \to \infty} \frac{x^2}{2}\Big|_{-\rho}^{\rho} = \lim_{\rho \to \infty} 0 = 0.$$

For this reason we introduce the following terminology: Given *any* function $f$ continuous on $(-\infty, \infty)$ the limit

$$\lim_{\rho \to \infty} \int_{-\rho}^{\rho} f(x)\, dx$$

(if it exists) is called the *Cauchy principal value* of the integral of $f$ over $(-\infty, \infty)$, and we write

$$\text{p.v.} \int_{-\infty}^{\infty} f(x)\, dx := \lim_{\rho \to \infty} \int_{-\rho}^{\rho} f(x)\, dx.$$

For example,

$$\text{p.v.} \int_{-\infty}^{\infty} x\, dx = 0.$$

We reiterate that whenever the improper integral $\int_{-\infty}^{\infty} f(x)\, dx$ exists, it must equal its principal value (p.v.).

We shall now show how the theory of residues can be used to compute p.v. *integrals* for certain functions $f$.

**Example 1**

Evaluate

$$I = \text{p.v.} \int_{-\infty}^{\infty} \frac{dx}{x^4 + 4} \left( = \lim_{\rho \to \infty} \int_{-\rho}^{\rho} \frac{dx}{x^4 + 4} \right).$$

**Solution.** As a first step, we recognize that the integral $I_\rho$ defined by

$$I_\rho := \int_{-\rho}^{\rho} \frac{dx}{x^4 + 4}$$

can be interpreted as a contour integral of an analytic function; in fact

$$I_\rho = \int_{\gamma_\rho} \frac{dz}{z^4 + 4},$$

where $\gamma_\rho$ is the directed segment of the real axis from $-\rho$ to $+\rho$. Now the key to using residue theory to find $I$ lies in constructing (for each sufficiently large value of $\rho$) a simple *closed* contour $\Gamma_\rho$ such that $\gamma_\rho$ is one of its components, i.e., $\Gamma_\rho = (\gamma_\rho, \gamma_\rho')$, and such that the integral of $1/(z^4 + 4)$ along the other component $\gamma_\rho'$ is somehow known. For then we will have

$$\int_{\Gamma_\rho} \frac{dz}{z^4 + 4} = I_\rho + \int_{\gamma_\rho'} \frac{dz}{z^4 + 4},$$

and if $\Gamma_\rho$ is positively oriented, the residue theorem yields

$$2\pi i \cdot \sum (\text{residues inside } \Gamma_\rho) = I_\rho + \int_{\gamma_\rho'} \frac{dz}{z^4 + 4}.$$

Consequently, the integral $I$ is evaluated by

$$I = \lim_{\rho \to \infty} I_\rho = \lim_{\rho \to \infty} 2\pi i \sum (\text{residues inside } \Gamma_\rho) - \lim_{\rho \to \infty} \int_{\gamma_\rho'} \frac{dz}{z^4 + 4}, \qquad (3)$$

provided the limits on the right exist. Actually, we see from Eq. (3) that it is only the *limiting value* of the integrals over $\gamma_\rho'$ which must be known in order to apply residue theory.

Now among the many curves that "close the contour $\gamma_\rho$," i.e., that start at $z = \rho$ and terminate at $z = -\rho$, how are we to find a suitable curve $\gamma_\rho'$? Observe that the integrand, $1/(z^4 + 4)$, is quite small in modulus when $|z|$ is large. This suggests that if we choose our curves $\gamma_\rho'$ far enough away from the origin, the integrals over them might well be negligible, i.e., approach zero as $\rho \to \infty$. Thus an obvious thing to try for $\gamma_\rho'$ is the half-circle $C_\rho^+$ parametrized by

$$C_\rho^+ : \quad z = \rho e^{it} \quad (0 \le t \le \pi) \qquad (4)$$

(see Fig. 6.4). To see if this works, we note that $|z| = \rho$ on $C_\rho^+$, so that by the triangle inequality

$$\left| \frac{1}{z^4 + 4} \right| \le \frac{1}{|z|^4 - 4} = \frac{1}{\rho^4 - 4} \qquad (\text{for } \rho^4 > 4),$$

and hence

$$\left| \int_{C_\rho^+} \frac{dz}{z^4 + 4} \right| \le \frac{1}{\rho^4 - 4} \cdot \pi\rho,$$

which certainly does go to zero as $\rho \to \infty$.

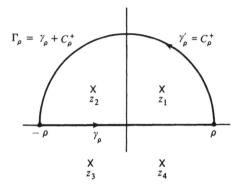

**Figure 6.4**   Closing the contour.

So now all we have to do in Eq. (3) is to evaluate the appropriate residues. First we locate the singularities of $1/(z^4 + 4)$. These occur at the zeros of $z^4 + 4$, i.e., at

$$z_1 = 1 + i, \qquad z_2 = -1 + i, \qquad z_3 = -1 - i, \qquad z_4 = 1 - i,$$

and the function

$$\frac{1}{z^4 + 4} = \frac{1}{(z - z_1)(z - z_2)(z - z_3)(z - z_4)}$$

has simple poles at these points. Since $z_3$ and $z_4$ lie in the lower half-plane, they are always excluded from the interior of the semicircular contour $\Gamma_\rho$ of Fig. 6.4, but $z_1$ and $z_2$ lie inside $\Gamma_\rho$ for every $\rho > \sqrt{2}$. Thus, for such $\rho$,

$$\int_{\Gamma_\rho} \frac{dz}{z^4 + 4} = 2\pi i [\text{Res}(z_1) + \text{Res}(z_2)]$$

$$= 2\pi i \left( \lim_{z \to z_1} \frac{z - z_1}{z^4 + 4} + \lim_{z \to z_2} \frac{z - z_2}{z^4 + 4} \right)$$

$$= 2\pi i \left[ \frac{1}{(z_1 - z_2)(z_1 - z_3)(z_1 - z_4)} + \frac{1}{(z_2 - z_1)(z_2 - z_3)(z_2 - z_4)} \right]$$

$$= 2\pi i \left[ \frac{1}{2(2 + 2i)2i} + \frac{1}{(-2)(2i)(-2 + 2i)} \right]$$

$$= 2\pi i \left[ \frac{-1 - i}{16} + \frac{1 - i}{16} \right] = \frac{\pi}{4}.$$

Putting this all together in Eq. (3) (with $\gamma_\rho' = C_\rho^+$) we have

$$I = \lim_{\rho \to \infty} \frac{\pi}{4} - \lim_{\rho \to \infty} \int_{C_\rho^+} \frac{dz}{z^4 + 4} = \frac{\pi}{4} - 0 = \frac{\pi}{4}. \quad \blacksquare$$

The technique of using expanding semicircular contours $\Gamma_\rho$ can readily be applied to a general class of integrands $f$. Indeed, the success of the procedure illustrated in Example 1 depends only on the following two conditions:

(i) $f$ is analytic on and above the real axis except for a *finite* number of isolated singularities in the open upper half-plane $\operatorname{Im} z > 0$ (this ensures that for $\rho$ sufficiently large, all the singularities in the upper half-plane will lie inside the contour $\Gamma_\rho$ of Fig. 6.4), and

(ii) $\lim_{\rho \to \infty} \int_{C_\rho^+} f(z)\, dz = 0$.

Whenever these conditions are satisfied, the value of the integral

$$\text{p.v.} \int_{-\infty}^{\infty} f(x)\, dx$$

is given by $2\pi i$ times the sum of the residues of $f$ at the singularities in the upper half-plane. (Of course, the lower half-plane can be used whenever analogous conditions hold there; see Prob. 8.)

A class of rational functions having property (ii) is given in the next lemma.

---

**Lemma 1.**   If $f(z) = P(z)/Q(z)$ is the quotient of two polynomials such that

$$\text{degree } Q \geq 2 + \text{degree } P, \tag{5}$$

then

$$\lim_{\rho \to \infty} \int_{C_\rho^+} f(z)\, dz = 0, \tag{6}$$

where $C_\rho^+$ is the upper half-circle of radius $\rho$ defined in Eq. (4).

---

*Proof.*   For large $|z|$, inequality (5) implies that

$$|f(z)| \leq \frac{K}{|z|^2},$$

where $K$ is some constant (recall Lemma 1 of Sec. 4.6, page 163); consequently,

$$\left| \int_{C_\rho^+} f(z)\, dz \right| \leq \frac{K}{\rho^2} \cdot \pi\rho = \frac{K\pi}{\rho} \to 0 \quad \text{as} \quad \rho \to \infty. \quad \blacksquare$$

We remark that the same proof shows that Eq. (6) remains valid if integration along $C_\rho^+$ is replaced by integration along the lower half-circle $C_\rho^-$.

**Example 2**

Compute

$$I = \text{p.v.} \int_{-\infty}^{\infty} \frac{x^2}{(x^2 + 1)^2}\, dx.$$

**Solution.**   Since the integrand has no singularities on the real axis and has numerator degree 2 and denominator degree 4, the expanding semicircular contour method is justified thanks to Lemma 1. On writing

$$f(z) := \frac{z^2}{(z^2+1)^2} = \frac{z^2}{(z-i)^2(z+i)^2},$$

we see that poles occur at $z = \pm i$. Thus, for any $\rho > 1$, the integral along the closed contour $\Gamma_\rho$ in Fig. 6.5 is given by

$$\int_{\Gamma_\rho} f(z)\,dz = 2\pi i \operatorname{Res}(f; +i).$$

Since $+i$ is a second-order pole, we have from the residue formula (Theorem 1, page 248)

$$\operatorname{Res}(f; +i) = \lim_{z \to i} \frac{1}{1!} \frac{d}{dz}\left[(z-i)^2 f(z)\right] = \lim_{z \to i} \frac{d}{dz}\left[\frac{z^2}{(z+i)^2}\right]$$

$$= \lim_{z \to i}\left[\frac{(z+i)2z - 2z^2}{(z+i)^3}\right] = \frac{1}{4i}.$$

Therefore,

$$\int_{\Gamma_\rho} f(z)\,dz = 2\pi i \cdot \frac{1}{4i} = \frac{\pi}{2} \qquad \text{(for all } \rho > 1). \tag{7}$$

On the other hand,

$$\int_{\Gamma_\rho} f(z)\,dz = \int_{-\rho}^{\rho} f(x)\,dx + \int_{C_\rho^+} f(z)\,dz,$$

and so on taking the limit as $\rho \to \infty$ in this last equation we deduce from Eq. (7) and Lemma 1 that

$$\frac{\pi}{2} = \lim_{\rho \to \infty} \int_{-\rho}^{\rho} f(x)\,dx + 0.$$

Hence

$$\frac{\pi}{2} = \text{p.v.} \int_{-\infty}^{\infty} \frac{x^2}{(x^2+1)^2}\,dx. \qquad \blacksquare$$

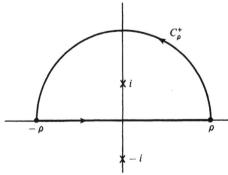

$$\Gamma_\rho = [-\rho, \rho] + C_\rho^+$$

**Figure 6.5**   Closed contour for Example 2.

As we shall see in the next section, semicircular contours are also useful in evaluating certain integrals involving trigonometric functions. Occasionally, a special integral may call for a specifically tailored contour, as in the following example.

**Example 3**

Compute

$$I = \text{p.v.} \int_{-\infty}^{\infty} \frac{e^{ax}}{1 + e^x} \, dx, \qquad \text{for } 0 < a < 1.$$

**Solution.** Observe that the function $e^{az}/(1 + e^z)$ has an infinite number of singularities in both the upper and lower half-planes; these occur at the points

$$z = (2n + 1)\pi i \qquad (n = 0, \pm 1, \pm 2, \ldots).$$

Hence if we employ expanding semicircles, the contribution due to the residues will result in an infinite series, which is undesirable. Moreover, there is no obvious way to estimate the contribution due to the semicircles themselves!

A better "return path" is revealed through careful examination of the integrand. The denominator of the function $e^{az}/(1 + e^z)$ is unchanged if $z$ is shifted by $2\pi i$, whereas the numerator changes by a factor $e^{2\pi a i}$. Thus if we consider the rectangular contour $\Gamma_\rho$ in Fig. 6.6, the contribution from $\gamma_3$ is easy to assess; it's merely $-e^{2\pi a i}$ times the contribution from $\gamma_1$ (negative because the path runs from right to left). Thus

$$\int_{\gamma_3} \frac{e^{az}}{1 + e^z} \, dz = -e^{2\pi a i} \int_{\gamma_1} \frac{e^{az}}{1 + e^z} \, dz.$$

For $\gamma_2$: $z = \rho + it$, $0 \le t \le 2\pi$, we have

$$\left| \int_{\gamma_2} \frac{e^{az}}{1 + e^z} \, dz \right| = \left| \int_0^{2\pi} \frac{e^{a(\rho + it)}}{1 + e^{\rho + it}} \, i \, dt \right|$$

$$\le \frac{e^{a\rho}}{e^{\rho} - 1} \cdot 2\pi,$$

which goes to zero as $\rho \to \infty$ since $a < 1$.

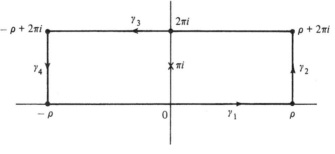

$$\Gamma_\rho = \gamma_1 + \gamma_2 + \gamma_3 + \gamma_4$$

**Figure 6.6**  Closed contour for Example 3.

Similarly, on $\gamma_4$: $z = -\rho + i(2\pi - t)$, $0 \le t \le 2\pi$, we have

$$\left| \int_{\gamma_4} \frac{e^{az}}{1+e^z} \, dz \right| = \left| \int_0^{2\pi} \frac{e^{a[-\rho+i(2\pi-t)]}}{1 + e^{-\rho+i(2\pi-t)}} (-i) \, dt \right|$$

$$\le \frac{e^{-a\rho}}{1 - e^{-\rho}} \cdot 2\pi,$$

again approaching zero as $\rho \to \infty$ since $a > 0$.

As a result, on taking the limit as $\rho \to \infty$ we have

$$\lim_{\rho \to \infty} \int_{\Gamma_\rho} \frac{e^{az}}{1+e^z} \, dz = (1 - e^{a2\pi i}) \text{ p.v.} \int_{-\infty}^{\infty} \frac{e^{ax}}{1+e^x} \, dx. \qquad (8)$$

Now we use residue theory to evaluate the contour integral in Eq. (8). For each $\rho > 0$, the function $e^{az}/(1 + e^z)$ is analytic inside and on $\Gamma_\rho$ except for a simple pole at $z = \pi i$, the residue there being given by

$$\text{Res}(\pi i) = \frac{e^{az}}{\dfrac{d}{dz}(1 + e^z)} \bigg|_{z = \pi i} = \frac{e^{a\pi i}}{e^{\pi i}} = -e^{a\pi i} \qquad (9)$$

(recall Example 2, Sec. 6.1, page 247). Consequently, putting Eqs. (8) and (9) together we obtain

$$\text{p.v.} \int_{-\infty}^{\infty} \frac{e^{ax}}{1+e^x} \, dx = \frac{1}{1 - e^{a2\pi i}} \cdot (2\pi i)(-e^{a\pi i})$$

$$= \frac{-2\pi i}{e^{-a\pi i} - e^{a\pi i}}$$

$$= \frac{\pi}{\sin a\pi}. \qquad \blacksquare$$

## EXERCISES 6.3

*Verify the integral formulas in Problems 1–7 with the aid of residues.*

**1.** p.v. $\displaystyle\int_{-\infty}^{\infty} \frac{dx}{x^2 + 2x + 2} = \pi$

**2.** p.v. $\displaystyle\int_{-\infty}^{\infty} \frac{x^2}{(x^2+9)^2} \, dx = \frac{\pi}{6}$

**3.** $\displaystyle\int_0^{\infty} \frac{x^2+1}{x^4+1} \, dx = \frac{\pi}{\sqrt{2}}$

**4.** p.v. $\displaystyle\int_{-\infty}^{\infty} \frac{dx}{(x^2+1)(x^2+4)} = \frac{\pi}{6}$

**5.** p.v. $\displaystyle\int_{-\infty}^{\infty} \frac{x}{(x^2+4x+13)^2} \, dx = -\frac{\pi}{27}$

**6.** $\displaystyle\int_0^{\infty} \frac{x^2}{(x^2+1)(x^2+4)} \, dx = \frac{\pi}{6}$

**7.** $\displaystyle\int_0^{\infty} \frac{x^6}{(x^4+1)^2} \, dx = \frac{3\pi\sqrt{2}}{16}$

**8.** Show that if $f(z) = P(z)/Q(z)$ is the quotient of two polynomials such that $\deg Q \geq 2 + \deg P$, where $Q$ has no real zeros, then

$$\text{p.v.} \int_{-\infty}^{\infty} f(x)\,dx = -2\pi i \cdot \sum [\text{residues of } f(z) \text{ at the poles in the lower half-plane}].$$

**9.** Show that

$$\text{p.v.} \int_{-\infty}^{\infty} \frac{e^{2x}}{\cosh(\pi x)}\,dx = \sec 1$$

by integrating $e^{2z}/\cosh(\pi z)$ around rectangles with vertices at $z = \pm \rho,\ \rho + i,\ -\rho + i$.

**10.** Given that

$$\int_{0}^{\infty} e^{-x^2}\,dx = \frac{\sqrt{\pi}}{2},$$

integrate $e^{-z^2}$ around a rectangle with vertices at $z = 0,\ \rho,\ \rho + \lambda i$, and $\lambda i$ (with $\lambda > 0$) and let $\rho \to \infty$ to derive

**(a)** $\displaystyle \int_{0}^{\infty} e^{-x^2} \cos(2\lambda x)\,dx = \frac{\sqrt{\pi}}{2}\,e^{-\lambda^2}$

**(b)** $\displaystyle \int_{0}^{\infty} e^{-x^2} \sin(2\lambda x)\,dx = e^{-\lambda^2} \int_{0}^{\lambda} e^{y^2}\,dy$

(The right-hand side of (b), as a function of $\lambda$, is known as the *Dawson integral* and is tabulated by Abramowitz and Stegun in Ref. [5].)

**11.** Show that

$$\int_{0}^{\infty} \frac{dx}{x^3 + 1} = \frac{2\pi\sqrt{3}}{9}$$

by integrating $1/(z^3 + 1)$ around the boundary of the circular sector $S_\rho: \{z = re^{i\theta}: 0 \leq \theta \leq 2\pi/3,\ 0 \leq r \leq \rho\}$ and letting $\rho \to \infty$.

**12.** Confirm the values of the integrals discussed in Prob. 18, Exercises 4.7, page 178.

**13.** Show that

$$\int_{-\infty}^{\infty} \frac{1}{(1 + x^2)^{n+1}}\,dx = \frac{\pi(2n)!}{2^{2n}(n!)^2}, \qquad \text{for } n = 0, 1, 2, \ldots.$$

## Summation of Series

**14.** Let $f(z)$ be a rational function of the form $P(z)/Q(z)$, where $\deg Q \geq 2 + \deg P$. Assume that no poles of $f(z)$ occur at the integer points $z = 0, \pm 1, \pm 2, \ldots$. Complete each of the following steps to establish the summation formula

$$\sum_{k=-\infty}^{\infty} f(k) = -\{\text{sum of the residues of } \pi f(z) \cot(\pi z) \text{ at the poles of } f(z)\}.$$

**(a)** Show that for the function $g(z) := \pi f(z) \cot(\pi z)$, we have $\text{Res}(g; k) = f(k)$, $k = 0$, $\pm 1, \pm 2, \ldots$.

**(b)** Let $\Gamma_N$ be the boundary of the square with vertices at $(N + \tfrac{1}{2})(1 + i)$, $(N + \tfrac{1}{2})(-1 + i)$, $(N + \tfrac{1}{2})(-1 - i)$, $(N + \tfrac{1}{2})(1 - i)$, taken in that order, where $N$ is a positive integer. Show that there is a constant $M$ independent of $N$ such that $|\pi \cot(\pi z)| \leq M$ for all $z$ on $\Gamma_N$.

(c) Prove that

$$\lim_{N \to +\infty} \int_{\Gamma_N} \pi f(z) \cot(\pi z)\, dz = 0,$$

where $\Gamma_N$ is as defined previously.

(d) Use the residue theorem and parts (a) and (c) to show that

$$\lim_{N \to +\infty} \sum_{k=-N}^{N} f(k) = -\{\text{sum of the residues of } \pi f(z) \cot(\pi z) \text{ at the poles of } f(z)\}.$$

**15.** Using the summation formula in Prob. 14 verify that

(a) $\displaystyle\sum_{k=-\infty}^{\infty} \frac{1}{k^2 + 1} = \pi \coth(\pi)$  [HINT: Take $f(z) = 1/(z^2 + 1)$.]

(b) $\displaystyle\sum_{k=-\infty}^{\infty} \frac{1}{(k - \frac{1}{2})^2} = \pi^2$

(c) $\displaystyle\sum_{k=1}^{\infty} \frac{1}{k^2} = \frac{\pi^2}{6}$  [HINT: The formula in Prob. 14 needs to be modified to compensate for the pole of $f(z) = 1/z^2$ at $z = 0$.]

**16.** Show that for $n$ a positive integer,

$$\sum_{k=1}^{\infty} \frac{1}{k^{2n}} = (-1)^{n-1} \pi^{2n} \frac{2^{2n-1}}{(2n)!} B_{2n},$$

where the constants $B_{2n}$ are the *Bernoulli numbers*, which are defined by the power series expansion

$$\frac{z}{e^z - 1} = \sum_{k=0}^{\infty} \frac{B_k}{k!} z^k.$$

[Compare Prob. 15(c).]  HINT: To determine the residue of $f(z) = 1/z^{2n}$ at $z = 0$, show that

$$\pi z \cot(\pi z) = \sum_{k=0}^{\infty} (-1)^k \frac{B_{2k}}{(2k)!} (2\pi z)^{2k}.$$

**17.** Show that if $a$ is real and noninteger and $0 < r < 1$,

(a) $\displaystyle\sum_{k=-\infty}^{\infty} \frac{1}{(k + a)^2} = \pi^2 \csc^2 \pi a$

(b) $\displaystyle\sum_{k=-\infty}^{\infty} \frac{1}{k^2 + a^2} = \frac{\pi}{a} \coth \pi a$

(c) $\displaystyle\sum_{k=-\infty}^{\infty} \frac{k^2 - a^2}{(k^2 + a^2)^2} = -\pi^2 \operatorname{csch}^2 \pi a$

(d) $\displaystyle\sum_{k=-\infty}^{\infty} \frac{1}{(k - r)^2 + a^2} = \frac{\pi}{2a} \frac{\sinh 2a\pi}{\sin^2 \pi r + \sinh^2 \pi a}$

(e) $\displaystyle\sum_{k=-\infty}^{\infty} \frac{(k - r)^2 - a^2}{[(k - r)^2 + a^2]^2} = \frac{\pi^2}{2} \frac{1 - \cos 2\pi r \cosh 2\pi a}{(\sin^2 \pi r + \sinh^2 \pi a)^2}$

(f) For which *complex* values of $a$ are the preceding identities valid?

**18.** To evaluate sums of the form $\sum_{k=-\infty}^{\infty} (-1)^k f(k)$ involving a sign alternation, we modify the approach of Prob. 14 by replacing $\pi f(z) \cot(\pi z)$ by $\pi f(z) \csc(\pi z)$. Again assuming that $f(z)$ is a rational function of the form $P/Q$, with deg $Q \geq 2 + \deg P$ and that $f$ has no poles at the integer points, derive the formula

$$\sum_{k=-\infty}^{\infty} (-1)^k f(k) = -\{\text{sum of residues of } \pi f(z) \csc(\pi z) \text{ at the poles of } f\}.$$

**19.** Use the formula of Prob. 18 to verify that

$$\sum_{k=1}^{\infty} \frac{(-1)^k}{k^2} = -\frac{\pi^2}{12}.$$

## 6.4 IMPROPER INTEGRALS INVOLVING TRIGONOMETRIC FUNCTIONS; JORDAN'S LEMMA

Our purpose here is to use residue theory to evaluate integrals of the general forms

$$\text{p.v.} \int_{-\infty}^{\infty} \frac{P(x)}{Q(x)} \cos mx \, dx, \qquad \text{p.v.} \int_{-\infty}^{\infty} \frac{P(x)}{Q(x)} \sin mx \, dx,$$

where $m$ is real and $P(x)/Q(x)$ denotes a certain rational function continuous on $(-\infty, \infty)$. As we shall show in the following example, the semicircular contour technique of the previous section can be applied, but some modifications are necessary.

**Example 1**

Compute

$$I = \text{p.v.} \int_{-\infty}^{\infty} \frac{\cos 3x}{x^2 + 4} \, dx.$$

**Solution.**    In utilizing semicircular contours our first inclination is to deal with the complex function

$$\frac{\cos 3z}{z^2 + 4}. \tag{1}$$

However, with this choice for $f(z)$ we are doomed to failure because the modulus of (1) does not go to zero in either the upper or lower half-plane. Indeed, when $z = \pm \rho i$ we have

$$\left| \frac{\cos 3z}{z^2 + 4} \right| = \frac{e^{-3\rho} + e^{3\rho}}{2|-\rho^2 + 4|},$$

which becomes infinite as $\rho \to \infty$.

To circumvent this difficulty we notice that since $\cos 3x$ is the real part of $e^{3ix}$, we have

$$I = \text{Re}(I_0), \quad \text{where } I_0 := \text{p.v.} \int_{-\infty}^{\infty} \frac{e^{3ix}}{x^2 + 4} \, dx.$$

Now if we deal with the function

$$f(z) := \frac{e^{3iz}}{z^2 + 4},$$

we encounter singularities at $z = \pm 2i$, and because

$$|f(z)| = |f(x + iy)| = \frac{|e^{3ix} \cdot e^{-3y}|}{|z^2 + 4|} = \frac{e^{-3y}}{|z^2 + 4|},$$

we have in the *upper* half-plane ($y \geq 0$)

$$|f(z)| \leq \frac{1}{|z^2 + 4|}.$$

Thus for any $\rho > 2$, the integral over the upper half-circle $C_\rho^+$ in Fig. 6.7 is bounded by

$$\left| \int_{C_\rho^+} f(z)\,dz \right| \leq \frac{\pi\rho}{\rho^2 - 4},$$

and so it goes to zero as $\rho \to \infty$. Furthermore, since $+2i$ is the only singularity in the upper half-plane, we have for $\rho > 2$

$$\int_{-\rho}^{\rho} f(x)\,dx + \int_{C_\rho^+} f(z)\,dz = 2\pi i\,\mathrm{Res}(f; 2i).$$

Hence on taking the limit as $\rho \to \infty$ we get

$$\text{p.v.} \int_{-\infty}^{\infty} \frac{e^{3ix}}{x^2 + 4}\,dx + 0 = 2\pi i\,\mathrm{Res}(f; 2i).$$

But

$$\mathrm{Res}(f; 2i) = \lim_{z \to 2i} (z - 2i)f(z) = \lim_{z \to 2i} \frac{e^{3iz}}{z + 2i} = \frac{e^{-6}}{4i}.$$

Thus

$$I_0 = 2\pi i \cdot \frac{e^{-6}}{4i} = \frac{\pi}{2e^6},$$

and finally

$$I = \mathrm{Re}(I_0) = \mathrm{Re}\left(\frac{\pi}{2e^6}\right) = \frac{\pi}{2e^6}. \quad \blacksquare$$

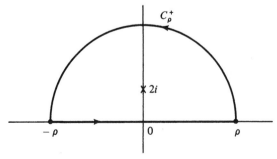

**Figure 6.7**  Contour for Example 1.

The technique of Example 1 can be used to evaluate any integral of the form

$$\text{p.v.} \int_{-\infty}^{\infty} e^{imx} \frac{P(x)}{Q(x)} \, dx \qquad (m > 0), \tag{2}$$

where $P$ and $Q$ are polynomials, $Q$ has no real zeros, and the degree of $Q$ exceeds that of $P$ by at least 2. Indeed, the function $e^{imz} P(z)/Q(z)$ has only a finite number of singularities and, for large $\rho$, its integral over $C_\rho^+$ is bounded by $1 \cdot (K/\rho^2) \cdot \pi\rho$, which goes to zero as $\rho \to \infty$. However, in applications it is sometimes necessary to evaluate integrals such as (2) where the degree of $Q$ is just *one* higher than that of $P$. For example, consider

$$\text{p.v.} \int_{-\infty}^{\infty} \frac{e^{ix} x}{1 + x^2} \, dx. \tag{3}$$

If we estimate the integral of $e^{iz} z/(1 + z^2)$ over $C_\rho^+$ as before, we find that it is bounded by $1 \cdot (K/\rho) \cdot \pi\rho = K\pi$, for some constant $K$. Since this does not go to zero, it is by no means obvious that the semicircular contour method will work in evaluating (3).

Surprisingly, it turns out that the integrals of such a function over $C_\rho^+$ *do* go to zero as $\rho \to \infty$; however, a much finer integral estimate is needed to show this. *Jordan's lemma* fills this need. First, we establish a rather obvious inequality.

---

**Lemma 2.**  Suppose that $f(t)$ and $M(t)$ are continuous functions on the real interval $a \leq t \leq b$, with $f$ complex and $M$ real-valued. If $|f(t)| \leq M(t)$ on this interval, then

$$\left| \int_a^b f(t) \, dt \right| \leq \int_a^b M(t) \, dt. \tag{4}$$

---

*Proof.*   For a careful proof of (4) we must revert to the integration theory presented in Chapter 4. Choose an arbitrary subdivision of the interval $[a, b]$, say,

$$a = \tau_0 < \tau_1 < \cdots < \tau_n = b,$$

and form a Riemann sum for $f$:

$$\sum_{k=1}^{n} f(c_k) \, \Delta\tau_k.$$

Since $|f(t)| \leq M(t)$ for all $t$, we have by the triangle inequality

$$\left| \sum_{k=1}^{n} f(c_k) \, \Delta\tau_k \right| \leq \sum_{k=1}^{n} M(c_k) \, \Delta\tau_k,$$

and we note that the right-hand side is a Riemann sum for $M(t)$ over $[a, b]$. Because the inequality holds for every partition of $[a, b]$, inequality (4) follows by letting the mesh tend to zero. ■

As an immediate consequence of this lemma we deduce that for any continuous function $f(t)$,

$$\left| \int_a^b f(t)\, dt \right| \le \int_a^b |f(t)|\, dt. \tag{5}$$

---

**Lemma 3.** *(Jordan's Lemma)* If $m > 0$ and $P/Q$ is the quotient of two polynomials such that

$$\text{degree } Q \ge 1 + \text{degree } P, \tag{6}$$

then

$$\lim_{\rho \to \infty} \int_{C_\rho^+} e^{imz} \frac{P(z)}{Q(z)}\, dz = 0, \tag{7}$$

where $C_\rho^+$ is the upper half-circle of radius $\rho$.

---

*Proof.*   Parametrizing $C_\rho^+$ we have

$$\int_{C_\rho^+} e^{imz} \frac{P(z)}{Q(z)}\, dz = \int_0^\pi g(t)\, dt,$$

where

$$g(t) := e^{im(\rho e^{it})} \frac{P(\rho e^{it})}{Q(\rho e^{it})} \rho i e^{it}.$$

Now

$$\left| e^{im(\rho e^{it})} \right| = \left| e^{im\rho \cos t - m\rho \sin t} \right| = e^{-m\rho \sin t}.$$

Furthermore, from (6) we know that there is some constant $K$ such that

$$\left| \frac{P(\rho e^{it})}{Q(\rho e^{it})} \right| \le \frac{K}{\rho} \qquad \text{(for } \rho \text{ large).}$$

Thus

$$|g(t)| \le e^{-m\rho \sin t} \cdot \frac{K}{\rho} \cdot \rho = K e^{-m\rho \sin t},$$

and so, by Lemma 2,

$$\left| \int_{C_\rho^+} e^{imz} \frac{P(z)}{Q(z)}\, dz \right| = \left| \int_0^\pi g(t)\, dt \right| \le K \int_0^\pi e^{-m\rho \sin t}\, dt. \tag{8}$$

To estimate the right-hand integral we first observe that the function $e^{-m\rho \sin t}$ on $[0, \pi]$ is symmetric about $t = \pi/2$. Consequently,

$$\int_0^\pi e^{-m\rho \sin t}\, dt = 2 \int_0^{\pi/2} e^{-m\rho \sin t}\, dt.$$

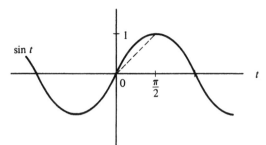

**Figure 6.8**   Graph of $\sin t$.

Furthermore, if we consider the graph of $\sin t$, we notice that it is concave downward on $[0, \pi/2]$; thus it lies above the dashed line in Fig. 6.8. In other words,

$$\sin t \geq \frac{2}{\pi} t, \qquad \text{for } 0 \leq t \leq \frac{\pi}{2}.$$

When $m\rho > 0$, this last inequality implies that

$$e^{-m\rho \sin t} \leq e^{-m\rho 2t/\pi} \quad \text{on} \quad [0, \pi/2],$$

and so

$$\int_0^\pi e^{-m\rho \sin t}\, dt = 2\int_0^{\pi/2} e^{-m\rho \sin t}\, dt \leq 2\int_0^{\pi/2} e^{-m\rho 2t/\pi}\, dt$$

$$= 2\left(\frac{-\pi}{m\rho 2}\right)[e^{-m\rho} - 1] < \frac{\pi}{m\rho}.$$

Hence from (8) we see that

$$\left| \int_{C_\rho^+} e^{imz} \frac{P(z)}{Q(z)}\, dz \right| \leq K \cdot \frac{\pi}{m\rho};$$

consequently, the integral goes to zero as $\rho \to \infty$. ∎

We remark that when $m$ is *negative* the function $e^{imz}P(z)/Q(z)$ is not bounded in the upper half-plane. However, under assumption (6), we do have

$$\lim_{\rho \to \infty} \int_{C_\rho^-} e^{imz} \frac{P(z)}{Q(z)}\, dz = 0 \qquad (m < 0),$$

where $C_\rho^-$ is the lower half-circle of radius $\rho$. To prove this result we need only make the change of variable $w = -z$ in (7).

**Example 2**

Evaluate

$$\text{p.v.} \int_{-\infty}^\infty \frac{x \sin x}{1 + x^2}\, dx.$$

**Solution.** We shall first compute the integral

$$I_0 := \text{p.v.} \int_{-\infty}^{\infty} \frac{xe^{ix}}{1 + x^2} \, dx$$

and then take its imaginary part as our answer. Since the hypotheses of Jordan's lemma are satisfied when $m = 1$ and $P/Q = x/(1 + x^2)$, the integral $I_0$ equals $2\pi i$ times the sum of the residues of $ze^{iz}/(1 + z^2)$ in the upper half-plane. Writing

$$\frac{ze^{iz}}{1 + z^2} = \frac{ze^{iz}}{(z + i)(z - i)},$$

we find for the residue at $z = +i$ the value

$$\lim_{z \to i} \frac{ze^{iz}}{z + i} = \frac{ie^{i^2}}{i + i} = \frac{e^{-1}}{2}.$$

Consequently,

$$I_0 = 2\pi i \cdot \frac{e^{-1}}{2},$$

and so

$$\text{p.v.} \int_{-\infty}^{\infty} \frac{x \sin x}{1 + x^2} \, dx = \text{Im}(I_0) = \text{Im}\left(\frac{2\pi i e^{-1}}{2}\right) = \frac{\pi}{e}. \quad \blacksquare$$

**Example 3**

Evaluate

$$I = \text{p.v.} \int_{-\infty}^{\infty} \frac{\sin x}{x + i} \, dx.$$

**Solution.** One may be tempted to say that this equals the imaginary part of

$$\text{p.v.} \int_{-\infty}^{\infty} \frac{e^{ix}}{x + i} \, dx.$$

This is wrong! (Why?) Moreover, we can't use $(\sin z)/(z + i)$ either, because it is unbounded in both the upper and lower half-planes. So let's try the substitution

$$\sin x = \frac{e^{ix} - e^{-ix}}{2i},$$

which leads to the representation

$$I = \frac{1}{2i}\left(\text{p.v.} \int_{-\infty}^{\infty} \frac{e^{ix}}{x + i} \, dx - \text{p.v.} \int_{-\infty}^{\infty} \frac{e^{-ix}}{x + i} \, dx\right). \tag{9}$$

Now we deal with each integral separately.

For

$$I_1 := \text{p.v.} \int_{-\infty}^{\infty} \frac{e^{ix}}{x + i} \, dx$$

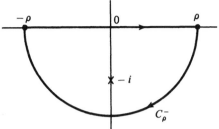

**Figure 6.9**  Contour for $I_2$ in Example 3.

we close the contour $[-\rho, \rho]$ with the half-circle $C_\rho^+$ in the upper half-plane. Then, by Jordan's lemma,

$$\lim_{\rho \to \infty} \int_{C_\rho^+} \frac{e^{iz}}{z+i}\, dz = 0,$$

and since the only singularity of the integrand is in the *lower* half-plane at $z = -i$, we deduce that $I_1 = 0$.

Now the second integral

$$I_2 := \text{p.v.} \int_{-\infty}^{\infty} \frac{e^{-ix}}{x+i}\, dx$$

involves the function $e^{-iz}$, which is unbounded in the upper half-plane, so we close the contour $[-\rho, \rho]$ in the lower half-plane with the semicircle $C_\rho^-: z = \rho e^{-it}$, $0 \le t \le \pi$ (see Fig. 6.9). Then by the analogue of Jordan's lemma for the case when $m < 0$, we deduce that

$$\lim_{\rho \to \infty} \int_{C_\rho^-} \frac{e^{-iz}}{z+i}\, dz = 0.$$

Observing that the closed contour in Fig. 6.9 is negatively oriented, we obtain

$$I_2 = -2\pi i \, \text{Res}\left(\frac{e^{-iz}}{z+i}; -i\right)$$

$$= -2\pi i \lim_{z \to -i} e^{-iz} = -2\pi i \cdot e^{-1}.$$

Consequently, from Eq. (9),

$$I = \frac{1}{2i}[I_1 - I_2] = \frac{1}{2i}[0 + 2\pi i e^{-1}] = \frac{\pi}{e}. \quad \blacksquare$$

## EXERCISES 6.4

*Using the method of residues, verify the integral formulas in Problems 1–3.*

**1.** p.v. $\displaystyle\int_{-\infty}^{\infty} \frac{\cos(2x)}{x^2+1}\, dx = \frac{\pi}{e^2}$

**2.** p.v. $\displaystyle\int_{-\infty}^{\infty} \frac{x \sin x}{x^2 - 2x + 10}\, dx = \frac{\pi}{3e^3}(3\cos 1 + \sin 1)$

**3.** $\displaystyle\int_{0}^{\infty} \frac{\cos x}{(x^2 + 1)^2}\, dx = \frac{\pi}{2e}$

*Compute each of the integrals in Problems 4–9.*

**4.** p.v. $\displaystyle\int_{-\infty}^{\infty} \frac{e^{3ix}}{x - 2i}\, dx$

**5.** p.v. $\displaystyle\int_{-\infty}^{\infty} \frac{x \sin(3x)}{x^4 + 4}\, dx$

**6.** p.v. $\displaystyle\int_{-\infty}^{\infty} \frac{e^{-2ix}}{x^2 + 4}\, dx$

**7.** p.v. $\displaystyle\int_{-\infty}^{\infty} \frac{\cos x}{(x^2 + 1)(x^2 + 4)}\, dx$

**8.** $\displaystyle\int_{0}^{\infty} \frac{x^3 \sin(2x)}{(x^2 + 1)^2}\, dx$

**9.** p.v. $\displaystyle\int_{-\infty}^{\infty} \frac{\cos(2x)}{x - 3i}\, dx$

**10.** Derive the formula

$$\text{p.v.} \int_{-\infty}^{\infty} \frac{\cos x}{x - w}\, dx = \begin{cases} \pi i\, e^{iw} & \text{if } \operatorname{Im} w > 0, \\ -\pi i\, e^{-iw} & \text{if } \operatorname{Im} w < 0. \end{cases}$$

[HINT: Express $\cos x$ in terms of exponentials and be judicious in your choice of closed contours.]

**11.** Give conditions under which the following formula is valid:

$$\text{p.v.} \int_{-\infty}^{\infty} e^{imx}\frac{P(x)}{Q(x)}\, dx = 2\pi i \cdot \sum [\text{residues of } e^{imz} P(z)/Q(z) \text{ at poles in the upper half-plane}].$$

**12.** Given that $\int_{0}^{\infty} e^{-x^2}\, dx = \sqrt{\pi}/2$, integrate $e^{iz^2}$ around the boundary of the circular sector $S_\rho : \{z = re^{i\theta} : 0 \le \theta \le \pi/4,\ 0 \le r \le \rho\}$, and let $\rho \to +\infty$ to prove that

$$\int_{0}^{\infty} e^{ix^2}\, dx = \frac{\sqrt{2\pi}}{4}(1 + i).$$

## 6.5 INDENTED CONTOURS

In the preceding sections the integrands $f$ were assumed to be defined and continuous over the whole interval of integration. We turn now to the problem of evaluating special integrals where $|f(x)| \to \infty$ as $x$ approaches certain finite points. Our first step is to give precise meaning to the integrals of $f$.

Let $f(x)$ be continuous on $[a, b]$ except at the point $c$, $a < c < b$. Then the *improper integrals* of $f$ over the intervals $[a, c]$, $[c, b]$, and $[a, b]$ are defined by

$$\int_a^c f(x)\,dx := \lim_{r\to 0^+} \int_a^{c-r} f(x)\,dx,$$

$$\int_c^b f(x)\,dx := \lim_{s\to 0^+} \int_{c+s}^b f(x)\,dx,$$

and

$$\int_a^b f(x)\,dx := \lim_{r\to 0^+} \int_a^{c-r} f(x)\,dx + \lim_{s\to 0^+} \int_{c+s}^b f(x)\,dx, \tag{1}$$

provided the appropriate limit(s) exists. For example,

$$\int_0^1 \frac{1}{\sqrt{x}}\,dx = \lim_{s\to 0^+} \int_s^1 \frac{1}{\sqrt{x}}\,dx = \lim_{s\to 0^+} 2\sqrt{x}\,\Big|_s^1$$

$$= \lim_{s\to 0^+}[2 - 2\sqrt{s}] = 2,$$

and therefore one can say that the area under the graph in Fig. 6.10 is *finite*, despite the vertical asymptote.

On the other hand, the areas on either side of the vertical asymptote in the graph of $f(x) = 1/(x - 2)$, depicted in Fig. 6.11, are both infinite, because

$$\int_{2+s}^4 \frac{dx}{x-2} = \text{Log}|x-2|\,\Big|_{x=2+s}^{x=4} = \text{Log}\,2 - \text{Log}\,s \to \infty \quad \text{as} \quad s \to 0^+,$$

and

$$\int_1^{2-r} \frac{dx}{x-2} = \text{Log}|x-2|\,\Big|_{x=1}^{x=2-r} = \text{Log}\,r - \text{Log}\,1 \to -\infty \quad \text{as} \quad r \to 0^+.$$

However if we close in on the asymptote from either side *symmetrically*, with $r = s$, the infinities "cancel each other" in the sense that

$$\int_1^{2-r} \frac{dx}{x-2} + \int_{2+r}^4 \frac{dx}{x-2} = \text{Log}|x-2|\,\Big|_{x=1}^{x=2-r} + \text{Log}|x-2|\,\Big|_{x=2+r}^{x=4}$$

$$= \text{Log}\,r - \text{Log}\,1 + \text{Log}\,2 - \text{Log}\,r$$

$$= \text{Log}\,2.$$

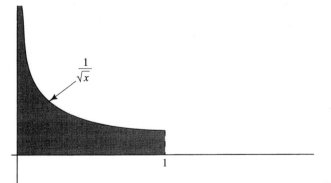

$$\frac{1}{\sqrt{x}}$$

**Figure 6.10**  Graph of $1/\sqrt{x}$.

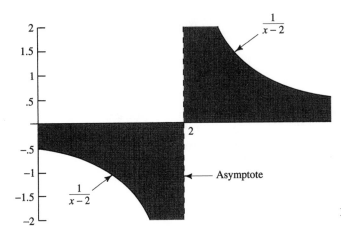

**Figure 6.11** Graph of $1/(x - 2)$.

To take advantage of this we again adopt the principal value notation for improper integrals when the limits in (1) are taken symmetrically:

$$\text{p.v.} \int_a^b f(x) \, dx := \lim_{r \to 0^+} \left\{ \int_a^{c-r} f(x) \, dx + \int_{c+r}^b f(x) \, dx \right\}.$$

The improper integral $\int_1^4 dx/(x - 2)$ does not exist, but its principal value is Log 2.

When the function $f(x)$ is continuous on the whole real line except at $c$, the principal value of its integral over $(-\infty, \infty)$ is defined by

$$\text{p.v.} \int_{-\infty}^\infty f(x) \, dx := \lim_{\substack{\rho \to \infty \\ r \to 0^+}} \left[ \int_{-\rho}^{c-r} f(x) \, dx + \int_{c+r}^\rho f(x) \, dx \right], \tag{2}$$

provided the limit exists as $\rho \to \infty$ and $r \to 0^+$ independently.[†] In the case of several discontinuities occurring at points $x = c_i$ we extend the definition of the p.v. integral over $(-\infty, \infty)$ in a natural way; namely, we remove a small symmetric interval $(c_i - r_i, c_i + r_i)$ about each $c_i$ and then take the limit of the integral as the variables $r_i \to 0^+$ and $\rho \to \infty$ independently. (See Fig. 6.12.)

Residue theory is useful in evaluating certain integrals of the form (2) when the integrand, considered as a function of $z$, has a simple pole at the exceptional point $c$. Assuming this to be the case, we must consider the integrals of $f$ along $[-\rho, c - r]$ and $[c + r, \rho]$, and to utilize residue theory we must form some closed contour which contains these segments. In the last two sections we discussed suitable ways to join $+\rho$ to $-\rho$. But now we also need to join $c - r$ to $c + r$. In so doing we cannot

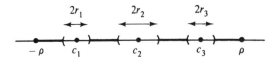

**Figure 6.12** Contour for p.v. integrals.

---

[†] More precisely, we say that the limit in (2) exists and equals $L$ if for every $\varepsilon > 0$ there exist positive constants $M$ and $\delta$ such that for $\rho > M$ and $0 < r < \delta$ the bracketed expression in (2) is within $\varepsilon$ of $L$.

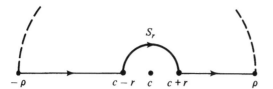

**Figure 6.13** Contour with detour.

proceed along the real axis, for such a segment would pass through the singularity at $c$. Instead, we detour around $c$ by forming, for example, the half-circle $S_r$ indicated in Fig. 6.13. Eventually we will let $r$ tend to zero, and so it will be necessary to determine the limit

$$\lim_{r \to 0^+} \int_{S_r} f(z) \; dz.$$

This is handled by the following lemma, which deals not only with half-circles, but with arbitrary circular arcs.

---

**Lemma 4.** If $f$ has a simple pole at $z = c$ and $T_r$ is the circular arc of Fig. 6.14 defined by

$$T_r: \quad z = c + re^{i\theta} \qquad (\theta_1 \le \theta \le \theta_2), \tag{3}$$

then

$$\lim_{r \to 0^+} \int_{T_r} f(z) \, dz = i(\theta_2 - \theta_1)\text{Res}(f; c). \tag{4}$$

Consequently, for the clockwise oriented half-circle $S_r$ of Fig. 6.13 we have

$$\lim_{r \to 0^+} \int_{S_r} f(z) \, dz = -i\pi \; \text{Res}(f; c). \tag{5}$$

---

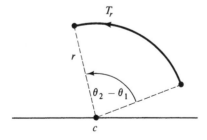

**Figure 6.14** Circular arc of Lemma 4.

*Proof.* Since $f$ has a simple pole at $c$, its Laurent expansion has the form

$$f(z) = \frac{a_{-1}}{z - c} + \sum_{k=0}^{\infty} a_k(z - c)^k,$$

valid in some punctured neighborhood of $c$, say $0 < |z - c| < R$. Thus if $0 < r < R$, we can write

$$\int_{T_r} f(z)\, dz = a_{-1} \int_{T_r} \frac{dz}{z-c} + \int_{T_r} g(z)\, dz, \qquad (6)$$

where

$$g(z) := \sum_{k=0}^{\infty} a_k (z-c)^k.$$

Now $g$ is analytic at $c$ (why?) and hence is bounded in some neighborhood of this point; i.e.,

$$|g(z)| \le M, \qquad \text{for } |z - c| < R_1.$$

Consequently, for $0 < r < R_1$, we have

$$\left| \int_{T_r} g(z)\, dz \right| \le M l(T_r) = M(\theta_2 - \theta_1) r,$$

and the last term goes to zero as $r \to 0^+$. Therefore,

$$\lim_{r \to 0^+} \int_{T_r} g(z)\, dz = 0.$$

To deal with the integral of $1/(z-c)$ we use the parametrization (3) to derive

$$\int_{T_r} \frac{dz}{z-c} = \int_{\theta_1}^{\theta_2} \frac{1}{re^{i\theta}} rie^{i\theta}\, d\theta = i \int_{\theta_1}^{\theta_2} d\theta = i(\theta_2 - \theta_1),$$

the value being independent of $r$. Hence from Eq. (6) we obtain

$$\lim_{r \to 0^+} \int_{T_r} f(z)\, dz = a_{-1} i(\theta_2 - \theta_1) + 0 = \mathrm{Res}(f; c)\, i(\theta_2 - \theta_1),$$

which is the desired limit (4).

In particular, when the $T_r$ are counterclockwise-oriented *half*-circles, we get the limiting value $i\pi \,\mathrm{Res}(f; c)$, and thus for the oppositely oriented half-circles $S_r$ in Fig. 6.13 we get minus this value. ∎

**Example 1**

Evaluate

$$I = \text{p.v.} \int_{-\infty}^{\infty} \frac{e^{ix}}{x}\, dx.$$

**Solution.**   First notice that the integrand is continuous except at $x = 0$. Hence

$$I = \lim_{\substack{\rho \to \infty \\ r \to 0^+}} \left( \int_{-\rho}^{-r} \frac{e^{ix}}{x}\, dx + \int_{r}^{\rho} \frac{e^{ix}}{x}\, dx \right).$$

Now we introduce the complex function

$$f(z) := \frac{e^{iz}}{z},$$

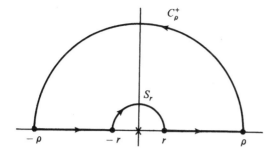

**Figure 6.15**  Contour for Example 1.

which has a simple pole at the origin but is analytic elsewhere. Next we must form a closed contour containing the segments $[-\rho, -r]$ and $[r, \rho]$. Observing that Jordan's lemma applies to $f(z)$ we join $+\rho$ to $-\rho$ by the half-circle $C_\rho^+$ in the upper half-plane. In joining $-r$ to $r$ we indent around the origin by using a half-circle $S_r$. This yields the closed contour of Fig. 6.15. Now since $e^{iz}/z$ has no singularities inside the closed contour, we have

$$\left(\int_{-\rho}^{-r} + \int_{S_r} + \int_r^\rho + \int_{C_\rho^+}\right)\frac{e^{iz}}{z}\,dz = 0;$$

i.e.,

$$\int_{-\rho}^{-r}\frac{e^{ix}}{x}\,dx + \int_r^\rho\frac{e^{ix}}{x}\,dx = -\int_{S_r}\frac{e^{iz}}{z}\,dz - \int_{C_\rho^+}\frac{e^{iz}}{z}\,dz. \tag{7}$$

By Jordan's lemma,

$$\lim_{\rho\to\infty}\int_{C_\rho^+}\frac{e^{iz}}{z}\,dz = 0,$$

and by Eq. (5) of Lemma 4

$$\lim_{r\to 0^+}\int_{S_r}\frac{e^{iz}}{z}\,dz = -i\pi\,\text{Res}(0)$$

$$= -i\pi\lim_{z\to 0} z\cdot\frac{e^{iz}}{z} = -i\pi.$$

Thus from Eq. (7) we obtain

$$\text{p.v.}\int_{-\infty}^{\infty}\frac{e^{ix}}{x}\,dx = -(-i\pi) - 0 = i\pi. \quad\blacksquare \tag{8}$$

**Example 2**

Find

$$\int_0^\infty\frac{\sin x}{x}\,dx = \lim_{\substack{\rho\to\infty \\ r\to 0^+}}\int_r^\rho\frac{\sin x}{x}\,dx.$$

**Solution.** Observe that the integrand $g(x) := (\sin x)/x$ is an even function of $x$; i.e., $g(-x) = g(x)$ for all $x$. Hence

$$2 \int_0^\infty \frac{\sin x}{x} \, dx = \text{p.v.} \int_{-\infty}^\infty \frac{\sin x}{x} \, dx.$$

Furthermore, the right-hand integral is the imaginary part of the integral of $e^{ix}/x$ over $(-\infty, \infty)$ and so, by Example 1, it equals $\text{Im}(i\pi) = \pi$. Thus

$$\int_0^\infty \frac{\sin x}{x} \, dx = \frac{\pi}{2}. \quad \blacksquare$$

We remark that as another consequence of Example 1 we have

$$\text{p.v.} \int_{-\infty}^\infty \frac{\cos x}{x} \, dx = \text{Re}(i\pi) = 0,$$

but this is scarcely surprising because the integrand $h(x) := (\cos x)/x$ is an odd function of $x$; i.e., $h(-x) = -h(x)$ for all $x$.

### Example 3

Compute

$$\text{p.v.} \int_{-\infty}^\infty \frac{xe^{2ix}}{x^2 - 1} \, dx.$$

**Solution.** Here the integrand is discontinuous at two real points, $x = \pm 1$. Thus we need to find

$$\lim_{\substack{\rho \to \infty \\ r_1, r_2 \to 0^+}} \left( \int_{-\rho}^{-1-r_1} + \int_{-1+r_1}^{1-r_2} + \int_{1+r_2}^{\rho} \right) \frac{xe^{2ix}}{x^2 - 1} \, dx.$$

For this purpose we work with

$$f(z) := \frac{ze^{2iz}}{z^2 - 1}$$

and indent around each of its simple poles, as indicated in Fig. 6.16. Then since

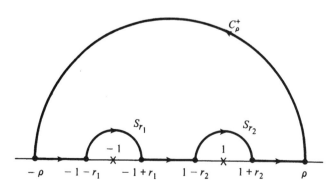

**Figure 6.16** Contour for Example 3.

$f(z)$ is analytic inside the closed contour, we obtain

$$\left(\int_{-\rho}^{-1-r_1} + \int_{-1+r_1}^{1-r_2} + \int_{1+r_2}^{\rho}\right)\frac{xe^{2ix}}{x^2-1}\,dx + J_{r_1} + J_{r_2} + J_\rho = 0, \qquad (9)$$

where $J_{r_1}, J_{r_2}, J_\rho$ are the integrals of $f(z)$ over $S_{r_1}, S_{r_2}, C_\rho^+$, respectively. Now by Jordan's lemma we have

$$\lim_{\rho \to \infty} J_\rho = 0,$$

and from Eq. (5) of Lemma 4,

$$\lim_{r_1 \to 0^+} J_{r_1} = -i\pi \, \mathrm{Res}(-1) = -i\pi \lim_{z \to -1} (z+1)f(z)$$

$$= -i\pi \lim_{z \to -1} \frac{ze^{2iz}}{z-1} = \frac{-i\pi e^{-2i}}{2},$$

and

$$\lim_{r_2 \to 0^+} J_{r_2} = -i\pi \, \mathrm{Res}(1) = -i\pi \lim_{z \to 1} (z-1)f(z)$$

$$= -i\pi \lim_{z \to 1} \frac{ze^{2iz}}{z+1} = \frac{-i\pi e^{2i}}{2}.$$

Hence on taking the limits in Eq. (9) we get

$$\mathrm{p.v.} \int_{-\infty}^{\infty} \frac{xe^{2ix}}{x^2-1}\,dx = \frac{i\pi e^{-2i}}{2} + \frac{i\pi e^{2i}}{2} - 0 = i\pi \cos 2. \qquad \blacksquare$$

# EXERCISES 6.5

1. Compute each of the following limits along the given circular arcs.

   (a) $\displaystyle\lim_{r \to 0^+} \int_{T_r} \frac{2z^2+1}{z}\,dz$, where $T_r: z = re^{i\theta}, \ 0 \le \theta \le \dfrac{\pi}{2}$

   (b) $\displaystyle\lim_{r \to 0^+} \int_{\Gamma_r} \frac{e^{3iz}}{z^2-1}\,dz$, where $\Gamma_r: z = 1 + re^{i\theta}, \ \dfrac{\pi}{4} \le \theta \le \pi$

   (c) $\displaystyle\lim_{r \to 0^+} \int_{\gamma_r} \frac{\mathrm{Log}\,z}{z-1}\,dz$, where $\gamma_r: z = 1 + re^{-i\theta}, \ \pi \le \theta \le 2\pi$

   (d) $\displaystyle\lim_{r \to 0^+} \int_{S_r} \frac{e^z-1}{z^2}\,dz$, where $S_r: z = re^{-i\theta}, \ \pi \le \theta \le 2\pi$

*Using the technique of residues, verify each of the integral formulas in Problems 2–8.*

2. $\mathrm{p.v.} \displaystyle\int_{-\infty}^{\infty} \frac{e^{2ix}}{x+1}\,dx = \pi i e^{-2i}$

3. $\mathrm{p.v.} \displaystyle\int_{-\infty}^{\infty} \frac{e^{ix}}{(x-1)(x-2)}\,dx = \pi i(e^{2i} - e^i)$

**4.** $\displaystyle\int_0^\infty \frac{\sin(2x)}{x(x^2+1)^2}\,dx = \pi\left(\frac{1}{2}-\frac{1}{e^2}\right)$

**5.** $\displaystyle\int_0^\infty \frac{\cos x - 1}{x^2}\,dx = -\frac{\pi}{2}$

**6.** p.v. $\displaystyle\int_{-\infty}^\infty \frac{\sin x}{(x^2+4)(x-1)}\,dx = \frac{\pi}{5}[\cos(1)-e^{-2}]$

**7.** p.v. $\displaystyle\int_{-\infty}^\infty \frac{x\cos x}{x^2-3x+2}\,dx = \pi[\sin(1)-2\sin(2)]$

**8.** p.v. $\displaystyle\int_{-\infty}^\infty \frac{\cos(2x)}{x^3+1}\,dx = \frac{\pi}{3}e^{-\sqrt{3}}[\sin(1)+\sqrt{3}\cos(1)] + \frac{\pi\sin(2)}{3}.$

**9.** Compute p.v. $\displaystyle\int_{-\infty}^\infty \frac{\sin^3 x}{x^3}\,dx.$ $\left[\text{HINT: } \sin^3 x = \text{Im}\left(\frac{3e^{ix}}{4}-\frac{e^{3ix}}{4}-\frac{1}{2}\right).\right]$

**10.** Verify that

$$\int_0^\infty \frac{\sin^2 x}{x^2}\,dx = \frac{\pi}{2}.$$

[HINT: $\sin^2 x = \frac{1}{2}(1-\cos 2x) = \frac{1}{2}\text{Re}(1-e^{2ix})$.]

**11.** Compute p.v. $\displaystyle\int_{-\infty}^\infty \frac{e^{ax}}{e^x-1}\,dx$ for $0 < a < 1$. [HINT: Indent the contour of Fig. 6.6, page 261, around the points $z = 0$ and $z = 2\pi i$.]

**12.** Verify that for $a > 0$ and $b > 0$

$$\int_0^\infty \frac{\sin(ax)}{x(x^2+b^2)}\,dx = \frac{\pi}{2b^2}(1-e^{-ab}).$$

## 6.6 INTEGRALS INVOLVING MULTIPLE-VALUED FUNCTIONS

In attempting to apply residue theory to compute an integral of $f(x)$, it may turn out that the complex function $f(z)$ is multiple-valued. If this happens, we need to modify our procedure by taking into account not only isolated singularities but also branch points and branch cuts. In fact we may find it necessary to integrate along a branch cut, so we turn first to a discussion of this technique.

To be specific let $\alpha$ denote a real number, but not an integer, and let $f(z)$ be the branch of $z^\alpha$ obtained by restricting the argument of $z$ to lie between 0 and $2\pi$; i.e.,

$$f(z) = e^{\alpha(\text{Log}\, r + i\theta)}, \qquad \text{where } z = re^{i\theta}, 0 < \theta < 2\pi. \tag{1}$$

As shown in Chapter 3 this function is analytic in the plane except along its branch cut, the nonnegative real axis. (See Fig. 6.17.) In fact, as $z$ approaches a point $x\,(>0)$ on the cut from the upper half-plane, $\theta$ goes to zero, and

$$f(z) \to e^{\alpha\,\text{Log}\, x} = x^\alpha \tag{2}$$

($x^\alpha$ being the principal value as in calculus); while if $z$ approaches $x$ from the lower

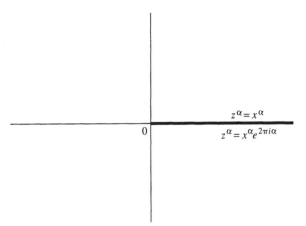

**Figure 6.17**    Branch cut for $z^\alpha$.

half-plane, $\theta$ tends to $2\pi$, and so

$$f(z) \to e^{\alpha(\text{Log } x + i2\pi)} = x^\alpha \cdot e^{2\pi i \alpha}. \tag{3}$$

In this sense we visualize $f$ as being equal to $x^\alpha$ on the "upper side" of the cut and being equal to $x^\alpha \cdot e^{2\pi i \alpha}$ on the "lower side."

Now if we are to integrate $f(z)$ along the cut, we avoid ambiguity by placing a direction arrow either above or below the cut to indicate which values of $f$ are to be used. For example, the integrals of $f$ along the segments $\gamma_1$ and $\gamma_2$ of Fig. 6.18 are given by

$$\int_{\gamma_1} f(z)\,dz = \int_\varepsilon^\rho x^\alpha\,dx, \qquad \int_{\gamma_2} f(z)\,dz = -\int_\varepsilon^\rho x^\alpha \cdot e^{2\pi i \alpha}\,dx$$

$$= -e^{2\pi i \alpha} \int_{\gamma_1} f(z)\,dz. \tag{4}$$

These same remarks apply to arbitrary functions with branch cuts—the values of the function on each side of the cut being determined by continuity from that side. We shall now illustrate how the residue theorem applies to such functions.

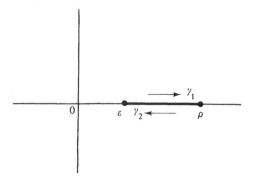

**Figure 6.18**    Integration along a branch cut.

**Example 1**

Let $\alpha$ be a real number (not an integer) and let $R(z)$ be a rational function having no poles on the closed contour $\Gamma$ of Fig. 6.19. Prove that if $f(z)$ is the branch of $z^\alpha R(z)$ obtained by restricting the argument of $z$ to be between 0 and $2\pi$, i.e.,

$$f(z) = \begin{cases} e^{\alpha(\text{Log } r + i\theta)} \cdot R(re^{i\theta}), & \text{for } z = re^{i\theta}, 0 < \theta < 2\pi, \\ x^\alpha R(x), & \text{for } z = x \text{ when integrating on } \gamma_1, \\ x^\alpha e^{2\pi i \alpha} R(x), & \text{for } z = x \text{ when integrating on } \gamma_2, \end{cases}$$

then

$$\int_\Gamma f(z)\, dz = 2\pi i \cdot \sum [\text{residues of } f(z) \text{ at the poles inside } \Gamma].^\dagger \qquad (5)$$

**Solution.** Notice that the residue theorem cannot be directly applied here because the integrand is multiple-valued on the portion $[\varepsilon, \rho]$ of the branch cut. (Notice also that the integrals along $\gamma_1$ and $\gamma_2$ do *not* cancel here.) To circumvent this difficulty we introduce a segment, indicated by the dashed line in Fig. 6.20, which joins the inner circle to the outer one and does not pass through any poles of $R(z)$. This creates two positively oriented closed contours $\Gamma_1$ and $\Gamma_2$ (as shown in Fig. 6.20) such that

$$\int_\Gamma f(z)\, dz = \int_{\Gamma_1} f(z)\, dz + \int_{\Gamma_2} f(z)\, dz \qquad (6)$$

(the integrals along the dashed segment cancel). Now on $\Gamma_1$ we *can* apply the residue theorem because $f(z)$ agrees with a function analytic on this contour; indeed, for $z$ on or inside $\Gamma_1$

$$f(z) = e^{\alpha \text{ Log } z} R(z)$$

(the *principal branch* Log $z$ having its branch cut along the negative real axis).

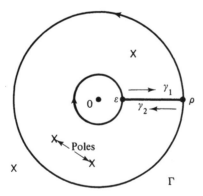

Figure 6.19   Contour for Example 1.

---

$^\dagger$ To be precise, by the *inside* of $\Gamma$ we mean the set of those points between the two circles but not lying on the segment $[\varepsilon, \rho]$.

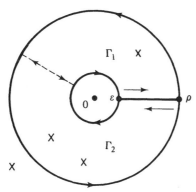

**Figure 6.20**   Modification of Fig. 6.19.

Hence

$$\int_{\Gamma_1} f(z)\, dz = 2\pi i \cdot \sum (\text{residues of } f \text{ at poles inside } \Gamma_1). \tag{7}$$

Similarly, on $\Gamma_2$ the function $f(z)$ again agrees with an analytic function; for instance, for $z = re^{i\theta}$ on $\Gamma_2$

$$f(z) = e^{\alpha(\text{Log } r + i\theta)} R(re^{i\theta}), \qquad \frac{\pi}{2} < \theta < \frac{5\pi}{2}$$

(with cut along the positive imaginary axis). Consequently

$$\int_{\Gamma_2} f(z)\, dz = 2\pi i \cdot \sum (\text{residues of } f \text{ at poles inside } \Gamma_2). \tag{8}$$

Therefore, since every pole inside $\Gamma$ lies either inside $\Gamma_1$ or inside $\Gamma_2$, adding Eqs. (7) and (8) gives the desired equation (5). ∎

Armed with this extension of the residue theorem we now can tackle problems of integrating certain functions involving fractional powers of $x$.

**Example 2**

Compute

$$I := \int_0^\infty \frac{dx}{\sqrt{x}(x + 4)},$$

where $\sqrt{x}$ denotes the principal value for $x > 0$.

**Solution.**   Observe that we are required here to find

$$I = \lim_{\substack{\rho \to \infty \\ \varepsilon \to 0^+}} \int_\varepsilon^\rho \frac{dx}{\sqrt{x}(x + 4)},$$

and for this purpose we take the branch of $z^{1/2}$ defined by

$$\sqrt{z} = e^{(\text{Log } r + i\theta)/2}, \qquad \text{for } z = re^{i\theta}, \qquad 0 < \theta < 2\pi,$$

which has the nonnegative real axis as its branch cut. With this choice of $\sqrt{z}$

we set

$$f(z) := \frac{1}{\sqrt{z}(z+4)}.$$

Then, according to our convention, for $x > 0$ on the upper side of the cut we have

$$f(x) = \frac{1}{\sqrt{x}(x+4)},$$

and, for $x > 0$ on the lower side,

$$f(x) = \frac{1}{\sqrt{x}e^{i2\pi/2}(x+4)} = \frac{-1}{\sqrt{x}(x+4)}.$$

Now we need to form a closed contour containing the segment $[\varepsilon, \rho]$, and in so doing we must take into account the branch point at the origin as well as the pole at $z = -4$. Consider then the closed contour of Fig. 6.21, where $\varepsilon$ is small enough and $\rho$ is large enough so that the pole at $-4$ lies inside the contour. Then for such $\varepsilon$ and $\rho$ we have, by Example 1,

$$\left( \int_{\Gamma_\varepsilon} + \int_{C_\rho} + \int_{\gamma_1} + \int_{\gamma_2} \right) f(z)\, dz = 2\pi i \operatorname{Res}(f; -4). \tag{9}$$

As discussed previously,

$$\int_{\gamma_1} f(z)\, dz + \int_{\gamma_2} f(z)\, dz = \int_\varepsilon^\rho \frac{1}{\sqrt{x}(x+4)}\, dx - \int_\varepsilon^\rho \frac{-1}{\sqrt{x}(x+4)}\, dx$$

$$= 2 \int_\varepsilon^\rho \frac{1}{\sqrt{x}(x+4)}\, dx,$$

and so we can identify

$$\lim_{\substack{\rho \to \infty \\ \varepsilon \to 0^+}} \left( \int_{\gamma_1} f(z)\, dz + \int_{\gamma_2} f(z)\, dz \right) = 2I.$$

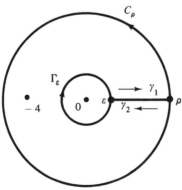

**Figure 6.21** Contour for Example 2.

Furthermore, on the circle of radius $\rho$ we have

$$|f(z)| = \frac{1}{|\sqrt{z}||z + 4|} \leq \frac{1}{\sqrt{\rho}(\rho - 4)} \qquad (\rho > 4),$$

which yields the estimate

$$\left| \int_{C_\rho} f(z)\, dz \right| \leq \frac{2\pi\rho}{\sqrt{\rho}(\rho - 4)}.$$

Consequently, the integral over $C_\rho$ tends to zero as $\rho \to \infty$. Similarly, on the inner circle of radius $\varepsilon$ we have

$$|f(z)| \leq \frac{1}{\sqrt{\varepsilon}(4 - \varepsilon)} \qquad (\varepsilon < 4),$$

which implies that

$$\left| \int_{\Gamma_\varepsilon} f(z)\, dz \right| \leq \frac{2\pi\varepsilon}{\sqrt{\varepsilon}(4 - \varepsilon)} = \frac{2\pi\sqrt{\varepsilon}}{4 - \varepsilon}.$$

As $\varepsilon \to 0^+$ this also goes to zero.

Hence on taking the limit as $\rho \to \infty$ and $\varepsilon \to 0^+$ in Eq. (9), we obtain

$$0 + 0 + 2I = 2\pi i \operatorname{Res}(f; -4). \tag{10}$$

Finally, since $z = -4$ is a simple pole of $f$,

$$\operatorname{Res}(f; -4) = \lim_{z \to -4} (z + 4)f(z) = \lim_{\substack{r \to 4 \\ \theta \to \pi}} \frac{1}{\sqrt{z}} = \lim_{\substack{r \to 4 \\ \theta \to \pi}} \frac{1}{e^{(\operatorname{Log} r + i\theta)/2}}$$

$$= e^{-(\operatorname{Log} 4)/2} e^{-i\pi/2} = \frac{1}{\sqrt{4}}(-i) = \frac{-i}{2}.$$

Therefore from Eq. (10) we get

$$I = \frac{2\pi i}{2}\left( \frac{-i}{2} \right) = \frac{\pi}{2}. \quad \blacksquare$$

A somewhat more complicated situation arises in the following example.

**Example 3**

Compute

$$I = \text{p.v.} \int_0^\infty \frac{dx}{x^\lambda(x - 4)}, \qquad \text{where } 0 < \lambda < 1.$$

**Solution.** There is a significant difference between this and the preceding example; here we have a singularity at $x = +4$ which lies *on* the interval of integration. Also notice that we have generalized the exponent of $x$ in the denominator. Thus we must compute

$$I = \lim_{\substack{\rho \to \infty \\ \varepsilon, \delta \to 0^+}} \left( \int_\varepsilon^{4-\delta} + \int_{4+\delta}^\rho \right) \frac{dx}{x^\lambda(x-4)}.$$

To do this we modify the approach of the preceding problem by indenting around the singularity. Choosing the branch

$$f(z) = \frac{1}{e^{\lambda(\operatorname{Log} r + i\theta)}(re^{i\theta} - 4)}, \quad \text{for } z = re^{i\theta}, \quad 0 < \theta < 2\pi,$$

we form the contour of Fig. 6.22. Since $f(x)$ has no singularities "inside" the closed contour, the integral over the latter must be zero. Utilizing different definitions for $f$ on the upper and lower sides of the branch cut, we can write this as

$$(1 - e^{-2\pi i\lambda}) \left( \int_\varepsilon^{4-\delta} + \int_{4+\delta}^\rho \right) \frac{dx}{x^\lambda(x-4)}$$

$$+ \left( \int_{\Gamma_\varepsilon} + \int_{S_\delta^+} + \int_{S_\delta^-} + \int_{C_\rho} \right) f(z)\, dz = 0, \quad (11)$$

where the contours are as indicated in Fig. 6.22.

Now for $0 < \lambda < 1$, it is easy to extend the estimates used in Example 2 to show that

$$\lim_{\varepsilon \to 0^+} \int_{\Gamma_\varepsilon} f(z)\, dz = 0 \quad \text{and} \quad \lim_{\rho \to \infty} \int_{C_\rho} f(z)\, dz = 0. \quad (12)$$

To compute the limits as $\delta \to 0^+$ of the integrals over $S_\delta^+$ and $S_\delta^-$, we apply the results of the preceding section concerning the behavior of integrals near simple poles. On the upper half-circles around $z = 4$, the function $f$ agrees with the principal branch

$$f_1(z) := \frac{1}{e^{\lambda \operatorname{Log} z}(z - 4)},$$

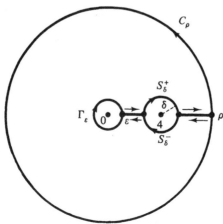

**Figure 6.22**  Contour for Example 3.

which is *analytic* on the positive real axis except for its simple pole at $z = 4$. Hence by Lemma 4 of Sec. 6.5, page 275,

$$\lim_{\delta \to 0^+} \int_{S_\delta^+} f(z)\, dz = -i\pi \operatorname{Res}(f_1; 4) = -i\pi \lim_{z \to 4} e^{-\lambda \operatorname{Log} z} = -i\pi 4^{-\lambda}. \tag{13}$$

However, on the lower half-circles $f(z)$ equals $e^{-2\pi i \lambda}$ times $f_1(z)$, and so

$$\lim_{\delta \to 0^+} \int_{S_\delta^-} f(z)\, dz = -i\pi 4^{-\lambda} e^{-2\pi i \lambda}. \tag{14}$$

Finally, on taking the limit as $\rho \to \infty$, $\varepsilon \to 0^+$, and $\delta \to 0^+$ in Eq. (11), we deduce from (12), (13), and (14) that

$$(1 - e^{-2\pi i \lambda})I + 0 - i\pi 4^{-\lambda} - i\pi 4^{-\lambda} e^{-2\pi i \lambda} + 0 = 0,$$

or, equivalently,

$$I = i\pi 4^{-\lambda} \frac{(1 + e^{-2\pi i \lambda})}{(1 - e^{-2\pi i \lambda})} = i\pi 4^{-\lambda} \frac{e^{i\pi\lambda} + e^{-i\pi\lambda}}{e^{i\pi\lambda} - e^{-i\pi\lambda}} = \pi 4^{-\lambda} \cot(\pi\lambda). \quad \blacksquare$$

## EXERCISES 6.6

*Use residue theory to verify each of the integral formulas in Problems 1–7.*

1. $\displaystyle \int_0^\infty \frac{\sqrt{x}}{x^2 + 1}\, dx = \frac{\pi}{\sqrt{2}}$

2. $\displaystyle \int_0^\infty \frac{x^{\alpha-1}}{x + 1}\, dx = \frac{\pi}{\sin(\pi\alpha)}, \quad 0 < \alpha < 1$

3. $\displaystyle \int_0^\infty \frac{x^\alpha}{(x + 9)^2}\, dx = \frac{9^{\alpha-1}\pi\alpha}{\sin(\pi\alpha)}, \quad -1 < \alpha < 1, \quad \alpha \neq 0$

4. $\displaystyle \int_0^\infty \frac{x^\alpha}{(x^2 + 1)^2}\, dx = \frac{\pi(1 - \alpha)}{4\cos(\alpha\pi/2)}, \quad -1 < \alpha < 3, \quad \alpha \neq 1$

5. $\displaystyle \int_0^\infty \frac{x^{\alpha-1}}{x^2 + x + 1}\, dx = \frac{2\pi}{\sqrt{3}}\cos\left(\frac{2\alpha\pi + \pi}{6}\right)\csc(\alpha\pi), \quad 0 < \alpha < 2, \quad \alpha \neq 1$

6. p.v. $\displaystyle \int_0^\infty \frac{x^\alpha}{x^2 - 1}\, dx = \frac{\pi}{2\sin(\pi\alpha)}[1 - \cos(\pi\alpha)], \quad -1 < \alpha < 1, \quad \alpha \neq 0$

7. $\displaystyle \int_0^\infty \frac{x^\alpha}{1 + 2x\cos\phi + x^2}\, dx = \frac{\pi}{\sin(\pi\alpha)}\frac{\sin(\phi\alpha)}{\sin\phi}, \quad -1 < \alpha < 1, \quad \alpha \neq 0, \quad -\pi < \phi < \pi, \quad \phi \neq 0$

8. Verify that p.v. $\displaystyle \int_{-\infty}^\infty \frac{\operatorname{Log}|x|}{x^2 + 4}\, dx = \frac{\pi}{2}\operatorname{Log} 2.$

[HINT: Integrate $(\operatorname{Log} z)/(z^2 + 4)$ around a semicircular contour indented at the origin (see Fig. 6.15, page 277) and note that $(\rho \operatorname{Log} \rho)/(\rho^2 - 4) \to 0$ as $\rho \to +\infty$ or as $\rho \to 0^+$.]

**9.** Verify that

$$\int_0^\infty \frac{\operatorname{Log} x}{x^2 + 1}\, dx = 0.$$

[HINT: See Prob. 8.]

**10.** Verify that

$$\int_0^\infty \frac{\operatorname{Log} x}{(x^2 + 1)^2}\, dx = -\frac{\pi}{4}.$$

[HINT: See Prob. 8.]

**11.** Verify that

$$\int_0^\infty x^{\alpha - 1} \sin x\, dx = \sin\left(\frac{\pi\alpha}{2}\right) \cdot \Gamma(\alpha) \qquad (0 < \alpha < 1),$$

where $\Gamma(\alpha) := \int_0^\infty e^{-x} x^{\alpha - 1}\, dx$ is the *Gamma function*. [HINT: Integrate $e^{-z} z^{\alpha - 1}$ around a quarter-circle indented at the origin.]

## *6.7 THE ARGUMENT PRINCIPLE AND ROUCHÉ'S THEOREM

In this section we shall use Cauchy's residue theorem to derive two theoretical results which have important practical applications. These results pertain to functions all of whose singularities are poles. Such functions are given a special name in the next definition.

---

**Definition 2.**    A function $f$ is said to be **meromorphic** in a domain $D$ if at every point of $D$ it is either analytic or has a pole.

---

In particular, we regard the analytic functions on $D$ as being special cases of meromorphic functions. The rational functions are examples of functions which are meromorphic in the whole plane.

Suppose now that we are given a function $f$ which is analytic and nonzero at each point of a simple closed contour $C$ and is meromorphic inside $C$. Under these conditions it can be shown that $f$ has at most a *finite* number of poles inside $C$. The proof of this depends on two facts: first, that the only singularities of $f$ are *isolated* singularities (poles), and, second, that every infinite sequence of points inside $C$ has a subsequence which converges to some point on or inside $C$. (The last fact is proved in advanced calculus texts under the name Bolzano-Weierstrass theorem.) Hence if $f$ had an infinite number of poles inside $C$, some subsequence of them would converge to a point which must be a singularity, but not an *isolated* singularity, of $f$. By contradiction, then, the number of poles must be finite.

By the same token, if $f$ had an infinite number of zeros inside $C$, a subsequence would converge to some point $z_0$. But if $f$ were analytic at $z_0$, then by Corollary 3, Sec. 5.6, page 219, the function $f$ would have to be identically zero in a neighborhood of $z_0$—and thus also on $C$, by analytic continuation; and if $f$ were *not* analytic at $z_0$, the deliberations of Sec. 5.6 would require that $z_0$ be an essential singularity. Both situations contradict our hypotheses; thus the number of zeros inside $C$ is also finite.

In counting the number of zeros or poles of a function it is common practice to include the multiplicity. For example, let $C: |z| = 4$ and take

$$f(z) = \frac{(z - 8)^2 z^3}{(z - 5)^4 (z + 2)^2 (z - 1)^5}. \tag{1}$$

Then the number $N_p(f)$ of poles of $f$ *inside* $C$ is to be interpreted as

$$N_p(f) := \sum_{\text{poles inside } C} (\text{order of each pole})$$

$$= (\text{order of pole at } z = -2) + (\text{order of pole at } z = 1)$$

$$= 2 + 5 = 7,$$

while the number $N_0(f)$ of its zeros inside $C$ is

$$N_0(f) := (\text{order of the zero at } z = 0) = 3.$$

**Example 1**

For the function in Eq. (1), evaluate $\int_C f'(z)/f(z)\, dz$, where $C: |z| = 4$ is positively oriented (Fig. 6.23).

**Solution.** To introduce a new concept we are going to attack this problem via a nonstandard approach. Observe that, formally speaking,

$$\frac{f'(z)}{f(z)} = \frac{d}{dz} \log f(z) = \frac{d}{dz} \{\text{Log} |f(z)| + i \arg f(z)\}.$$

Now let $\gamma$ be a subarc of $C$, sufficiently short so that (some branch of) arg $f(z)$ varies by less than $2\pi$ along $\gamma$. Then there is a branch of log $f(z)$ that is analytic

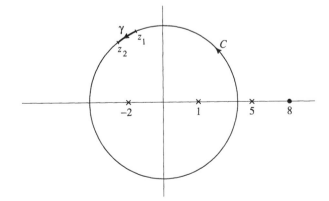

**Figure 6.23**  Contour for Example 1.

on $\gamma$. It follows from Theorem 6, Sec. 4.3 (page 124), that

$$\int_\gamma \frac{f'(z)}{f(z)}\, dz = \{\mathrm{Log}\,|f(z)| + i \arg f(z)\}\Big|_{z_1}^{z_2} \qquad (2)$$

for this branch, where $z_1$ and $z_2$ are the endpoints of $\gamma$ (see Fig. 6.23). If we break $C$ into such subarcs and piece together the contributions (2), we can write the result as

$$\int_C \frac{f'(z)}{f(z)}\, dz = \Delta_C \mathrm{Log}|f(z)| + i\Delta_C \arg f(z) = i\Delta_C \arg f(z), \qquad (3)$$

where we interpret $\Delta_C \arg f(z)$ to be the *net* excursion in $\arg f(z)$ as we go around $C$. (Since $\mathrm{Log}\,|f(z)|$ is single-valued, its net excursion on a closed contour is zero.)
From (1) we see that

$$\arg f(z) = 2 \arg(z - 8) + 3 \arg z - 4 \arg(z - 5) - 2 \arg(z + 2) - 5 \arg(z - 1).$$

The net excursion in $\arg z$, as we go around $C$, is $2\pi$, since $C$ encircles the origin. Similarly, the net excursion in $\arg(z + 2)$ and $\arg(z - 1)$ is $2\pi$, since $C$ encircles $-2$ and $+1$. But the excursion in $\arg(z - 8)$ and $\arg(z - 5)$ is zero; 8 and 5 lie outside $C$. Thus

$$\int_C \frac{f'(z)}{f(z)}\, dz = i2\pi(3 - 2 - 5) = -4(2\pi i) = -8\pi i. \quad \blacksquare$$

In the preceding example, every zero of $f(z)$ inside $C$ contributed $2\pi i$ times its multiplicity, and every pole contributed $-2\pi i$ times its multiplicity, to the integral. The *Principle of the Argument* generalizes this result for all meromorphic functions.

---

**Theorem 3.**   (*Argument Principle*) If $f$ is analytic and nonzero at each point of a simple closed positively oriented contour $C$ and is meromorphic inside $C$, then

$$\frac{1}{2\pi i} \int_C \frac{f'(z)}{f(z)}\, dz = N_0(f) - N_p(f), \qquad (4)$$

where $N_0(f)$ and $N_p(f)$ are, respectively, the number of zeros and poles of $f$ inside $C$ (multiplicity included).

---

*Proof.*   The strategy is quite straightforward; we locate the singularities and compute the residues of the integrand

$$G(z) := \frac{f'(z)}{f(z)}.$$

Notice that this function is analytic at each point *on* $C$ because $f$ is analytic and nonzero there. Inside $C$ the singularities of $G$ occur at those points where $f$ has a zero or a pole.

Consider first a point $z_0$ inside $C$ which is a zero of $f$ of order $m$. Then we know $f$ can be written in the form

$$f(z) = (z - z_0)^m \cdot h(z),$$

where $h(z)$ is analytic and not zero at $z = z_0$ (recall Theorem 16, Sec. 5.6, page 219). Hence in some punctured neighborhood of $z_0$ we compute that

$$G(z) = \frac{f'(z)}{f(z)} = \frac{m}{z - z_0} + \frac{h'(z)}{h(z)}.$$

Since the function $h'/h$ is analytic at $z_0$, this representation shows that $G$ has a simple pole at $z_0$ with residue equal to $m$.

On the other hand, if $f$ has a pole of order $k$ at $z_p$, then

$$f(z) = \frac{H(z)}{(z - z_p)^k},$$

where $H(z)$ is analytic at $z_p$ and $H(z_p) \neq 0$ (recall Lemma 7, Sec. 5.6, page 222). This time we derive that in a punctured neighborhood of $z_p$

$$G(z) = \frac{f'(z)}{f(z)} = \frac{-k}{z - z_p} + \frac{H'(z)}{H(z)},$$

and since $H'/H$ is analytic at $z_p$, we find that $G$ has a simple pole at $z_p$ with residue equal to *minus k*.

Finally, by the residue theorem, the integral of $G$ around $C$ must equal $2\pi i$ times the sum of the residues at the singularities inside $C$. This, from the preceding deliberations, equals $2\pi i$ times the sum of the orders of the zeros of $f$ inside $C$ plus the sum of the negatives of the orders of the poles of $f$ inside $C$; that is,

$$\int_C G(z)\, dz = \int_C \frac{f'(z)}{f(z)}\, dz = 2\pi i [N_0(f) - N_p(f)],$$

which is the same as Eq. (4).  ∎

Of course if the function $f$ of Theorem 3 has no poles inside $C$, then $N_p(f) = 0$, and we have the following.

---

**Corollary 1.**   If $f$ is analytic inside and on a simple closed positively oriented contour $C$ and if $f$ is nonzero on $C$, then

$$\frac{1}{2\pi i} \int_C \frac{f'(z)}{f(z)}\, dz = N_0(f),$$

where $N_0(f)$ is the number of zeros of $f$ inside $C$ (multiplicity included).

---

As we have discussed, the conclusion of the argument principle can be written in the form ·

$$\frac{1}{2\pi} \Delta_C \arg f(z) = N_0(f) - N_p(f). \tag{5}$$

There is yet another way to express the variation in the argument of $f$ along $C$; it involves the *image curve* $f(C)$. This is simply the image (in the $w$-plane) of the curve $C$ under the mapping $w = f(z)$; i.e., if $C$ is parametrized by $z = z(t)$, $a \leq t \leq b$, then $f(C)$ is the curve parametrized by

$$w = f(z(t)) \qquad (a \leq t \leq b). \tag{6}$$

Obviously the image curve is closed, but unlike $C$ it need not be simple or positively oriented.

Now if we sketch the image curve as in Fig. 6.24, it is easy to follow the net change in the argument of $f(z)$. Every time $f(z(t))$ encircles the origin $w = 0$ in the positive (counterclockwise) direction, arg $f$ increases by $2\pi$, while it decreases by $2\pi$ for a negative circuit. Hence, since $f(C)$ is closed, $\Delta_C$ arg $f(z)$ equal $2\pi$ multiplied by the net number of times $f(C)$ winds around $w = 0$ *in the positive sense*, i.e., counterclockwise minus clockwise.[†]

As a specific illustration, consider

$$f(z) = z^3 \quad \text{and} \quad C: z = e^{it} \qquad (0 \leq t \leq 2\pi).$$

Notice that the image of this circle under $w = f(z)$ winds around the origin three times in the positive sense (see Fig. 6.25). Hence the net change in the argument of $f$ is $6\pi$, and Eq. (5) correctly predicts $N_0(f) - N_p(f) = 6\pi/2\pi = 3$.

Next, suppose we have a function $f(z)$ analytic on $C$ and meromorphic inside, and we know how many times $f(C)$ winds around the origin $w = 0$. Now we are interested in *perturbing* $f(z)$ by some analytic function $h(z)$ to form $g(z) := f(z) + h(z)$. We would like to know how small the perturbation $h$ must be to guarantee that $g(C)$ winds around $w = 0$ the same number of times as $f(C)$.

The problem is a familiar one to anyone who has tried to walk a dog on a leash in a big city. If the pair encounters a lamppost and the leash is long, the canine will inevitably tangle the leash around the post. But if the human continually adjusts the length of the leash as they walk so that it never quite extends to the post, then

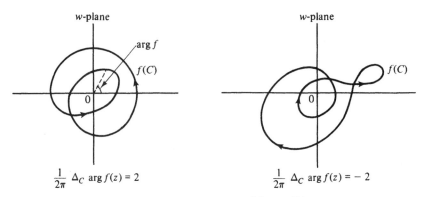

$$\frac{1}{2\pi} \Delta_C \text{ arg } f(z) = 2 \qquad\qquad \frac{1}{2\pi} \Delta_C \text{ arg } f(z) = -2$$

**Figure 6.24**    Visualization of $\Delta_C$ arg $f(z)$.

---

[†] Some authors call this the *winding number* of $f(C)$ about the origin.

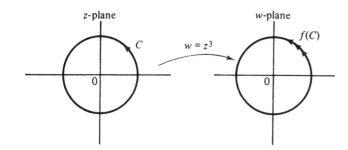

**Figure 6.25**   $\Delta_C \arg(z^3)$.

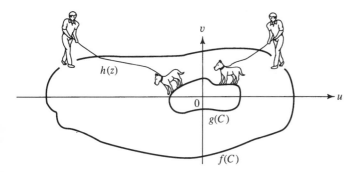

**Figure 6.26**   Contours for Rouché's theorem.

both dog and human will wind around the post an equal number of times and avoid entanglement.

Let us interpret the origin in the $w$-plane as the lamppost, the contour $f(C)$ as the path of the human, and $g(C)$ as the dog's path. (See Fig. 6.26.) Thus $h(z) = g(z) - f(z)$ becomes the leash, and the condition that the leash never extends from the human to the lamppost—and thus that $f(C)$ and $g(C)$ wind around the origin the same number of times—is expressed as

$$|h(z)| < |f(z)|, \qquad z \text{ on } C.$$

*Rouché's theorem* results when we combine these deliberations with the argument principle, for functions $f$ with no poles inside $C$.

---

**Theorem 4.**   (*Rouché's Theorem*) If $f$ and $h$ are each functions that are analytic inside and on a simple closed contour $C$ and if the strict inequality

$$|h(z)| < |f(z)| \tag{7}$$

holds at each point on $C$, then $f$ and $f + h$ must have the same total number of zeros (counting multiplicity) inside $C$.

---

Observe that the inequality (7) need only hold *on* $C$, not inside, and that (7) prevents $f$ (as well as $g = f + h$) from being zero on $C$. See Prob. 15 for an extension to the case when $f$ is meromorphic inside $C$.

One typically uses Rouché's theorem to deduce some information about the location of zeros of a complicated analytic function $g$ by comparing it with an analytic function $f$ whose zeros are known.

**Example 2**

Prove that all five zeros of the polynomial

$$g(z) = z^5 + 3z + 1$$

lie in the disk $|z| < 2$.

**Solution.** We take $C$ as the circle $|z| = 2$, and we regard $g$ as a perturbation of the function $f(z) = z^5$, which clearly has five zeros inside $C$. To test condition (7) we estimate the perturbation $h(z) = 3z + 1$ on $C$ by

$$|h(z)| = |3z + 1| \leq 3|z| + 1 = 3 \cdot 2 + 1 = 7,$$

which sure enough, is strictly less than $|f(z)| = |z^5| = 2^5 = 32$. Therefore, $g$ also has five zeros inside $|z| < 2$. ∎

**Example 3**

Prove that the equation

$$z + 3 + 2e^z = 0$$

has precisely one root in the left half-plane.

**Solution.** Since Rouché's theorem refers to domains bounded by contours, we cannot apply it directly to an unbounded set such as a half-plane. But by this time the reader will probably be able to anticipate our strategy; we choose $C_\rho$ as in Fig. 6.27 and regard the function

$$g(z) = z + 3 + 2e^z$$

as a perturbation of

$$f(z) = z + 3,$$

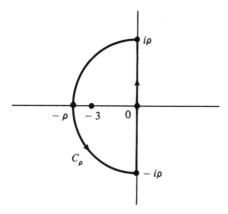

**Figure 6.27**    Contour for Example 3.

which has exactly one zero in the left half-plane. Then for $z$ on $C_\rho$ we have

$$|h(z)| = |g(z) - f(z)| = |2e^z| = 2e^{\text{Re } z} \leq 2e^0 = 2,$$

while $f(z)$ is bounded below on $C_\rho$ by (see Fig. 6.27)

$$|f(z)| = |z + 3| \geq \begin{cases} 3, & \text{for } z = iy, \\ |z| - 3 = \rho - 3, & \text{for } |z| = \rho. \end{cases}$$

Thus when $\rho > 5$ we have $|h(z)| < |f(z)|$ for all $z$ on $C_\rho$. This implies that $g$ also has precisely one (simple) zero inside $C_\rho$, and hence (letting $\rho \to \infty$) in the left half-plane. ■

Rouché's theorem can also be used to give an alternative proof of the Fundamental Theorem of Algebra.

**Example 4**

Prove that every polynomial of degree $n$ has $n$ zeros.

**Solution.**  Consider the polynomial $g(z)$ of degree $n$

$$g(z) = a_n z^n + a_{n-1} z^{n-1} + \cdots + a_1 z + a_0 \qquad (a_n \neq 0),$$

as a perturbation of

$$f(z) = a_n z^n,$$

which has $n$ zeros (at the origin). The difference

$$h(z) = g(z) - f(z) = a_{n-1} z^{n-1} + \cdots + a_1 z + a_0$$

is then a polynomial of degree at most $n - 1$ and hence does not grow as rapidly as the polynomial $f$ of degree $n$. In more precise terms, on the circle $C: |z| = R$ we have

$$|f(z)| = |a_n| R^n,$$

and

$$|h(z)| \leq |a_{n-1}| R^{n-1} + \cdots + |a_1| R + |a_0|.$$

Hence if we choose $R \; (>1)$ so large that

$$\frac{|a_{n-1}|}{|a_n|} + \cdots + \frac{|a_1|}{|a_n|} + \frac{|a_0|}{|a_n|} < R,$$

then the inequality $|h(z)| < |f(z)|$ will be valid on $C$. Therefore, $g(z)$ also has $n$ zeros. ■

Now suppose $f(z)$ is analytic in some open neighborhood of a point $z_0$ and $f(z_0) = 0$. Then from Sec. 5.6 we know that unless $f$ is identically zero, there is some circle $C$, centered at $z_0$ and lying in the neighborhood, such that $f$ is nonzero on $C$. Let $\sigma$ be the minimum value of $|f(z)|$ for $z$ on $C$, and perturb $f$ by $h(z) = -c$, where $c$ is *any* complex number smaller in magnitude than $\sigma$ (continuity implies $\sigma > 0$). It follows from Rouché's theorem that $f(z) - c$ also achieves the value 0, and hence $f$ achieves the value $c$, inside the circle.

In other words, the values taken on by $w = f(z)$ in this neighborhood of $z$ completely cover the open disk $|w| < \sigma$ in the $w$-plane. So the image of *every* open neighborhood of $z_0$ contains an open neighborhood of $w_0 = f(z_0) = 0$, unless $f$ is identically zero.

If $f(z_0) = w_0$ is not zero, we can apply this same argument to the function $f(z_0) - w_0$ and conclude that the image of every open neighborhood of $z_0$ contains an open neighborhood of $f(z_0)$, unless $f$ is constant. This is the *open mapping* property of analytic functions, which we state as follows.

---

**Theorem 5.**  If $f$ is nonconstant and analytic in a domain $D$, then its range

$$f(D) := \{w \mid w = f(z) \text{ for some } z \text{ in } D\}$$

is an open set.

---

## EXERCISES 6.7

1. Which of the following functions are meromorphic in the whole plane?

   (a) $2z + z^3$

   (b) $\text{Log } z$

   (c) $\dfrac{\sin z}{z^3 + 1}$

   (d) $e^{1/z}$

   (e) $\tan z$

   (f) $\dfrac{2i}{(z - 3)^2} + \cos z$

2. Let $P(z) = a_n z^n + a_{n-1} z^{n-1} + \cdots + a_1 z + a_0$, where $a_n \neq 0$. Explain why for each sufficiently large value of $R$

$$\oint_{|z| = R} \frac{P'(z)}{P(z)} \, dz = 2n\pi i.$$

3. Evaluate

$$\frac{1}{2\pi i} \oint_{|z| = 3} \frac{f'(z)}{f(z)} \, dz,$$

   where $f(z) = \dfrac{z^2(z - i)^3 e^z}{3(z + 2)^4(3z - 18)^5}$.

4. Let $f(z)$ be analytic on the closed disk $|z| \le \rho$, and suppose that $f(z) \neq w_0$ for all $z$ on the circle $|z| = \rho$. Explain why the value of the integral

$$\frac{1}{2\pi i} \oint_{|z| = \rho} \frac{f'(z)}{f(z) - w_0} \, dz$$

   equals the number of solutions of $f(z) = w_0$ inside the disk.

5. Prove that if $f(z)$ is analytic inside and on a simple closed contour $C$ and is one-to-one on $C$, then $f(z)$ is one-to-one inside $C$. [HINT: Consider the image curve $f(C)$.]

6. Use Rouché's theorem to show that the polynomial $z^6 + 4z^2 - 1$ has exactly two zeros in the disk $|z| < 1$.

7. Prove that the equation $z^3 + 9z + 27 = 0$ has no roots in the disk $|z| < 2$.

8. Prove that all the roots of the equation $z^6 - 5z^2 + 10 = 0$ lie in the annulus $1 < |z| < 2$.

9. Find the number of roots of the equation $6z^4 + z^3 - 2z^2 + z - 1 = 0$ in the disk $|z| < 1$.

10. Prove that the equation $z = 2 - e^{-z}$ has exactly one root in the right half-plane. Why must this root be real?

11. Prove that the polynomial $P(z) = z^4 + 2z^3 + 3z^2 + z + 2$ has exactly two zeros in the right half-plane. HINT: Write $P(iy) = (y^2 - 2)(y^2 - 1) + iy(1 - 2y^2)$, and show that

$$\lim_{R \to \infty} \arg P(iy)\Big|_{-R}^{R} = 0.$$

12. Suppose that $f(z)$ is analytic on $|z| \leq 1$ and satisfies $|f(z)| < 1$ for $|z| = 1$.
    (a) Prove that the equation $f(z) = z$ has exactly one root (counting multiplicity) in $|z| < 1$. (This root is called a *fixed point* of $f$.)
    (b) Prove that if $|z_0| \leq 1$, then the sequence $z_n$ defined recursively by $z_n = f(z_{n-1})$, $n = 1$, $2, \ldots$, converges to the fixed point of $f$.

13. Give an example to show that the conclusion of Rouché's theorem may be false if the strict inequality $|h(z)| < |f(z)|$ is replaced by $|h(z)| \leq |f(z)|$ for $z$ on $C$.

14. State and prove a generalization of Rouché's theorem for meromorphic functions $f$ and $h$, which concludes that $N_0(f) - N_p(f) = N_0(f + h) - N_p(f + h)$.

15. Prove: If $f$ is analytic and nonzero at each point of a simple closed contour $C$ and is *meromorphic* inside $C$ and if $h$ is *analytic* inside and on $C$ and satisfies $|h(z)| < |f(z)|$ on $C$, then $f$ and $f + h$ have the same number of zeros inside $C$ (counting multiplicity).

16. Let $\lambda > 0$ be fixed, and let $g(z) = \tan z - \lambda z$. The zeros of $g$ are important in certain problems related to heat flow. By completing each of the following steps, show that for $n$ large, $g(z)$ has exactly $2n + 1$ zeros inside the square $\Gamma_n$ with vertices at $n\pi(1 \pm i)$, $n\pi(-1 \pm i)$.
    (a) Show that

    $$\tan(x + iy) = \left[\frac{\sin(2x)}{\cosh(2y) + \cos(2x)}\right] + i\left[\frac{\sinh(2y)}{\cosh(2y) + \cos(2x)}\right].$$

    (b) Prove that for all large integers $n$, the inequality $|\tan z| \leq 2$ holds on the boundary of $\Gamma_n$. [HINT: Use the formula in part (a) for the horizontal segments.]
    (c) Show for all large integers $n$, the inequality $|g(z) + \lambda z| < \lambda|z|$ holds on the boundary of $\Gamma_n$.
    (d) Show that $g(z)$ has exactly $2n$ poles inside $\Gamma_n$, $n = 1, 2, \ldots$.
    (e) Conclude from the general form of Rouché's theorem (Prob. 14) that $g(z)$ has $2n + 1$ zeros inside $\Gamma_n$ for large integers $n$.

17. Let $f(z)$ be analytic in a domain $D$, and suppose that $f(z) - f(z_0)$ has a zero of order $n$ at $z_0$ in $D$. Prove that for $\varepsilon > 0$ sufficiently small, there exists a $\delta > 0$ such that for all $w$ in $|w - f(z_0)| < \delta$ the equation $f(z) - w$ has exactly $n$ roots in $|z - z_0| < \varepsilon$.

18. Use the open mapping property (Theorem 5) to give a quick proof of the following familiar facts: If $f$ is analytic in a domain $D$, then $f$ is identically constant in $D$ if any of the following conditions holds.
    (a) Re $f(z)$ is constant in $D$.
    (b) Im $f(z)$ is constant in $D$.
    (c) $|f(z)|$ is constant in $D$.

19. Let $f_n(z)$, $n = 1, 2, \ldots$, be a sequence of functions analytic in the disk $D: |z| < R$ which converges uniformly to the analytic function $f(z)$ on each closed subset of $D$. Prove that if $f(z) \neq 0$ on $|z| = \delta$, $0 < \delta < R$, then for each $n$ sufficiently large $f_n(z)$ has the same number of zeros in $|z| < \delta$ as does $f(z)$.

20. Let $P(z) = a_n z^n + \cdots + a_1 z + a_0$ and let

$$P^*(z) = z^n \overline{P(1/\bar{z})} = \bar{a}_0 z^n + \bar{a}_1 z^{n-1} + \cdots + \bar{a}_n.$$

Prove that if $|a_0/a_n| > 1$, then $P(z)$ has the same number of zeros in $|z| < 1$ as does the polynomial $\bar{a}_0 P(z) - a_n P^*(z)$. [HINT: $|P(z)| = |P^*(z)|$ for $|z| = 1$.]

21. To establish the stability of feedback control systems one often has to ensure that a certain meromorphic function of the form $F(z) = 1 + P(z)$ has all its zeros in the left half-plane. The *Nyquist stability criterion* proceeds as follows: First consider the contour $\Gamma_r$ shown in Fig. 6.28, and let $m$ equal the net number of times that the image contour $P(\Gamma_r)$ encircles the point $w_0 = -1$ in a counterclockwise direction. Then let $n$ equal the number of poles of $P(z)$ with positive real parts. Argue that if $m$ equals $n$ for all sufficiently large $r$, then all the zeros of $F(z)$ lie in the left half-plane (and thus the system is stable). [If for $r$ large, $P(\Gamma_r)$ passes through the point $w_0 = -1$, then of course $F(z)$ has a zero on the imaginary axis; in such a case stability cannot be guaranteed.]

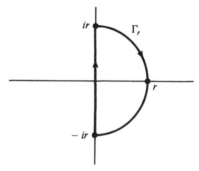

**Figure 6.28**  Contour for Prob. 21.

22. A stronger version (in that the hypothesis is weaker) of Rouché's theorem was discovered by Glicksberg. With reference to the dog-and-human situation in Fig. 6.26, let $\tau$ denote the ray extending from the lamppost in the direction away from the human, as in Fig. 6.29. (Obviously $\tau$ turns as the human traverses the path $f(C)$.) Now, if the dog is restricted to stay on one side or the other of $\tau$—and never to cross it—then the leash will not tangle around the lamppost and both dog and human will encircle the post the same number of times.

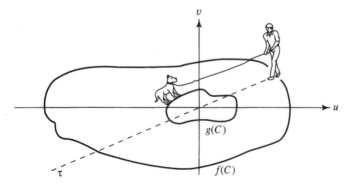

**Figure 6.29**  Stronger Rouché theorem.

**(a)** From this consideration argue that condition (7) in Rouché's theorem can be replaced by

$$|h(z)| < |f(z)| + |f(z) + h(z)|. \tag{8}$$

[HINT: Inequality (8) is a *strict* triangle inequality, and as such it ensures that the points $h(z)$, $f(z)$, and $f(z) + h(z)$ do not align unfavorably; recall Sec. 1.3.]

**(b)** Give an alternative derivation based on the observation that (8) implies $f/(f + h)$ is never negative or zero, and thus

$$\int_C \left[ \frac{f'}{f} - \frac{(f+h)'}{(f+h)} \right] dz = \int_C \left[ \mathrm{Log}\, \frac{f}{f+h} \right]' dz = 0;$$

now apply Corollary 1, page 291.

# SUMMARY

A useful way to evaluate certain contour integrals is by means of residues. The residue of a function $f(z)$ at an isolated singularity $z_0$ is the coefficient $a_{-1}$ of $1/(z - z_0)$ in the Laurent expansion for $f(z)$ about $z_0$. Simple formulas exist for computing the residues at the poles of $f(z)$. When all the singularities of $f(z)$ are isolated, its integral along a simple closed positively oriented contour is equal to $2\pi i$ times the sum of the residues at the singularities inside the contour.

Residue theory can be employed to evaluate certain integrals which arise in the calculus of functions of a real variable. For example, definite integrals over $[0, 2\pi]$ which involve $\sin \theta$ and $\cos \theta$ can be rewritten as contour integrals around the unit circle $C: |z| = 1$ after making the identification

$$\cos \theta = \frac{1}{2}\left( z + \frac{1}{z} \right), \qquad \sin \theta = \frac{1}{2i}\left( z - \frac{1}{z} \right) \qquad \text{(for } z = e^{i\theta}\text{)}.$$

Also certain improper integrals over infinite intervals, say over the real axis $(-\infty, +\infty)$, can be computed with the aid of expanding closed contours (such as semicircles or rectangles) which have a segment of the real axis as a component. For this approach to be successful the contours must be selected so that the sum of residues inside is easily computed, and so that the limiting values of the integral over the nonreal portions of the contours are known. Some specific results such as Jordan's lemma are useful in this regard.

Sometimes these contours must be modified by indention to compensate for singularities on the original interval of integration. In the case when the complex version of an integrand is multiple-valued, it might be necessary to integrate along a branch cut. For this purpose the cut is regarded as having two distinct sides, with the integrand being defined differently on each side.

When the only singularities of $f(z)$ are poles, the variation in the argument of $f(z)$ along a simple closed contour is related to the difference in the number of its zeros and poles inside the contour. A consequence of this fact is Rouché's theorem, which provides a comparison technique for counting the number of zeros of an analytic function in a certain domain.

## SUGGESTED READING

### *Residue Calculus*

[1] Conway, J. B. *Functions of One Complex Variable*, 2nd ed. Springer-Verlag Inc., New York, 1978.

[2] Copson, E. T. *An Introduction to the Theory of Functions of a Complex Variable*. Oxford University Press, Inc., New York, 1962.

[3] Eves, H. W. *Functions of a Complex Variable*, Vol. 2. Prindle, Weber & Schmidt, Inc., Boston, 1966.

[4] Henrici, P. *Applied and Computational Analysis*, Vol. 1. John Wiley & Sons, Inc., New York, 1974.

### *Special Functions*

[5] Abramowitz, M. and Stegun, I. A. *Handbook of Mathematical Functions with Formulas, Graphs, and Mathematical Tables*. Dover Publications, New York, 1965.

### *Stability and Control*

[6] Dorf, R. C. *Modern Control Systems*, 6th ed. Addison-Wesley Publishing Company, Inc., Reading, Mass., 1992.

### *Rouché's Theorem*

[7] Glicksberg, I. "A Remark on Rouché's Theorem," *Amer. Math. Mon.* 83 (1976), 186–87.

# :7:

# Conformal Mapping

In this chapter we shift our point of view somewhat; rather than dealing with the algebraic properties of an analytic function $f(z)$, we are going to regard $f$ as a *mapping* from its domain to its range and consider its *geometric* properties. The ability to map one region onto another via an analytic function proves invaluable in applied mathematics, as we shall see in Sec. 7.1.

## 7.1 INVARIANCE OF LAPLACE'S EQUATION

One of the most valuable aspects of mappings generated by analytic functions is the persistence of Laplace's equation; roughly, this means that if $\phi(x, y)$ is harmonic in a certain domain $D$ of the $xy$-plane (so that $\phi$ satisfies Laplace's equation

$$\frac{\partial^2 \phi}{\partial x^2} + \frac{\partial^2 \phi}{\partial y^2} = 0$$

in $D$), and if

$$w = f(z) \tag{1}$$

is an analytic function mapping $D$ onto a domain $D'$ in the $uv$-plane, then $\phi$ is "carried over" by the mapping to a function which is harmonic in $D'$.

To express this fact precisely we must elaborate somewhat on the nature of the mapping. We assume that relation (1) provides a *one-to-one* correspondence between the points of $D$ and those of $D'$.[†] Recall that this means $f(z_1) = f(z_2)$ only if $z_1 = z_2$.

---

[†] Such a mapping is sometimes called *univalent*, or *schlicht*.

We also assume that the derivative $df/dz$ is never zero in $D$; actually the latter is a consequence of the one-to-one assumption, but we won't prove it here.

Now since the mapping is one-to-one, it has an inverse; i.e., with each point of $D'$ there can be associated a point of $D$, namely its preimage under $f$. This relationship, which is the inverse of $w = f(z)$, is suggestively written

$$z = f^{-1}(w) \tag{2}$$

(recall, for instance, $\sin^{-1} w$). Figure 7.1 depicts both the relationships. Observe that $f^{-1}$ is a single-valued function (since only one $z$ is mapped to a particular $w$). The reader will probably not be surprised to learn that it is, in fact, an *analytic* function, and that its derivative is given by

$$\frac{df^{-1}(w)}{dw} = \frac{1}{\dfrac{df(z)}{dz}} \qquad [\text{where } w = f(z)]. \tag{3}$$

This equation can also be written in the form

$$\frac{dz}{dw} = \frac{1}{\dfrac{dw}{dz}}, \tag{4}$$

keeping in mind the functional relationships. We have already proved a special case of Eq. (3) for the function $w = e^z$ and its inverse $z = \text{Log } w$, taking $D$ to be the strip $|\text{Im } z| < \pi$ and $D'$ to be the entire plane slit along the negative real axis (cf. Sec. 3.2). We shall invite the reader to prove Eq. (3) in general at the end of the next section.

It is sometimes helpful to indicate the mappings (1) and (2) in terms of real variables; thus Eq. (1) becomes

$$u = u(x, y),$$
$$v = v(x, y), \tag{5}$$

and its inverse, Eq. (2), becomes

$$x = x(u, v),$$
$$y = y(u, v). \tag{6}$$

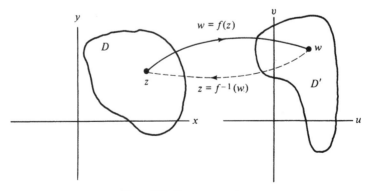

**Figure 7.1**  Inverse mapping.

Now we are prepared to demonstrate the property stated before about the persistence of Laplace's equation. Observe that if $\phi(x, y)$ is a function defined on $D$, then the domain $D'$ "inherits" $\phi$ through the one-to-one mapping; that is, the function $\psi(u, v)$ defined by

$$\psi(u, v) := \phi(x(u, v), y(u, v)),$$

or, equivalently,

$$\psi(w) := \phi(f^{-1}(w)),$$

agrees with $\phi(x, y)$ are corresponding points. Our claim is that if $\phi$ is harmonic in $D$, then $\psi$ is harmonic in $D'$.

The proof is quite simple. Consider any point $w_0$ in $D'$, say $w_0 = f(z_0)$. Now $D$ is an open set, so there exists an open disk $\mathcal{N}$ centered at $z_0$ and lying entirely in $D$. Since $f^{-1}(w)$ is an analytic function, and thus continuous, there must be a sufficiently small neighborhood $\mathcal{N}'$ around $w_0$ whose image under $f^{-1}$ lies entirely inside $\mathcal{N}$ (this is depicted in Fig. 7.2). To see that $\psi(u, v)$ is harmonic in the neighborhood $\mathcal{N}'$, recall that in Sec. 2.5 we proved that any harmonic function can be taken as the real part of an analytic function in "nice enough" domains—in particular, this is true for disks. Hence $\phi(x, y)$ is the real part of an analytic function $g(z)$ on $\mathcal{N}$. But then $\psi(u, v)$ is the real part of the composite function $g(f^{-1}(w))$ on $\mathcal{N}'$, and since the composition of analytic functions is again analytic, $\psi$ must be harmonic in the neighborhood $\mathcal{N}'$ of $w_0$. Consequently, as $w_0$ is an arbitrary point of $D'$, the function $\psi$ must be harmonic everywhere on $D'$.

Another proof of this fact can be based upon the direct verification of Laplace's equation in the $w$-plane, using the Cauchy-Riemann equations (see Prob. 2).

One can readily see why the preceding result is so useful in applications. Consider the *Dirichlet problem*, which requires us to find a function harmonic in a domain and taking specified values on its boundary (cf. Sec. 4.7). Once we have solved this problem on a particular domain, we immediately have solutions on all the domains which we can map onto the original one via a one-to-one analytic function, as long as the boundary values correspond. So we select whichever domain renders the problem simplest.

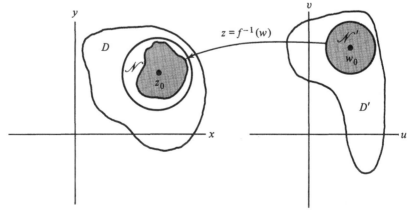

**Figure 7.2**  Mapping of neighborhoods.

**Example 1**

Find a function $\phi(x, y)$ harmonic inside the unit disk $|z| < 1$ and satisfying the boundary conditions

$$\phi(x, y) \to +1 \qquad \text{on the upper half-circle,}$$

$$\phi(x, y) \to -1 \qquad \text{on the lower half-circle}$$

(see Fig. 7.3). (This function gives the temperature profile inside an infinitely long right circular cylinder whose outer wall is partitioned into sections maintained at different temperatures. Thermal insulation, of course, must be provided at the points of discontinuity $z = \pm 1$, and the temperature is not specified there.)

**Solution.** This complicated-looking problem is solved by means of a mapping that carries the disk onto the right half-plane. In Sec. 7.3 the reader will be instructed on how to construct such mappings, but for now we simply want to illustrate the power of the technique. So we state without proof that the function

$$w = f(z) = \frac{1 + z}{1 - z} \tag{7}$$

maps the unit circle onto the imaginary axis and its interior onto the right half-plane. (The exterior maps to the left half-plane.) The correspondences are illustrated in Fig. 7.4. Observe that $|w|$ increases without bound as $z \to 1$, whereas

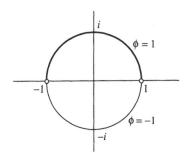

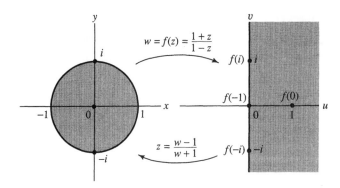

**Figure 7.3** Dirichlet problem for Example 1.

**Figure 7.4** Mapping for Example 1.

$z = -1$ maps to $w = 0$. Moreover, the mapping (7) is one-to-one. In fact, the inverse is easily computed by solving for $z$ in terms of $w$:

$$z = \frac{w - 1}{w + 1}. \tag{8}$$

The pair of functions (7) and (8) are employed in microwave engineering to construct the *Smith chart*, which is discussed in Exercises 7.3, page 326. Figure 7.5, from the paper "Conformal Image Warping" by Frederick and Schwartz (see Ref. [9]), offers a somewhat whimsical depiction of the mapping.

Notice that the upper half-circle, where the unknown function $\phi$ equals 1, is mapped to the positive imaginary axis, whereas the lower half-circle (where $\phi = -1$) corresponds to the negative imaginary axis. Consequently, we can solve the given problem if we can solve the following simpler problem:

Find a function $\psi(u, v)$ that is harmonic in the right half-plane and approaches 1 on the positive $v$-axis and $-1$ on the negative $v$-axis.

The experience we have acquired by now tells us immediately that

$$\psi(u, v) = \frac{2}{\pi} \operatorname{Arg}(w).$$

Hence the solution to the original problem is derived from $\psi$ by the mapping (7):

$$\phi(x, y) = \psi(u(x, y), v(x, y)) = \frac{2}{\pi} \operatorname{Arg}(f(z)) = \frac{2}{\pi} \operatorname{Arg}\left(\frac{1 + z}{1 - z}\right).$$

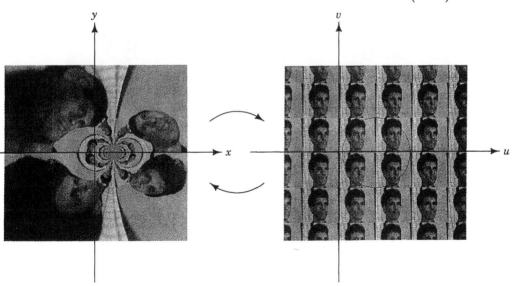

**Figure 7.5**  Correspondences for mapping Eqs. (7) and (8). (From C. Frederick and E. L. Schwartz, "Conformal Image Warping," *IEEE Computer Graphics and Applications*, March 1990, p. 56. © 1990 IEEE.)

A little algebra results in the expression[†]

$$\phi(x, y) = \frac{2}{\pi} \tan^{-1} \frac{2y}{1 - x^2 - y^2},$$

where the value of the arctangent is taken to be between $-\pi/2$ and $\pi/2$. Note that $\phi(x, 0) = 0$, as we would expect from symmetry.  ∎

With this example as motivation we devote the next few sections to a study of mappings given by analytic functions. The final two sections of the chapter will return us to applications, illustrating the power of this technique in handling many different situations. A table of some of the more useful mappings appears as Appendix II, for the reader's future convenience.

## EXERCISES 7.1

1. Verify that the function $w = e^z$ maps the half-strip $x > 0$, $-\pi/2 < y < \pi/2$ onto the portion of the right half $w$-plane that lies outside the unit circle (see Fig. 7.6). What harmonic function $\psi(w)$ does the $w$-plane "inherit," via this mapping, from the harmonic function $\phi(z) = x + y$? What harmonic function $\phi(z)$ is inherited from $\psi(w) = u + v$?

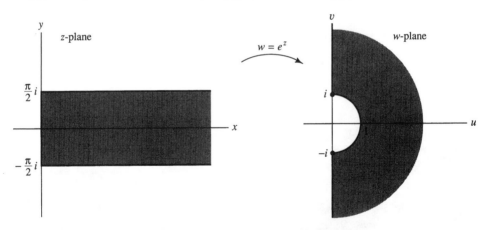

**Figure 7.6**  Exponential mapping of half-strip.

2. Suppose that Eqs. (5) and (6) describe a one-to-one analytic mapping. Let $\phi(x, y)$ be a real-valued twice-continuously differentiable function that is carried over in the $w$-plane to the function

$$\psi(u, v) := \phi(x(u, v), y(u, v)).$$

---

[†] This solution can also be obtained from the Poisson integral formula described in (optional) Sec. 4.7.

**(a)** The *gradient* of $\phi(x, y)$ is the vector $(\partial\phi/\partial x, \partial\phi/\partial y)$; it corresponds to the complex number (recall Sec. 1.3) $\partial\phi/\partial x + i(\partial\phi/\partial y)$. Similarly, the gradient of $\psi$ corresponds to $\partial\psi/\partial u + i(\partial\psi/\partial v)$. Use the chain rule and the Cauchy-Riemann equations to show that these gradients are related by

$$\frac{\partial\psi}{\partial u} + i\frac{\partial\psi}{\partial v} = \left(\frac{\partial\phi}{\partial x} + i\frac{\partial\phi}{\partial y}\right)\overline{\left(\frac{dz}{dw}\right)}.$$

**(b)** Show that the *Laplacians* of $\psi$ and $\phi$ are related by

$$\left\{\frac{\partial^2\psi}{\partial u^2} + \frac{\partial^2\psi}{\partial v^2}\right\} = \left\{\frac{\partial^2\phi}{\partial x^2} + \frac{\partial^2\phi}{\partial y^2}\right\}\left|\frac{dz}{dw}\right|^2.$$

**(c)** Show that if $\phi(x, y)$ satisfies Laplace's equation in the $z$-plane, then $\psi$ satisfies Laplace's equation in the $w$-plane.

**(d)** Show that if $\phi$ satisfies *Helmholtz's* equation,

$$\frac{\partial^2\phi}{\partial x^2} + \frac{\partial^2\phi}{\partial y^2} = \Lambda\phi$$

($\Lambda$ is a constant), in the $z$-plane, then $\psi$ satisfies

$$\frac{\partial^2\psi}{\partial u^2} + \frac{\partial^2\psi}{\partial v^2} = \Lambda\left|\frac{dz}{dw}\right|^2\psi$$

in the $w$-plane. (The Helmholtz equation arises in transient thermal analysis.)

**3.** Find a function $\phi$ harmonic in the upper half-plane and taking the boundary values specified in Fig. 7.7. [HINT: Use a combination of $\arg(z - 2)$ and $\arg(z + 1)$.] Generalize this to the case of a function harmonic in the upper half-plane taking boundary values as indicated in Fig. 7.8.

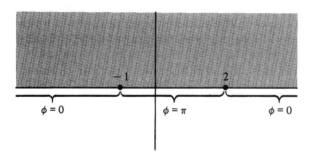

**Figure 7.7**  Dirichlet problem in Prob. 3.

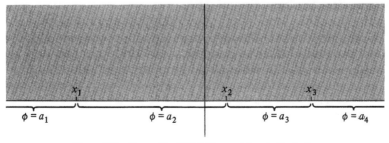

**Figure 7.8**  Generalized Dirichlet problem in Prob. 3.

**4.** Consider the problem of finding a function $\phi$ that is harmonic in the right half-plane and takes the values $\phi(0, y) = y/(1 + y^2)$ on the imaginary axis. Observe that the obvious first guess,

$$\phi(z) = \operatorname{Im} \frac{z}{1 - z^2},$$

fails because $z/(1 - z^2)$ is *not* analytic at $z = 1$. However, the following strategy can be used.

**(a)** According to the text, the mappings (7) and (8) provide a correspondence between the right half-plane and the unit disk. (Of course, one should interchange the roles of $z$ and $w$ in the formulas.) Thus the $w$-plane inherits from $\phi(z)$ a function $\psi(w)$ harmonic in the unit disk. Show that the values of $\psi(w)$ on the unit circle $w = e^{i\theta}$ must be given by

$$\psi(e^{i\theta}) = \frac{\sin \theta}{2}.$$

**(b)** Argue that the harmonic function $\psi(w)$ must be given by

$$\psi(w) = \tfrac{1}{2} \operatorname{Im} w$$

throughout the unit disk.

**(c)** Use the mappings to carry $\psi(w)$ back to the $z$-plane, producing the function

$$\phi(z) = \frac{y}{y^2 + (x + 1)^2}$$

as a solution of the problem.

**5.** Use the strategy of Prob. 4 to find a function $\phi$ harmonic in the right half-plane such that $\phi(0, y) = 1/(y^2 + 1)$.

**6.** Suppose that the harmonic function $\phi(x, y)$ in the domain $D$ is carried over to the harmonic function $\psi(u, v)$ in the domain $D'$ via the one-to-one analytic mapping $w = f(z)$. Prove that if the normal derivative $\partial\phi/\partial n$ is zero on a curve $\Gamma$ in $D$, then the normal derivative $\partial\psi/\partial n$ is zero on the image curve of $\Gamma$ under $f$. (The boundary condition $\partial\phi/\partial n = 0$ is known as a *Neumann* condition.) [HINT: $\partial\phi/\partial n$ is the projection of the gradient $(\partial\phi/\partial x) + i(\partial\phi/\partial y)$ onto the normal, and the gradient is orthogonal to the level curves $\phi(x, y) = $ constant.]

**7.** Suppose that $f(z)$ is analytic and one-to-one. Then, according to the text, you may presume that $f^{-1}$ is also one-to-one. If $x, y, u, v$ are as in Eqs. (5) and (6), explain the identities

$$\frac{\partial x}{\partial u} = \frac{\partial y}{\partial v}, \qquad \frac{\partial x}{\partial v} = -\frac{\partial y}{\partial u}.$$

## 7.2 GEOMETRIC CONSIDERATIONS

The geometric aspects of analytic mappings split rather naturally into two categories: *local* properties and *global* properties. Local properties need only hold in sufficiently small neighborhoods, while global properties hold throughout a domain. For example, consider the function $e^z$. It is one-to-one in any disk of diameter less than $2\pi$, and hence it is locally one-to-one, but since $e^{z_1} = e^{z_2}$ when $z_1 - z_2 = 2\pi i$, the function is not globally one-to-one. On the other hand, sometimes local properties can be extended to global properties; in fact, this is the essence of *analytic continuation* (see Sec. 5.8).

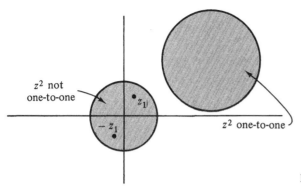

$z^2$ not
one-to-one

$z^2$ one-to-one

**Figure 7.9**    Locally one-to-one mapping.

Let us begin our study of local properties by considering "one-to-oneness." As the example $e^z$ shows, a function may be locally one-to-one without being globally one-to-one. (Of course the opposite situation is impossible.) Furthermore, an analytic function may be locally one-to-one at some points but not at others. Indeed, consider

$$f(z) = z^2.$$

In any open set which contains the origin there will be distinct points $z_1$ and $z_2$ such that $z_2 = -z_1$, and hence (since $z_2^2 = z_1^2$) the function $f$ will not be one-to-one. However, around any point other than the origin, we *can* find a neighborhood in which $z^2$ is one-to-one (any disk that excludes the origin will do; see Fig. 7.9). Thus $f(z) = z^2$ is locally one-to-one at every point other than the origin. An explanation of the exceptional nature of $z = 0$ in this example is provided by the following.

---

**Theorem 1.**    If $f$ is analytic at $z_0$ and $f'(z_0) \neq 0$, then there is some neighborhood $\mathcal{N}$ of $z_0$ such that $f$ is one-to-one on $\mathcal{N}$.

---

*Proof.*    If no such neighborhood existed, i.e., if inside *every* disk around $z_0$ one could find distinct points $\alpha$ and $\beta$ such that $f(\alpha) = f(\beta)$, then by choosing a sequence of disks shrinking to $z_0$ we could construct an infinite number of pairs of complex numbers $\{\alpha_n, \beta_n\}_{n=1}^{\infty}$ such that for each $n$

$$\alpha_n \neq \beta_n \quad \text{but} \quad f(\alpha_n) = f(\beta_n), \tag{1}$$

while the sequences $\{\alpha_n\}$ and $\{\beta_n\}$ each converged to $z_0$. (See Fig. 7.10.) But this contradicts the fact that $f'(z_0) \neq 0$, as we now show:

Let $C$ be a circle centered at $z_0$ such that $f$ is analytic inside and on $C$. Clearly each $\alpha_n$ and $\beta_n$ will lie interior to $C$ for $n$ sufficiently large. Hence, for such $n$, we can use the Cauchy integral formula to write

$$\frac{f(\beta_n) - f(\alpha_n)}{\beta_n - \alpha_n} = \frac{1}{(\beta_n - \alpha_n)} \left[ \frac{1}{2\pi i} \oint_C \frac{f(z)}{z - \beta_n} \, dz - \frac{1}{2\pi i} \oint_C \frac{f(z)}{z - \alpha_n} \, dz \right]$$

$$= \frac{1}{2\pi i} \oint_C \frac{f(z)}{(z - \beta_n)(z - \alpha_n)} \, dz.$$

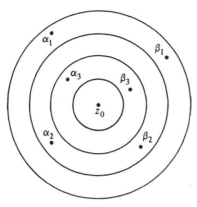

**Figure 7.10** Geometry for Theorem 1.

Now since each of the sequences $\{\alpha_n\}$ and $\{\beta_n\}$ converges to $z_0$, it is easy to see that the integrands

$$\frac{f(z)}{(z - \beta_n)(z - \alpha_n)}$$

converge uniformly (as $n \to \infty$) on $C$ to $f(z)/(z - z_0)^2$. Hence the integrals themselves converge (cf. Theorem 8 in Chapter 5, page 199); i.e.,

$$\lim_{n \to \infty} \frac{f(\beta_n) - f(\alpha_n)}{\beta_n - \alpha_n} = \lim_{n \to \infty} \frac{1}{2\pi i} \oint_C \frac{f(z)}{(z - \beta_n)(z - \alpha_n)} \, dz$$

$$= \frac{1}{2\pi i} \oint_C \frac{f(z)}{(z - z_0)^2} \, dz.$$

But by (1) the left-hand side is zero, and by the generalized Cauchy formula the last integral equals $f'(z_0)$. Thus $f'(z_0) = 0$, contrary to the given hypothesis. ∎

Theorem 1 says that an analytic function is locally one-to-one at points where its derivative does not vanish. In advanced texts it is shown more generally that if $z_0$ is a zero of order $m$ for $f'$, then $f$ is locally "$(m + 1)$-to-one" around (but excluding) $z_0$; in other words, each value of $f$ is taken on $m + 1$ times. This is reinforced by the observation that $f(z) = z^2$ is two-to-one in any punctured neighborhood of the origin.

The next local property we shall discuss is *conformality*. Consider the following situation: $f(z)$ is analytic and one-to-one in a neighborhood of the point $z_0$, and $\gamma_1$ and $\gamma_2$ are two directed smooth curves (in this neighborhood) intersecting at $z_0$. Under the mapping $f$ the images of these curves, $\gamma_1'$ and $\gamma_2'$, are also directed smooth curves, and they will intersect at $w_0 = f(z_0)$. At the point $z_0$ we construct vectors $\mathbf{v}_1$ and $\mathbf{v}_2$ tangent to $\gamma_1$ and $\gamma_2$, respectively, and pointing in the directions consistent with the orientations of the curves (see Fig. 7.11). Then the *angle from $\gamma_1$ to $\gamma_2$* is the angle $\theta$ through which $\mathbf{v}_1$ must be rotated counterclockwise in order to lie along $\mathbf{v}_2$. The angle $\theta'$ from $\gamma_1'$ to $\gamma_2'$ is defined similarly.

Now the mapping $f$ is said to be *conformal at $z_0$* if these angles are preserved; i.e., $\theta = \theta'$ for every pair of directed smooth curves that intersect at $z_0$. For analytic mappings we have the following theorem.

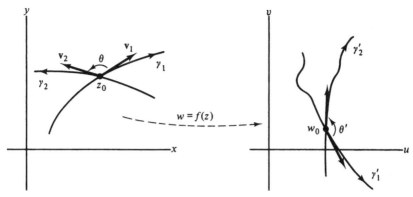

**Figure 7.11** Conformality.

---

**Theorem 2.** An analytic function $f$ is conformal at every point $z_0$ for which $f'(z_0) \neq 0$.

---

*Proof.* By Theorem 1 we know that there is some open disk containing $z_0$ in which $f$ is one-to-one. We will argue that every directed smooth curve through $z_0$ has its tangent (at $z_0$) turned through the same angle, under the mapping $w = f(z)$. Consequently, the angle between any two curves intersecting at $z_0$ will be preserved.

So let $\gamma$ be any directed smooth curve through $z_0$ parametrized, say, by $z = z(t)$ with $z(t_0) = z_0$. The vector $z'(t_0)$ is then tangent to $\gamma$ at $z_0$. Under the mapping the image, $\gamma'$, of $\gamma$ has the parametrization

$$w = w(t) = f(z(t))$$

with

$$w_0 := f(z_0) = f(z(t_0)),$$

and the vector $w'(t_0)$ (if it is nonzero) is tangent to $\gamma'$ at $w_0$. But the chain rule implies

$$w'(t_0) = f'(z_0)z'(t_0), \tag{2}$$

so $w'(t_0)$ is nonzero since $f'(z_0) \neq 0$. Furthermore, we see from Eq. (2) that the angles which the tangent vectors $z'(t_0)$, $w'(t_0)$ make with the horizontal are related by

$$\arg w'(t_0) = \arg f'(z_0) + \arg z'(t_0).$$

Hence every curve through $z_0$ is rotated through the same angle $\arg f'(z_0)$ which, of course, is a constant independent of the particular curve. ∎

The condition that $f'(z_0)$ not vanish in Theorem 2 is crucial; the function $f(z) = z^2$, for which $f'(0) = 0$, does not preserve angles at the origin—it *doubles* them. However, it is common practice to call any mapping generated by a nonconstant analytic function a "conformal map," overlooking the violations occurring at the points where $f'$ is zero. (Incidentally, these exceptional points will be reexamined in Sec. 7.5; they turn out to be quite important.)

Moving now to the global aspects of conformal mapping, we begin with a property which has both local and global ramifications: the *open mapping property*. A function is said to be an *open mapping* if the image of every open set in its domain is, itself, open; that is, the function maps open sets to open sets. For analytic functions we have the following theorem.

> **Theorem 3.**   Any analytic function that is nonconstant on domains is an open mapping.

A proof of this theorem was given in Sec. 6.7. Note that the theorem prohibits, for instance, a nonconstant analytic mapping of a disk onto a portion of a line.

It is very useful in investigating conformal maps to exploit the concept of *connectivity*. For example, one can show that any nonconstant analytic function takes domains, i.e., open connected sets, to domains. Openness is preserved because of Theorem 3, and as for connectivity, we argue as follows. Let us say $f$ maps the domain $D$ onto the open set $\mathcal{O}$. To join $w_1 = f(z_1)$ to $w_2 = f(z_2)$ by a polygonal path in $\mathcal{O}$, first join $z_1$ to $z_2$ in $D$. Then the image of this path is a path joining $w_1$ to $w_2$ in $\mathcal{O}$. Of course, the image path need not be polygonal, but any competent topologist can prove that such a path, lying inside an *open* set, can be deformed into a polygonal path without leaving the set. Hence $\mathcal{O}$ is connected. (See Fig. 7.12.)

The remaining topic from the global theory of analytic functions which we shall consider here is the *Riemann mapping theorem*. Because it is primarily an *existence* theorem, its usefulness in applied mathematics is somewhat limited.

> **Theorem 4.**   (*Riemann Mapping Theorem*) Let $D$ be any simply connected domain in the plane other than the entire plane itself. Then there is a one-to-one analytic function that maps $D$ onto the interior of the unit circle. Moreover, one can prescribe an arbitrary point of $D$ and a direction through that point which are to be mapped to the origin and the direction of the positive real axis, respectively. Under such restrictions the mapping is unique.

A direction through a point $z_0$ is specified, of course, by an angle $\phi$, as in Fig. 7.13. The Riemann mapping theorem allows us to specify $z_0$ and $\phi$ so that all curves through $z_0$ with tangent in the direction $\phi$ are mapped to curves through the origin with tangent along the positive real axis. This yields three "degrees of freedom," or three choices to be made, in fixing the map: the real and the imaginary parts of the point which goes to 0, and the direction.

Again, we appeal to the references for a complete treatment of this theorem. (As an ominous note, we tantalize the reader by pointing out that the theorem makes no predictions about the boundary values of the function.)

From the Riemann mapping theorem we can conclude that any simply con-

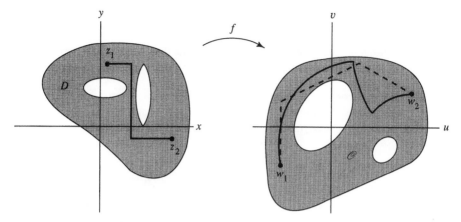

**Figure 7.12**   Conformal maps preserve connectivity.

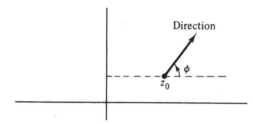

**Figure 7.13**   Three degrees of freedom.

nected domain $D_1$ can be analytically mapped one-to-one onto any other simply connected domain $D_2$, assuming that neither $D_1$ nor $D_2$ is the whole plane. Indeed, let $f$ map $D_1$ onto the unit disk, and let $g$ map $D_2$ onto this disk, in accordance with the theorem. Then $g^{-1}(f(z))$ maps $D_1$ onto $D_2$ and is one-to-one and analytic. (See Fig. 7.14.)

The remainder of this chapter will deal with constructing and applying *specific* conformal mappings.

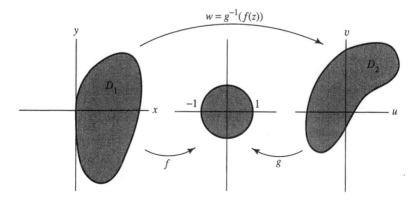

**Figure 7.14**   Mapping of simply connected domains.

## EXERCISES 7.2

1. For each of the following functions, determine the order $m$ of the zero of the derivative $f'$ at $z_0$ and show explicitly that the function is not one-to-one in any neighborhood of $z_0$.
   (a) $f(z) = z^2 + 2z + 1$,   $z_0 = -1$
   (b) $f(z) = \cos z$,   $z_0 = 0$, $\pm\pi$, $\pm 2\pi, \ldots$
   (c) $f(z) = e^{z^3}$,   $z_0 = 0$

2. Prove that if $w = f(z)$ is analytic at $z_0$ and $f'(z_0) \neq 0$, then $z = f^{-1}(w)$ is analytic at $w_0 = f(z_0)$, and

$$\frac{df^{-1}(w)}{dw} = \frac{1}{\dfrac{df(z)}{dz}}$$

   for $w = w_0$, $z = z_0$. [HINT: Theorem 1 guarantees that $f^{-1}(w)$ exists near $w_0$ and Theorem 3 implies that $f^{-1}(w)$ is continuous. Now generalize the proof in Sec. 3.2 (page 84).]

3. What happens to angles at the origin under the mapping $f(z) = z^\alpha$ for $\alpha > 1$? For $0 < \alpha < 1$?

4. Use the open mapping theorem to prove the maximum-modulus principle.

5. Find all functions $f(z)$ analytic in $D: |z| < 1$ which assume only pure imaginary values in $D$.

6. If $f$ is analytic at $z_0$ and $f'(z_0) \neq 0$, show that the function $g(z) = \overline{f(z)}$ preserves the magnitude, but reverses the orientation, of angles at $z_0$.

7. Show that the mapping $w = z + 1/z$ maps circles $|z| = \rho$ $(\rho \neq 1)$ onto ellipses

$$\frac{u^2}{\left(\rho + \dfrac{1}{\rho}\right)^2} + \frac{v^2}{\left(\rho - \dfrac{1}{\rho}\right)^2} = 1.$$

8. Let $f$ be analytic at $z_0$ with $f'(z_0) \neq 0$. By considering the difference quotient, argue that "infinitesimal" lengths of segments drawn from $z_0$ are magnified by the factor $|f'(z_0)|$ under the mapping $w = f(z)$.

9. Let $w = f(z)$ be a one-to-one analytic mapping of the domain $D$ onto the domain $D'$, and let $A' = \text{area}(D')$. Using Prob. 8, argue the plausibility of the formula

$$A' = \iint_D |f'(z)|^2 \, dx \, dy.$$

10. Why is it impossible for $D$ to be the whole plane in the Riemann mapping theorem? [HINT: Appeal to Liouville's theorem.]

11. Describe the image of each of the following domains under the mapping $w = e^z$.
    (a) the strip $0 < \operatorname{Im} z < \pi$
    (b) the slanted strip between the two lines $y = x$ and $y = x + 2\pi$
    (c) the half-strip $\operatorname{Re} z < 0$, $0 < \operatorname{Im} z < \pi$
    (d) the half-strip $\operatorname{Re} z > 0$, $0 < \operatorname{Im} z < \pi$
    (e) the rectangle $1 < \operatorname{Re} z < 2$, $0 < \operatorname{Im} z < \pi$
    (f) the half-planes $\operatorname{Re} z > 0$ and $\operatorname{Re} z < 0$

12. Let $P(z) = (z - \alpha)(z - \beta)$, and let $L$ be any straight line through $(\alpha + \beta)/2$. Prove that $P$ is one-to-one on each of the open half-planes determined by $L$.

**13.** Describe the image of each of the following domains under the mapping
$w = \cos z = \cos x \cosh y - i \sin x \sinh y$. [HINT: Consider the image of the boundary in each case.]

**(a)** the half-strip $0 < \operatorname{Re} z < \pi$, $\operatorname{Im} z < 0$

**(b)** the half-strip $0 < \operatorname{Re} z < \dfrac{\pi}{2}$, $\operatorname{Im} z > 0$

**(c)** the strip $0 < \operatorname{Re} z < \pi$

**(d)** the rectangle $0 < \operatorname{Re} z < \pi$, $-1 < \operatorname{Im} z < 1$

**14.** Prove that if $f$ has a simple pole at $z_0$, then there exists a punctured neighborhood of $z_0$ on which $f$ is one-to-one.

**15.** A domain $D$ is said to be *convex* if for any two points $z_1$, $z_2$ in $D$, the line segment joining $z_1$ and $z_2$ lies entirely in $D$. Prove the *Noshiro-Warschawski theorem*: Let $f$ be analytic in a convex domain $D$. If $\operatorname{Re} f'(z) > 0$ for all $z$ in $D$, then $f$ is one-to-one in $D$. [HINT: Write $f(z_2) - f(z_1)$ as an integral of $f'$.]

**16.** Let $f$ be analytic at $z_0$ and suppose that $f'(z_0) = \lambda \neq 0$. Use the Noshiro-Warschawski theorem (Prob. 15) to give another proof of the fact that there exists a neighborhood of $z_0$ on which $f$ is one-to-one. [HINT: Choose a real number $\theta$ such that $e^{i\theta}f'(z_0)$ is real and positive. Then show that $e^{i\theta}f(z)$ is one-to-one in a neighborhood of $z_0$.]

**17.** (For students who have read Sec. 4.4a)  Argue that a one-to-one analytic function will map simply connected domains to simply connected domains.

## 7.3 MÖBIUS TRANSFORMATIONS

The problem of finding a one-to-one analytic function that maps one domain onto another can be quite perplexing, so it is worthwhile to investigate a few elementary mappings in order to compile some rules of thumb which we can draw upon. The basic properties of *Möbius transformations*, which we shall investigate in this section, constitute an essential portion of every analyst's bag of tricks. (Some of these mappings were previewed in Exercises 2.1.)

First let's consider the simplest mapping of all, the *translation* defined by the function

$$w = f(z) = z + c, \tag{1}$$

where $c$ is a fixed complex number. Under this mapping every point is shifted by the vector corresponding to $c$. Its properties are quite apparent: The entire complex plane is mapped one-to-one onto itself, and every geometric object is mapped onto a congruent object. (See Fig. 7.15 on page 316.)

*Rotations* are quite simple also. Observe that under the transformation

$$w = f(z) = e^{i\phi}z, \tag{2}$$

with $\phi$ real, every point is rotated about the origin through the angle $\phi$. Such transformations are also one-to-one mappings of the complex plane onto itself and map geometric objects onto congruent objects. (See Fig. 7.16.)

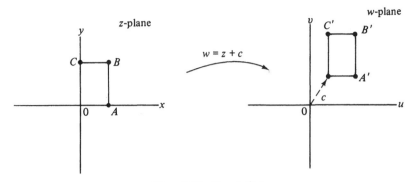

**Figure 7.15**   Translation.

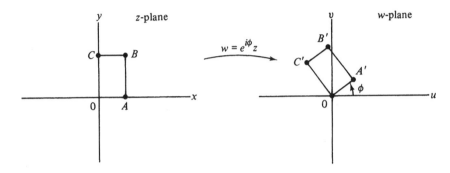

**Figure 7.16**   Rotation.

The mapping defined by

$$w = f(z) = \rho z, \tag{3}$$

where $\rho$ is a positive real constant, simply enlarges (or contracts) the distance of every point from the origin by the factor $\rho$; hence such a transformation is called a *magnification*. Observe that the distance between *any* two points is multiplied by this same constant, since

$$|w_1 - w_2| = |f(z_1) - f(z_2)| = |\rho z_1 - \rho z_2| = \rho |z_1 - z_2|.$$

Magnifications thus rescale distances and (since they are conformal) preserve angles; consequently any geometric object is mapped onto an object that is *similar* to the original. And again, the complex plane is mapped one-to-one onto itself. (See Fig. 7.17.)

A *linear transformation*† is any mapping of the form

$$w = f(z) = az + b, \tag{4}$$

where $a$ and $b$ are complex constants with $a \neq 0$. Such a transformation can be considered as the composition of a rotation, a magnification, and a translation (each

---

† This is, unfortunately, bad terminology, because in other branches of mathematics a "linear transformation" has the property $f(z_1 + z_2) = f(z_1) + f(z_2)$, while this is true of Eq. (4) only when $b = 0$. Worse yet, *Möbius* transformations are called linear transformations by some authors.

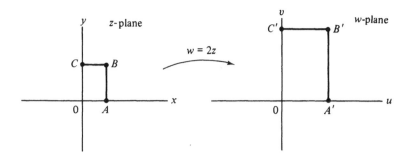

**Figure 7.17**   Magnification.

of which is, of course, a special case of the linear transformation): Writing $a$ in polar form as $a = \rho e^{i\phi}$, we express the linear transformation (4) by the composition of

$$w_1 = e^{i\phi}z,$$

$$w_2 = \rho w_1,$$

and, finally,

$$w = w_2 + b.$$

Hence the linear transformation is, once again, one-to-one in the complex plane, and the image of any object is geometrically similar to the original. (See Fig. 7.18, page 318.)

**Example 1**

Find the linear transformation that rotates the entire complex plane through an angle $\theta$ about a given point $z_0$.

**Solution.**   We know that the mapping

$$w_1 = e^{i\theta}z$$

rotates the plane through the angle $\theta$ about the origin. In particular, the point $z_0$ is mapped to the point $e^{i\theta}z_0$. If we now shift the whole plane so that the latter point is carried *back to* $z_0$, the net result will be a rotation of the plane about $z_0$. (Think about this; every straight line gets rotated through the angle $\theta$, and $z_0$ is left fixed.) Thus the required answer is

$$w = w_1 + (z_0 - e^{i\theta}z_0) = e^{i\theta}z + (1 - e^{i\theta})z_0. \quad \blacksquare$$

**Example 2**

Find a linear transformation that maps the circle $C_1: |z - 1| = 1$ onto the circle $C_2: |w - \frac{3}{2}i| = 2$.

**Solution.**   Refer to Fig. 7.19. First we translate by $-1$ so that $C_1$ becomes a unit circle centered at the origin. Then we magnify by the factor 2. Finally we translate $\frac{3}{2}$ units up the imaginary axis, bringing us to $C_2$. The mappings are

$$w_1 = z - 1,$$

$$w_2 = 2w_1 = 2z - 2,$$

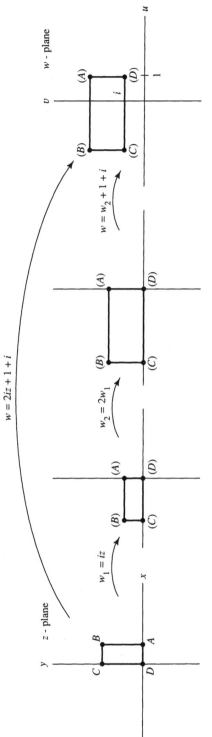

**Figure 7.18** Linear transformation $w = 2iz + 1 + i$.

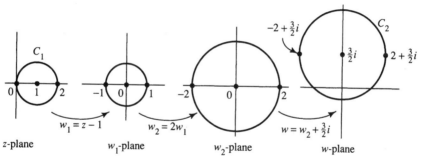

**Figure 7.19**   Mappings for Example 2.

and, last,

$$w = w_2 + \tfrac{3}{2}i = 2z - 2 + \tfrac{3}{2}i.$$

Moreover, any subsequent rotation about the point $\tfrac{3}{2}i$ can be permitted.   ∎

Now we consider the *inversion* transformation defined by

$$w = f(z) = \frac{1}{z}. \tag{5}$$

It is easy to see that the inversion is a one-to-one mapping of the *extended* complex plane onto itself. Although it fails to preserve the shape of objects, this transformation does have an important geometrical property, which will reveal itself if we look at what inversion does to lines and circles. For this purpose we regard $\infty$ as a member of every line.

**(a)** *Lines Passing through the Origin.*   The point $z = \rho e^{i\theta}$ is mapped to

$$w = \frac{1}{\rho e^{i\theta}} = \frac{1}{\rho} e^{-i\theta}.$$

Letting the real number $\rho$ vary we see that the line making an angle $\theta$ with the real axis is mapped onto the line making the angle $-\theta$ with the real axis; the point at $\infty$ goes to the origin and vice versa. (See Fig. 7.20, page 320.)

**(b)** *Lines Not Passing through the Origin.*   In the $xy$-plane, the equation of such a line is given by

$$Ax + By = C, \qquad \text{with } C \neq 0. \tag{6}$$

Now since $w = 1/z$, we have

$$z = \frac{1}{w} = \frac{\bar{w}}{|w|^2} = \frac{u - iv}{u^2 + v^2}, \qquad \text{for } w = u + iv,$$

and so

$$x = \frac{u}{u^2 + v^2}, \qquad y = \frac{-v}{u^2 + v^2}. \tag{7}$$

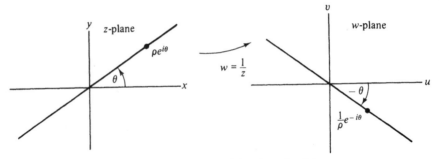

**Figure 7.20**    Inversion of line through origin.

Making these substitutions into (6) we find

$$A\left(\frac{u}{u^2 + v^2}\right) + B\left(\frac{-v}{u^2 + v^2}\right) = C,$$

which can be written in the equivalent form

$$u^2 + v^2 - \frac{A}{C}u + \frac{B}{C}v = 0. \tag{8}$$

Eq. (8) describes a circle through the origin in the $w$-plane. Hence we conclude that the image points of the line (6) all lie on this circle. In fact, since both "ends" of the line (i.e., $\infty$) are mapped to $w = 0$, the image set is simple and closed and hence must be the *whole* circle having equation (8), as is indicated in Fig. 7.21.

(c) *Circles Passing through the Origin.* Because the inverse of the map $w = 1/z$ is itself an inversion, $z = 1/w$, this case is just the reverse of case (b). In other words, from case (b) the image of the circle

$$x^2 + y^2 - \frac{A}{C}x + \frac{B}{C}y = 0 \tag{9}$$

under the map $w = 1/z$ is the line

$$Au + Bv = C.$$

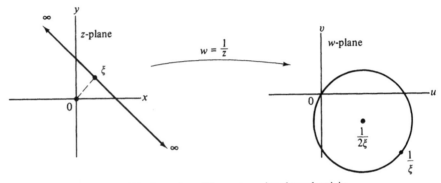

**Figure 7.21**    Inversion of line not passing through origin.

This is all we need because any circle through $z = 0$ can be written in the form (9) for suitable choices of $A$, $B$, and $C$.

**(d)** *Circles Not Passing through the Origin.* In the $xy$-plane, the equation of such a circle is of the form

$$x^2 + y^2 + Ax + By = C, \qquad \text{with} \quad C \neq 0.$$

Using the relations (7), this equation becomes

$$\frac{u^2}{(u^2 + v^2)^2} + \frac{v^2}{(u^2 + v^2)^2} + \frac{Au}{u^2 + v^2} - \frac{Bv}{u^2 + v^2} = C,$$

which simplifies to $1 + Au - Bv = C(u^2 + v^2)$. As $C \neq 0$, we therefore obtain

$$u^2 + v^2 - \frac{A}{C}u + \frac{B}{C}v = \frac{1}{C}$$

for the equation satisfied by the image of the given circle. Such an equation describes a circle in the $w$-plane that does not pass through the origin and, as before, the image is the whole circle. (See Fig. 7.22.)

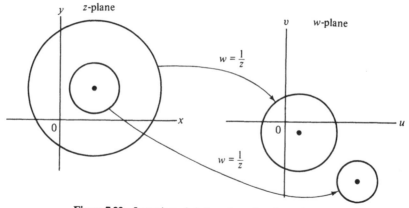

**Figure 7.22** Inversion of circle not passing through origin.

Summarizing, we see that the inversion mapping is one-to-one and carries the class of straight lines and circles into itself, a property shared with translations, rotations, and magnifications.

Now we are ready to define the *Möbius transformations*.

---

**Definition 1.** A **Möbius** (or Moebius) **transformation** (sometimes known as a **linear fractional transformation** or **bilinear transformation**) is any function of the form

$$w = f(z) = \frac{az + b}{cz + d} \tag{10}$$

with the restriction that $ad \neq bc$ (so that $w$ is not a constant function).

Notice that since

$$f'(z) = \frac{ad - bc}{(cz + d)^2}$$

does not vanish, the Möbius transformation $f(z)$ is conformal at every point except its pole $z = -d/c$.

It is easy to see that the Möbius transformations include the previous elementary transformations of this section as special cases. More important is the fact that any Möbius transformation can be decomposed into a succession of these elementary transformations. If $c = 0$, we have the linear transformation which was treated earlier. For $c \neq 0$, the decomposition can be seen by writing

$$\frac{az + b}{cz + d} = \frac{\frac{a}{c}(cz + d) - \frac{ad}{c} + b}{cz + d} = \frac{a}{c} + \frac{b - \frac{ad}{c}}{cz + d},$$

which shows that the Möbius transformation can be expressed as a linear transformation (rotation + magnification + translation)

$$w_1 = cz + d, \tag{11}$$

followed by an inversion

$$w_2 = \frac{1}{w_1}, \tag{12}$$

and then another linear transformation

$$w = \left(b - \frac{ad}{c}\right)w_2 + \frac{a}{c}. \tag{13}$$

As a result of this decomposition and of our previous deliberations, we can summarize some properties of Möbius transformations.

---

**Theorem 5.** Let $f$ be any Möbius transformation. Then

(i) $f$ can be expressed as the composition of a finite sequence of translations, magnifications, rotations, and inversions.

(ii) $f$ maps the extended complex plane one-to-one onto itself.

(iii) $f$ maps the class of circles and lines to itself.

(iv) $f$ is conformal at every point except its pole.

---

One way of distinguishing the possibilities in property (iii) is as follows. If a line or circle passes through the pole ($z = -d/c$) of the Möbius transformation, it gets mapped to an unbounded figure. Hence its image is a straight line. A line or circle that avoids the pole, then, must be mapped to a circle.

**Example 3**

Find the image of the *interior* of the circle $C: |z - 2| = 2$ under the Möbius transformation

$$w = f(z) = \frac{z}{2z - 8}.$$

**Solution.** First we find the image of the circle $C$. Since $f$ has a pole at $z = 4$ and this point lies on $C$, the image has to be a straight line. To specify this line all we need is to determine two of its finite points. The points $z = 0$ and $z = 2 + 2i$ which lie on $C$ have, as their images,

$$w = f(0) = 0 \quad \text{and} \quad w = f(2 + 2i) = \frac{2 + 2i}{2(2 + 2i) - 8} = -\frac{i}{2}.$$

Thus the image of $C$ is the imaginary axis in the $w$-plane. From our discussion of connectivity in Sec. 7.2, we know that the interior of $C$ is therefore mapped either onto the right half-plane Re $w > 0$ or onto the left half-plane Re $w < 0$. Since $z = 2$ lies inside $C$ and

$$w = f(2) = \frac{2}{4 - 8} = -\frac{1}{2}$$

lies in the left half-plane, we conclude that the image of the interior of $C$ is the left half-plane (see Figure 7.23). ∎

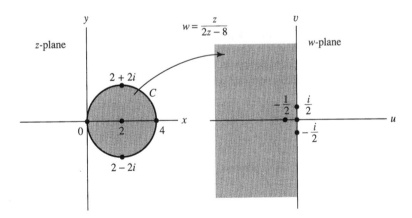

**Figure 7.23**   Mapping for Example 3.

Now we shall present an example showing how to *construct* a conformal map of one region onto another.

**Example 4**

Find a conformal map of the unit disk $|z| < 1$ onto the right half-plane Re $w > 0$.

**Solution.**    We are naturally led to look for a Möbius transformation that maps the circle $|z| = 1$ onto the imaginary axis (Fig. 7.24). The transformation must therefore have a pole on the circle, according to our earlier remarks. Moreover, the origin $w = 0$ must also lie on the image of the circle. As a first step, let's look at

$$w = f_1(z) = \frac{z + 1}{z - 1}, \tag{14}$$

which maps 1 to $\infty$ and $-1$ to 0.

From the geometric properties of Möbius transformations we have learned, we can conclude that (14) maps $|z| = 1$ onto *some* straight line through the origin. To see *which* straight line, we plug in $z = i$ and find that the point

$$w = \frac{i + 1}{i - 1} = -i$$

also lies on the line. Hence the image of the circle under $f_1$ must be the imaginary axis.

To see which half-plane is the image of the interior of the circle, we check the point $z = 0$. It is mapped by (14) to the point $w = -1$ in the *left* half-plane. This is not what we want, but it can be corrected by a final rotation of $\pi$, yielding

$$w = f(z) = -\frac{z + 1}{z - 1} = \frac{1 + z}{1 - z} \tag{15}$$

as an answer to the problem. (Of course, any subsequent vertical translation or magnification can be permitted.) Observe that (15) is precisely the mapping that was introduced in Example 1, Sec. 7.1, to solve a thermal problem, and we have thus verified the claims made there. ∎

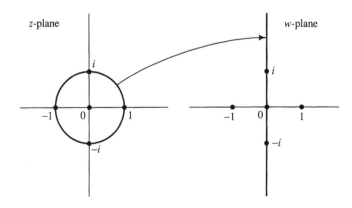

**Figure 7.24**    Mapping for Example 4.

# EXERCISES 7.3

1. Find a linear transformation mapping the circle $|z| = 1$ onto the circle $|w - 5| = 3$ and taking the point $z = i$ to $w = 2$.

2. What is the image of the strip $0 < \text{Im } z < 1$ under the mapping $w = (z - i)/z$?

3. Discuss the image of the circle $|z - 2| = 1$ and its interior under the following transformations.

   **(a)** $w = z - 2i$          **(b)** $w = 3iz$          **(c)** $w = \dfrac{z - 2}{z - 1}$

   **(d)** $w = \dfrac{z - 4}{z - 3}$          **(e)** $w = \dfrac{1}{z}$

4. Find a Möbius transformation mapping the lower half-plane to the disk $|w + 1| < 1$. [HINT: Do it in steps.]

5. Find a Möbius transformation mapping the unit disk $|z| < 1$ onto the right half-plane and taking $z = -i$ to the origin.

6. A *fixed point* of a function $f(z)$ is a point $z_0$ satisfying $f(z_0) = z_0$. Show that a Möbius transformation $f(z)$ can have at most two fixed points in the complex plane unless $f(z) \equiv z$.

7. Find the Möbius transformation that maps $0, 1, \infty$ to the following respective points.

   **(a)** $0, i, \infty$      **(b)** $0, 1, 2$      **(c)** $-i, \infty, 1$      **(d)** $-1, \infty, 1$

8. What is the image of the third quadrant under the mapping $w = (z + i)/(z - i)$?

9. What is the image of the sector $-\pi/4 < \text{Arg } z < \pi/4$ under the mapping $w = z/(z - 1)$?

10. Find a conformal map of the semidisk $|z| < 1$, $\text{Im } z > 0$, onto the upper half-plane. [HINT: Combine a Möbius transformation with the mapping $w = z^2$. Make sure you cover the entire upper half-plane.]

11. Map the shaded region in Fig. 7.25 conformally onto the upper half-plane. [HINT: Use a Möbius transformation to map the point 2 to $\infty$. Argue that the image region will be a *strip*. Then use the exponential map.]

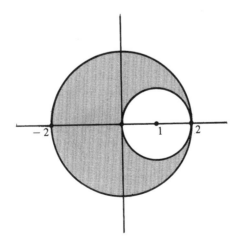

**Figure 7.25**   Region for Prob. 11.

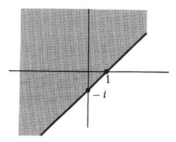

**Figure 7.26**   Region for Prob. 12.

**12.** Find a Möbius transformation which takes the half-plane depicted in Fig. 7.26 onto the unit disk $|w| < 1$.

**13.** (*Smith Chart*)   The *impedance Z* of an electrical circuit oscillating at a frequency $\omega$ is a complex number, denoted $Z = R + iB$, which characterizes the voltage-current relationship of the circuit; recall Sec. 3.4. In practice $R$ can take any value from 0 to $\infty$ and $B$ can take any value from $-\infty$ to $\infty$. Thus the usual representation of $Z$ as a point in the complex plane becomes unwieldy (inasmuch as the entire right half-plane comes into play). The *Smith chart* provides a more compact graphical description, displaying the entire range of impedances within the unit circle. The impedance $Z$ is depicted as the point

$$W = \frac{Z - 1}{Z + 1}.$$

This mapping (its inverse, actually) is portrayed in Figs. 7.4 and 7.5. $W$ is also known as the *reflection coefficient* corresponding to $Z$.

**(a)** Show that the circles in the Smith chart depicting the lines Re $Z = R = $ constant, indicating constant-resistance contours, have the equations

$$\left(u - \frac{R}{1 + R}\right)^2 + v^2 = \frac{1}{(1 + R)^2}.$$

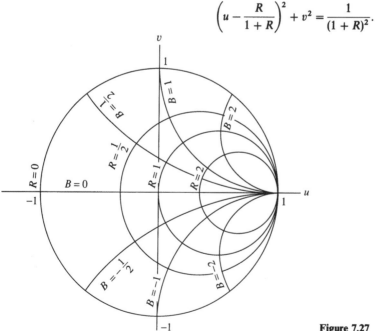

**Figure 7.27**   Smith chart.

**(b)** Show that the circles in the Smith chart depicting the lines Im $Z = B =$ constant, indicating constant-reactance contours, have the equations

$$(u - 1)^2 + \left(v - \frac{1}{B}\right)^2 = \frac{1}{B^2}.$$

(See Fig. 7.27.)

14. If a circuit with impedance $Z$ is connected to a length $\ell$ of *transmission line* with "phase constant" $\beta$ and a "characteristic impedance" of unity, then the new configuration has a transformed impedance $Z'$ given by

$$Z' = \frac{Z \cos \beta\ell + i \sin \beta\ell}{\cos \beta\ell + iZ \sin \beta\ell}.$$

Show that the Smith chart point depicting $Z'$ can be obtained from the Smith chart point depicting $Z$ by a clockwise rotation of $2\beta\ell$ radians about the origin.

## 7.4 MÖBIUS TRANSFORMATIONS, CONTINUED

We shall now explore some additional properties of Möbius transformations which enhance their usefulness as conformal mappings. These are the group properties, the cross-ratio formula, and the symmetry property.

Given any Möbius transformation

$$w = f(z) = \frac{az + b}{cz + d} \qquad (ad \neq bc), \tag{1}$$

its inverse $f^{-1}(w)$ can be found by simply solving Eq. (1) for $z$ in terms of $w$. This computation yields

$$z = f^{-1}(w) = \frac{dw - b}{-cw + a},$$

and we see that *the inverse of any Möbius transformation is again a Möbius transformation.* Furthermore, if we take the composition of two Möbius transformations, say

$$w = f_1(z) = \frac{a_1 z + b_1}{c_1 z + d_1} \quad \text{and} \quad \zeta = f_2(w) = \frac{a_2 w + b_2}{c_2 w + d_2},$$

it can be readily shown that

$$\zeta = f_2(f_1(z)) = \frac{(a_2 a_1 + b_2 c_1)z + (a_2 b_1 + b_2 d_1)}{(c_2 a_1 + d_2 c_1)z + (c_2 b_1 + d_2 d_1)}.$$

Hence *the composition of any two Möbius transformations is also a Möbius transformation.* [Rigorously speaking, we should make this claim only after verifying that $f_2(f_1)$ cannot reduce to a constant, but this is obviously true because the one-to-oneness of $f_1$ and $f_2$ implies that $f_2(f_1)$ is one-to-one.] Of course when we take $f_1^{-1}(f_1)$ we get the identity function $I(z) \equiv z$, which is certainly a Möbius transformation. These facts are known as the *group properties* of the Möbius transformations (see Prob. 19).

We have already seen that Möbius transformations map the class of circles and lines to itself. Now we turn to the problem of finding a specific transformation which maps a *given* circle (or line) $C_z$ in the z-plane to a *given* circle (or line) $C_w$ in the w-plane. Recall from geometry that any three distinct noncollinear points uniquely determine a circle; if these points are collinear, then, of course, they uniquely determine a line (in particular this will be the case when one of them is $\infty$). Hence if we choose three points $z_1$, $z_2$, $z_3$ on $C_z$ and three points $w_1$, $w_2$, $w_3$ on $C_w$ and find a Möbius transformation $f$ satisfying

$$f(z_1) = w_1, \qquad f(z_2) = w_2, \qquad f(z_3) = w_3, \tag{2}$$

then $f$ must map $C_z$ onto $C_w$.

It is not difficult to write down a Möbius transformation that satisfies Eqs. (2) in the case when $w_1 = 0$, $w_2 = 1$, and $w_3 = \infty$; this corresponds to the problem of mapping $C_z$ onto the real axis. If all the points $z_1$, $z_2$, $z_3$ are finite, it is easy to check that

$$T(z) = \frac{(z - z_1)(z_2 - z_3)}{(z - z_3)(z_2 - z_1)} \tag{3}$$

satisfies

$$T(z_1) = 0, \qquad T(z_2) = 1, \qquad T(z_3) = \infty, \tag{4}$$

while if one of the $z_i$ is $\infty$, the conditions (4) will be satisfied by

$$T(z) = \frac{z_2 - z_3}{z - z_3} \quad (z_1 = \infty), \qquad T(z) = \frac{z - z_1}{z - z_3} \quad (z_2 = \infty),$$

$$\text{or} \qquad T(z) = \frac{z - z_1}{z_2 - z_1} \quad (z_3 = \infty).^\dagger \tag{5}$$

We remark that the right-hand side of Eq. (3) [or Eqs. (5)] is called the *cross-ratio* of the four points $z$, $z_1$, $z_2$, $z_3$ and is abbreviated by writing $(z, z_1, z_2, z_3)$; that is,

$$(z, z_1, z_2, z_3) := \frac{(z - z_1)(z_2 - z_3)}{(z - z_3)(z_2 - z_1)} \tag{6}$$

in the case of finite points. Notice that the order in which the points are listed is crucial in this notation. For example,

$$(z, 3, 0, i) = \frac{(z - 3)(0 - i)}{(z - i)(0 - 3)} = \frac{-iz + 3i}{-3z + 3i},$$

but

$$(z, i, 3, 0) = \frac{(z - i)(3 - 0)}{(z - 0)(3 - i)} = \frac{3z - 3i}{(3 - i)z}.$$

---

† Note that the Möbius transformations in (5) can be obtained immediately from (3) by simply deleting the factors involving $\infty$.

Now if we wish to solve the general problem of finding a Möbius transformation $f$ that maps

$$z_1 \text{ to } w_1, \quad z_2 \text{ to } w_2, \quad z_3 \text{ to } w_3,$$

where $w_1$, $w_2$, $w_3$ are any three distinct points, just discussed, we can proceed as follows: Let $T(z)$ be the Möbius transformation just discussed, taking

$$z_1 \text{ to } 0, \quad z_2 \text{ to } 1, \quad z_3 \text{ to } \infty,$$

and let $S(w)$ be the analogous transformation that maps

$$w_1 \text{ to } 0, \quad w_2 \text{ to } 1, \quad w_3 \text{ to } \infty.$$

Then the desired Möbius transformation $f$ is given by the composition

$$w = f(z) = S^{-1}(T(z)), \tag{7}$$

because

$$f(z_1) = S^{-1}(T(z_1)) = S^{-1}(0) = w_1,$$

$$f(z_2) = S^{-1}(T(z_2)) = S^{-1}(1) = w_2,$$

$$f(z_3) = S^{-1}(T(z_3)) = S^{-1}(\infty) = w_3.$$

Notice that Eq. (7) is equivalent to the equation

$$S(w) = T(z);$$

in other words, to map $z_1$, $z_2$, $z_3$ to the respective points $w_1$, $w_2$, $w_3$, we need merely equate the two cross-ratios

$$(w, w_1, w_2, w_3) = (z, z_1, z_2, z_3) \tag{8}$$

and solve for $w$ in terms of $z$.

**Example 1**

Find a Möbius transformation that maps 0 to $i$, 1 to 2, and $-1$ to 4.

**Solution.** The appropriate cross-ratios are given by

$$(z, 0, 1, -1) = \frac{(z - 0)[1 - (-1)]}{[z - (-1)](1 - 0)} = \frac{2z}{z + 1}$$

and

$$(w, i, 2, 4) = \frac{(w - i)(2 - 4)}{(w - 4)(2 - i)} = \frac{-2(w - i)}{(w - 4)(2 - i)}.$$

Hence, solving the equation

$$\frac{-2(w - i)}{(w - 4)(2 - i)} = \frac{2z}{z + 1}$$

for $w$ yields the desired transformation

$$w = \frac{(16 - 6i)z + 2i}{(6 - 2i)z + 2}. \quad \blacksquare$$

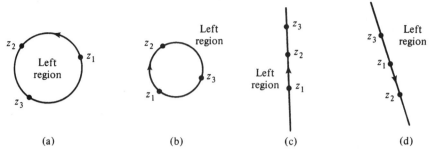

**Figure 7.28** Left regions determined by three-point sequence.

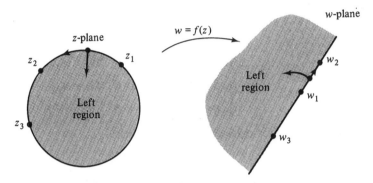

**Figure 7.29** Correspondence of left regions.

It is important to note that the circle or line $\Gamma$ determined by the three points $z_1, z_2, z_3$ is also *oriented* by the order of these points. That is, $\Gamma$ acquires the direction obtained by proceeding through the points $z_1, z_2, z_3$ in succession. [Notice that lines are regarded as "closed" at $\infty$ in the present context. Hence they, like circles, require a sequence of *three* points to determine a direction. See Fig. 7.28(d).] This orientation, in turn, uniquely specifies the "left region," the region that lies to the left of an observer traversing $\Gamma$ (Fig. 7.28). Since Möbius transformations are conformal, it can be shown that a Möbius transformation that takes $z_1, z_2, z_3$ to the respective points $w_1, w_2, w_3$ must map the left region of the circle (or line) oriented by $z_1, z_2, z_3$ onto the left region of the circle (or line) oriented by $w_1, w_2, w_3$. To see this in the special case depicted in Fig. 7.29 imagine a short directed segment drawn from a point on the circle in the $z$-plane into the left region. The image of this segment is, by conformality, a curve from the image line into its left region. By connectivity, then, we conclude that the left region is mapped to the left region. The other situations of mapping a circle to a circle, a line to a circle, and a line to a line can be treated in a similar manner.

Thus by judiciously selecting ordered points we can quickly write down an algebraic formula for a mapping of one "circular" region onto another.

**Example 2**

Find a Möbius transformation that maps the region $D_1 \colon |z| > 1$ onto the region $D_2 \colon \operatorname{Re} w < 0$.

**Solution.**   We shall take both $D_1$ and $D_2$ to be left regions. This is accomplished for $D_1$ by choosing any three points on the circle $|z| = 1$ that give it a negative (clockwise) orientation, say

$$z_1 = 1, \qquad z_2 = -i, \qquad z_3 = -1.$$

Similarly the three points

$$w_1 = 0, \qquad w_2 = i, \qquad w_3 = \infty$$

on the imaginary axis make $D_2$ its left region. Hence a solution to the problem is given by the transformation that takes

$$1 \text{ to } 0, \qquad -i \text{ to } i, \qquad -1 \text{ to } \infty.$$

This we obtain by setting

$$(w, 0, i, \infty) = (z, 1, -i, -1),$$

that is,

$$\frac{w - 0}{i - 0} = \frac{(z - 1)(-i + 1)}{(z + 1)(-i - 1)},$$

which yields

$$w = \frac{(z - 1)(1 + i)}{(z + 1)(-i - 1)} = \frac{1 - z}{1 + z}. \quad \blacksquare$$

Another important aspect of Möbius transformations is their symmetry-preserving property. First recall that two points $z_1$ and $z_2$ are symmetric with respect to a straight line $\mathscr{L}$ if $\mathscr{L}$ is the perpendicular bisector of the line segment joining $z_1$ and $z_2$ [see Fig. 7.30(a)]. From elementary geometry this is equivalent to saying that *every* line or circle through $z_1$ and $z_2$ intersects $\mathscr{L}$ orthogonally, i.e., at right angles. See Fig. 7.30(b). (Remember that a circle is orthogonal to $\mathscr{L}$ if and only if its center lies on $\mathscr{L}$.)

These considerations suggest the following definition of symmetry with respect to a circle $C$.

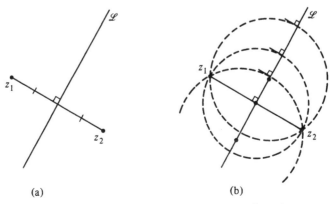

(a)                                        (b)

**Figure 7.30**   $z_1$, $z_2$ symmetric with respect to line $\mathscr{L}$.

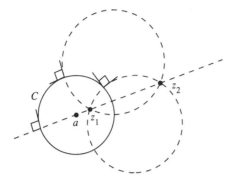

**Figure 7.31**  $z_1$, $z_2$ symmetric with respect to circle C.

**Definition 2.**  Two points $z_1$ and $z_2$ are said to be **symmetric with respect to a circle** $C$ if every straight line or circle passing through $z_1$ and $z_2$ intersects $C$ orthogonally (Fig. 7.31).

In particular the center $a$ of the circle $C$, and the point $\infty$, are symmetric with respect to $C$; there are no circles through these two points, and any line containing $a$ (and necessarily $\infty$) is orthogonal to $C$, so the condition in Definition 2 holds.

Now we are in a position to state the symmetry-preserving property of Möbius transformations.

**Theorem 6.**  (*Symmetry Principle*) Let $C_z$ be a line or circle in the $z$-plane, and let $w = f(z)$ be any Möbius transformation. Then two points $z_1$ and $z_2$ are symmetric with respect to $C_z$ if and only if their images $w_1 = f(z_1)$, $w_2 = f(z_2)$ are symmetric with respect to the image of $C_z$ under $f$.

The theorem is illustrated in Fig. 7.32 for the special case when $C_z$ is a circle that maps to a straight line.

*Proof.*  This is easy; think about it. Two points are symmetric with respect to a circle (line) if every circle or line containing the points intersects the given circle

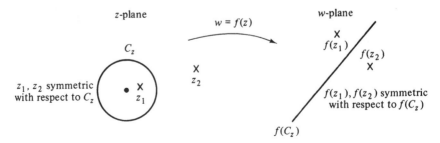

**Figure 7.32**  Symmetry Principle.

(line) orthogonally. But Möbius transformations preserve the class of circles and lines, and they also preserve the orthogonality; hence they preserve the symmetry condition.  ■

Now given a circle $C$ with center $a$ and radius $R$, and given a point $\alpha$, it would be convenient to have a formula for the point $\alpha^*$ symmetric to $\alpha$ with respect to $C$. For this purpose observe that the transformation

$$T(z) = (z, a - R, a + Ri, a + R)$$

$$= \frac{[z - (a - R)](Ri - R)}{[z - (a + R)](Ri + R)} = i\frac{z - (a - R)}{z - (a + R)} \tag{9}$$

maps three points of $C$, and hence all of $C$, onto the real axis. Thus, by the symmetry principle, $\alpha^*$ is symmetric to $\alpha$ with respect to $C$ if and only if $T(\alpha^*)$ is symmetric to $T(\alpha)$ with respect to the real axis. But the latter condition is clearly equivalent to saying that $T(\alpha^*)$ and $T(\alpha)$ must be conjugate points, i.e.,

$$T(\alpha^*) = \overline{T(\alpha)},$$

or, using Eq. (9),

$$i\frac{\alpha^* - (a - R)}{\alpha^* - (a + R)} = \overline{\left[i\frac{\alpha - (a - R)}{\alpha - (a + R)}\right]} = -i\frac{\bar{\alpha} - (\bar{a} - R)}{\bar{\alpha} - (\bar{a} + R)}. \tag{10}$$

Solving Eq. (10) for $\alpha^*$ yields the formula

$$\alpha^* = \frac{R^2}{\bar{\alpha} - \bar{a}} + a. \tag{11}$$

(Notice that this also shows that the point symmetric to $\alpha$ with respect to $C$ is *unique*.) From representation (11) we see that

$$\arg(\alpha^* - a) = \arg\left(\frac{R^2}{\bar{\alpha} - \bar{a}}\right) = \arg\left[\frac{R^2(\alpha - a)}{|\alpha - a|^2}\right] = \arg(\alpha - a),$$

and

$$|\alpha^* - a| = \frac{R^2}{|\bar{\alpha} - \bar{a}|} = \frac{R^2}{|\alpha - a|},$$

implying that symmetric points $\alpha^*$ and $\alpha$ lie on the same ray from the center $a$ and that the product of their distances from the center ($|\alpha^* - a| \cdot |\alpha - a|$) is equal to the radius squared. Figure 7.33, page 334, suggests a construction of symmetric points.

## Example 3

Find all Möbius transformations that map $|z| < 1$ onto $|w| < 1$.

**Solution.**  Let $f(z)$ be any such Möbius transformation. Then $f$ maps the circle $C_z$: $|z| = 1$ onto $C_w$: $|w| = 1$. Furthermore, there must be some point $\alpha$, $|\alpha| < 1$, that is mapped to the origin, i.e., $f(\alpha) = 0$. According to formula (11) (with $a = 0$, $R = 1$) the point

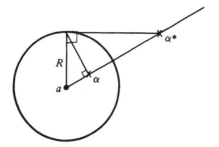

**Figure 7.33** Similar triangles give $R/(|\alpha - a|) = (|\alpha^* - a|)/R$.

$$\alpha^* = \frac{1^2}{\bar{\alpha} - \bar{0}} + 0 = \frac{1}{\bar{\alpha}}$$

is symmetric to $\alpha$ with respect to $C_z$. Hence $f(1/\bar{\alpha})$ must be symmetric to $f(\alpha) = 0$ with respect to $C_w$. But since the origin is the center of $C_w$, its symmetric point is $\infty$, i.e.,

$$f\left(\frac{1}{\bar{\alpha}}\right) = \infty.$$

Consequently, $f$ has a zero at $\alpha$ and a pole at $1/\bar{\alpha}$, so $f$ is of the form

$$f(z) = k \cdot \frac{z - \alpha}{z - \dfrac{1}{\bar{\alpha}}} = k\bar{\alpha}\,\frac{z - \alpha}{\bar{\alpha}z - 1}$$

for some constant $k$. Moreover, since $f(1)$ lies on $C_w$, we have

$$1 = |f(1)| = |k\bar{\alpha}| \cdot \left|\frac{1 - \alpha}{\bar{\alpha} - 1}\right| = |k\bar{\alpha}|.$$

Thus $k\bar{\alpha} = e^{i\theta}$ for some real $\theta$, and we find

$$f(z) = e^{i\theta} \cdot \frac{z - \alpha}{\bar{\alpha}z - 1} \qquad (|\alpha| < 1). \tag{12}$$

Conversely, the reader can easily show (Prob. 14) that any transformation of the form (12) maps $|z| < 1$ onto $|w| < 1$. ∎

More generally, it can be shown that the functions in Eq. (12) are the only one-to-one *analytic* mappings of the unit disk onto itself.

## EXERCISES 7.4

1. Let $f_1(z) = (z + 2)/(z + 3)$, $f_2(z) = z/(z + 1)$. Find $f_1^{-1}(f_2(z))$.
2. Argue why the Möbius transformation defined by $(w, -i, 1, i) = (z, -i, i, 1)$ maps the unit circle onto itself but maps the interior onto the exterior. [HINT: Consider orientation.]

3. Find the point symmetric to $4 - 3i$ with respect to each of the following circles.

  (a) $|z| = 1$          (b) $|z - 1| = 1$          (c) $|z - 1| = 2$

4. Prove that if $z_2$, $z_3$, and $z_4$ are distinct points in the extended complex plane and $T$ is any Möbius transformation, then

$$(z_1, z_2, z_3, z_4) = (T(z_1), T(z_2), T(z_3), T(z_4))$$

for any point $z_1$ in the extended plane. That is, *the cross-ratio is invariant under Möbius transformations.* [HINT: Consider translations, rotations, magnifications, and inversions.]

5. Let $w = f(z)$ be the Möbius transformation mapping the points $0$, $\lambda$, $\infty$ to $-i$, $1$, $i$, respectively, where $\lambda$ is real. For what values of $\lambda$ is the upper half-plane mapped onto $|w| < 1$?

6. Using the cross-ratio notation, write an equation defining a Möbius transformation that maps the half-plane below the line $y = 2x - 3$ onto the interior of the circle $|w - 4| = 2$. Repeat for the exterior of this circle.

7. Does there exist a Möbius transformation $f$ that maps the real axis onto the unit circle $|w| = 1$ and satisfies $f(i) = 2$, $f(-i) = -\tfrac{1}{2}$?

8. Prove that if $z_1$, $z_2$, $z_3$ are distinct points and $w_1$, $w_2$, $w_3$ are distinct points, then the Möbius transformation $T$ satisfying $T(z_1) = w_1$, $T(z_2) = w_2$, $T(z_3) = w_3$ is *unique*. [HINT: Suppose that $S$ is another such Möbius transformation and consider the fixed points of the composition $T^{-1} \circ S$ (see Prob. 6, Exercises 7.3, page 325).]

9. Let $f$ be a Möbius transformation such that $f(1) = \infty$ and $f$ maps the imaginary axis onto the unit circle $|w| = 1$. What is the value of $f(-1)$?

10. By completing the following steps, prove that given any two nonintersecting circles $C_1$ and $C_2$ there always exist two distinct points $z_1$ and $z_2$ that are symmetric with respect to $C_1$ and $C_2$ *simultaneously.*

  (a) Argue that there exists a Möbius transformation that maps $C_1$ onto the real axis $\mathbf{R}$ and $C_2$ onto some circle $C$ of the form $|w - \lambda i| = R$ with $\lambda$ real and $R < |\lambda|$. (See Fig. 7.34.)

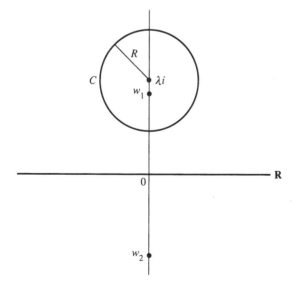

**Figure 7.34**  $w_1$, $w_2$ symmetric in $\mathbf{R}$ and $C$.

**(b)** Show that $w_1$ and $w_2$ are symmetric with respect to both **R** and $C$ if and only if

$$w_2 = \overline{w_1} \quad \text{and} \quad w_2 = \frac{R^2}{\overline{w_1} + \lambda i} + \lambda i.$$

Solve this pair of equations to obtain

$$w_1 = i\sqrt{\lambda^2 - R^2}, \qquad w_2 = -i\sqrt{\lambda^2 - R^2}$$

as the simultaneously symmetric points.

**(c)** Use the results of parts (a) and (b) and the symmetry principle to conclude that there are points $z_1$ and $z_2$ symmetric in both $C_1$ and $C_2$.

**11.** Use the result of Prob. 10 to show that for any two nonintersecting circles $C_1$ and $C_2$ there always exists a Möbius transformation that maps $C_1$ and $C_2$ onto *concentric circles*. [HINT: Map $z_1$ to the origin and $z_2$ to infinity, where $z_1, z_2$ are points symmetric with respect to both circles.]

**12.** Let $z_1$, $z_2$, and $z_3$ be three distinct points that lie on a circle (or line) $C$. Prove that $z$ and $z^*$ are symmetric with respect to $C$ if and only if $(z^*, z_1, z_2, z_3) = \overline{(z, z_1, z_2, z_3)}$.

**13.** Show that the distinct points $w_1$, $w_2$, $w_3$, and $w_4$ all lie on the same circle or line if and only if the cross-ratio $(w_1, w_2, w_3, w_4)$ is real. [HINT: Consider the Möbius transformation defined by

$$(w, w_2, w_3, w_4) = (z, 0, 1, \infty)$$

and observe that $(z, 0, 1, \infty) \equiv z$.]

**14.** Verify that any transformation of the form (12) maps $|z| < 1$ onto $|w| < 1$.

**15.** Find a conformal map of the unit disk onto itself, taking the point $i/2$ to the origin.

**16.** Show that the Möbius transformation taking $z_i$ to $w_i$ ($i = 1, 2, 3$) can be expressed in determinant form as

$$\begin{vmatrix} 1 & z & w & zw \\ 1 & z_1 & w_1 & z_1 w_1 \\ 1 & z_2 & w_2 & z_2 w_2 \\ 1 & z_3 & w_3 & z_3 w_3 \end{vmatrix} = 0.$$

**17.** Find all Möbius transformations that map the upper half-plane onto itself.

**18.** Show that every Möbius transformation that maps the upper half-plane onto the open unit disk must be of the form

$$f(z) = e^{i\theta} \frac{z - z_0}{z - \bar{z}_0}, \qquad \text{where } \text{Im}(z_0) > 0.$$

**19.** A set $\mathcal{G}$ of mathematical objects (such as numbers or mappings) together with an operation $*$ defined on ordered pairs of objects in $\mathcal{G}$ is called a *group* if it satisfies the following conditions:

(i) $a * b$ is a unique element of $\mathcal{G}$ for every ordered pair $(a, b)$ of elements of $\mathcal{G}$.

(ii) The operation $*$ is associative; i.e. for any three elements $a, b, c$ of $\mathcal{G}$,

$$(a * b) * c = a * (b * c).$$

(iii) There exists an element $e$ in $\mathcal{G}$ (called the *identity* element) with the property that

$$e * a = a * e = a \qquad \text{for all } a \text{ in } \mathcal{G}.$$

(iv) For each $a$ in $\mathscr{G}$ there exists an element $a^{-1}$ in $\mathscr{G}$ (called the *inverse* of $a$) such that

$$a^{-1} * a = a * a^{-1} = e.$$

(a) Prove that the set $\mathscr{M}$ of Möbius transformations forms a group under the operation of composition $\circ$ of mappings.

(b) Is the group of Möbius transformations commutative? (That is, is $T \circ S = S \circ T$ for all $S,\ T \in \mathscr{M}$?)

20. Let $\mathscr{S}$ be the set of all two-by-two (complex number) matrices having determinant 1:

$$\begin{pmatrix} a & b \\ c & d \end{pmatrix}, \qquad ad - bc = 1.$$

(a) Prove that $\mathscr{S}$ forms a group under ordinary multiplication of matrices (see Prob. 19 for the definition of *group*).

(b) Show that on multiplying numerator and denominator by a suitable number, any Möbius transformation $T$ can be written in the form

$$T(z) = \frac{\alpha z + \beta}{\gamma z + \delta}$$

with $\alpha\delta - \beta\gamma = 1$. Thus $T$ can be associated with the element

$$\begin{pmatrix} \alpha & \beta \\ \gamma & \delta \end{pmatrix}$$

of $\mathscr{S}$.

(c) Show that if the Möbius transformations $T_1$ and $T_2$ are associated as in part (b) with the elements

$$S_1 = \begin{pmatrix} \alpha_1 & \beta_1 \\ \gamma_1 & \delta_1 \end{pmatrix} \quad \text{and} \quad S_2 = \begin{pmatrix} \alpha_2 & \beta_2 \\ \gamma_2 & \delta_2 \end{pmatrix}$$

of $\mathscr{S}$, then the composition $T_1 \circ T_2$ is associated with the product matrix $S_1 S_2$.

21. Let $z$ be fixed with Re $z \geq 0$, and let

$$T_0(w) = \frac{a_0}{z + a_0 + b_1 + w}, \qquad T_k(w) = \frac{a_k}{z + b_{k+1} + w} \qquad (k = 1, 2, \ldots, n-1)$$

be a sequence of Möbius transformations such that each $a_k$ is real and positive and each $b_k$ is pure imaginary or zero. Prove, by induction, that the composition

$$\zeta = S(w) := T_0 \circ T_1 \circ \cdots \circ T_{n-2} \circ T_{n-1}(w)$$

maps the half-plane Re $w > 0$ onto a region contained in the disk $|\zeta - \tfrac{1}{2}| < \tfrac{1}{2}$.

22. Let $P(z) = z^n + c_1 z^{n-1} + c_2 z^{n-2} + \cdots + c_n$ be a polynomial of degree $n > 0$ with complex coefficients $c_k = p_k + iq_k$, $k = 1, 2, \ldots, n$. Set $Q(z) := p_1 z^{n-1} + iq_2 z^{n-2} + p_3 z^{n-3} + iq_4 z^{n-4} + \cdots$. Prove *Wall's criterion* that if $Q(z)/P(z)$ can be written in the form

$$\frac{Q(z)}{P(z)} = \cfrac{a_0}{z + a_0 + b_1 + \cfrac{a_1}{z + b_2 + \cfrac{a_2}{z + b_3 + \cfrac{\ddots}{+\cfrac{a_{n-1}}{z + b_n}}}}},$$

where each $a_k$ is real and positive and each $b_k$ is pure imaginary or zero, then all the zeros of $P(z)$ have negative real parts. [HINT: Write $Q(z)/P(z) = T_0 \circ T_1 \circ \cdots \circ T_{n-1}(0)$, where the transformations $T_k$ are defined as in Prob. 21.]

**23.** Prove that $P(z) = z^3 + 3z^2 + 6z + 6$ has all its zeros in the left half-plane by applying the result of Prob. 22. [HINT: Use ordinary long division to obtain the representation for $Q(z)/P(z)$.]

## 7.5 THE SCHWARZ-CHRISTOFFEL TRANSFORMATION

We have seen that a function $f(z)$ is conformal at every point at which it is analytic and its derivative is nonzero. It is instructive to analyze what happens at certain isolated points where these conditions are not met. For concreteness, let $x_1$ be a fixed point on the real axis and let $f(z)$ be a function whose *derivative* $f'(z)$ is given by $(z - x_1)^\alpha$ for some real $\alpha$ satisfying $-1 < \alpha < 1$. [To be precise, we shall take the argument of $z - x_1$ to lie between $-\pi/2$ and $3\pi/2$, introducing a branch cut vertically downward from $x_1$; see Fig. 7.35(a).] We are going to use the equation

$$f'(z) = (z - x_1)^\alpha \tag{1}$$

to determine certain features of the image of the real axis under the mapping $f$.

If $z$ lies on the $x$-axis to the left of $x_1$, as in Fig. 7.35(a), then two observations follow from Eq. (1); namely, $f$ is conformal at $z$ (since $f'$ exists and is nonzero), and the argument of $f'(z)$ is constant for all such $z$:

$$\arg f'(z) = \arg(z - x_1)^\alpha = \alpha \arg(z - x_1) = \alpha\pi$$

(we ignore multiples of $2\pi$ in this derivation). From this we can conclude that $f$ maps the interval $(-\infty, x_1)$ onto a portion of a *straight line* terminating at $f(x_1)$; after all, if we view $(-\infty, x_1)$ as a curve whose tangents are all parallel to the real axis, then according to the discussion of Sec. 7.2 its image must be a curve, all of whose tangents make an angle $\alpha\pi$ with the real axis—i.e., a straight line. See Fig. 7.35(b).

For $z$ on the real axis to the right of $x_1$, we have

$$\arg f'(z) = \alpha \arg(z - x_1) = \alpha \cdot 0 = 0.$$

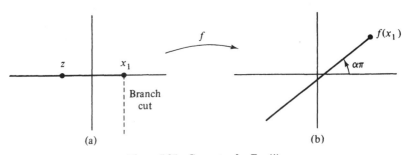

**Figure 7.35**  Geometry for Eq. (1).

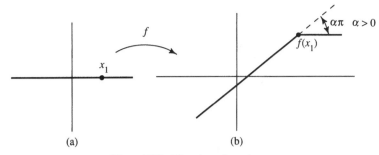

**Figure 7.36**   Mapping of x-axis.

Hence, by similar reasoning, the interval $(x_1, \infty)$ is mapped to a *horizontal* straight line, and the whole picture looks like Fig. 7.36(b).

For the special case where $f(z) = \frac{2}{3}z^{3/2}$, which has $f'(z) = z^{1/2}$, the mapping (for the branch described earlier) is sketched in Fig. 7.37.

Now we start to generalize this model. If, instead of Eq. (1) we have

$$f'(z) = A(z - x_1)^{\alpha} \tag{2}$$

for some complex constant $A \ (\neq 0)$, then

$$\arg f'(z) = \arg A + \alpha \arg(z - x_1),$$

and the mapping can be visualized by rotating Fig. 7.36(b) by an amount $\arg A$; see Fig. 7.38. In particular, the angle made by the image of the interval $(x_1, \infty)$ is now $\arg A$, but the angle of the turn at $f(x_1)$ is unchanged.

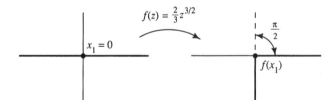

**Figure 7.37**   Mapping of x-axis by $f(z) = \frac{2}{3}z^{3/2}$.

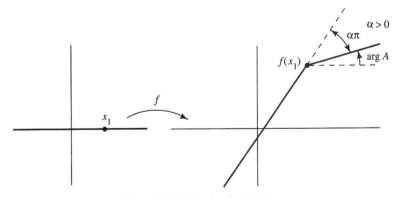

**Figure 7.38**   Mapping for Eq. (2).

The next generalization is to consider a mapping given by a function $f$ with a derivative of the form

$$f'(z) = A(z - x_1)^{\alpha_1}(z - x_2)^{\alpha_2} \cdots (z - x_n)^{\alpha_n}; \tag{3}$$

here $A$ $(\neq 0)$ is a complex constant, each $\alpha_i$ lies between $-1$ and $+1$, and the (real) $x_i$ satisfy

$$x_1 < x_2 < \cdots < x_n.$$

(As before we take the argument of each $z - x_i$ to be between $-\pi/2$ and $3\pi/2$.) What does this mapping $f$ do to the real axis?

From the equation

$$\arg f'(z) = \arg A + \alpha_1 \arg(z - x_1) + \alpha_2 \arg(z - x_2) + \cdots + \alpha_n \arg(z - x_n)$$

and the previous discussion we see that the images of the intervals $(-\infty, x_1)$, $(x_1, x_2), \ldots, (x_n, \infty)$ are each portions of straight lines, making angles measured counterclockwise from the horizontal in accordance with the following prescription:

| Interval | Angle of image |
|---|---|
| $(-\infty, x_1)$ | $\arg A + \alpha_1\pi + \alpha_2\pi + \cdots + \alpha_n\pi$ |
| $(x_1, x_2)$ | $\arg A + \alpha_2\pi + \cdots + \alpha_n\pi$ |
| $\vdots$ | $\vdots$ |
| $(x_{n-1}, x_n)$ | $\arg A + \alpha_n\pi$ |
| $(x_n, \infty)$ | $\arg A.$ |

Hence as $z$ traverses the real axis from left to right $f(z)$ generates a polygonal path whose tangent at the point $f(x_i)$ *makes a right turn through the angle* $\alpha_i\pi$; see Fig. 7.39.

Now if the function $f(z)$ satisfies Eq. (3) it is, a priori, differentiable and hence analytic on the complex plane with the exception of the (downward) branch cuts from the points $x_i$. So for any $z$ in the upper half-plane we can set

$$g(z) := \int_\Gamma f'(\zeta)\, d\zeta, \tag{4}$$

where $\Gamma$ is, for definiteness, the straight line segment from 0 to $z$, and conclude then

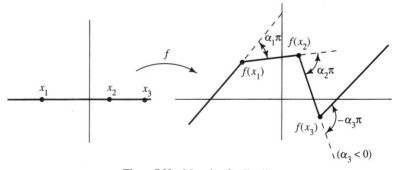

**Figure 7.39**  Mapping for Eq. (3).

that $f(z) = g(z) + B$ for some constant $B$. In particular, we can write

$$f(z) = A \int_0^z (\zeta - x_1)^{\alpha_1}(\zeta - x_2)^{\alpha_2} \cdots (\zeta - x_n)^{\alpha_n} \, d\zeta + B. \qquad (5)$$

Functions of the form (5) are known as *Schwarz-Christoffel transformations*. We have seen that such transformations map the real axis onto a polygonal path. Now one of the most important problems in conformal mapping applications is the construction of a one-to-one analytic function carrying the upper half-plane to the interior of a *given* polygon. We thus turn to the task of tailoring a Schwarz-Christoffel transformation to accomplish this.

To be specific, let the polygon $P$ have vertices at the consecutive points $w_1, w_2, \ldots, w_n$ taken in *counterclockwise* order, giving $P$ a positive orientation, as in Fig. 7.40. In traversing the polygon we make a right turn at vertex $w_i$ through the angle $\theta_i$. Thus each angle lies between $-\pi$ and $\pi$ and a negative value of $\theta_i$ indicates a left turn. The *net* rotation for a counterclockwise tour must be $2\pi$ radians to the left:

$$\theta_1 + \theta_2 + \cdots + \theta_n = -2\pi. \qquad (6)$$

To map the $x$-axis onto $P$ with a Schwarz-Christoffel transformation $w = g(z)$ we begin by picking real points $x_1, x_2, \ldots, x_{n-1}$ as the preimages of the vertices $w_1, w_2, \ldots, w_{n-1}$, and presume that both $x = -\infty$ and $x = \infty$ are the preimages of $w_n$; see Fig. 7.41, page 342. From the discussion of Eq. (5) it follows that the function

$$g(z) := \int_0^z (\zeta - x_1)^{\theta_1/\pi}(\zeta - x_2)^{\theta_2/\pi} \cdots (\zeta - x_{n-1})^{\theta_{n-1}/\pi} \, d\zeta \qquad (7)$$

maps the real axis onto *some* polygon $P'$. Although $P'$ may not be the desired polygon $P$, it does have the proper right-turn angles $\alpha_i \pi = \theta_i$ at the corners $g(x_i)$ for $i = 1, 2, \ldots, n-1$; and since the initial and final segments intersect at $g(\pm\infty)$, the right turn at this final vertex must match the angle $\theta_n$ (because both are given by $-2\pi - \theta_1 - \theta_2 - \cdots - \theta_{n-1}$).

Now because $P'$ has the same angles as $P$, by adjusting the lengths of the sides of $P'$ we can make it *geometrically similar* to $P$. And it seems quite plausible that we

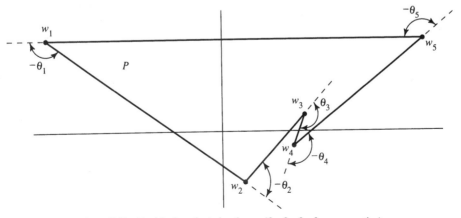

**Figure 7.40**    Positively oriented polygon ($\theta_2, \theta_3, \theta_4, \theta_5$ are negative).

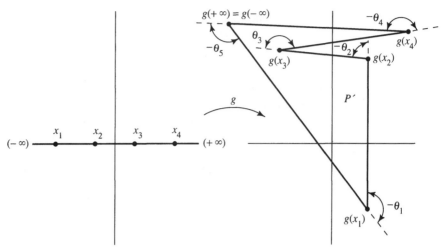

**Figure 7.41**    Mapping for Eq. (7).

could accomplish this by adjusting the points $x_1, x_2, \ldots, x_{n-1}$; after all, they determine where the corners of $P'$ lie. Then, with the use of a rotation, a magnification, and a translation—in other words, a linear transformation—we could make these similar polygons coincide.

Summarizing, we are led to speculate that with an appropriate choice of the constants we can construct a function

$$f(z) = Ag(z) + B$$

$$= A \int_0^z (\zeta - x_1)^{\theta_1/\pi}(\zeta - x_2)^{\theta_2/\pi} \cdots (\zeta - x_{n-1})^{\theta_{n-1}/\pi} \, d\zeta + B, \tag{8}$$

i.e., a Schwarz-Christoffel transformation, which maps the real axis onto the perimeter of a given polygon $P$, with the correspondences

$$f(x_1) = w_1, \quad f(x_2) = w_2, \quad \ldots, \quad f(x_{n-1}) = w_{n-1}, \quad f(\infty) = w_n. \tag{9}$$

Moreover, if our speculations are valid, we can use conformality and connectivity arguments to show that $f$ maps the upper half-plane to the interior of $P$, as was requested; for observe that if $\gamma$ is a segment as indicated in Fig. 7.42, conformality requires that its image, $\gamma'$, have a tangent that initially points inward as shown, and connectivity completes the argument (assuming one-to-oneness). The whole story about Schwarz-Christoffel transformations is given in Theorem 7, whose proof can be found in the references.

---

**Theorem 7.**    Let $P$ be a positively oriented polygon having consecutive corners at $w_1, w_2, \ldots, w_n$ with corresponding right turn angles $\theta_i$ $(i = 1, 2, \ldots, n)$. Then there exists a function of the form (8) that is a one-to-one conformal map from the upper half-plane onto the interior of $P$. Furthermore, the correspondences (9) hold.

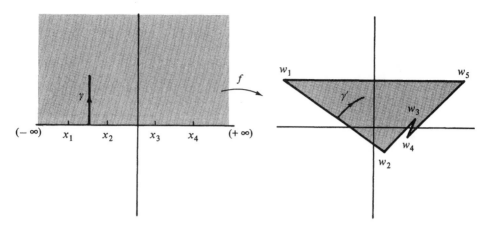

**Figure 7.42**    The upper half-plane is mapped to the interior of $P$.

Before we illustrate the technique, we must make two remarks. First, recall that in constructing the map we have three "degrees of freedom" at our disposal (from the Riemann mapping theorem). Thus we can specify three points on the real axis to be the preimages of three of the $w_i$. However, formula (9) already designates $\infty$ as the preimage of $w_n$, so we are free to choose only, say, $x_1$ and $x_2$, and the other $x_i$ are then determined.

Second, to get a closed-form expression for the mapping we must be able to compute the integral in Eq. (8). A glance through a standard table of integrals shows that this is hopeless for $n > 4$ and not always possible even for smaller $n$. Numerical integration, however, is always feasible. Appendix I describes Trefethen's implementation of the computations.

**Example 1**

Derive a Schwarz-Christoffel transformation mapping the upper half-plane onto the triangle in Fig. 7.43, page 344.

**Solution.**    The right turns are through angles $\theta_1 = \theta_2 = -3\pi/4$, $\theta_3 = -\pi/2$. Hence, choosing $x_1 = -1$ and $x_2 = 1$ we have

$$f(z) = A \int_0^z (\zeta + 1)^{-3/4}(\zeta - 1)^{-3/4} \, d\zeta + B$$

$$= A \int_0^z (\zeta^2 - 1)^{-3/4} \, d\zeta + B.$$

The integration must be performed numerically. To evaluate the constants we compute

$$f(x_1) = f(-1) = A \int_0^{-1} (\zeta^2 - 1)^{-3/4} \, d\zeta + B = A\eta + B,$$

where

$$\eta := \int_0^{-1} (\zeta^2 - 1)^{-3/4} \, d\zeta \approx 1.85(1 + i)$$

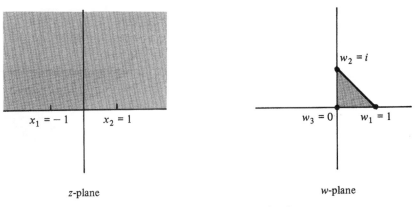

z-plane · w-plane

**Figure 7.43** Mapping onto a triangle.

and

$$f(x_2) = f(1) = A \int_0^1 (\zeta^2 - 1)^{-3/4} \, d\zeta + B = -A\eta + B.$$

Setting these equal to $w_1$ and $w_2$, respectively, we find

$$A\eta + B = 1,$$

$$-A\eta + B = i.$$

Consequently,

$$A = \frac{1-i}{2\eta}, \qquad B = \frac{1+i}{2}. \quad \blacksquare$$

## Example 2

Determine a Schwarz-Christoffel transformation which maps the upper half-plane onto the semi-infinite strip $|\text{Re } w| < 1$, $\text{Im } w > 0$ (Fig. 7.44).

**Solution.** We return to the analysis surrounding Eq. (3) for mapping the real axis onto a polygonal path. To have the upper half-plane map onto the interior of the strip we choose the orientation indicated by the arrows in Fig. 7.44. Left

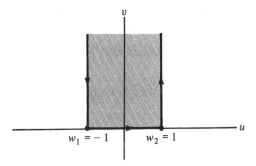

**Figure 7.44** Semi-infinite strip for Example 2.

turns of $\pi/2$ radians at $w_1$ and $w_2$ can be accommodated by a mapping whose derivative is of the form

$$f'(z) = A(z - x_1)^{-1/2}(z - x_2)^{-1/2}.$$

Choosing $x_1 = -1$ and $x_2 = 1$ again, we compute

$$f(z) = A \int_0^z (\zeta + 1)^{-1/2}(\zeta - 1)^{-1/2} \, d\zeta + B = \frac{A}{i} \int_0^z \frac{d\zeta}{\sqrt{1 - \zeta^2}} + B$$

$$= \frac{A}{i} \sin^{-1} z + B.$$

Setting $f(-1) = w_1 = -1$ and $f(1) = w_2 = 1$, we have

$$-iA \sin^{-1}(-1) + B = -1,$$

$$-iA \sin^{-1}(1) + B = 1,$$

which implies that $B = 0$ and $A = 2i/\pi$. Hence

$$f(z) = \frac{2}{\pi} \sin^{-1} z. \quad \blacksquare$$

## Example 3

Map the upper half-plane onto the domain consisting of the fourth quadrant plus the strip $0 < v < 1$.

**Solution.**  The boundary of this domain consists of the line $v = 1$, the negative $u$-axis, and the negative $v$-axis. We shall regard this as the limiting form of the polygonal path indicated in Fig. 7.45, again choosing the orientation so that the specified domain lies to the left. A left turn of $\pi$ radians is called for at the

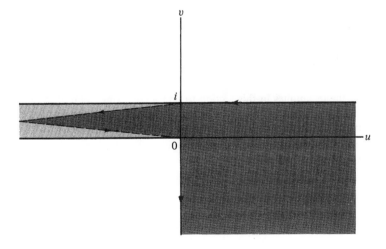

**Figure 7.45**  Domain for Example 3.

corner "near $w = -\infty$," and a right turn of $\pi/2$ radians occurs at $w = 0$. Selecting $x_1 = -1$ and $x_2 = 1$ as the respective preimages of these points we write, in accordance with Eq. (3),

$$f'(z) = A(z+1)^{-1}(z-1)^{1/2}.$$

Using integral tables, with some labor one arrives at

$$f(z) = Ai\left\{2\sqrt{1-z} + \sqrt{2}\log\frac{\sqrt{1-z}-\sqrt{2}}{\sqrt{1-z}+\sqrt{2}}\right\} + B.$$

The selection of branches is quite involved in this case, so we shall leave it to the industrious reader (Prob. 6) to verify that with the choice

$$\log \zeta = \text{Log } |\zeta| + i \arg \zeta, \qquad -\frac{3}{2}\pi < \arg \zeta \le \frac{\pi}{2},$$

$$\sqrt{\zeta} = e^{(\log \zeta)/2}, \qquad\qquad \log \zeta \text{ as above,}$$

one finds that

$$f(z) = \frac{\sqrt{2}}{\pi}\sqrt{1-z} + \frac{1}{\pi}\log\frac{\sqrt{1-z}-\sqrt{2}}{\sqrt{1-z}+\sqrt{2}} + i$$

satisfies the required conditions

$$\text{Re } f(x) \to +\infty, \quad \text{Im } f(x) \to 1 \quad \text{as} \quad x \to -\infty,$$
$$\text{Re } f(x) \to -\infty, \quad \text{Im } f(x) \to 1 \quad \text{as} \quad x \to (-1)^-,$$
$$\text{Re } f(x) \to -\infty, \quad \text{Im } f(x) \to 0 \quad \text{as} \quad x \to (-1)^+,$$
$$f(1) = 0,$$
$$\text{Re } f(x) \to 0, \qquad \text{Im } f(x) \to -\infty \quad \text{as} \quad x \to +\infty. \quad \blacksquare$$

## EXERCISES 7.5

1. Use the techniques in this section to find a conformal map of the upper half-plane onto the whole plane slit along the negative real axis up to the point $-1$. [HINT: Consider the slit as the limiting form of the wedge indicated in Fig. 7.46.]

2. Use the Schwarz-Christoffel formula to derive the mapping $w = \sqrt{z}$ of the upper half-plane onto the first quadrant.

3. Map the upper half-plane onto the semi-infinite strip $u > 0, 0 < v < 1$, indicated in Fig. 7.47.

4. Show that the transformation

$$w = \int_0^z \frac{d\zeta}{(1-\zeta^2)^{2/3}}$$

maps the upper half-plane onto the interior of an equilateral triangle.

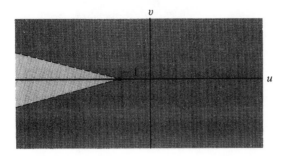

**Figure 7.46**  Region for Prob. 1.

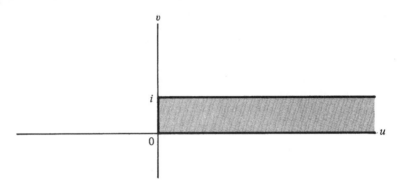

**Figure 7.47**  Region for Prob. 3.

5. Map the upper half-plane onto the *exterior* of the semi-infinite strip in Fig. 7.44, page 344.

6. Verify that the choice of branches indicated in Example 3 yields the appropriate correspondences for $f(-\infty)$, $f(-1)$, $f(1)$, and $f(+\infty)$. [HINT: Argue that if $z$ stays in the upper half-plane, $1 - z$ stays in the lower half-plane, $\sqrt{1 - z}$ stays in the fourth quadrant, and $(\sqrt{1 - z} - \sqrt{2})/(\sqrt{1 - z} + \sqrt{2})$ stays in the lower half-plane.]

7. Map the upper half-plane onto the shaded region in Fig. 7.48.

8. Derive the expression

$$w = f(z) = \int_0^z \frac{d\zeta}{\sqrt{(1 - \zeta^2)(k^2 - \zeta^2)}}$$

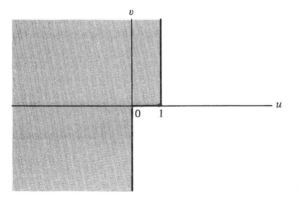

**Figure 7.48**  Region for Prob. 7.

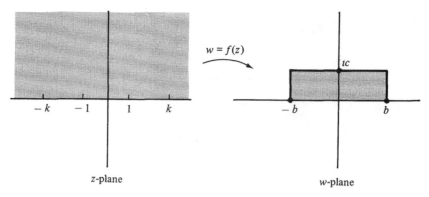

**Figure 7.49**   Mapping onto a rectangle.

for a conformal map of the upper half-plane onto a rectangle, as indicated in Fig. 7.49. Show that the rectangular dimensions $b$ and $c$ must be related, through $k$, by the equations

$$b = \frac{1}{k} \int_0^1 \frac{dx}{\sqrt{(1 - x^2)\left(1 - \dfrac{x^2}{k^2}\right)}},$$

$$c = \frac{1}{k} \int_1^k \frac{dx}{\sqrt{(x^2 - 1)\left(1 - \dfrac{x^2}{k^2}\right)}}.$$

(These are so-called *elliptic integrals*; see Ref. [2].)

9. Map the upper half-plane onto the strip $0 < v < 1$, considered as the limiting form of Fig. 7.50.

10. Argue that a conformal mapping of the unit disk $|z| < 1$ onto the interior of a positively oriented polygon with consecutive corners at $w_1, w_2, \ldots, w_n$ should have the form

$$w = A \int_0^z (\zeta - z_1)^{\theta_1/\pi}(\zeta - z_2)^{\theta_2/\pi} \cdots (\zeta - z_n)^{\theta_n/\pi} \, d\zeta + B,$$

where the $\theta_i$ are the corresponding right-turn angles and the points $z_i$ on the unit circle are the preimages of the corresponding $w_i$.

11. What does the Schwarz reflection principle (Probs. 13 and 14 in Exercises 5.8, page 241) say about the image of the lower half-plane under the Schwarz-Christoffel transformations? (Consider, in turn, Example 2, then Example 1, and then the general case of Fig. 7.42.)

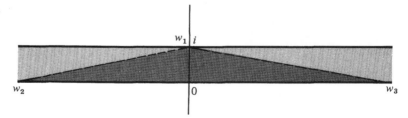

**Figure 7.50**   Region for Prob. 9.

## 7.6 APPLICATIONS IN ELECTROSTATICS, HEAT FLOW, AND FLUID MECHANICS

The next two sections in this chapter are devoted to the solution of certain physical problems involving Laplace's equation

$$\frac{\partial^2 \phi}{\partial x^2} + \frac{\partial^2 \phi}{\partial y^2} = 0, \tag{1}$$

using conformal mapping techniques.

We remind the reader that Eq. (1) governs the temperature distribution in two-dimensional steady-state heat flow (Sec. 2.6). The function $\phi(x, y)$ is the temperature, and the curves $\phi = $ constant are the *isotherms*. Usually one assumes that idealized heat sources or heat sinks are used to maintain fixed (specified) values of $\phi$ on certain parts of the boundary of a domain and that the rest of the boundary is thermally insulated. The latter condition is expressed mathematically by saying that the normal derivative of $\phi$ is zero; i.e., $\partial \phi / \partial n = 0$, where $n$ is a coordinate measured perpendicular to the boundary. The problem, of course, is to find $\phi$ inside the domain.

Eq. (1) also arises in electrical applications. In electrostatics $\phi(x, y)$ is interpreted as the electric potential, or voltage, at the point $(x, y)$, and its partial derivatives $\partial \phi / \partial x$ and $\partial \phi / \partial y$ are the components of the electric field intensity. Typically one specifies either the potential or the *normal* component of the intensity vector on the boundary of a domain and asks for the values of the potential inside (or outside) the domain. The curves defined by the equation $\phi = $ constant are called *equipotentials*.

Flow patterns of fluids can also be analyzed through Eq. (1). The fluid mechanical interpretation of $\phi$ that we shall adopt is the following: The curves given by

$$\phi(x, y) = \text{constant}$$

are the paths which the fluid particles follow. In other words, they are *streamlines*. Thus in studying flow around a nonporous obstacle, the perimeter of that obstacle must constitute part of a streamline, and we specify $\phi = $ constant there. Sometimes $\phi(x, y)$ is known as the *stream function*. For details of these physical interpretations of solutions of Eq. (1), the reader is directed to the references at the end of this chapter.

As we indicated in Sec. 7.1, the basic strategy in solving these problems is to map the given domain conformally onto a simpler domain, to determine the harmonic function which satisfies the "transplanted" boundary conditions, and to carry this function back via the conformal map.

### Example 1

Find the function $\phi$ that is harmonic in the lens-shaped domain of Fig. 7.51(a) and takes the values 0 and 1 on the bounding circular arcs, as illustrated. Here $\phi$ can be interpreted as the steady-state temperature inside an infinitely long strip of material having this lens-shaped region as its cross section, with its sides maintained at the given temperatures.

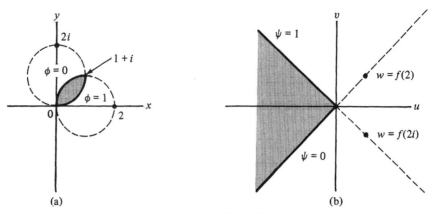

**Figure 7.51**   Lens region and its image.

**Solution.**  Because the domain is bounded by circular arcs, we are naturally inclined to see what we can do with Möbius transformations. If we choose the pole of the transformation to be at $z = 1 + i$, both circles will become straight lines—orthogonal lines, in fact, because conformality will preserve the right angle at $z = 0$. So let's consider

$$w = f(z) = \frac{z}{z - (1 + i)}, \tag{2}$$

which takes $z = 0$ to $w = 0$ and $z = 1 + i$ to $w = \infty$. To determine the image of the lens, we observe that since $z = 2$ goes to $w = 2/(1 - i) = 1 + i$ and $z = 2i$ goes to $w = 2i/(-1 + i) = 1 - i$, the lens is mapped onto the shaded region in Fig. 7.51(b), bounded by the rays Arg $w = 3\pi/4$ (the image of the arc where $\phi = 1$) and Arg $w = -3\pi/4$ (the image of the arc where $\phi = 0$). The corresponding harmonic function $\psi(w)$ in the $w$-plane is easily seen to be

$$\psi(w) = \frac{2}{\pi}\left(\frac{5\pi}{4} - \arg w\right),$$

taking the branch $0 < \arg w < 2\pi$. Carrying this back to the $z$-plane via Eq. (2), we find

$$\phi(x, y) = \frac{2}{\pi}\left(\frac{5\pi}{4} - \arg \frac{z}{z - (1 + i)}\right),$$

which can be expressed as

$$\phi(x, y) = \frac{2}{\pi}\left(\frac{\pi}{4} - \tan^{-1}\frac{x - y}{x(x - 1) + y(y - 1)}\right);$$

here $-\pi/2 < \tan^{-1}\theta < \pi/2$.  ∎

## Example 2

Find the function $\phi$ that is harmonic in the shaded domain depicted in Fig. 7.52(a) and takes the value 0 on the inner circle and 1 on the outer circle. One might interpret $\phi$ as the electrostatic potential inside a capacitor formed by two nested parallel (but nonconcentric) cylindrical conductors.

**Solution.** This problem would be trivial if the circles were concentric; recall that $\text{Log} |z|$ is harmonic (except at $z = 0$) and constant on circles centered at the origin, so we could construct a solution of the form $a \, \text{Log} |bz|$. Thus it seems we should try to map the given region onto an annulus, as in Fig. 7.52(b).

The key to constructing such a map is the fact that one can find a pair of (real) points $z = x_1$ and $z = x_2$ that are *symmetric with respect to both circles simultaneously*. To see this, note that the condition that $x_1$ and $x_2$ are symmetric with respect to the outer circle reads [Sec. 7.4, Eq. (11), page 333]

$$x_2 = \frac{1}{x_1},$$

while symmetry with respect to the inner circle is expressed by

$$x_2 - 0.3 = \frac{(0.3)^2}{x_1 - 0.3}.$$

The solution to these equations is easily seen to be

$$x_1 = \tfrac{1}{3}, \qquad x_2 = 3.$$

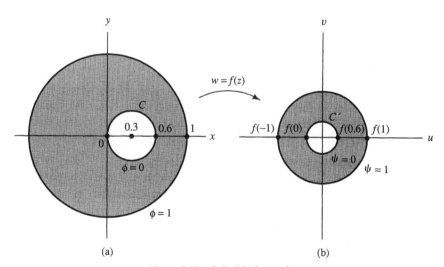

**Figure 7.52**  Cylindrical capacitor.

Now what happens if we perform a Möbius transformation sending $x_1$ to 0 and $x_2$ to $\infty$ via

$$w = f(z) = \frac{z - \frac{1}{3}}{z - 3}? \tag{3}$$

Since Möbius mappings preserve symmetries of this nature, we end up with a pair of circles for which 0 and $\infty$ are symmetric points. But the point at infinity is symmetric to the *center* of a circle [see Sec. 7.4, Eq. (11)]. Therefore, the centers of the image circles coincide and the circles are concentric, as depicted in Fig. 7.52(b)!

The radius of the image of the inner circle can be calculated from

$$|w| = |f(0)| = \tfrac{1}{9},$$

and for the image of the outer circle,

$$|w| = |f(1)| = \tfrac{1}{3}.$$

The solution to the problem in the $w$-plane is thus seen to be

$$\psi(w) = \frac{\text{Log} \, |9w|}{\text{Log} \, 3}.$$

Transforming back, we find

$$\phi(z) = \psi\left(\frac{z - \frac{1}{3}}{z - 3}\right) = \frac{\text{Log} \left|\dfrac{9z - 3}{z - 3}\right|}{\text{Log} \, 3}$$

$$= \frac{1}{\text{Log} \, 3} \left\{ \text{Log} \, 3 + \frac{1}{2} \text{Log}[(3x - 1)^2 + 9y^2] \right.$$

$$\left. - \frac{1}{2} \text{Log}[(x - 3)^2 + y^2] \right\}. \quad \blacksquare$$

**Example 3**

Find a function $\phi(x, y)$ that is harmonic in the portion of the upper half-plane exterior to the circle $C: |z - 5i| = 4$ and which takes the value $+1$ on the circle and 0 on the real axis (Fig. 7.53). The solution can be interpreted as the electric potential due to a charged conducting cylinder lying above a conducting plane.

**Solution.** Actually, this configuration is similar to that of Example 2, if we interpret a straight line to be a circle with a center at infinity. Thus once again we look for a pair of points that are simultaneously symmetric with respect to both the line and the circle.

If $z_p$ denotes a point in the upper half-plane, its symmetric point with respect to the $x$-axis is obviously $\bar{z}_p$. Now $z_p$ and $\bar{z}_p$ will also be symmetric with respect to the circle if (recall Eq. (11), Sec. 7.4)

$$\bar{z}_p = \frac{4^2}{(\bar{z}_p + 5i)} + 5i.$$

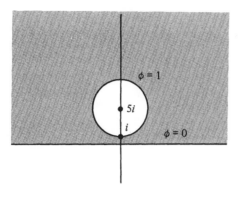

**Figure 7.53**   Charged cylinder over conducting plane.

Solving, we find $z_p = 3i$.

Reasoning as in Example 2, we argue that the Möbius transformation

$$w = f(z) = \frac{z - 3i}{z + 3i},$$

which carries $3i$ to the origin and $-3i$ to infinity, will map the two conductors onto concentric circles *centered at the origin*. The radius of the image of the real axis is

$$r_1 = |f(0)| = 1,$$

and the radius of the image of $C$ is

$$r_2 = |f(i)| = \frac{1}{2}.$$

The function that is harmonic in the annulus $r_2 < |w| < r_1$ and takes the proper boundary values is

$$\psi(w) = \frac{\text{Log } |w|}{\text{Log } \frac{1}{2}} = \frac{-\text{Log } |w|}{\text{Log } 2},$$

so the solution to the problem is

$$\phi(x, y) = \frac{-1}{\text{Log } 2} \text{Log} \left| \frac{z - 3i}{z + 3i} \right|.$$

Note that the equipotentials are circles in the $z$-plane (Fig. 7.54).   ∎

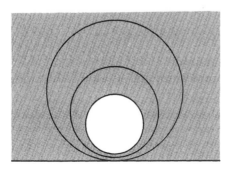

**Figure 7.54**   Equipotentials for Example 3.

*w*-plane

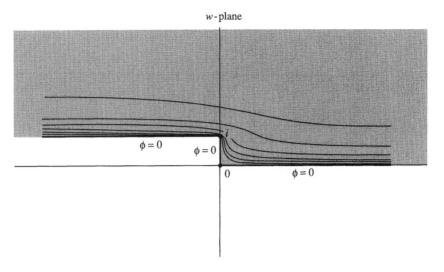

$\phi = 0$            $\phi = 0$

0            $\phi = 0$

**Figure 7.55**   Discontinuous streambed.

### Example 4

Find a *nonconstant* function $\phi$ harmonic inside the infinite domain depicted in Fig. 7.55 and taking the value 0 on the indicated polygonal path. As the sketched lines indicate, the curves $\phi =$ constant will be streamlines for the flow of a deep river over a discontinuous streambed.

**Solution.**    The Schwarz-Christoffel transformation is the tool for this geometry. Using the analysis surrounding Eq. (3) of Sec. 7.5, page 340, we map the *x*-axis onto the discontinuous streambed in the *w*-plane. (Notice that the Schwarz-Christoffel transformation maps the *simple* region onto the *complicated* one.) We assume that the corners $w_1 = i$ and $w_2 = 0$ are the images of $z = -1$ and $z = 1$, respectively. Since the streambed has a right-turn angle of $\pi/2$ at $i$ and a left-turn angle of $\pi/2$ at 0, we must have

$$\frac{dw}{dz} = f'(z) = A(z + 1)^{+1/2}(z - 1)^{-1/2};$$

thus with the aid of an integral table we find

$$w = f(z) = A\{(z^2 - 1)^{1/2} + \log[z + (z^2 - 1)^{1/2}]\} + B. \tag{4}$$

We take branches of these functions that are real and positive for large real $z$ and which are analytic in the upper half-plane.[†] Then the correspondences

---

[†] We are omitting many important details here. Branching is often a very subtle business when Schwarz-Christoffel transformations are used. A painstaking analysis would reveal that here $(z^2 - 1)^{1/2}$ is positive for large positive $z$ and negative for large negative $z$ and that the log function is handled by the restriction $-\pi/2 < \arg \zeta < 3\pi/2$.

$f(-1) = i$, $f(1) = 0$ require

$$A \log(-1) + B = Ai\pi + B = i,$$

$$A \log(1) + B = B = 0,$$

with solution $A = 1/\pi$ and $B = 0$. Hence the mapping from the $z$-plane to the flow region is

$$w = f(z) = \frac{1}{\pi} \{(z^2 - 1)^{1/2} + \log[z + (z^2 - 1)^{1/2}]\}. \tag{5}$$

Now we must find a nonconstant harmonic function of $z$ in the upper half-plane that vanishes on the real axis. One answer is obvious:

$$\psi(x, y) = y = \text{Im}(z). \tag{6}$$

To complete the problem we must carry $\psi(x, y)$ back to the flow region in the $w$-plane. But since we have constructed the map from the simple domain to the complicated domain, we need the inverse of the function (5) to complete the problem. Rather than going through the details of solving Eq. (5) for $z$ in terms of $w$, let's simply abbreviate the answer by stating that the harmonic function $\phi(u, v)$ is given by

$$\phi(u, v) = \text{Im}(f^{-1}(w)).$$

Actually, this is not a serious "cop-out." We have an explicit expression for the streamlines $\phi = $ constant simply by holding $y$ constant and regarding $x$ as a parameter in Eq. (5).  ∎

One of the classic problems in elementary physics involves the parallel-plate capacitor. Here we have two oppositely charged flat conducting sheets separated by a fixed distance, and we must determine the electrostatic potential in the region between them. In the simple case of infinite (square) plates, $\phi$ is proportional to $y$ and the equipotentials are as in Fig. 7.56. This is a good approximation to the more realistic problem of two large plates separated by a relatively small distance; however, near the edges of the plates the potential behaves in a more complicated manner, and this can be computed by conformal mapping. Example 5 should thus be interpreted as finding the potential in the region around two *semi*-infinite conducting plates holding opposite charges.

### Example 5

Find a function $\phi$ that is harmonic in the doubly slit plane and takes the values $-1$ and $+1$ on the two slits indicated in Fig. 7.57.

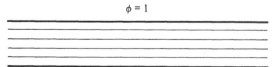

$\phi = 1$

$\phi = -1$                        **Figure 7.56**  Infinite charged plates.

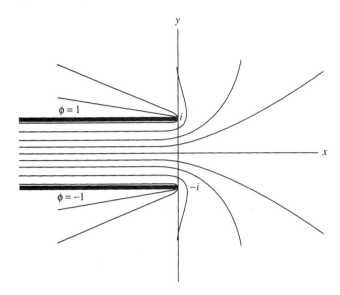

**Figure 7.57** Semi-infinite charged plates.

**Solution.** We shall use the Schwarz-Christoffel transformation again, regarding the domain as the limiting form of the region sketched in Fig. 7.58, with the point $w_0$ going to $-\infty$. The limiting right-turn angles are $\pi$ at $w = i$, $-\pi$ at $w = w_0$, and $\pi$ at $w = -i$. If we select the preimages of $w = i$ and $w = w_0$ to be $z = -1$ and $z = 0$ respectively, the symmetry of the configuration clearly dictates that $z = +1$ will be the preimage of $w = -i$. Thus employing Eq. (3), Sec. 7.5 again, we have

$$\frac{dw}{dz} = A(z + 1)z^{-1}(z - 1) = A\left(z - \frac{1}{z}\right).$$

Consequently,

$$w = f(z) = A\left(\frac{z^2}{2} - \log z\right) + B.$$

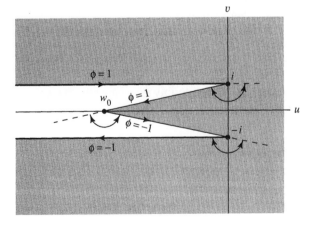

**Figure 7.58** Approximate geometry for semi-infinite charged plates.

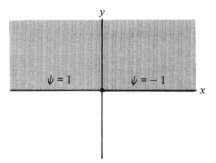

**Figure 7.59**  Transformed problem for semi-infinite charged plates.

Enforcing $f(-1) = i$, $f(1) = -i$ and choosing a branch of $\log z$ that is positive for large positive $z$, we get

$$A\left(\frac{1}{2} - i\pi\right) + B = i,$$

$$A\left(\frac{1}{2} - 0\right) + B = -i,$$

yielding $A = -2/\pi$, $B = 1/\pi - i$. Hence the mapping is

$$w = f(z) = -\frac{2}{\pi}\left(\frac{z^2}{2} - \log z\right) + \frac{1}{\pi} - i. \tag{7}$$

Notice that we have not checked the condition at $z = 0$. This is unnecessary because of the symmetry of the situation. At any rate, it is easy to see that $|w| \to \infty$ as $z \to 0$.

The transformed problem is depicted in Fig. 7.59. The obvious solution is

$$\psi(z) = \frac{2}{\pi} \operatorname{Arg} z - 1.$$

Again the labor involved in inverting Eq. (7) is prohibitive, but the curves $\psi = $ constant are given parametrically by writing Eq. (7) as

$$w = -\frac{2}{\pi}\left(\frac{r^2}{2} e^{2i\theta} - \operatorname{Log} r - i\theta\right) + \frac{1}{\pi} - i$$

and holding $\theta$ constant while $r$ varies from 0 to $\infty$.  ∎

## Example 6

Find a nonconstant function $\phi$ that is harmonic in the slit upper half-plane of Fig. 7.60(a), taking the value $\phi = 0$ on the slit and the real axis. The lines $\phi = $ constant can be interpreted as the streamlines for fluid flow past a simple obstacle.

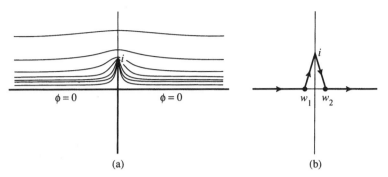

(a)                                                        (b)

**Figure 7.60**    Flow past a simple obstacle.

**Solution.** Regarding the boundary as the limiting form of the polygonal path in Fig. 7.60(b), we construct a Schwarz-Christoffel transformation that maps $-1$ to $w_1$, 0 to $i$, and $+1$ to $w_2$, with limiting right-turn angles $-\pi/2$, $\pi$, and $-\pi/2$, respectively, as $w_1$ and $w_2$ approach 0. Hence

$$\frac{dw}{dz} = A(z + 1)^{-1/2}z(z - 1)^{-1/2} = \frac{Az}{(z^2 - 1)^{1/2}},$$

and

$$w = f(z) = A(z^2 - 1)^{1/2} + B. \tag{8}$$

Taking a branch which is positive for large positive $z$, we make the correspondences

$$f(-1) = B = 0,$$

$$f(1) = B = 0,$$

$$f(0) = Ai + B = i.$$

Hence $A = 1$, $B = 0$. (Again we are able to satisfy three conditions with only two constants in this case because of the symmetry of the region.)

In the $z$-plane the problem becomes, just as in Example 4, that of finding a nonconstant harmonic function vanishing on the real axis, so again we have $\psi(z) = \text{Im } z = y$. In this case we can invert the map (8) to find

$$z = (w^2 + 1)^{1/2},$$

so that the required function is

$$\phi(u, v) = \text{Im}\{(w^2 + 1)^{1/2}\},$$

taking a branch that is positive for large positive $w$. ∎

# EXERCISES 7.6

1. Find the electrostatic potential $\phi$ in the semidisk with the boundary values as shown in Fig. 7.61.

2. Find the electrostatic potential in the upper half-plane exterior to the unit circle under the conditions shown in Fig. 7.62.

3. Find the temperature distribution in Fig. 7.63.

4. Find the temperature distribution in the unit disk with boundary values as shown in Fig. 7.64. [HINT: Map to the upper half-plane; then use Prob. 3 of Exercises 7.1.]

5. Find the electrostatic potential in the slit upper half-plane with the boundary values as depicted in Fig. 7.65.

6. Find the electrostatic potential $\phi$ in the region between two conducting cylinders under the conditions shown in Fig. 7.66.

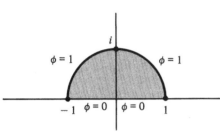

**Figure 7.61**   Region for Prob. 1.

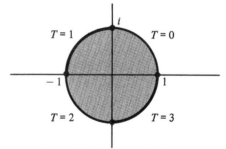

**Figure 7.62**   Region for Prob. 2.

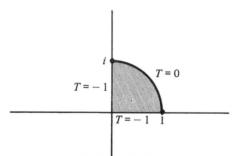

**Figure 7.63**   Region for Prob. 3.

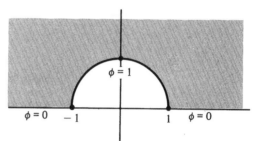

**Figure 7.64**   Region for Prob. 4.

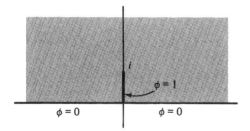

**Figure 7.65**   Region for Prob. 5.

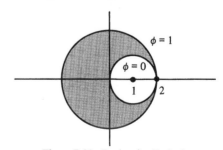

**Figure 7.66**   Region for Prob. 6.

7. Find the temperature inside the infinite regions depicted in Fig. 7.67.

8. Find the potential in the region exterior to two conducting cylinders charged as shown in Fig. 7.68. [HINT: A Möbius transformation can be used to reduce this problem to the situation in Example 2, page 351.]

9. Find the potential between two nested nonconcentric conducting cylinders charged as shown in Fig. 7.69.

10. Find the temperature distribution in the crescent-shaped region given in Fig. 7.70.

11. Find the streamlines for fluid flow in the region indicated in Prob. 7 in Exercises 7.5, page 347.

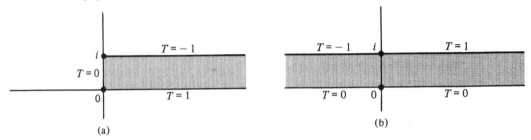

**Figure 7.67** Regions for Prob. 7.

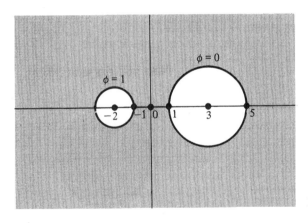

**Figure 7.68** Region for Prob. 8.

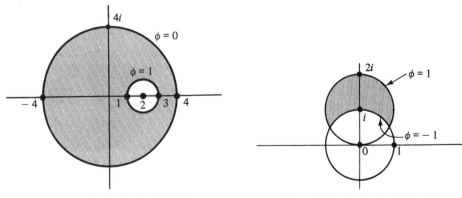

**Figure 7.69** Region for Prob. 9.

**Figure 7.70** Region for Prob. 10.

# 7.7 FURTHER PHYSICAL APPLICATIONS OF CONFORMAL MAPPING

The examples in the previous section were rather straightforward. We were assigned the task of solving a boundary value problem in an irregular region, and we constructed a mapping to a simpler region by techniques which we had learned earlier. In the present section we shall study some problems wherein the mappings are not found by straightforward methods but rather by techniques which may seem to arise from divine inspiration, dumb luck, or simply "experience."

The situation might be described as follows. When one is dealing with an area of mathematics which is so complicated that the direct solution of problems is not feasible, it is useful to try to gain some "feel" for the area by *postulating* a solution and then finding out what problem it solves. The mathematical community calls such procedures *inverse methods*, as opposed to the *direct methods* illustrated in Sec. 7.6. As our first experiment with an inverse method, let us consider what situations can be analyzed with the mapping $w = \sin z$ and its inverse $z = \sin^{-1} w$. This transformation turns out to be quite a versatile tool, and we shall utilize it to solve four very different physical problems.

We saw in Sec. 7.5 that $z = \sin^{-1} w$ maps the upper half-plane onto the semi-infinite strip depicted in Fig. 7.71, with the real axis mapping to the sides and bottom of the strip. We can thus use the harmonic function $\psi_1(u, v) = v = \text{Im } w$, which is zero on the real axis, to find the streamlines for flow in a blocked channel, as illustrated in the figure. They are given by the level curves

$$\phi_1(x, y) = \text{Im}(\sin z) = \text{constant}.$$

On the other hand, the harmonic function $\phi_2(x, y) = y = \text{Im } z$ in the $z$-plane is zero on the bottom of the strip, and the lines $\phi_2 = \text{constant}$ are perpendicular to the sides of the strip. Therefore, their images, the curves

$$\psi_2(u, v) = \text{Im}(\sin^{-1} w) = \text{constant}, \qquad (1)$$

intersect the $u$-axis orthogonally for $|u| > 1$, while $\psi_2 = 0$ on the segment $-1 \leq u \leq 1$;

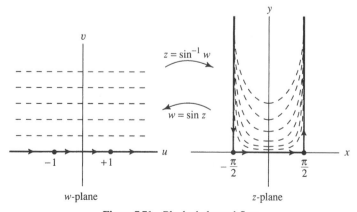

**Figure 7.71**  Blocked channel flow.

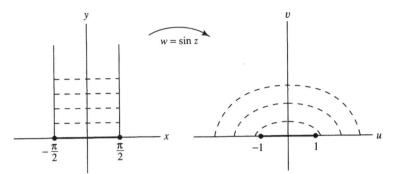

**Figure 7.72**   Charged conducting strip.

see Fig. 7.72. Hence Eq. (1) can be interpreted as the equipotentials around an infinitely long charged conducting strip of width 2.

The harmonic function which is conjugate to this $\psi_2(u, v)$ is, obviously,

$$\eta(u, v) = \text{Re}(\sin^{-1} w) \tag{2}$$

(to be accurate, we should say that $\eta$ is *one* of the harmonic conjugates of $\psi_2$; recall Sec. 2.5). The level curves of $\eta(u, v)$ intersect those of $\psi_2(u, v)$ orthogonally, so they look like the dashed curves in Fig. 7.73. In particular, $\eta(u, v) = -\pi/2$ on the $u$-axis to the left of $-1$, and $\eta = \pi/2$ on the $u$-axis to the right of $+1$. Thus (2) can be interpreted as the potential due to two oppositely charged semi-infinite conducting plates lying side by side.

Finally, it is easy to verify that the mapping $w = \sin z$ as depicted in Fig. 7.71 is symmetric about the $y$-axis, so that we can restrict it to determine a one-to-one map of the strip in Fig. 7.74 onto the first quadrant. Consider the harmonic function $\phi_3(x, y) = 2x/\pi = (2 \text{ Re } z)/\pi$. It is zero on the $y$-axis and $+1$ on the line $x = \pi/2$, and its level curves intersect the $x$-axis orthogonally. Hence the "inherited" function

$$\psi_3(u, v) = \frac{2}{\pi} \text{Re}(\sin^{-1} w)$$

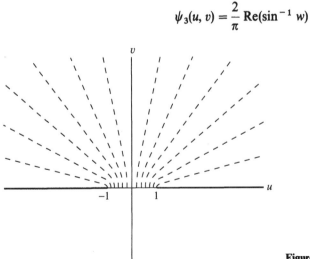

**Figure 7.73**   Coplanar charged plates.

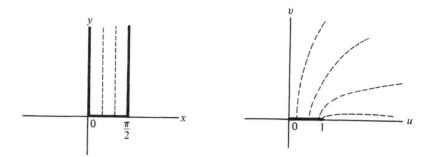

**Figure 7.74**   Thermal configuration with insulation.

is zero on the $v$-axis and $+1$ on the $u$-axis for $u > 1$, while the curves $\psi_3 = $ constant intersect the segment $0 < u < 1$ orthogonally. The latter condition implies that the normal derivative of $\psi_3$ is zero on the segment. As a result, we can interpret $\psi_3$ as the steady-state temperature in the first quadrant when the $v$-axis is held at 0 degrees, the $u$-axis is held at 1 degree for $u > 1$, and the portion $0 < u < 1$ of the $u$-axis is thermally insulated.

Continuing in this vein, observe that we can solve an interesting variety of problems with the function $f(z) = z^\alpha$. For $\alpha > 0$, we have seen that a suitable branch of $f$ maps the wedge $0 < \arg z < \pi/\alpha$ onto the upper half-plane, and thus $\operatorname{Im} z^\alpha$ is the stream function for fluid flow in a wedge. But of course *any* of the streamlines $\operatorname{Im} z^\alpha = $ constant could be considered as the perimeter of an obstacle placed in the flow. For example, if $\alpha = 2$,

$$\operatorname{Im} z^2 = 2xy,$$

and we have a stream function for flow inside, say, the rectilinear hyperbola $xy = 1$ [cf. Fig. 2.6(b), page 66]. For $\alpha = -1$, the level curves (see Figure 7.75)

$$\operatorname{Im} z^{-1} = \frac{-y}{x^2 + y^2} = \text{constant}$$

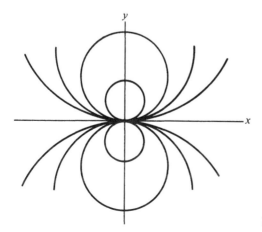

**Figure 7.75**   Dipole equipotentials.

are all circles through the origin. One can visualize these curves as equipotentials for a (two-dimensional) dipole at $z = 0$.

The level curves for the harmonic function Log $|z|$ have an interesting interpretation. Recall that they are concentric circles (Fig. 7.76). Considered as streamlines, they represent the flow due to a *vortex*, or whirlpool, at the origin. On the other hand, if they are interpreted as isotherms or equipotentials, we infer the existence of a heat source or point charge[†] at $x = y = 0$. As we shall illustrate shortly, it is often instructive to study the effect of superimposing vortices or sources on a given pattern.

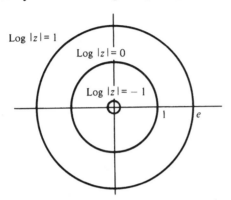

**Figure 7.76** Vortex streamlines.

A very important application of the inverse method, which we shall describe presently, springs from considerations involving the following simple example.

**Example 1**

Find the streamlines for flow around a cylindrical obstacle as depicted in Fig. 7.77.

**Solution.** We assume for the moment that the flow is symmetric with respect to the x-axis. (We shall see later that other interpretations are possible.) Then we only have to deal with $y \geq 0$.

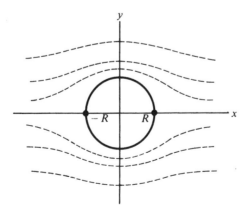

**Figure 7.77** Flow around a cylinder.

---

† Remember that a point charge in two dimensions corresponds to a *line* charge in three dimensions.

To map the flow region onto the upper-half $w$-plane we need a (nonconstant) analytic function $f(z)$ that is real on the $x$-axis (at least for $|x| > R$) and on the circle $|z| = R$. The first condition is satisfied by a large class of functions, e.g., rational functions with real coefficients. To handle the second condition, we observe that since the circle can be described by $z\bar{z} = R^2$, we have $\bar{z} = R^2/z$ there; hence the rational function $z + R^2/z$ is equal to $z + \bar{z} = 2\,\mathrm{Re}\,z$ on the circle. Consequently we are led to the mapping

$$w = f(z) = z + \frac{R^2}{z}. \tag{3}$$

It is easily verified that (3) produces a one-to-one map of the flow region (for $y > 0$) onto the upper half-plane. Thus the appropriate stream function corresponds to $\psi(u, v) = v = \mathrm{Im}\,w$, yielding the streamlines

$$\phi(x, y) = \mathrm{Im}\left(z + \frac{R^2}{z}\right) = \text{constant.} \quad \blacksquare$$

If we drop the symmetry assumption, we require only that the circle itself be a streamline. Hence we can add any constant multiple of $\mathrm{Log}\,|z|$ (since it has $|z| = R$ as a streamline) to the stream function and obtain "circulating" flow patterns as in Fig. 7.78.

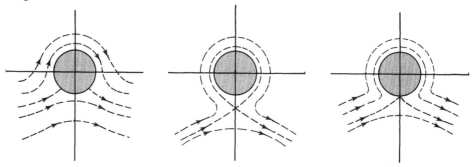

**Figure 7.78**    Circulating flows around cylinder.

In 1908 the mathematician N. Joukowski had the inspiration to see what the mapping (3) would do, not to the circle $|z| = R$, but to an off-center circle such as $C$ in Fig. 7.79(a). The result was an *airfoil*, as indicated in Fig. 7.79(b). By starting with different circles $C$ we can generate a variety of these so-called Joukowski airfoils. Furthermore, since we have already found a wide class of flows around cylinders (Figs. 7.77 and 7.78), we can use the "Joukowski transformation" (3) to carry them over and compute flows around these airfoils! For instance, we can adjust the "point of attachment" $P$ in Fig. 7.80 by modifying the constant multiple of the Log term in the "original" stream function. And we can shape the airfoil to meet certain specifications by trying different choices for $C$ and introducing modifications into the mapping; e.g.,

$$w = f(z) = z + \frac{R^2}{z} + \frac{a}{z^2} + \frac{b}{z^3} + \cdots.$$

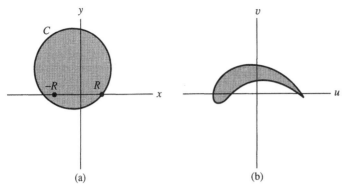

(a)

(b)

**Figure 7.79**   Joukowski airfoil.

**Figure 7.80**   Point of attachment.

This technique has been extremely useful in aircraft design, and we direct the interested reader to the specialized literature for further study.

## EXERCISES 7.7

1. Analyze the temperature distribution in the plate depicted in Fig. 7.81. [HINT: Use the solution to Prob. 7 in Exercises 7.5, page 347, and the sine function.]

2. Consider the problem of fluid flow around a straight obstacle inclined at an angle $\alpha$, as in Fig. 7.82. The stream function $\psi$ must be constant on the obstacle, and the streamlines ($\psi$ = constant) must tend to horizontal straight lines at large distances from the origin. Show that

$$\psi(x, y) = \text{Im}\left[ e^{-i\alpha}z\left( \cos \alpha + i \sin \alpha \sqrt{1 - \frac{e^{2i\alpha}}{z^2}} \right) \right]$$

satisfies these conditions.

3. Find the temperature distribution in the first quadrant under the boundary conditions indicated in Fig. 7.83.

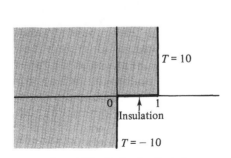

**Figure 7.81**   Region for Prob. 1.

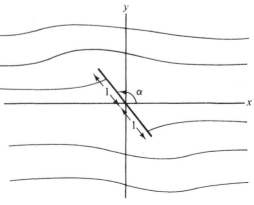

**Figure 7.82**   Region for Prob. 2.

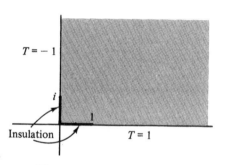

**Figure 7.83**  Region for Prob. 3.               **Figure 7.84**  Region for Prob. 4.

4. Another feasible approach to Example 1, page 364, is to map the shaded region in Fig. 7.84 to the upper half-plane as follows: First use a Möbius transformation to map $R$ to $\infty$ and $-R$ to 0. Argue that the shaded region then maps onto a 90° wedge, which can be rotated if necessary to coincide with the first quadrant. Squaring then maps onto the upper half-plane, and taking the imaginary part of the whole transformation should solve the problem of finding the stream function. Show the implementation of this scheme leads to the mapping

$$w = -\left(\frac{z+R}{z-R}\right)^2$$

and the stream function

$$\psi(z) = \mathrm{Im}\ w = \frac{4yR(x^2 + y^2 - R^2)}{[(x-R)^2 + y^2]^2}.$$

Although this function is harmonic in the shaded region and zero on the boundary, we reject it on the physical basis that the flow we seek must have nearly horizontal streamlines for large $y$. That is, the curves $\psi(x, y) = $ constant must approximate $y = $ constant far away from the obstacle. This is obviously not the case for the above solution. (This problem illustrates an additional complication which we have ignored in our elementary treatment, namely, consideration of boundary conditions "at infinity" for unbounded domains.)

5. Analyze the temperature distribution in the slab $0 < y < 1$ under the conditions shown in Fig. 7.85.

6. Show that the mapping (3) takes two concentric circles, $|z| = R$ and $|z| = R' > R$, onto a line segment and an ellipse as shown in Fig. 7.86, page 368. Use this to find the electrostatic potential between a conducting elliptic cylinder surrounding a conducting strip.

7. Using the mapping (3), find the streamlines for the flow indicated in Fig. 7.87 on page 368.

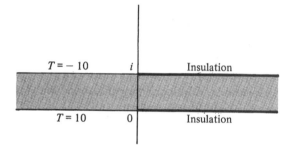

**Figure 7.85**  Region for Prob. 5.

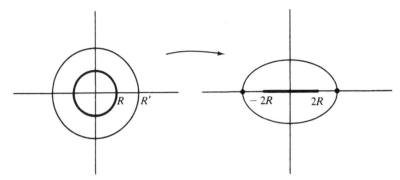

**Figure 7.86**   Region for Prob. 6.

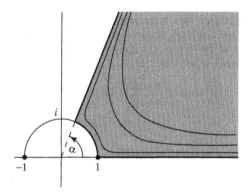

**Figure 7.87**   Region for Prob. 7.

## SUMMARY

One of the most important aspects of mappings generated by analytic functions is the persistence of solutions of Laplace's equation. That is, if $w = f(z)$ is a one-to-one analytic function mapping one domain onto another [which implies that $z = f^{-1}(w)$ is analytic also] and if $\phi(z)$ is harmonic in the first domain, then $\psi(w) := \phi(f^{-1}(w))$ is harmonic in the second domain. This leads to a useful technique for solving boundary value problems for Laplace's equation; one performs a preliminary mapping to a domain such as a quadrant, half-plane, or annulus, where the corresponding problem is easy to solve, and then one carries the solution function back via the mapping.

An analytic mapping is conformal, i.e., it preserves angles, at all points where the derivative is nonzero. Using this property and connectivity considerations one can often determine the mapped image of a domain from the image of its boundary.

One important category of conformal mappings is the Möbius transformations. They are functions of the form $(az + b)/(cz + d)$, with $ad \neq bc$, and they can be expressed as compositions of translations, rotations, magnifications, and inversions. Because they preserve the class of straight lines and circles and their associated symmetries, they are usually the method of choice for solving problems in domains bounded by such figures.

On the other hand, Schwarz-Christoffel transformations map half-planes to polygons, and thus they are the appropriate tool for such geometries. These mappings are computed by determining the conditions imposed on the derivative $f'$ at the corners of the polygon.

With these devices one can solve many two-dimensional problems in electrostatics, heat flow, and fluid dynamics. Sometimes insight into the nature of more complicated situations is provided by experimenting with inverse methods, where one first postulates solutions and then analyzes what problems they solve.

## SUGGESTED READING

Most of the references listed previously treat conformal mapping, but the following are particularly useful:

### Riemann Mapping Theorem and Geometric Considerations

[1] Goluzin, G. M. *Geometric Theory of Functions of a Complex Variable.* American Mathematical Society, 1969.

[2] Nehari, Z. *Conformal Mapping.* Dover Publications, Inc., New York, 1975.

### Schwarz-Christoffel Transformation

Reference 2

[3] Levinson, N., and Redheffer, R. *Complex Variables.* Holden-Day, Inc., San Francisco, 1970.

[4] Henrici, P. *Applied and Computational Complex Analysis*, Vol. I. John Wiley & Sons, Inc., New York, 1974.

### Compendium of Conformal Maps

[5] Kober, H. *Dictionary of Conformal Representations*, 2nd ed. Dover Publications, Inc., New York, 1957.

### Applications

[6] Courant, R. *Dirichlet's Principle, Conformal Mapping, and Minimal Surfaces.* John Wiley & Sons, Inc. (Interscience Division), New York, 1950.

[7] Dettman, J. W. *Applied Complex Variables.* Dover Publications, Inc., New York, 1984.

[8] England, A. H. *Complex Variable Methods in Elasticity.* John Wiley & Sons, Inc. (Interscience Division), New York, 1971.

[9] Frederick, C., and Schwartz, E. L. "Conformal Image Warping," *IEEE Computer Graphics and Applications*, March 1990, pp. 54–61.

[10] Kyrala, A. *Applied Functions of a Complex Variable.* John Wiley & Sons, Inc., New York, 1972.

[11] Marsden, J. E., and Hoffman, M. J. *Basic Complex Analysis*, 2nd ed. W. H. Freeman and Company, Publishers, New York, 1987.

[12] Milne-Thomson, L. M. *Theoretical Hydrodynamics*, 2nd ed. The Macmillan Company, New York, 1968.

[13] Rothe, R., Ollendorff, F., and Pohlhausen, K. *Theory of Functions*. Dover Publications, Inc., New York, 1961.

[14] Smythe, W. R. *Static and Dynamic Electricity*, 3rd ed. Hemisphere Publications Corporation, New York, 1989.

# :8:

# The Transforms
# of Applied
# Mathematics

In Sec. 3.4 we gave some indication of why, when analyzing linear time-invariant systems, it is particularly advantageous to deal with sinusoidal functions as inputs. Briefly, the virtues of employing an input of the form $Ae^{i\omega t}$ are as follows:

1. Compactness of notation—a real expression such as $\alpha \cos(\omega t + \phi) + \beta \sin(\omega t + \psi)$ can be represented simply by $\text{Re}(Ae^{i\omega t})$.
2. The fact that differentiation amounts to multiplication by $i\omega$—thus, in a sense, replacing calculus by algebra.
3. The fact that the steady-state response of the system to this input will have the same form, a complex constant times $e^{i\omega t}$.

   For these reasons it would be very helpful if a general input function $F(t)$ could be expressed as a sum of these sinusoids. One could then determine the output by finding the response to each sinusoidal component (which is an easier problem) and then adding these responses together (recall that superposition of solutions is permissible in a linear system).

   *Fourier analysis*, as implemented through the Fourier series and the Fourier transform, is devoted to the decomposition of a function into these sinusoids. Other transforms—notably, the Mellin, Laplace, and $z$ transforms—have been developed with the same objective: the decomposition of *arbitrary* functions into superpositions of elementary forms that are convenient for a particular analytical task at hand. Another transform, named for Hilbert, is intimately related to the others both theoretically and in applications, although it does not address the specific objective of functional decomposition.

The range of validity and applicability of these mathematical operations extends well beyond the domain of *analytic* functions, but the derivations of the key properties are much more transparent if we restrict ourselves. Thus we devote the final chapter of this book to a survey of the analytic-functional aspects of these transforms.

## 8.1 FOURIER SERIES (THE FINITE FOURIER TRANSFORM)

As indicated in the introduction, the main goal of this chapter is to establish the possibility of expressing a (possibly complex-valued) function of a real variable, $F(t)$, as a sum of sinusoidal functions of the form $e^{i\omega t}$. The present section is devoted to the special case when $F(t)$ is periodic with period $L$; i.e., $F(t) = F(t + L)$ for all $t$.

Naturally we are inclined to seek a decomposition of $F$ into sinusoids with the same period; i.e., only those values of $\omega$ should occur such that $e^{i\omega(t+L)} = e^{i\omega t}$. This implies that $e^{i\omega L} = 1$, so that $\omega$ must be one of the numbers

$$\omega_n = \frac{2\pi n}{L} \qquad (n = 0, \pm 1, \pm 2, \ldots).$$

To be specific, we assume that $L = 2\pi$ (one can always rescale to achieve this condition). Our problem is thus to find (complex) numbers $c_n$ such that

$$F(t) = \sum_{n=-\infty}^{\infty} c_n e^{int}. \tag{1}$$

Suppose, for the moment, that the series in Eq. (1) converges *uniformly* to $F(t)$ for $-\pi \leq t \leq \pi$ (and hence for all $t$). For any fixed integer $m$ we can multiply by $e^{-imt}$ to obtain

$$F(t)e^{-imt} = \sum_{n=-\infty}^{\infty} c_n e^{i(n-m)t}, \tag{2}$$

again converging uniformly, from which it follows that $F(t)e^{-imt}$ is a continuous function and that termwise integration of the series is valid [recall Theorem 8 of Chapter 5 (page 199)]. Integrating Eq. (2) over the interval $[-\pi, \pi]$ yields

$$\int_{-\pi}^{\pi} F(t)e^{-imt}\, dt = \sum_{n=-\infty}^{\infty} c_n \int_{-\pi}^{\pi} e^{i(n-m)t}\, dt; \tag{3}$$

however,

$$\int_{-\pi}^{\pi} e^{i(n-m)t}\, dt = \begin{cases} \dfrac{e^{i(n-m)t}}{i(n-m)}\bigg|_{-\pi}^{\pi} = 0 & \text{if } n \neq m, \\[2mm] t\bigg|_{-\pi}^{\pi} = 2\pi & \text{if } n = m. \end{cases}$$

Hence only the term $2\pi c_m$ survives on the right-hand side of Eq. (3). As a result we have the following formula for the coefficient $c_m$:

$$c_m = \frac{1}{2\pi} \int_{-\pi}^{\pi} F(t)e^{-imt}\, dt, \tag{4}$$

valid whenever the series in Eq. (1) is uniformly convergent.

Whether or not the series is convergent, we use the following terminology.

---

**Definition 1.** If $F$ has period $2\pi$ and is integrable over $[-\pi, \pi]$, the (formal) series $\sum_{n=-\infty}^{\infty} c_n e^{int}$ with coefficients given by Eq. (4) is called the **Fourier series** for $F$; the numbers $c_n$ are called the **Fourier coefficients** of $F$.

---

More generally, if $F(t)$ has period $L$, the Fourier series looks like $\sum_{n=-\infty}^{\infty} c_n e^{in2\pi t/L}$, and the Fourier coefficients become

$$c_n = \frac{1}{L} \int_{-L/2}^{L/2} F(t)e^{-in2\pi t/L}\, dt.$$

What we have shown is that under the assumption that $F(t)$ has a representation of the form (1) which is known to be uniformly convergent, then the series in question must be the Fourier series. Now we must investigate this assumption and try to determine *under what conditions the Fourier series will converge to F*.

A partial answer to this question can be derived from analytic function theory. Consider a function $f(z)$ analytic in some annulus, such as $D$ in Fig. 8.1, which contains the unit circle. Then, of course, $f$ can be represented by a Laurent series:

$$f(z) = \sum_{n=-\infty}^{\infty} a_n z^n \qquad (z \text{ in } D). \tag{5}$$

We shall be particularly concerned with the values of $f$ on the unit circle where the series converges uniformly. Parametrizing this circle by $z = e^{it}$, $-\pi \le t \le \pi$, we introduce the notation $F(t) := f(e^{it})$ and rewrite Eq. (5) as

$$F(t) = \sum_{n=-\infty}^{\infty} a_n e^{int}. \tag{6}$$

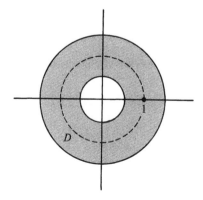

**Figure 8.1** Annulus of analyticity.

Observe our good fortune; the function $F(t)$ has period $2\pi$, and Eq. (6) is a decomposition of $F$ into a series of sinusoids, converging uniformly! Thus Eq. (6) must be the Fourier series for $F(t)$.

In fact, we can even present an independent derivation of formula (4) for the Fourier coefficients in this case. According to Theorem 14 of Chapter 5 (page 211) the coefficients in Eq. (5) are given by

$$a_n = \frac{1}{2\pi i} \oint_{|z|=1} \frac{f(\zeta)}{\zeta^{n+1}} \, d\zeta,$$

and inserting the parametrization we find

$$a_n = \frac{1}{2\pi i} \int_{-\pi}^{\pi} f(e^{it}) e^{-it(n+1)} i e^{it} \, dt$$

$$= \frac{1}{2\pi} \int_{-\pi}^{\pi} F(t) e^{-int} \, dt,$$

in agreement with Eq. (4).

Of course, we have not proved a great deal; we have only shown that *the Fourier series of a function F converges uniformly to F in those cases when the values of F(t) coincide with the values of an analytic function f(z) for $z = e^{it}$*. Furthermore, this technique of finding a Fourier series by way of a Laurent expansion is usually of more theoretical than practical value.

## Example 1

Find the Fourier series for the periodic function

$$F(t) = e^{2\cos t}$$

using the preceding technique.

**Solution.** First we must find an analytic function $f(z)$ that matches the values of $F(t)$ for $z$ on the unit circle. This is easy; since

$$\cos t = \frac{e^{it} + e^{-it}}{2},$$

we see that

$$F(t) = e^{(z+1/z)} =: f(z)$$

when $z = e^{it}$. Hence the Fourier series for $F$ can be obtained from the Laurent series for $f$. We have

$$e^{(z+1/z)} = e^z e^{1/z}$$

$$= \left( \sum_{m=0}^{\infty} \frac{z^m}{m!} \right) \left( \sum_{l=0}^{\infty} \frac{z^{-l}}{l!} \right),$$

and we can multiply these series termwise. (Termwise multiplication of Laurent series is valid, although we have not proved it in this text.) The term involving

$z^n$ in the result comes from the sum of products of terms $z^m/m!$ times $z^{-l}/l!$ with $m - l = n$; collecting these we have

$$e^{(z + 1/z)} = \sum_{n=-\infty}^{\infty} z^n \left( \sum_{m=n}^{\infty} \frac{1}{m!} \cdot \frac{1}{(m-n)!} \right).$$

Hence the Fourier series for $F$ is

$$F(t) = \sum_{n=-\infty}^{\infty} c_n e^{int}$$

with

$$c_n = \sum_{m=n}^{\infty} \frac{1}{m!(m-n)!}. \quad \blacksquare$$

## Example 2

Show that when $F(t) = f(e^{it})$ with $f$ analytic, termwise differentiation of the Fourier series for $F$ is valid.

**Solution.** We know that the Laurent series (5) can be differentiated termwise:

$$\frac{df(z)}{dz} = \sum_{n=-\infty}^{\infty} n a_n z^{n-1}. \tag{7}$$

For $z = e^{it}$, the chain rule yields

$$\frac{df}{dt} = \frac{df}{dz} \frac{dz}{dt} = \frac{df}{dz} i e^{it}.$$

Inserting Eq. (7) for $df/dz$ and identifying $f(e^{it})$ as $F(t)$, we find

$$\frac{d}{dt} f(e^{it}) = \frac{dF(t)}{dt} = \sum_{n=-\infty}^{\infty} n a_n e^{i(n-1)t} i e^{it},$$

or

$$\frac{dF(t)}{dt} = \sum_{n=-\infty}^{\infty} i n a_n e^{int},$$

which agrees with termwise differentiation of Eq. (6).    $\blacksquare$

As another illustration of the fertility of this approach, we shall present a heuristic derivation of *Poisson's formula* for harmonic functions on the unit disk. Since the validity of the formula has been stated in Sec. 4.7, we shall proceed formally and not worry about the rigorous justification of each detail.

We are given a continuous real-valued function $U(\theta)$ having period $2\pi$, and we want to find a function $u(z)$ that is harmonic for $|z| < 1$ and approaches the value $U(\theta)$ as $z \to e^{i\theta}$; in other words, we want to solve the Dirichlet problem for the unit disk (see Sec. 4.7). First we assume that $U(\theta)$ has a Fourier expansion

$$U(\theta) = \sum_{n=-\infty}^{\infty} c_n e^{in\theta} = \sum_{n=-\infty}^{\infty} \left[ \frac{1}{2\pi} \int_{-\pi}^{\pi} U(\phi) e^{-in\phi} \, d\phi \right] e^{in\theta},$$

where we have inserted the coefficient formula (4). If we combine the terms for $n$ and $-n$, we derive (observe that $n = 0$ is exceptional)

$$U(\theta) = \frac{1}{2\pi} \int_{-\pi}^{\pi} U(\phi)\, d\phi + \sum_{n=1}^{\infty} \frac{1}{2\pi} \int_{-\pi}^{\pi} U(\phi)(e^{in(\theta - \phi)} + e^{-in(\theta - \phi)})\, d\phi$$

$$= \frac{1}{2\pi} \int_{-\pi}^{\pi} U(\phi)\, d\phi + 2 \sum_{n=1}^{\infty} \frac{1}{2\pi} \int_{-\pi}^{\pi} U(\phi) \cos n(\theta - \phi)\, d\phi.$$

Now we use a device known to mathematicians as *Abel-Poisson summation* to sum the series. First we artificially introduce the variable $r$ to obtain a function $g(r, \theta)$:

$$g(r, \theta) := \frac{1}{2\pi} \int_{-\pi}^{\pi} U(\phi)\, d\phi + \frac{2}{2\pi} \sum_{n=1}^{\infty} \int_{-\pi}^{\pi} U(\phi) r^n \cos n(\theta - \phi)\, d\phi. \qquad (8)$$

This yields three dividends; first, observe that the series

$$1 + 2 \sum_{n=1}^{\infty} r^n \cos n(\theta - \phi) \qquad (9)$$

converges uniformly in $\phi$, if $0 \le r < 1$. Hence it can be multiplied by $U(\phi)$ and integrated termwise. But this results in $2\pi$ times the right-hand side of Eq. (8). Thus we can rewrite Eq. (8) as

$$g(r, \theta) = \frac{1}{2\pi} \int_{-\pi}^{\pi} U(\phi) \left\{ 1 + 2 \sum_{n=1}^{\infty} r^n \cos n(\theta - \phi) \right\} d\phi. \qquad (10)$$

Second, observe that the series (9) is, in fact, the real part of the series

$$1 + 2 \sum_{n=1}^{\infty} r^n e^{in\theta} e^{-in\phi} = 1 + 2 \sum_{n=1}^{\infty} z^n e^{-in\phi}, \qquad (11)$$

a power series in $z = re^{i\theta}$. Since the latter series converges for $|z| < 1$, it defines an analytic function inside the unit disk, and consequently its real part, (9), is harmonic! As a result, $g(r, \theta)$ is the real part of an analytic function [since $U(\phi)$ is real], and hence $g$ is a harmonic function of $z = re^{i\theta}$ for $r < 1$. The formal substitution $r = 1$ in Eq. (10) yields the Fourier series for $U(\theta)$, so we are led to postulate that Eq. (10) solves the Dirichlet problem; i.e., $u(z) = u(re^{i\theta}) = g(r, \theta)$ is a function which is harmonic for $|z| < 1$ and approaches $U(\theta)$ as $|z| \to 1$.

Finally, the third dividend of our labors follows from the equality

$$1 + 2 \sum_{n=1}^{\infty} r^n \cos n(\theta - \phi) = \frac{1 - r^2}{1 - 2r \cos(\theta - \phi) + r^2}, \qquad (12)$$

which we invite the reader to prove as Prob. 4. Using this in Eq. (10), we arrive at the *Poisson formula*

$$u(re^{i\theta}) = \frac{1 - r^2}{2\pi} \int_{-\pi}^{\pi} \frac{U(\phi)}{1 - 2r \cos(\theta - \phi) + r^2}\, d\phi,$$

expressing a harmonic function inside the unit disk in terms of its "boundary values."

As we indicated earlier, we refer the reader to Sec. 4.7 for a more precise statement of the validity of Poisson's formula.

At this point our achievements can be summarized as follows: Subject to some fairly restrictive analyticity assumptions, the equation

$$F(t) = \sum_{n=-\infty}^{\infty} c_n e^{int} \tag{13}$$

is valid when

$$c_n = \frac{1}{2\pi} \int_{-\pi}^{\pi} F(t)e^{-int}\, dt \qquad \text{(for all } n\text{)}. \tag{14}$$

Now notice that there is nothing in Eqs. (13) or (14) that would indicate the necessity of any analytic properties of $F$. Indeed, the coefficients (14) can be evaluated for any integrable $F$. So, we speculate, why should the validity of Eq. (13) hinge on analyticity? Shouldn't we expect that the Fourier series converges under weaker conditions? The answer is yes, but the proofs of the more general convergence theorems lie outside analytic function theory. We shall simply quote some of these results without proof.

The first theorem is more or less in line with our speculations. It postulates only the integrability of $|F|^2$, but it pays the price in that a much weaker type of convergence occurs.

---

**Theorem 1.** If the integral $\int_{-\pi}^{\pi} |F(t)|^2\, dt$ exists, then the Fourier series defined by Eqs. (13) and (14) exists and converges to $F$ in the mean square sense; i.e.,

$$\lim_{N\to\infty} \int_{-\pi}^{\pi} \left| F(t) - \sum_{n=-N}^{N} c_n e^{int} \right|^2 dt = 0.$$

---

**Example 3**

Prove *Parseval's identity* for the Fourier coefficients:

$$\int_{-\pi}^{\pi} |F(t)|^2\, dt = \lim_{N\to\infty} 2\pi \sum_{n=-N}^{N} |c_n|^2, \tag{15}$$

if $|F|^2$ is integrable over $[-\pi, \pi]$.

**Solution.** We have

$$\int_{-\pi}^{\pi} \left| F(t) - \sum_{n=-N}^{N} c_n e^{int} \right|^2 dt = \int_{-\pi}^{\pi} \left[ F(t) - \sum_{n=-N}^{N} c_n e^{int} \right]\left[ \overline{F(t)} - \sum_{n=-N}^{N} \bar{c}_n e^{-int} \right] dt.$$

Since the conjugate of $F(t) \cdot \sum_{n=-N}^{N} \bar{c}_n e^{-int}$ is $\overline{F(t)} \cdot \sum_{n=-N}^{N} c_n e^{int}$, the right-hand side becomes

$$\int_{-\pi}^{\pi} |F(t)|^2\, dt - 2\,\mathrm{Re} \sum_{n=-N}^{N} \bar{c}_n \int_{-\pi}^{\pi} F(t)e^{-int}\, dt$$

$$+ \int_{-\pi}^{\pi} \left( \sum_{n=-N}^{N} c_n e^{int} \right)\left( \sum_{n=-N}^{N} \bar{c}_n e^{-int} \right) dt. \tag{16}$$

Recognizing the expression for the Fourier coefficient [Eq. (14)] in the preceding, we can write the second term as $-2(2\pi)\sum_{m=-N}^{N}|c_n|^2$. The third term can be expanded, but we must change one of the summation indices to avoid confusion; this term then becomes

$$\sum_{n=-N}^{N} c_n \sum_{m=-N}^{N} \bar{c}_m \int_{-\pi}^{\pi} e^{i(n-m)t}\, dt.$$

Recalling our previous evaluation of this integral, we see that this reduces to $2\pi\sum_{m=-N}^{N}c_n\bar{c}_n$. Thus we have shown

$$\int_{-\pi}^{\pi}\left|F(t) - \sum_{n=-N}^{N} c_n e^{int}\right|^2 dt = \int_{-\pi}^{\pi}|F(t)|^2\, dt - 2\pi\sum_{n=-N}^{N}|c_n|^2.$$

According to Theorem 1, the left-hand side approaches zero as $N \to \infty$. Hence

$$\int_{-\pi}^{\pi}|F(t)|^2\, dt - \lim_{N\to\infty} 2\pi\sum_{n=-N}^{N}|c_n|^2 = 0,$$

and Eq. (15) results. ∎

The next Fourier convergence theorem is valuable in electrical engineering applications, where switching circuits may produce (theoretically) discontinuous input functions, such as the periodic step function illustrated in Fig. 8.2.

We restrict ourselves to periodic functions $F$ with a finite number of discontinuities in any period. Specifically, we assume that $F$ has period $2\pi$ and that there is a finite subdivision of the interval $[-\pi, \pi]$ given by

$$-\pi = \tau_0 < \tau_1 < \tau_2 < \cdots < \tau_{n-1} < \tau_n = \pi$$

such that

1. $F(t)$ is continuously differentiable on each open subinterval $(\tau_j, \tau_{j+1})$ for $j = 0, 1, 2, \ldots, n-1$,
2. As $t$ approaches any subdivision point $\tau_j$ from the left, $F(t)$ and $F'(t)$ approach limiting values denoted by $F(\tau_j-)$ and $F'(\tau_j-)$, respectively, and
3. As $t$ approaches any $\tau_j$ from the right, $F(t)$ and $F'(t)$ again approach limiting values denoted $F(\tau_j+)$ and $F'(\tau_j+)$, respectively.

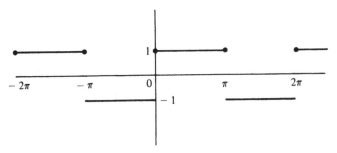

**Figure 8.2**    Periodic step function.

Such a function is said to be *piecewise smooth*.

Of course, if $F$ is continuous at $\tau_j$, then $F(\tau_j-) = F(\tau_j+) = F(\tau_j)$. For the step function in Fig. 8.2, $F(0-) = F(\pi+) = -1$, and $F(0+) = F(\pi-) = +1 = F(0) = F(\pi)$. Moreover, $F'(0-) = F'(0+) = 0$, but $F'(0)$ does not exist.

---

**Theorem 2.**    Suppose that $F$ is periodic and piecewise smooth. Then the Fourier series for $F$ converges to $F(t)$ at all points $t$ where $F$ is continuous and converges to $\frac{1}{2}[F(\tau_j+) + F(\tau_j-)]$ at the subdivision points.

---

### Example 4

Compute the Fourier series for the step function in Fig. 8.2, and state its convergence properties.

**Solution.**    The Fourier coefficients are given by

$$c_n = \frac{1}{2\pi} \int_{-\pi}^{\pi} F(t)e^{-int}\, dt = \frac{1}{2\pi} \int_{-\pi}^{0} (-1)e^{-int}\, dt + \frac{1}{2\pi} \int_{0}^{\pi} (1)e^{-int}\, dt$$

$$= \begin{cases} 0 & \text{if } n = 0, \\ \dfrac{i\{(-1)^n - 1\}}{\pi n} & \text{otherwise.} \end{cases}$$

Hence the Fourier series is

$$\frac{i}{\pi} \sum_{\substack{n=-\infty \\ n \neq 0}}^{\infty} \left[\frac{(-1)^n - 1}{n}\right] e^{int}. \tag{17}$$

According to Theorem 2, it converges to $+1$ for $0 < t < \pi$, to $-1$ for $-\pi < t < 0$, and to the average, 0, for $t = 0$ and $t = \pi$. ∎

When Fourier analysis (or *frequency analysis*, as it is sometimes called) is used to solve linear systems governed by differential equations, the question naturally arises as to whether or not a Fourier series can legitimately be differentiated termwise. (Obviously, the result of Example 2 is much too restrictive.) The following argument seems to cover a great many cases of interest to engineers: Suppose that $F$ has a convergent Fourier series expansion

$$F(t) = \sum_{n=-\infty}^{\infty} c_n e^{int}, \tag{18}$$

and suppose furthermore that the termwise-differentiated series

$$\sum_{n=-\infty}^{\infty} inc_n e^{int} \tag{19}$$

can be shown (by, say, the $M$-test) to be uniformly convergent on $[-\pi, \pi]$. Under such circumstances we know the "derived series" (19) can be legitimately *integrated* from $-\pi$ to $t$, termwise. But the result of this integration is the original series (18), up to a constant. Hence the sum function of (19) must be the *derivative* of $F(t)$, and we have proved the following.

---

**Theorem 3.**  Suppose that the Fourier expansion (18) is valid and that the derived series (19) converges uniformly on $[-\pi, \pi]$. Then

$$\sum_{n=-\infty}^{\infty} inc_n e^{int} = \frac{d}{dt} \sum_{n=-\infty}^{\infty} c_n e^{int}.$$

---

**Example 5**

Find the Fourier series for the periodic function

$$F(t) = \left| \sin \frac{t}{2} \right|^5,$$

and state the convergence properties for the derived series.

**Solution.**    (Observe that $F$ has period $2\pi$.) The Fourier coefficients are given by

$$c_n = \frac{1}{2\pi} \int_{-\pi}^{\pi} \left| \sin \frac{t}{2} \right|^5 e^{-int} \, dt.$$

These integrals can be evaluated by standard techniques after application of the identity

$$\sin^5 \theta = \tfrac{5}{8} \sin \theta - \tfrac{5}{16} \sin 3\theta + \tfrac{1}{16} \sin 5\theta.$$

With some labor one finds that the Fourier series for $F(t)$ is given by

$$\sum_{n=-\infty}^{\infty} \frac{240/\pi}{225 - 1036n^2 + 560n^4 - 64n^6} \, e^{int}, \tag{20}$$

and, according to Theorem 2, it converges to $F(t)$. Differentiating termwise we derive

$$\sum_{n=-\infty}^{\infty} \frac{in(240/\pi)}{225 - 1036n^2 + 560n^4 - 64n^6} \, e^{int}. \tag{21}$$

This series converges uniformly, as can be seen by comparing its increasing and decreasing parts with the (convergent) series $\sum_{n=1}^{\infty} 2 \cdot 240/\pi 64n^5$ [the factor 2 ensures that the terms of this series dominate those of (21) for large $n$]. Hence (21) represents $F'(t)$. Moreover, termwise differentiation of (21) can be justified by comparing the result with $\sum_{n=1}^{\infty} 2 \cdot 240/\pi 64n^4$; hence

$$F''(t) = \sum_{n=-\infty}^{\infty} \frac{-n^2(240/\pi)}{225 - 1036n^2 + 560n^4 - 64n^6} \, e^{int}.$$

Clearly two more termwise differentiations are justified, leading to Fourier series for $F^{(3)}(t)$ and $F^{(4)}(t)$. In fact, the student should verify that the original function $F(t)$ is continuously differentiable exactly four times! (The fifth derivative jumps from $-\tfrac{15}{4}$ to $+\tfrac{15}{4}$ as $t$ increases through $0, \pm 2\pi, \pm 4\pi$, etc.) ∎

The next example illustrates how Fourier series are used in practice to solve linear problems.

**Example 6**

Find a function $f$ that satisfies the differential equation

$$\frac{d^2f(t)}{dt^2} + 2\frac{df(t)}{dt} + 2f(t) = F(t),$$    (22)

where $F$ is the periodic "sawtooth" function prescribed by

$$F(t) := \begin{cases} -1 - \dfrac{2t}{\pi}, & -\pi \le t \le 0, \\[2mm] -1 + \dfrac{2t}{\pi}, & 0 \le t \le \pi \end{cases}$$

(see Fig. 8.3).

**Solution.**    First we show how a solution can be found to the simpler equation

$$\frac{d^2g(t)}{dt^2} + 2\frac{dg(t)}{dt} + 2g(t) = e^{i\omega t},$$    (23)

where the "forcing function" on the right-hand side has been replaced by a simple sinusoid. The considerations outlined in the introduction to this chapter indicate that Eq. (23) has a solution of the form $g(t) = Ae^{i\omega t}$. To find $A$, we insert this expression into Eq. (23) and obtain

$$-\omega^2 Ae^{i\omega t} + 2i\omega Ae^{i\omega t} + 2Ae^{i\omega t} = e^{i\omega t},$$

or (dividing by $e^{i\omega t}$)

$$(-\omega^2 + 2i\omega + 2)A = 1.$$

Solving for $A$, we deduce that

$$g(t) = \frac{e^{i\omega t}}{-\omega^2 + 2i\omega + 2}$$    (24)

solves Eq. (23).

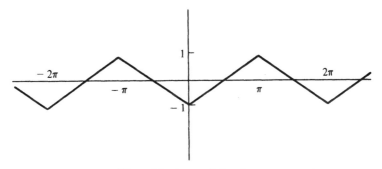

**Figure 8.3**  Sawtooth function.

Next we expand the given function $F(t)$ into a Fourier series. Using formula (14) for the coefficients, we find

$$c_n = \frac{1}{2\pi} \int_{-\pi}^{\pi} F(t) e^{-int} \, dt$$

$$= \frac{1}{2\pi} \int_{-\pi}^{0} \left( -1 - \frac{2t}{\pi} \right) e^{-int} \, dt + \frac{1}{2\pi} \int_{0}^{\pi} \left( -1 + \frac{2t}{\pi} \right) e^{-int} \, dt$$

$$= \begin{cases} 0 & \text{if } n = 0, \\ \dfrac{2}{\pi^2 n^2} \{ (-1)^n - 1 \} & \text{if } n \neq 0. \end{cases}$$

Hence

$$F(t) = \sum_{\substack{n=-\infty \\ (n \neq 0)}}^{\infty} \frac{2\{(-1)^n - 1\}}{\pi^2 n^2} e^{int}, \tag{25}$$

which is valid by Theorem 2. Now we argue as follows: We have Eq. (25) expressing $F(t)$ as a linear combination, albeit infinite, of sinusoids, and we have Eq. (24) expressing a solution for a single sinusoid. By linearity, then, we are led to postulate that the same linear combination of these solutions ought to solve the given equation; i.e.,

$$f(t) = \sum_{\substack{n=-\infty \\ (n \neq 0)}}^{\infty} \frac{2\{(-1)^n - 1\}}{\pi^2 n^2} \frac{e^{int}}{-n^2 + 2in + 2} \tag{26}$$

should be valid. To complete the argument, we first make the observation that Eq. (26) certainly solves Eq. (22) termwise. Furthermore, by comparing the series in Eq. (26) with the convergent series $\sum_{n=-\infty, n \neq 0}^{\infty} (8/\pi^2 n^4)$, we conclude that the former converges and that it can be legitimately differentiated termwise twice. Since Eq. (22) involves no derivatives higher than the second, we are done.

∎

[The reader should observe that termwise differentiation of the sawtooth series (25) yields $2/\pi$ times the step function series (17), which is consistent with the fact that the derivative of the sawtooth is $2/\pi$ times the step function, except at the "break points" 0, $\pm \pi, \pm 2\pi, \dots$. This phenomenon, which is not predicted by Theorem 3, reflects the fact that there are more powerful convergence results for Fourier series. Some of these can be found in the references.]

Formula (14) for the coefficients in the Fourier series is sometimes known as the *finite Fourier transform* (the "infinite" Fourier transform is covered in Sec. 8.2). The efficient computation of this transform is of crucial importance in engineering applications. But practically speaking one usually has to evaluate the integral numerically, for the following reasons:

1. $F(t)$ may be known only through measured data—no formula is available.
2. Even if an analytic formula for $F(t)$ is given, there may be no closed-form expression for the indefinite integral.

Now, for $n$ fixed, a Riemann sum approximating the integral

$$\int_{-\pi}^{\pi} \frac{F(t)}{2\pi} e^{-int} \, dt$$

takes the form (Sec. 4.2)

$$S_{n,N} = F(\tau_1)e^{-in\tau_1}\frac{t_1 - t_0}{2\pi} + F(\tau_2)e^{-in\tau_2}\frac{t_2 - t_1}{2\pi} + \cdots + F(\tau_N)e^{-in\tau_N}\frac{t_N - t_{N-1}}{2\pi},$$

where $-\pi = t_0 < t_1 < t_2 < \cdots < t_N = \pi$ and $t_{j-1} \leq \tau_j \leq t_j$. Let us choose the partition so that there are $N$ equal intervals, giving $t_j = -\pi + 2\pi j/N$; and let us choose the "sample" points $\tau_j$ to be the left endpoints of each respective interval, $\tau_j = t_{j-1}$. Then the sum can be abbreviated

$$S_{n,N} = \sum_{j=0}^{N-1} F\left(-\pi + \frac{2\pi j}{N}\right)\frac{e^{-in(-\pi + 2\pi j/N)}}{N} = \sum_{j=0}^{N-1} A_j e^{-i2\pi nj/N}, \tag{27}$$

where $A_j := F(-\pi + 2\pi j/N)e^{in\pi}/N$. As $N$ is increased, the sum $S_{n,N}$ converges to the coefficient $c_n$; thus error is controlled by choosing $N$ large.

Of course, larger values of $N$ also imply more computational effort. To estimate a single coefficient $c_n$ by Eq. (27) requires $N$ multiplications, and typically one evaluates the finite Fourier transform for $N$ such coefficients—calling for a total of $N^2$ (complex) multiplications. In applications it is often desirable to take $N$ to be several thousand, and the algorithm in this form is too computation-intensive. However, by judicious grouping of the terms in (27), the work can be reduced considerably.

Suppose, for instance, that $N = 16$, so that the computation of, say, $S_{1,16}$ takes the symbolic form

$$S_{1,16} = \sum_{j=0}^{15} A_j e^{-i2\pi j/16} = \sum_{j=0}^{15} A_j e^{-i\pi j/8}.$$

The numerical values of $\{e^{-ij\pi/8}: j = 0, 1, 2, \ldots, 15\}$ are quite redundant:

| $j$ | $e^{-ij\pi/8}$ |
|---|---|
| 0 | 1.000 |
| 1 | $.924 - .383i$ |
| 2 | $.707 - .707i$ |
| 3 | $.383 - .924i$ |
| 4 | $-1.000i$ |
| 5 | $-.383 - .924i$ |
| 6 | $-.707 - .707i$ |
| 7 | $-.924 - .383i$ |
| 8 | $-1.000$ |
| 9 | $-.924 + .383i$ |
| 10 | $-.707 + .707i$ |
| 11 | $-.383 + .924i$ |
| 12 | $+1.000i$ |
| 13 | $.383 + .924i$ |
| 14 | $.707 + .707i$ |
| 15 | $.924 + .383i$ |

So the computation of $S_{1,16}$ can, in fact, be carried out with only three complex multiplications:

$$S_{1,16} = A_0 - A_8 - i(A_4 - A_{12})$$
$$+ .924[A_1 - A_7 - A_9 + A_{15} - i(A_3 + A_5 - A_{11} - A_{13})]$$
$$+ .707[A_2 - A_6 - A_{10} + A_{14} - i(A_2 + A_6 - A_{10} - A_{14})]$$
$$+ .383[A_3 - A_5 - A_{11} + A_{13} - i(A_1 + A_7 - A_9 - A_{15})].$$

If we could achieve this same savings for 16 values of $S_{n,16}$, the number of multiplications would be reduced from $16^2 = 256$ to $16 \times 3 = 48$.

The *fast Fourier transform* (FFT) is an algorithm that systematically exploits these rearrangements of terms in the evaluation of (27). The emergence of the FFT in the late 1960s was a major milestone in modern system analysis and signal processing. For values of $N$ of the form $2^m$, the total number of multiplications required for $N$ values of $S_{n,N}$ is reduced to roughly $Nm/2 = (N/2) \log_2 N$. Codes are readily available, and small computers can perform a 4096-point transform in seconds. An outline of the basic strategy of the FFT is given in Problem 12; applications and error analyses are discussed in the references.

### EXERCISES 8.1

1. Compute the Fourier series for the following functions.
   (a) $F(t) = \sin^3 t$
   (b) $F(t) = \left|\cos^3 \dfrac{t}{3}\right|$
   (c) $F(t) = t^2$ $\quad (-\pi < t < \pi)$
   (d) $F(t) = t|t|$ $\quad (-\pi < t < \pi)$
2. Verify the Fourier representations of the indicated functions and state the convergence properties on the interval $[-\pi, \pi]$.
   (a) $\displaystyle\sum_{\substack{n=-\infty \\ n \text{ even}}}^{\infty} \frac{-2}{\pi(n^2 - 1)} e^{int} = |\sin t|$
   (b) $\displaystyle\sum_{n=-\infty}^{\infty} \frac{(-1)^n \sinh \pi}{(1 - in)\pi} e^{int} = e^t$
   (c) $\dfrac{1}{\pi} + \dfrac{1}{2} \sin t - \dfrac{2}{\pi} \displaystyle\sum_{n=1}^{\infty} \frac{\cos 2nt}{4n^2 - 1} = \begin{cases} 0, & -\pi \le t \le 0, \\ \sin t, & 0 \le t \le \pi \end{cases}$
   (d) $\dfrac{\pi}{4} + \displaystyle\sum_{n=1}^{\infty} \frac{(-1)^n - 1}{\pi n^2} \cos nt - \sum_{n=1}^{\infty} \frac{(-1)^n}{n} \sin nt = \begin{cases} 0, & -\pi \le t \le 0, \\ t, & 0 \le t < \pi \end{cases}$
3. Which of the series in Prob. 1 can be differentiated termwise?
4. Prove Eq. (12). [HINT: Use Eq. (11).]
5. Rewrite the series

$$\sum_{n=-\infty}^{\infty} c_n e^{int}$$

as a *trigonometric series* of the form

$$\sum_{n=0}^{\infty} \alpha_n \cos nt + \sum_{n=1}^{\infty} \beta_n \sin nt,$$

deriving the relations

$$\alpha_0 = c_0,$$

$$\alpha_n = c_n + c_{-n} \qquad (n \geq 1),$$

$$\beta_n = i(c_n - c_{-n}) \qquad (n \geq 1).$$

What are the conditions on the coefficients $c_n$ such that the sum of the series is a real function?

**6. (a)** If $F(t)$ is defined only for $0 \leq t \leq \pi$, show that by defining $F(-t) := -F(t), 0 < t \leq \pi$, and constructing the Fourier series for this function over the interval $[-\pi, \pi]$, one arrives at a *Fourier sine series*

$$\sum_{n=1}^{\infty} \beta_n \sin nt$$

for $F$, with coefficients given by

$$\beta_n = \frac{2}{\pi} \int_0^{\pi} F(t) \sin nt \, dt.$$

State conditions for the Fourier sine series to converge to $F(t)$ for $0 \leq t \leq \pi$.

**(b)** As in part (a), show that the definition $F(-t) := F(t), 0 < t \leq \pi$, produces a *Fourier cosine series*

$$\sum_{n=0}^{\infty} \alpha_n \cos nt$$

with coefficients

$$\alpha_0 = \frac{1}{\pi} \int_0^{\pi} F(t) \, dt, \qquad \alpha_n = \frac{2}{\pi} \int_0^{\pi} F(t) \cos nt \, dt \quad (n \geq 1).$$

State the conditions for convergence on $[0, \pi]$.

**7.** Find the Fourier representation for the periodic solutions of the following equations.

**(a)** $\dfrac{d^2 f}{dt^2} + 3f = \sin^4 t$

**(b)** $\dfrac{d^2 f}{dt^2} + \dfrac{df}{dt} + f = t^2, \ -\pi \leq t \leq \pi$, continued with period $2\pi$

**(c)** $\dfrac{d^2 f}{dt^2} + 4 \dfrac{df}{dt} + 2f = $ (the step function in Fig. 8.2)

**8.** Show that the Fourier sine and cosine series for the function $F(t) = t, 0 \leq t \leq \pi$, are given by

$$\sum_{n=1}^{\infty} \frac{2(-1)^{n+1}}{n} \sin nt$$

and

$$\frac{\pi}{2} - \frac{4}{\pi} \sum_{\substack{n=1 \\ n \text{ odd}}}^{\infty} \frac{\cos nt}{n^2},$$

respectively. (See Prob. 6.)

9. Suppose that we wish to solve the Dirichlet problem for the unit disk and that the boundary values of the desired harmonic function for $z = e^{i\theta}$ are represented by the series

$$U(\theta) = \sum_{n=0}^{\infty} \alpha_n \cos n\theta + \sum_{n=1}^{\infty} \beta_n \sin n\theta.$$

Argue that the solution to the problem is given by

$$u(re^{i\theta}) = \sum_{n=0}^{\infty} \alpha_n r^n \cos n\theta + \sum_{n=1}^{\infty} \beta_n r^n \sin n\theta.$$

10. As an illustration of the power of Fourier methods in solving partial differential equations, consider the *nonstatic* problem of heat flow along a uniform rod of length $\pi$, whose ends are maintained at zero degrees temperature. The temperature $T$ is now a function of position $x$ along the rod ($0 \le x \le \pi$) and time $t$. If the initial ($t = 0$) temperature distribution is specified to be $f(x)$, the equations that $T$ must satisfy are

$$\frac{\partial T(x, t)}{\partial t} = \frac{\partial^2 T(x, t)}{\partial x^2}$$

$$T(0, t) = T(\pi, t) = 0$$

$$T(x, 0) = f(x)$$

for $0 < x < \pi$, $t > 0$. Assuming the validity of termwise differentiation and Fourier expansions, show that

$$T(x, t) = \sum_{n=1}^{\infty} a_n \sin nx \, e^{-n^2 t}$$

solves the equations, where $a_n$ is defined by

$$a_n = \frac{2}{\pi} \int_0^{\pi} f(\xi) \sin n\xi \, d\xi.$$

[HINT: You will need the Fourier sine series, Prob. 6.] What is the limiting value of $T(x, t)$ as $t \to \infty$? Interpret this.

11. Another illustration of the power of Fourier methods is provided by the vibrating string problem. A taut string fastened at $x = 0$ and $x = \pi$ is initially distorted into the shape $u = f(x)$, where $u$ is the displacement of the string at the point $x$, and then the string is released. The equations governing the displacement $u(x, t)$ of the string are

$$\frac{\partial^2 u(x, t)}{\partial x^2} = \frac{\partial^2 u(x, t)}{\partial t^2}$$

$$u(0, t) = u(\pi, t) = 0$$

$$u(x, 0) = f(x)$$

$$\frac{\partial u(x, 0)}{\partial t} = 0$$

for $0 < x < \pi$, $t > 0$. Again assuming the validity of termwise differentiation and Fourier expansions, show that

$$u(x, t) = \sum_{n=1}^{\infty} b_n \sin nx \cos nt$$

solves the equations, where $b_n$ is defined by

$$b_n = \frac{2}{\pi} \int_0^\pi f(\xi) \sin n\xi \, d\xi.$$

[HINT: Use the Fourier sine series again.] How would you modify this representation if the "initial conditions" were interchanged to read

$$u(x, 0) = 0$$

$$\frac{\partial u(x, 0)}{\partial t} = f(x)?$$

Combine these formulas to satisfy the more general set of initial conditions

$$u(x, 0) = f_1(x)$$

$$\frac{\partial u(x, 0)}{\partial t} = f_2(x).$$

**12.** (*Fast Fourier Transform*)   Consider the evaluation of (27). As mentioned in the text, the computation of $N$ values of $S_{n,N}$ apparently entails $N^2$ complex multiplications.
   **(a)** Suppose that $N$ is even: $N = 2N_1$. Show that the formula for $S_{n,N}$ can be rewritten as

$$S_{n,N} = \sum_{j=0}^{N_1-1} A_j e^{-i2\pi nj/N} + \sum_{j=0}^{N_1-1} A_{j+N_1} e^{-i2\pi n(j+N_1)/N}$$

$$= \sum_{j=0}^{N_1-1} \{A_j + (-1)^n A_{j+N_1}\} e^{-i2\pi nj/N} = \sum_{j=0}^{N_1-1} B_j e^{-i2\pi nj/N}$$

   **(b)** Now how many complex multiplications will it take to compute $N$ values of $S_{n,N}$? [ANSWER: $N$ coefficients times $N_1 = N/2$ multiplications per coefficient $= N^2/2$; the multiplications by $(-1)$, of course, are not counted.] We seek to iterate this process, halving the number of multiplications again (assuming $N_1$ is even). However the sum in (a) does not have the same form as that in (27)—the $N$ in the exponent does not match the $N_1$ in the summation limits. So we have to back up.
   **(c)** Show that if $n$ is even, $n = 2n_1$, then the sum formula in (a) takes the form

$$S_{n,N} = \sum_{j=0}^{N_1-1} B_j e^{-i2\pi nj/N} = \sum_{j=0}^{N_1-1} B_j e^{-i2\pi n_1 j/N_1},$$

whereas if $n$ is odd, $n = 2n_1 + 1$, the formula can be written

$$S_{n,N} = \sum_{j=0}^{N_1-1} \{B_j e^{-i2\pi j/N}\} e^{-i2\pi n_1 j/N_1}$$

$$= \sum_{j=0}^{N_1-1} C_j e^{-i2\pi n_1 j/N_1}.$$

   **(d)** Noting that in (c) the computation of the coefficients $C_j$ requires a one-time "overhead" of $N_1 = N/2$ multiplications, how much work does it take to compute $N$ values of $S_{n,N}$? [ANSWER: $N/2$ multiplications plus $N$ coefficients times $N/2$ multiplications per coefficient $= N/2 + N^2/2$.]
   **(e)** At this point the sums in (c) have exactly the same form as the sum in (27), with $N$ replaced by $N_1 = N/2$. If each of the two new sums is manipulated as before, it will be replaced by a sum of $N_2 = N_1/2 = N/4$ terms, with an overhead of $N_2$ multiplications per sum to form the new coefficients. Now how much work is required to com-

pute $N$ values $S_{n,N}$? [ANSWER: $N_1$ multiplications overhead to form the coefficients for (c) plus 2 times $N_2$ multiplications to perform the same overhead for each sum in (c) plus $N$ coefficients times $N_2$ multiplications per coefficient $= N/2 + 2(N/4) + N^2/4 = 2(N/2) + N^2/4$.]

(f) If the trick in (c) is implemented yet again for the sums therein, how much work will be required to compute $N$ values of $S_{n,N}$? [ANSWER: $3(N/2) + N^2/8$.]

(g) If $N$ is a power of 2 and the trick in (c) is implemented to reduce the sums down to one term each, what is the net computational load to compute $N$ Fourier coefficients? [ANSWER: $(\log_2 N)(N/2) + N$ multiplications.]

## 8.2 THE FOURIER TRANSFORM

We move on to the next stage in our program of decomposing arbitrary functions into sinusoids. We have seen how a periodic function can be expressed as a Fourier series, so now we seek a similar representation for nonperiodic functions.

To begin with, let's assume we are given a nonperiodic function $F(t)$, $-\infty < t < \infty$, which is, say, continuously differentiable. Then if we pick an interval of the form $(-L/2, L/2)$ we can represent $F(t)$ by a Fourier series *for t in this interval*:

$$F(t) = \sum_{n=-\infty}^{\infty} c_n e^{in2\pi t/L}, \qquad \frac{-L}{2} < t < \frac{L}{2}, \tag{1}$$

with coefficients given by

$$c_n = \frac{1}{L} \int_{-L/2}^{L/2} F(t) e^{-in2\pi t/L}\, dt \qquad (n = 0, \pm 1, \pm 2, \ldots). \tag{2}$$

Actually the series in Eq. (1) defines a *periodic* function $F_L(t)$, $-\infty < t < \infty$, which coincides with $F(t)$ on $(-L/2, L/2)$; see Fig. 8.4. [Notice that $F_L(t)$ may be discontinuous even though $F(t)$ is smooth.]

Thus we have a sinusoidal representation of $F(t)$ over an interval of length $L$. If we now let $L \to \infty$, it seems reasonable to conjecture that this might evolve into a sinusoidal representation of $F(t)$ valid for *all* $t$. Let's explore this possibility.

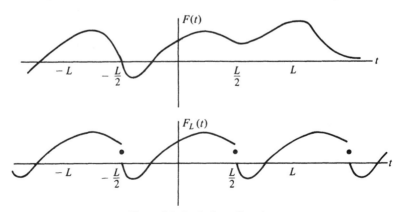

**Figure 8.4**   Periodic replica of $F(t)$.

We are going to rewrite these equations in what will seem at first like a rather bizarre form, but it will aid in interpreting them as $L \to \infty$. We define $g_n$ to be $c_n L/2\pi$, and introduce the factor $[(n + 1) - n] \equiv 1$ into the series in Eq. (1). Then we have

$$F_L(t) = \sum_{n=-\infty}^{\infty} g_n e^{in2\pi t/L} \frac{[(n + 1) - n]2\pi}{L} \tag{3}$$

and

$$g_n = \frac{1}{2\pi} \int_{-L/2}^{L/2} F(t)e^{-in2\pi t/L}\, dt. \tag{4}$$

Now write $\omega_n = n2\pi/L$, producing

$$F_L(t) = \sum_{n=-\infty}^{\infty} G_L(\omega_n)e^{i\omega_n t}(\omega_{n+1} - \omega_n), \tag{5}$$

where the function $G_L(\omega)$ is defined for *any* real $\omega$ by

$$G_L(\omega) := \frac{1}{2\pi} \int_{-L/2}^{L/2} F(t)e^{-i\omega t}\, dt. \tag{6}$$

As $L$ goes to infinity, $G_L(\omega)$ evolves rather naturally into a function $G(\omega)$ which is known as the *Fourier transform* of $F$:

$$G(\omega) := \frac{1}{2\pi} \int_{-\infty}^{\infty} F(t)e^{-i\omega t}\, dt. \tag{7}$$

Moreover, since $\Delta\omega_n := \omega_{n+1} - \omega_n$ goes to zero as $L \to \infty$ and since $\omega_n$ ranges from $-\infty$ to $+\infty$, Eq. (5) begins to look very much like a Riemann sum for the integral

$$\int_{-\infty}^{\infty} G(\omega)e^{i\omega t}\, d\omega.$$

Thus we are led to propose the equality

$$F(t) = \int_{-\infty}^{\infty} G(\omega)e^{i\omega t}\, d\omega \tag{8}$$

for nonperiodic $F$, when $G$ is defined by Eq. (7). Equation (8) is called the *Fourier inversion formula*.

Equations (7) and (8) are the essence of Fourier transform theory. As is suggested by this discussion, it is often profitable to indulge one's whimsy and think of the integral in Eq. (8) as a generalized "sum" of sinusoids, summed over a *continuum* of frequencies $\omega$. Equation (7) then dictates the "coefficients," $G(\omega)\, d\omega$, in the sum.

**Example 1**

Find the Fourier transform and verify the inversion formula for the function

$$F(t) = \frac{1}{t^2 + 4}.$$

**Solution.**    Observe that

$$F(t) = \frac{1}{t^2 + 4} = \frac{1}{(t - 2i)(t + 2i)}$$

is analytic except for simple poles at $t = \pm 2i$. We shall use residue theory to evaluate the Fourier transform, interpreting the integral as a principal value:

$$G(\omega) = \frac{1}{2\pi} \text{ p.v. } \int_{-\infty}^{\infty} \frac{e^{-i\omega t}}{t^2 + 4} \, dt.$$

If $\omega \geq 0$, we close the contour with expanding semicircles in the lower half-plane; by the techniques of Chapter 6 we find

$$G(\omega) = \frac{1}{2\pi} (-2\pi i) \text{Res} \left( \frac{e^{-i\omega t}}{t^2 + 4}; -2i \right)$$

$$= -i \cdot \lim_{t \to -2i} \frac{e^{-i\omega t}}{t - 2i} = \frac{e^{-2\omega}}{4} \qquad (\omega \geq 0).$$

Similarly, for $\omega < 0$ we close in the upper half-plane and find

$$G(\omega) = \frac{1}{2\pi} (2\pi i) \text{Res} \left( \frac{e^{-i\omega t}}{t^2 + 4}; 2i \right) = \frac{e^{2\omega}}{4} \qquad (\omega < 0).$$

In short,

$$G(\omega) = \frac{e^{-2|\omega|}}{4}.$$

To verify the Fourier inversion formula we compute

$$\int_{-\infty}^{\infty} G(\omega) e^{i\omega t} \, d\omega = \int_{-\infty}^{\infty} \frac{e^{-2|\omega|}}{4} \cdot e^{i\omega t} \, d\omega.$$

By symmetry, the imaginary part vanishes, and this integral equals

$$\text{Re} \int_{-\infty}^{\infty} \frac{e^{-2|\omega|}}{4} e^{i\omega t} \, d\omega = 2 \cdot \text{Re} \int_{0}^{\infty} \frac{e^{-2\omega}}{4} e^{i\omega t} \, d\omega$$

$$= \frac{1}{2} \text{Re} \left. \frac{e^{(-2+it)\omega}}{-2+it} \right|_{\omega=0}^{\infty} = \frac{1}{t^2 + 4}.$$

Hence

$$\frac{1}{t^2 + 4} = \int_{-\infty}^{\infty} \frac{e^{-2|\omega|}}{4} \cdot e^{i\omega t} \, d\omega. \quad \blacksquare \qquad (9)$$

As in the case of Fourier series, a wealth of theorems have been discovered stating conditions under which the Fourier integral representations (7) and (8) are valid. A very useful one for applications deals with piecewise smooth functions $F(t)$ like those in Theorem 2 of the previous section; that is, on every bounded interval $F(t)$ is continuously differentiable for all but the finite number of values $t = \tau_1$, $\tau_2, \ldots, \tau_n$, and at each $\tau_j$ the "one-sided limits" of $F(t)$ and $F'(t)$ exist.

**Theorem 4.** Suppose that $F(t)$ is piecewise smooth on every bounded interval and that $\int_{-\infty}^{\infty} |F(t)|\, dt$ exists. Then the Fourier transform, $G(\omega)$, of $F$ exists and

$$\int_{-\infty}^{\infty} G(\omega)e^{i\omega t}\, d\omega = \begin{cases} F(t) & \text{where } F \text{ is continuous,} \\ \dfrac{F(t+) + F(t-)}{2} & \text{otherwise.} \end{cases}$$

## Example 2

Find the Fourier transform of the function

$$F(t) = \begin{cases} 1, & -\pi \le t \le \pi, \\ 0, & \text{otherwise} \end{cases}$$

(Fig. 8.5), and confirm the inversion formula.

**Solution.** We have

$$G(\omega) = \frac{1}{2\pi} \int_{-\pi}^{\pi} (1)e^{-i\omega t}\, dt = \frac{\sin \omega\pi}{\omega\pi}.$$

Hence Theorem 4 tells us that

$$\int_{-\infty}^{\infty} \frac{\sin \omega\pi}{\omega\pi} e^{i\omega t}\, d\omega = \begin{cases} 1, & |t| < \pi, \\ 0, & |t| > \pi, \\ \dfrac{1}{2}, & t = \pm\pi. \end{cases} \tag{10}$$

To confirm this, rewrite the left-hand side of (10) as

$$\frac{1}{2\pi i} \int_{-\infty}^{\infty} \frac{e^{i\omega(\pi + t)} - e^{i\omega(-\pi + t)}}{\omega}\, d\omega. \tag{11}$$

Now recall from Example 1, Sec. 6.5 (page 276) that

$$\text{p.v.} \int_{-\infty}^{\infty} \frac{e^{ix}}{x}\, dx = i\pi$$

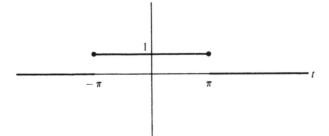

**Figure 8.5** "Boxcar" function.

which, with the change of variables $x = C\omega$, generalizes to

$$\text{p.v.} \int_{-\infty}^{\infty} \frac{e^{iC\omega}}{\omega} \, d\omega = \begin{cases} i\pi & \text{if } C > 0, \\ -i\pi & \text{if } C < 0. \end{cases}$$

Of course,

$$\text{p.v.} \int_{-\infty}^{\infty} \frac{1}{x} \, dx = 0.$$

Therefore, by introducing principal values we derive

if $t < -\pi$, (11) becomes $\dfrac{1}{2\pi i} \left[ -i\pi - (-i\pi) \right] = 0$;

if $t = -\pi$, (11) becomes $\dfrac{1}{2\pi i} \left[ 0 - (-i\pi) \right] = \dfrac{1}{2}$;

if $-\pi < t < \pi$, (11) becomes $\dfrac{1}{2\pi i} \left[ i\pi - (-i\pi) \right] = 1$;

if $t = \pi$, (11) becomes $\dfrac{1}{2\pi i} \left[ i\pi - 0 \right] = \dfrac{1}{2}$;

if $\pi < t$, (11) becomes $\dfrac{1}{2\pi i} \left[ i\pi - i\pi \right] = 0$. ∎

## Example 3

Find the Fourier transform of the function

$$F(t) = \begin{cases} \sin t, & |t| \leq 6\pi, \\ 0, & \text{otherwise} \end{cases}$$

(Fig. 8.6), and confirm the inversion formula. (Physicists call this function a *finite wave train*.)

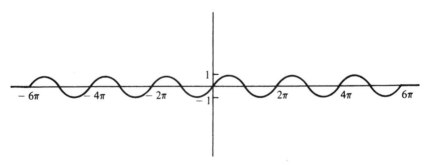

**Figure 8.6**    Finite wave train.

**Solution.**  We have

$$G(\omega) = \frac{1}{2\pi} \int_{-6\pi}^{6\pi} (\sin t) e^{-i\omega t}\, dt$$

$$= \frac{i \sin 6\pi\omega}{\pi(1 - \omega^2)}.$$

Since $F(t)$ is continuous everywhere, the inversion formula implies

$$\int_{-\infty}^{\infty} \frac{i \sin 6\pi\omega}{\pi(1 - \omega^2)} e^{i\omega t}\, d\omega = F(t). \tag{12}$$

To confirm this rewrite the left-hand side as

$$\frac{i}{2i\pi} \int_{-\infty}^{\infty} \frac{e^{i\omega(6\pi + t)} - e^{i\omega(-6\pi + t)}}{(1 - \omega^2)}\, d\omega = \text{p.v.} \frac{-1}{2\pi} \int_{-\infty}^{\infty} \frac{e^{i\omega(6\pi + t)} - e^{i\omega(-6\pi + t)}}{(\omega - 1)(\omega + 1)}\, d\omega$$

(because of the removable singularities at $\omega = \pm 1$).

Now the integral

$$\text{p.v.} \frac{-1}{2\pi} \int_{-\infty}^{\infty} \frac{e^{i\omega(6\pi + t)}}{(\omega - 1)(\omega + 1)}\, d\omega$$

can be evaluated using the indented-contour techniques of Sec. 6.5. For $t \geq -6\pi$ we employ the contour shown in Fig. 8.7(a) and invoke Lemmas 3 and 4 of

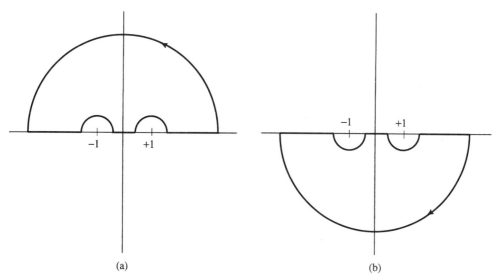

(a)                    (b)

**Figure 8.7**   Contours for Example 3.

Chapter 6 to obtain

$$\text{p.v.} \frac{-1}{2\pi} \int_{-\infty}^{\infty} \frac{e^{i\omega(6\pi+t)}}{(\omega-1)(\omega+1)} \, d\omega = \frac{-1}{2\pi} (\pi i)\{\text{Res}(-1) + \text{Res}(1)\}$$

$$= \frac{-i}{2} \left[ \frac{e^{-i(6\pi+t)}}{-2} + \frac{e^{i(6\pi+t)}}{2} \right]$$

$$= \frac{\sin(6\pi+t)}{2} = \frac{\sin t}{2}.$$

Similarly, for $t \le -6\pi$ we use the contour of Fig. 8.7(b) and find

$$\text{p.v.} \frac{-1}{2\pi} \int_{-\infty}^{\infty} \frac{e^{i\omega(6\pi+t)}}{(\omega-1)(\omega+1)} \, d\omega = \frac{-1}{2\pi}(-\pi i)\{\text{Res}(-1) + \text{Res}(1)\}$$

$$= -\frac{\sin t}{2}.$$

By the same reasoning one obtains

$$\text{p.v.} \frac{-1}{2\pi} \int_{-\infty}^{\infty} \frac{e^{i\omega(-6\pi+t)}}{(\omega-1)(\omega+1)} \, d\omega = \begin{cases} \dfrac{\sin t}{2} & \text{if } t \ge 6\pi, \\ -\dfrac{\sin t}{2} & \text{if } t \le 6\pi. \end{cases}$$

Piecing this together we validate (12). ∎

The Fourier transform equations are used just like Fourier series in solving linear systems; as an illustration, consider the following.

### Example 4

Find a function that satisfies the differential equation

$$\frac{d^2 f(t)}{dt^2} + 2 \frac{df(t)}{dt} + 2f(t) = \begin{cases} \sin t, & |t| \le 6\pi, \\ 0 & \text{otherwise.} \end{cases} \tag{13}$$

**Solution.** In Example 6, Sec. 8.1, page 381, we learned that a solution to $f'' + 2f' + 2f = e^{i\omega t}$ is

$$\frac{e^{i\omega t}}{-\omega^2 + 2i\omega + 2}.$$

Now the right-hand side of Eq. (13) is the function $F(t)$ in the previous example, and Eq. (12) can be interpreted as expressing $F$ as a "superposition" of sinusoids of the form $e^{i\omega t}$. Hence we propose that the corresponding superposition of solutions

$$f(t) = \int_{-\infty}^{\infty} \frac{i \sin 6\pi\omega}{\pi(1-\omega^2)} \left( \frac{e^{i\omega t}}{-\omega^2 + 2i\omega + 2} \right) d\omega \tag{14}$$

solves the given equation. As before, we should establish that this expression converges and can be differentiated twice under the integral sign, but, instead, we invite the student to verify that residue theory yields the expression (see Prob. 2)

$$f(t) = \begin{cases} 0, & t \le -6\pi, \\ \frac{2}{5}(e^{-6\pi-t} - 1)\cos t + \frac{1}{5}(e^{-6\pi-t} + 1)\sin t, & -6\pi \le t \le 6\pi, \\ -(e^{6\pi} - e^{-6\pi})e^{-t}(\frac{2}{5}\cos t + \frac{1}{5}\sin t), & t \ge 6\pi, \end{cases}$$

which, as direct computation shows, solves the differential equation (13). ∎

As an amusing exercise in the manipulation of contour integrals we now present an informal derivation of an identity that can be considered as a Fourier expansion theorem, if we are lenient in interpreting relations (7) and (8) between the function and its transform.

### Example 5

Suppose that the function $F(t)$ is analytic and bounded by a constant $M$ in an open strip $|\text{Im } t| < \delta$, and define $G_L(\omega)$ as in Eq. (6). Argue that, as $L \to \infty$,

$$\text{p.v.} \int_{-\infty}^{\infty} G_L(\omega)e^{i\omega t}\, d\omega \to F(t) \tag{15}$$

for each real $t$.

**Solution.**    Notice, first of all, that if $F(t)$ has a Fourier transform, it will be given by the limit of $G_L(\omega)$ as $L \to \infty$. Hence (15) looks very much like a Fourier inversion formula. In fact, we shall argue that the members of (15) are equal whenever $L > 2|t|$.

For this purpose we define $I_r$ via

$$\text{p.v.} \int_{-\infty}^{\infty} G_L(\omega)e^{i\omega t}\, d\omega$$

$$= \lim_{r \to \infty} \frac{1}{2\pi} \int_{-r}^{r} \left[ \int_{-L/2}^{L/2} F(\tau)e^{-i\omega\tau}\, d\tau \right] e^{i\omega t}\, d\omega =: \lim_{r \to \infty} I_r. \tag{16}$$

We state without proof that the order of integration can legitimately be reversed under these circumstances, producing

$$I_r = \frac{1}{2\pi} \int_{-r}^{r} \left[ \int_{-L/2}^{L/2} F(\tau)e^{-i\omega\tau}\, d\tau \right] e^{i\omega t}\, d\omega$$

$$= \frac{1}{2\pi} \int_{-L/2}^{L/2} F(\tau) \left[ \int_{-r}^{r} e^{i\omega(t-\tau)}\, d\omega \right] d\tau$$

$$= \frac{1}{2\pi} \int_{-L/2}^{L/2} \frac{F(\tau)}{i(t-\tau)} \left( e^{ir(t-\tau)} - e^{-ir(t-\tau)} \right) d\tau.$$

We write this as

$$I_r = \frac{1}{2\pi i} \int_{-L/2}^{L/2} \frac{F(\tau)}{t - \tau} (e^{ir(t-\tau)} - 1) \, d\tau$$

$$+ \frac{1}{2\pi i} \int_{-L/2}^{L/2} \frac{F(\tau)}{t - \tau} (1 - e^{-ir(t-\tau)}) \, d\tau, \tag{17}$$

for reasons that will become apparent shortly. Observe that each integrand is analytic in $\tau$ as long as we stay in the strip $|\operatorname{Im} \tau| < \delta$, because the singularity at $\tau = t$ is removable. Hence the integrals are independent of path. We choose to evaluate the first integral in Eq. (17) along the contour $\Gamma^-$ in Fig. 8.8 and the second along $\Gamma^+$:

$$I_r = \frac{1}{2\pi i} \int_{\Gamma^-} \frac{F(\tau)}{t - \tau} (e^{ir(t-\tau)} - 1) \, d\tau + \frac{1}{2\pi i} \int_{\Gamma^+} \frac{F(\tau)}{t - \tau} (1 - e^{-ir(t-\tau)}) \, d\tau.$$

If we rewrite this as

$$I_r = \frac{1}{2\pi i} \int_{\Gamma^-} \frac{F(\tau)}{\tau - t} \, d\tau - \frac{1}{2\pi i} \int_{\Gamma^+} \frac{F(\tau)}{\tau - t} \, d\tau$$

$$+ \frac{1}{2\pi i} \int_{\Gamma^-} \frac{F(\tau)}{t - \tau} e^{ir(t-\tau)} \, d\tau - \frac{1}{2\pi i} \int_{\Gamma^+} \frac{F(\tau)}{t - \tau} e^{-ir(t-\tau)} \, d\tau,$$

we recognize that the first two integrals combine to give the integral of $F(\tau)/2\pi i(\tau - t)$ around the simple closed positively oriented contour $(\Gamma^-, -\Gamma^+)$, and according to Cauchy's integral formula, this is precisely $F(t)$. Thus (15) will follow if we show that the last two integrals go to zero as $r \to \infty$. This is most encouraging, since $\exp[ir(t - \tau)] \to 0$ for $\tau$ in the lower half-plane and $\exp[-ir(t - \tau)] \to 0$ for $\tau$ in the upper half-plane.

To fill in the details, consider first the part of the integration along $\gamma_1$: $\tau = -L/2 - iy$, $0 \le y \le \varepsilon$ (see Fig. 8.8). We have

$$\left| \frac{1}{2\pi i} \int_{\gamma_1} \frac{F(\tau)}{t - \tau} e^{ir(t-\tau)} \, d\tau \right| \le \frac{1}{2\pi} \int_0^\varepsilon \frac{M}{L/2 - |t|} \left| e^{irt} e^{irL/2} e^{-ry} \right| \, dy$$

$$= \frac{1}{\pi} \frac{M}{L - 2|t|} \int_0^\varepsilon e^{-ry} \, dy = \frac{1}{\pi} \frac{M}{L - 2|t|} \frac{1 - e^{-r\varepsilon}}{r},$$

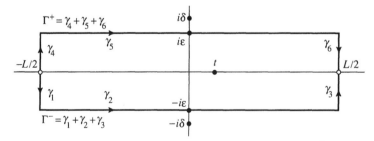

Figure 8.8   Contour for Example 5.

where $M$ is the bound for $|F|$. This certainly approaches zero as $r \to \infty$. For the integration along $\gamma_2$: $\tau = x - i\varepsilon$, $-L/2 \le x \le L/2$ (Fig. 8.8 again),

$$\left| \frac{1}{2\pi i} \int_{\gamma_2} \frac{F(\tau)}{t - \tau} e^{ir(t-\tau)} \, d\tau \right| \le \frac{1}{2\pi} \int_{-L/2}^{L/2} \frac{M}{\varepsilon} \left| e^{ir(t-x+i\varepsilon)} \right| \, dx$$

$$= \frac{1}{2\pi} \frac{Me^{-r\varepsilon}}{\varepsilon} L,$$

again vanishing as $r \to \infty$. The integrals over $\gamma_3$, $\gamma_4$, and $\gamma_6$ in Fig. 8.8 are similar to that over $\gamma_1$, and the integral over $\gamma_5$ is handled like that over $\gamma_2$. ∎

# EXERCISES 8.2

1. Find the Fourier transform of the following functions and express the inversion formula.
   (a) $F(t) = e^{-|t|}$
   (b) $F(t) = e^{-t^2}$  [HINT: Complete the square and use the fact that $\int_{-\infty}^{\infty} e^{-x^2} \, dx = \sqrt{\pi}$.]
   (c) $F(t) = te^{-t^2}$  [HINT: Integrate by parts and use part (b).]

   (d) $F(t) = \dfrac{\sin t}{t}$  [HINT: Use Eq. (10), page 391.]

   (e) $F(t) = \dfrac{\sin \pi t}{1 - t^2}$  [HINT: Exploit Example 3, page 392.]

2. Use residue theory to evaluate the integral in Eq. (14).

3. Find the Fourier transform and confirm the inversion formula (by residue theory or some other method) for the following functions.

   (a) $F(t) = \dfrac{1}{t^4 + 1}$  (b) $F(t) = \dfrac{t}{t^4 + 1}$

   (c) $F(t) = e^{-t^2}$  [HINT: See Prob. 1(b).]

4. Find the Fourier integral representations for solutions to the following differential equations.

   (a) $\dfrac{d^2f}{dt^2} + \dfrac{df}{dt} + f = e^{-t^2}$

   (b) $\dfrac{d^2f}{dt^2} + 4\dfrac{df}{dt} + f = \begin{cases} 0, & t < 0 \\ e^{-t}, & t \ge 0 \end{cases}$

   (c) $\dfrac{d^2f}{dt^2} + 2\dfrac{df}{dt} + 3f = \begin{cases} 1, & |t| < 1 \\ 0, & \text{otherwise} \end{cases}$

5. If $f(x)$ is a given function whose Fourier transform is $G(\omega)$, show that the function

$$T(x, t) = \int_{-\infty}^{\infty} G(\omega)e^{i\omega x}e^{-\omega^2 t} \, d\omega$$

solves the partial differential equation

$$\frac{\partial T(x, t)}{\partial t} = \frac{\partial^2 T(x, t)}{\partial x^2}$$

and the initial condition

$$T(x, 0) = f(x)$$

for $t > 0$ and $-\infty < x < \infty$. [These equations describe the flow of heat in an infinite rod heated initially to the temperature $T = f(x)$. Recall Prob. 10 in Exercises 8.1, page 386.] Assume the validity of the Fourier representations and differentiation under the integral sign.

Insert the expression for the transform $G(\omega)$, interchange the order of integration, and use the hint accompanying Prob. 1(b) to derive the formula

$$T(x, t) = \frac{1}{2\sqrt{\pi t}} \int_{-\infty}^{\infty} f(\xi) e^{-(x - \xi)^2/4t} \, d\xi.$$

6. If $f(x)$ is a given function whose Fourier transform is $G(\omega)$, show that the function

$$u(x, t) = \int_{-\infty}^{\infty} G(\omega) e^{i\omega x} \cos \omega t \, d\omega$$

solves the partial differential equation

$$\frac{\partial^2 u(x, t)}{\partial x^2} = \frac{\partial^2 u(x, t)}{\partial t^2}$$

and the initial conditions

$$u(x, 0) = f(x)$$

$$\frac{\partial u}{\partial t}(x, 0) = 0$$

for $t > 0$ and $-\infty < x < \infty$. [These equations govern the motion of an infinite taut string initially displaced to the configuration $u = f(x)$ and released at $t = 0$. Recall Prob. 11 in Exercises 8.1, page 386.] Assume the validity of the Fourier representations and differentiation under the integral sign.

How would you modify this representation if the initial conditions were

$$u(x, 0) = 0$$

$$\frac{\partial u(x, 0)}{\partial t} = f(x)?$$

Combine these results to handle the general set of initial conditions

$$u(x, 0) = f_1(x)$$

$$\frac{\partial u}{\partial t}(x, 0) = f_2(x).$$

7. The *Mellin transform* can be obtained from the Fourier transform by a change of variables. Suppose $f(r)$ is defined for $0 < r < \infty$. Let $x = -\text{Log}\, r$ (so $r = e^{-x}$) and set $F(x) := f(e^{-x})$; then $x$ runs from $\infty$ down to $-\infty$.

   (a) Write the Fourier transform equations for $F(x)$ and recast them in terms of $r$ and $f$ to obtain

$$f(r) = \int_{-\infty}^{\infty} g(\omega) e^{-i\omega \, \text{Log}\, r} \, d\omega, \qquad g(\omega) = \frac{1}{2\pi} \int_{0}^{\infty} f(r) e^{i\omega \, \text{Log}\, r} r^{-1} \, dr.$$

   (b) The Mellin transform of $f$ is formally defined by

$$M[s; f] := \int_{0}^{\infty} f(r) r^{s-1} \, dr.$$

Show that the inverse transform can be expressed as

$$f(r) = \frac{1}{2\pi} \int_{-\infty}^{\infty} M[i\omega; f] r^{-i\omega} \, d\omega.$$

**8.** The two-dimensional Laplace equation for a harmonic function $\phi$, expressed in polar coordinates as $\phi = \phi(r, \theta)$, is given by

$$\frac{1}{r} \frac{\partial}{\partial r} \left( r \frac{\partial \phi}{\partial r} \right) + \frac{1}{r^2} \frac{\partial^2 \phi}{\partial \theta^2} = 0.$$

**(a)** Presuming that the order of differentiation and integration can be interchanged freely, use the Mellin transform to show that

$$\phi(r, \theta) = \frac{1}{2\pi} \int_{-\infty}^{\infty} \frac{M[i\omega; f]}{\sinh \omega \theta_0} r^{-i\omega} \sinh \omega \theta \, d\omega$$

solves Laplace's equation in the wedge $0 < \theta < \theta_0$ and takes the boundary values $\phi(r, 0) = 0$, $\phi(r, \theta_0) = f(r)$.
**(b)** Construct a solution to Laplace's equation in the wedge, taking the boundary values $\phi(r, 0) = f(r)$, $\phi(r, \theta_0) = 0$.
**(c)** Construct a solution to Laplace's equation in the wedge, taking the boundary values $\phi(r, 0) = f_1(r)$, $\phi(r, \theta_0) = f_2(r)$.
**(d)** Construct a solution to Laplace's equation in the wedge satisfying the boundary conditions $\partial \phi(r, 0)/\partial \theta = 0$, $\phi(r, \theta_0) = f(r)$.

## 8.3 THE LAPLACE TRANSFORM

In the two previous sections we were motivated by the desire to solve linear systems by means of frequency analysis. The strategy we were employing can be stated as follows: If a linear system is forced by a sinusoidal input function, $e^{i\omega t}$, then we expect that there ought to be a solution which is a sinusoid having the same frequency.

Now this is probably not the *only* solution; e.g., consider the problem of finding a function $f(t)$ that satisfies the differential equation

$$\frac{d^2 f(t)}{dt^2} + 2 \frac{df(t)}{dt} + f(t) = e^{i2t}. \tag{1}$$

It has a solution of the form $Ae^{i2t}$ with $A = 1/(4i - 3)$. But if $g(t)$ is a solution of the so-called *associated homogeneous equation*

$$\frac{d^2 g(t)}{dt^2} + 2 \frac{dg(t)}{dt} + g(t) = 0,$$

then the function $g$ may be added to a solution $f$ of Eq. (1) to produce another solution of Eq. (1). For example, the function

$$\frac{1}{4i - 3} e^{i2t} + 7e^{-t} \tag{2}$$

also solves Eq. (1), since $7e^{-t}$ is a "homogeneous solution." The reader should verify the solution (2) by direct computation to see exactly what's going on.

Now for most *physical* systems, these homogeneous solutions are transient in nature; i.e., they die out as time increases [like $e^{-t}$ in (2)]. This is evidenced by the fact that most physical systems, if not forced, eventually come to rest due to dissipative phenomena such as resistance, damping, radiation loss, etc. Such systems are called *asymptotically stable*. In these cases, we argue that the analysis of the preceding sections provides the *unique* solutions for the types of problems formulated there, because for both the periodic functions and the functions integrable over the whole real line the inputs have been driving the system "since $t = -\infty$" and hence the transients must have died out by any (finite) time.

Now it is time to become more flexible and to develop some mathematical machinery that will handle the transients. That is, we must take into account two considerations: The input is "turned on" at $t = 0$ and has *not* been driving the system for all time, and the system starts in some "initial configuration" at $t = 0$ that probably does *not* coincide with the steady-state solution. The *Laplace transform*, as we shall see, handles both these effects. It also accomodates nondissipative systems.

Let us begin by dealing with the input function. We have $F(t)$ defined for all $t \geq 0$. For this discussion it is convenient to extend the domain to the whole line, so we set $F(t) = 0$ for $t < 0$ (such a function is commonly called "causal") and then consider the Fourier transform of $F$:

$$G(\omega) = \frac{1}{2\pi} \int_{-\infty}^{\infty} F(t)e^{-i\omega t}\, dt.$$

In our case,

$$G(\omega) = \frac{1}{2\pi} \int_{0}^{\infty} F(t)e^{-i\omega t}\, dt. \tag{3}$$

Now if $F$ is sufficiently well behaved near infinity (we shall not be precise here), one can show that Eq. (3) defines a function of $\omega$ that is analytic in the lower half-plane $\text{Im}\,\omega < 0$. Indeed, the derivative is given, as expected, by

$$\frac{dG(\omega)}{d\omega} = \frac{-i}{2\pi} \int_{0}^{\infty} tF(t)e^{-i\omega t}\, dt;$$

the *lower* half-plane is appropriate because

$$\left| e^{-i\omega t} \right| = e^{(\text{Im}\,\omega)t}$$

is bounded there. If we let $\omega$ be pure imaginary, say $\omega = -is$ with $s$ nonnegative, we create

$$g(s) := 2\pi G(-is), \tag{4}$$

a function called the *Laplace transform* of $F(t)$:

$$g(s) = \int_{0}^{\infty} F(t)e^{-st}\, dt. \tag{5}$$

It is often useful to indicate the relation between $g(s)$ and $F(t)$ by employing the notation

$$g(s) = \mathscr{L}\{F\}(s).$$

As an example, consider $F(t) = e^{-t}$; its Laplace transform is

$$g(s) = \mathscr{L}\{e^{-t}\}(s) = \int_0^\infty e^{-t}e^{-st}\, dt = -\frac{e^{-(s+1)t}}{s+1}\bigg|_0^\infty = \frac{1}{s+1}.$$

We remark that the integral in Eq. (5) may converge even if $F$ does not approach zero as $t \to \infty$, provided that $s$ is sufficiently large. Indeed, for the function $F(t) = e^{7t}$, which might characterize a nondissipative "runaway" physical system, we have

$$\int_0^b e^{7t}e^{-st}\, dt = \frac{e^{(7-s)b} - 1}{7 - s},$$

and if $s > 7$, this approaches $(s - 7)^{-1}$ as $b \to \infty$. In fact, whenever there exist two positive numbers $M$ and $\alpha$ such that

$$|F(t)| \le Me^{\alpha t}, \qquad \text{for all} \quad t \ge 0,$$

one can show that the integral in Eq. (5) converges for any *complex* $s$ satisfying $\text{Re}(s) > \alpha$. Accordingly, we shall say that Eq. (5) defines the Laplace transform $\mathscr{L}\{F\}(s)$ for any (complex) value of $s$ for which the integral converges. In essence the Laplace transform is able to encompass more functions than the Fourier transform, by allowing the frequency variable $\omega$ to be complex.

As a simple extension of the preceding computation shows, the Laplace transform of the function $e^{at}$ is $1/(s - a)$ for $\text{Re}(s) > \text{Re}(a)$. By interpreting this statement with various choices of the constant $a$, we are able to derive the first eight entries in the Laplace transform table on page 402. Entry (ix) is obtained by integration by parts, and it leads immediately to entries (x), (xi), and (xii).

The derivation of (xiii) proceeds as follows:

$$\mathscr{L}\{F(t)e^{-at}\}(s) = \int_0^\infty F(t)e^{-at}e^{-st}\, dt = \int_0^\infty F(t)e^{-(s+a)t}\, dt = \mathscr{L}\{F\}(s + a).$$

Entry (xiv) says, of course, that the Laplace transform is linear.

By looking at the transform of the derivative $F'(t)$, we can see how the Laplace transform takes initial configurations into account; we have

$$\mathscr{L}\{F'\}(s) = \int_0^\infty e^{-st}F'(t)\, dt.$$

Now if $F'$ is sufficiently well behaved so that integration by parts is permitted, this becomes

$$\mathscr{L}\{F'\}(s) = -\int_0^\infty (-s)e^{-st}F(t)\, dt + e^{-st}F(t)\bigg|_0^\infty,$$

and assuming that $e^{-st}F(t) \to 0$ as $t \to \infty$, we find

$$\mathscr{L}\{F'\}(s) = s\mathscr{L}\{F\}(s) - F(0). \tag{6}$$

## TABLE OF LAPLACE TRANSFORMS

(i) $\mathcal{L}\{e^{at}\} = \dfrac{1}{s-a}$     $[\text{Re}(s) > \text{Re}(a)]$

(ii) $\mathcal{L}\{1\} = \mathcal{L}\{e^{0t}\} = \dfrac{1}{s}$     $[\text{Re}(s) > 0]$

(iii) $\mathcal{L}\{\cos \omega t\} = \text{Re}\,\mathcal{L}\{e^{i\omega t}\} = \dfrac{s}{s^2 + \omega^2}$     $[\omega \text{ real, } \text{Re}(s) > 0]$

(iv) $\mathcal{L}\{\sin \omega t\} = \text{Im}\,\mathcal{L}\{e^{i\omega t}\} = \dfrac{\omega}{s^2 + \omega^2}$     $[\omega \text{ real, } \text{Re}(s) > 0]$

(v) $\mathcal{L}\{\cosh \omega t\} = \mathcal{L}\{\cos i\omega t\} = \dfrac{s}{s^2 - \omega^2}$     $[\omega \text{ real, } \text{Re}(s) > |\omega|]$

(vi) $\mathcal{L}\{\sinh \omega t\} = \mathcal{L}\{-i \sin i\omega t\} = \dfrac{\omega}{s^2 - \omega^2}$     $[\omega \text{ real, } \text{Re}(s) > |\omega|]$

(vii) $\mathcal{L}\{e^{-\lambda t} \cos \omega t\} = \text{Re}\,\mathcal{L}\{e^{(-\lambda + i\omega)t}\} = \dfrac{s + \lambda}{(s + \lambda)^2 + \omega^2}$     $[\omega, \lambda \text{ real, } \text{Re}(s) > -\lambda]$

(viii) $\mathcal{L}\{e^{-\lambda t} \sin \omega t\} = \text{Im}\,\mathcal{L}\{e^{(-\lambda + i\omega)t}\} = \dfrac{\omega}{(s + \lambda)^2 + \omega^2}$     $[\omega, \lambda \text{ real, } \text{Re}(s) > -\lambda]$

(ix) $\mathcal{L}\{t^n e^{at}\} = \dfrac{n!}{(s - a)^{n+1}}$     $[\text{Re}(s) > \text{Re}(a)]$

(x) $\mathcal{L}\{t^n\} = \dfrac{n!}{s^{n+1}}$     $[\text{Re}(s) > 0]$

(xi) $\mathcal{L}\{t \cos \omega t\} = \text{Re}\,\mathcal{L}\{te^{i\omega t}\} = \dfrac{s^2 - \omega^2}{(s^2 + \omega^2)^2}$     $[\omega \text{ real, } \text{Re}(s) > 0]$

(xii) $\mathcal{L}\{t \sin \omega t\} = \text{Im}\,\mathcal{L}\{te^{i\omega t}\} = \dfrac{2s\omega}{(s^2 + \omega^2)^2}$     $[\omega \text{ real, } \text{Re}(s) > 0]$

(xiii) $\mathcal{L}\{F(t)e^{-at}\}(s) = \mathcal{L}\{F\}(s + a)$

(xiv) $\mathcal{L}\{aF(t) + bH(t)\} = a\mathcal{L}\{F(t)\} + b\mathcal{L}\{H(t)\}$

Iterating this equation results in

$$\mathcal{L}\{F''\}(s) = s\mathcal{L}\{F'\}(s) - F'(0)$$
$$= s^2 \mathcal{L}\{F\}(s) - sF(0) - F'(0), \tag{7}$$

and, in general,

$$\mathcal{L}\{F^{(k)}\}(s) = s^k \mathcal{L}\{F\}(s) - s^{k-1}F(0) - s^{k-2}F'(0) - \cdots - F^{(k-1)}(0). \tag{8}$$

Sufficient conditions for the validity of these equations are given in the following theorem.

**Theorem 5.** Suppose that the function $F(t)$ and its first $n - 1$ derivatives are continuous for $t \geq 0$ and that $F^{(n)}(t)$ is piecewise smooth on every finite interval $[0, b]$. Also, suppose that there are positive constants $M$, $\alpha$ such that for $k = 0, 1, \ldots, n - 1$

$$\left| F^{(k)}(t) \right| \leq M e^{\alpha t} \qquad (t \geq 0).$$

Then the Laplace transforms of $F, F', F'', \ldots, F^{(n)}$ exist for $\mathrm{Re}(s) > \alpha$, and Eq. (8) is valid for $k = 1, 2, \ldots, n$.

The reader is invited to prove this theorem in Prob. 2.

To illustrate how the Laplace transform is used in solving the so-called initial-value problems, we consider an example.

### Example 1

Find the function $f(t)$ which satisfies

$$\frac{d^2 f(t)}{dt^2} + 2 \frac{df(t)}{dt} + f(t) = \sin t \tag{9}$$

for $t \geq 0$ and which at $t = 0$ has the properties

$$f(0) = 1, \qquad f'(0) = 0. \tag{10}$$

**Solution.** We begin by taking the transform of the Eq. (9). Thanks to the linearity property (xiv) we have

$$\mathscr{L}\{f''(t)\} + 2\mathscr{L}\{f'(t)\} + \mathscr{L}\{f(t)\} = \mathscr{L}\{\sin t\}.$$

Using Eq. (8) and the initial conditions (10) we find

$$\mathscr{L}\{f'(t)\} = s\mathscr{L}\{f(t)\} - 1,$$
$$\mathscr{L}\{f''(t)\} = s^2 \mathscr{L}\{f(t)\} - s \cdot 1 - 0.$$

Thus our equation is transformed to

$$(s^2 + 2s + 1)\mathscr{L}\{f(t)\} - s - 2 = \mathscr{L}\{\sin t\},$$

or, from entry (iv) of the table,

$$\mathscr{L}\{f(t)\} = \frac{s + 2}{s^2 + 2s + 1} + \frac{1}{(s^2 + 2s + 1)(s^2 + 1)}.$$

Writing the first term on the right as

$$\frac{s + 2}{s^2 + 2s + 1} = \frac{s + 1}{(s + 1)^2} + \frac{1}{(s + 1)^2} = \frac{1}{s + 1} + \frac{1}{(s + 1)^2},$$

we find that, according to entries (i) and (ix), it is the Laplace transform of the function

$$e^{-t} + te^{-t}.$$

To analyze the second term, we use partial fractions to express

$$\frac{1}{(s^2 + 2s + 1)(s^2 + 1)} = \frac{1}{2}\frac{1}{s+1} + \frac{1}{2}\frac{1}{(s+1)^2} - \frac{1}{2}\frac{s}{s^2 + 1},$$

which is the Laplace transform of

$$\frac{1}{2}e^{-t} + \frac{1}{2}te^{-t} - \frac{1}{2}\cos t$$

[see entries (i), (ix), and (iii)].

Hence the solution is

$$f(t) = e^{-t} + te^{-t} + \frac{1}{2}e^{-t} + \frac{1}{2}te^{-t} - \frac{1}{2}\cos t$$

$$= \frac{3}{2}e^{-t} + \frac{3}{2}te^{-t} - \frac{1}{2}\cos t,$$

which can be directly verified. ∎

In the example we found it fairly easy to solve for the Laplace transform of the solution; to find the solution itself we had to invert the transform. Often, as illustrated, this can be done by referring to a table of Laplace transforms. However, since this transform was derived from the Fourier transform, which has an inversion formula, we suspect that a formula also exists for the inverse Laplace transform. To see this, recall that by Theorem 4, page 391, the Fourier inversion formula for a continuously differentiable, integrable $F$ is

$$F(t) = \int_{-\infty}^{\infty} G(\omega)e^{i\omega t}\, d\omega.$$

Recall, also, that the Laplace transform $\mathscr{L}\{F\}$ was expressed in terms of the Fourier transform by formula (4), or, equivalently,

$$G(\omega) = \frac{1}{2\pi}\,\mathscr{L}\{F\}(i\omega).$$

Hence we have immediately

$$F(t) = \frac{1}{2\pi}\int_{-\infty}^{\infty} \mathscr{L}\{F\}(i\omega)e^{i\omega t}\, d\omega$$

for such functions. This formula is often written (substituting $-is$ for $\omega$) as

$$F(t) = \frac{1}{2\pi i}\int_{-i\infty}^{i\infty} \mathscr{L}\{F\}(s)e^{st}\, ds, \tag{11}$$

with the obvious interpretation of these imaginary limits of integration.

We would like to generalize formula (11) to cover *nonintegrable* functions whose Laplace transforms are defined only for Re(*s*) sufficiently large. This is easy to achieve if we can find a positive number *a* sufficiently large so that $F(t)e^{-at}$ is integrable. Then we write the inversion formula (11) for the function $F(t)e^{-at}$:

$$F(t)e^{-at} = \frac{1}{2\pi i} \int_{-i\infty}^{i\infty} \mathcal{L}\{F(t)e^{-at}\}(s)e^{st}\, ds. \qquad (12)$$

Inserting entry (xiii) of the table in Eq. (12) we multiply by $e^{at}$ to derive

$$F(t) = \frac{1}{2\pi i} \int_{-i\infty}^{i\infty} \mathcal{L}\{F\}(s + a)e^{(s+a)t}\, ds, \qquad (13)$$

which we interpret as the so-called *Bromwich integral*

$$F(t) = \frac{1}{2\pi i} \int_{a-i\infty}^{a+i\infty} \mathcal{L}\{F\}(s)e^{st}\, ds.$$

A rigorous analysis produces the following generalization:

---

**Theorem 6.**   Suppose that $F(t)$ is piecewise smooth on every finite interval $[0, b]$ and that $|F(t)|$ is bounded by $Me^{\alpha t}$ for $t \geq 0$. Then $\mathcal{L}\{F\}(s)$ exists for Re(*s*) > α, and for all $t > 0$ and any $a > \alpha$,

$$\frac{F(t+) + F(t-)}{2} = \frac{1}{2\pi i} \int_{a-i\infty}^{a+i\infty} \mathcal{L}\{F\}(s)e^{st}\, ds.$$

---

**Example 2**

Find the piecewise smooth function whose Laplace transform is

$$\frac{1}{s^4 - 1}.$$

**Solution.**   It is possible to employ partial fractions and the transform table to solve this problem, but we shall illustrate the use of the inversion formula. Observe that this function is certainly analytic for Re(*s*) > 1. To get the inverse transform, let us evaluate the integral

$$I := \int_{a-i\infty}^{a+i\infty} \frac{1}{s^4 - 1} e^{st}\, ds \qquad (t > 0)$$

with, say, $a = 2$. This can be done by residue theory. $I$ is the limit, as $\rho \to \infty$, of the contour integral

$$I_\rho := \int_{\gamma_\rho} \frac{e^{zt}}{z^4 - 1}\, dz,$$

where $\gamma_\rho$ is the vertical segment from $2 - i\rho$ to $2 + i\rho$. For $t > 0$ we close the contour with the half-circle $C_\rho : z = 2 + \rho e^{i\theta}$, $\pi/2 \leq \theta \leq 3\pi/2$; see Fig. 8.9(a). The

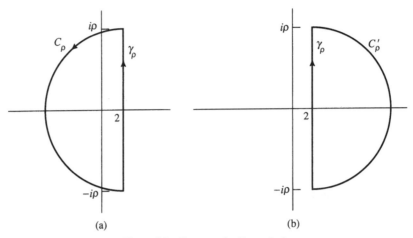

**Figure 8.9**   Contours for Example 2.

integral over $C_\rho$ is bounded by

$$\max_{\pi/2 \le \theta \le 3\pi/2} \frac{\left|e^{(2+\rho\cos\theta)t}e^{i\rho t \sin\theta}\right|}{(\rho-2)^4-1}\,\pi\rho = \frac{e^{2t}\pi\rho}{(\rho-2)^4-1},$$

which goes to zero as $\rho \to \infty$.

Now the integrand has four simple poles at $\pm 1, \pm i$, all of which eventually lie inside the semicircular contour of Fig. 8.9(a); in fact,

$$\frac{e^{zt}}{z^4-1} = \frac{e^{zt}}{(z-1)(z+1)(z-i)(z+i)}.$$

Hence

$$I = \lim_{\rho\to\infty} I_\rho = 2\pi i[\text{Res}(1)+\text{Res}(-1)+\text{Res}(i)+\text{Res}(-i)]$$

$$= 2\pi i\left[\frac{e^t}{2(1-i)(1+i)} + \frac{e^{-t}}{(-2)(-1-i)(-1+i)}\right.$$

$$\left. + \frac{e^{it}}{(i-1)(i+1)(2i)} + \frac{e^{-it}}{(-i-1)(-i+1)(-2i)}\right]$$

$$= \pi i \sinh t - \pi i \sin t.$$

So the inverse transform is

$$F(t) = \frac{1}{2\pi i}I = \frac{\sinh t - \sin t}{2}, \qquad t > 0.$$

For $t \le 0$ we close the contour as in Fig. 8.9(b); the integral over $C'_\rho$ is characterized by $-\pi/2 \le \theta \le \pi/2$, and for negative $t$ it goes to zero. Since this contour encloses no singularities, we confirm $F(t) = 0$ for $t \le 0$. ∎

## EXERCISES 8.3

1. Compute the Laplace transforms of the following functions.
   (a) $F(t) = 3\cos 2t - 8e^{-2t}$

   (b) $F(t) = 2 - e^{4t}\sin \pi t$

   (c) $F(t) = \begin{cases} 1, & t < 1 \\ 0, & t \geq 1 \end{cases}$

   (d) $F(t) = \begin{cases} 0, & t < 1 \\ 1, & 1 \leq t \leq 2 \\ 0, & t > 2 \end{cases}$

   (e) $F(t) = \sin^2 t$

   (f) $F(t) = \dfrac{1}{\sqrt{t}}$

   [HINT: In (f) let $\chi = \sqrt{st}$ and use the fact that $\int_0^\infty e^{-\chi^2}\,d\chi = \tfrac{1}{2}\sqrt{\pi}$.]

2. Prove Theorem 5.

3. Find the inverse transform of the following functions.
   (a) $\dfrac{1}{s^2 + 4}$

   (b) $\dfrac{4}{(s-1)^2}$

   (c) $\dfrac{s+1}{s^2 + 4s + 4}$

   (d) $\dfrac{1}{s^3 + 3s^2 + 2s}$

   (e) $\dfrac{s+3}{s^2 + 4s + 7}$

4. The effect of a time delay in a physical system is described mathematically by replacing a function $f(t)$ by the *delayed function*

$$f_\tau(t) := \begin{cases} 0, & 0 \leq t < \tau, \\ f(t - \tau), & \tau \leq t < \infty. \end{cases}$$

   Show that $\mathscr{L}\{f_\tau(t)\} = e^{-\tau s}\mathscr{L}\{f(t)\}$. [Compare Prob. 1(c) and (d).]

5. Use the Laplace transform to solve the following initial-value problems.
   (a) $\dfrac{df}{dt} - f = e^{3t}, \qquad f(0) = 3$

   (b) $\dfrac{d^2f}{dt^2} - 5\dfrac{df}{dt} + 6f = 0, \qquad f(0) = 1, f'(0) = -1$

   (c) $\dfrac{d^2f}{dt^2} - \dfrac{df}{dt} - 2f = e^{-t}\sin 2t, \qquad f(0) = 0, f'(0) = 2$

   (d) $\dfrac{d^2f}{dt^2} - 3\dfrac{df}{dt} + 2f = \begin{cases} 0, & 0 \leq t < 3 \\ 1, & 3 \leq t \leq 6 \\ 0, & t > 6 \end{cases}, \qquad f(0) = 0, f'(0) = 0$

   [HINT: See Prob. 4]

6. Verify the inversion formula for the following functions.
   (a) $F(t) = e^{-t}$

   (b) $F(t) \equiv 1$

7. In control theory the differential equation

$$a_n\frac{d^nf}{dt^n} + a_{n-1}\frac{d^{n-1}f}{dt^{n-1}} + \cdots + a_1\frac{df}{dt} + a_0 f = u$$

   is interpreted as a relation between the (known) input $u(t)$ and the (unknown) output $f(t)$.
   (a) Show that the Laplace transforms $U(s)$ and $F(s)$ of the input and output are related by

$$F(s) = \frac{U(s)}{a_n s^n + \cdots + a_1 s + a_0} + \frac{P(s)}{a_n s^n + \cdots + a_1 s + a_0},$$

where $P(s)$ is a polynomial in $s$ whose coefficients depend on the $a_i$ and the initial values of $f(t)$ and its derivatives. The coefficient of $U(s)$ in this expression is called the *transfer function* of the system.

**(b)** Show that every solution of the equation with $u(t) \equiv 0$ will go to zero as $t \to +\infty$ if all the poles of the transfer function are simple and lie in the left half-plane. (In other words, the system is *stable*.)

**8.** Identify the transfer functions and determine the stability of the differential equations in Prob. 5.

**9.** Consider the mass-spring system shown in Fig. 8.10. Each spring has the same natural (unstretched) length $L$, but when it is compressed or elongated it exerts a force proportional to the amount of compression or elongation (Hooke's law); the constant of proportionality is denoted by $K$. If we let $x$ and $y$ be the respective displacements of the masses $m_1$ and $m_2$ from equilibrium, we have the situation depicted in Fig. 8.11 (for $x$ and $y$ both positive.)

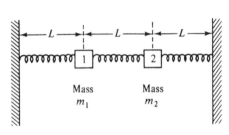

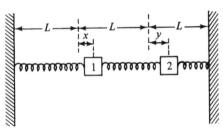

**Figure 8.10**  Mass-spring system unstretched.          **Figure 8.11**  Mass-spring system stretched.

**(a)** By writing Newton's law (force equals mass times acceleration) for each mass, derive the equations of motion

$$m_1 \frac{d^2x}{dt^2} = -Kx - K(x - y),$$

$$m_2 \frac{d^2y}{dt^2} = -Ky + K(x - y).$$

**(b)** Setting $K = 1$, $m_1 = m_2 = 1$, use Laplace transforms to find the solutions $x(t)$, $y(t)$ of these equations if the masses are released from rest (zero velocity) with each of the following initial displacements:

(i) $x(0) = 1$, $y(0) = -1$.

(ii) $x(0) = 1$, $y(0) = 1$.

(iii) $x(0) = 1$, $y(0) = 0$.

**(c)** For which initial conditions in part (b) is the *system's* response periodic [note that this requires *both* $x(t + T) = x(t)$ *and* $y(t + T) = y(t)$]? Can you visualize these responses? (They are called *normal modes* for the system.)

## 8.4 THE z-TRANSFORM

As we have seen, the application of transforms to various families of functions can often bring about certain computational advantages—e.g., the replacement of differentiation by multiplication. For continuous functions on a finite interval, or periodic functions, we use the Fourier series; if the interval is semi-infinite, we use the Laplace transform; and the Fourier transform is employed when the interval comprises the whole real line.

Often in practice we encounter "functions" that are *discrete* data structures. For example, when a continuous function $f(t)$ is measured in a laboratory it is sampled at a discrete set of points $\{t_j\}$. The transform tool that facilitates the mathematical manipulation of such discrete data streams is known as the *z-transform*.

Let us denote a discrete sequence of numbers by $a(n)$; we assume $n$ to take integer values from $-\infty$ to $\infty$, and we allow the possibility that $a(n)$ is complex. As examples consider

$$a(0)$$
$$\downarrow$$
$$\ldots, \tfrac{1}{16}, \tfrac{1}{8}, \tfrac{1}{4}, \tfrac{1}{2}, 1, \tfrac{1}{2}, \tfrac{1}{4}, \tfrac{1}{8}, \tfrac{1}{16}, \ldots \tag{1}$$

(sampled values of $f(x) = 2^{-|x|}$ at integers $x = n$);

$$a(0)$$
$$\downarrow$$
$$\ldots, -1, 1, -1, 1, -1, 1, -1, 1, \ldots \tag{2}$$

(sampled values of $\cos \pi x$);

$$a(0)$$
$$\downarrow$$
$$\ldots, 0, 0, 0, 3, 1, 4, 1, 5, 9, \ldots \tag{3}$$

(digits in decimal representation of $\pi$);

$$a(1976)$$
$$\downarrow$$
$$\ldots, 0, 0, 0, 1.2\text{K}, 2.5\text{K}, 4\text{K}, 10\text{K}, 12\text{K}, 12\text{K}, 13\text{K}, 12\text{K}, 3\text{K}, 0, 0, \ldots \tag{4}$$

(annual royalties from sales of a book first published in 1976).

The z-transform of the sequence $a(n)$ is defined as the sum of the series

$$A(z) := \sum_{n=-\infty}^{\infty} a(n)z^{-n}$$
$$= \cdots + a(-2)z^2 + a(-1)z + a(0) + a(1)z^{-1} + a(2)z^{-2} + \cdots \tag{5}$$

at all points where (5) converges. Note the exponent for $z$ is *minus n*.

For the sequence $a(n) = 2^{-|n|}$ in (1) the z-transform can be reorganized as follows:

$$\sum_{n=-\infty}^{\infty} 2^{-|n|} z^{-n} = \sum_{n=-\infty}^{0} 2^n z^{-n} + \sum_{n=1}^{\infty} 2^{-n} z^{-n}$$

$$= \sum_{n=0}^{\infty} \left(\frac{z}{2}\right)^n + \sum_{n=1}^{\infty} \left(\frac{1}{2z}\right)^n. \tag{6}$$

Both "sums" are geometric series. The first converges to $1/(1 - z/2)$ for $|z| < 2$, whereas the second converges to

$$\sum_{n=1}^{\infty} \left(\frac{1}{2z}\right)^n = \left(\frac{1}{2z}\right) \sum_{n=0}^{\infty} \left(\frac{1}{2z}\right)^n = \frac{\dfrac{1}{2z}}{1 - \dfrac{1}{2z}} = \frac{1}{2z - 1}$$

for $|z| > \frac{1}{2}$. Thus, in the common annulus of convergence $\frac{1}{2} < |z| < 2$, the z-transform of the sequence $2^{-|n|}$ is the analytic function

$$A(z) = \frac{1}{1 - z/2} + \frac{1}{2z - 1} = \frac{-3z}{2(z - 2)(z - \frac{1}{2})}. \tag{7}$$

Clearly the z-transform is the Laurent series of $A(z)$ in this annulus.

The z-transform for the oscillating sequence (2) has the form $\sum_{n=-\infty}^{\infty} (-1)^n z^{-n}$; the sum for $n \geq 0$ converges for $|z| > 1$, whereas the sum for $n < 0$ converges only for $|z| < 1$. These regions are disjoint, so the z-transform converges nowhere.

Since the negatively indexed elements of the "pi" sequence (3) are all zero, the corresponding portion of the series converges for all $z$. The terms of the positively indexed portion of the series are bounded by $9|z|^{-n}$ and thus by the M-test (Theorem 13, Sec. 5.4) this subseries converges for $|z| > 1$, which is then the common region of convergence and the domain of definition of the z-transform.

The sequence (4) has only a finite number of nonzero terms, so its z-transform is simply a polynomial in $1/z$ converging for all values of $z \neq 0$.

From these examples we can see the general nature of the z-transform of a sequence; it is an analytic function defined by the Laurent series whose coefficients are the terms of the sequence taken in reverse order (since $a(n)$ multiplies $z^{-n}$). From the convergence theory surveyed in Sec. 5.4 we deduce that the positively indexed portion of the series converges for

$$|z| > \limsup_{n>0} \sqrt[n]{|a(n)|},$$

and the negatively indexed portion converges for

$$|z| < \frac{1}{\limsup_{n>0} \sqrt[n]{|a(-n)|}}.$$

The z-transform is thus well defined for $z$ in the annulus

$$\limsup_{n>0} \sqrt[n]{|a(n)|} < |z| < \frac{1}{\limsup_{n>0} \sqrt[n]{|a(-n)|}}, \tag{8}$$

if this set is nonempty. The transform is analytic in this annulus, and the Laurent series enjoys the usual properties of termwise differentiation, integration, and multiplication.

If $a(n) = 0$ for $n < 0$ [as in (3)], the sequence is said to be *causal*. From (8) we see that the z-transform of a causal sequence converges *outside a circle* (i.e., the outer radius of the annulus is infinite).

A given analytic function can be the z-transform of more than one sequence, since its Laurent series representation is not unique (it depends on the region of convergence). The computations in Example 2 of Sec. 5.5, page 216, show that the function $1/[(z - 1)(z - 2)]$ is the transform of each of the following sequences:

$$a(n) = \begin{cases} 1 - 2^{n-1}, & n \leq 0 \\ 0, & n > 0 \end{cases} \quad \text{for } |z| < 1;$$

$$a(n) = \begin{cases} -2^{n-1}, & n \leq 0 \\ -1, & n > 0 \end{cases} \quad \text{for } 1 < |z| < 2;$$

$$a(n) = \begin{cases} 0, & n < 1 \\ 2^{n-1} - 1, & n \geq 1 \end{cases} \quad \text{for } |z| > 2.$$

The third of these sequences is causal; in general any Laurent series converging in the exterior of a circle is the z-transform of a causal sequence.

When the z-transform of a sequence can be written in closed form, it provides a very compact representation of the sequence. Also, as we shall see, it facilitates the solution of recursion relations, or "difference equations," involving sequences. The tools for recovering a sequence from its z-transform are precisely the tools for constructing Laurent series, which we explored in Sec. 5.5; one employs Maclaurin series such as the geometric series, partial fractions, etc. In this regard, the following version of Theorem 14, page 211, of that section can be interpreted as an inverse z-transform formula:

---

**Theorem 7.**   Let $A(z)$ be the z-transform of the sequence $\{a(n): -\infty < n < \infty\}$ in the annulus $a < |z| < b$. Then

$$a(n) = \frac{1}{2\pi i} \oint_\Gamma A(\zeta)\zeta^{n-1} \, d\zeta, \qquad (n = 0, \pm 1, \pm 2, \ldots)$$

where $\Gamma$ is any positively oriented simple closed contour lying in the annulus and encircling the origin.

---

The key to most applications of the z-transform is the following property.

---

**Theorem 8.**   Let $A(z)$ be the z-transform of the sequence $\{a(n): -\infty < n < \infty\}$ in the annulus $a < |z| < b$. Then the corresponding z-transform of the shifted sequence $\{b(n) = a(n + 1): -\infty < n < \infty\}$ is given by $zA(z)$. More generally, the z-transform of the sequence $\{c(n) = a(n + N): -\infty < n < \infty\}$ equals $z^N A(z)$ for any $N$ (positive or negative).

---

The proof is transparent: the z-transform of $b(n)$ is

$$\sum_{n=-\infty}^{\infty} b(n)z^{-n} = \sum_{n=-\infty}^{\infty} a(n + 1)z^{-n} = z \sum_{n=-\infty}^{\infty} a(n + 1)z^{-(n+1)} = zA(z),$$

and the generalization follows easily.

Some examples will demonstrate how this property is exploited in the solution of difference equations.

## Example 1

Let $a(n)$ represent the balance in a savings account at the beginning of month $n$. Starting from month 1 a monthly deposit of $t$ dollars is made and compound interest is accrued at the rate $r \cdot 100$ percent (per month). If the account holds $P$ dollars at the beginning of month 0 and no monies are withdrawn, develop the formula for $a(n)$.

**Solution.**    The balance in the account on successive months has the pattern

$$\vdots$$
$$a(-2) = 0$$
$$a(-1) = 0$$
$$a(0) = P$$
$$a(1) = P(1 + r) + t$$
$$a(2) = [P(1 + r) + t](1 + r) + t$$
$$\vdots$$

From this display we can formulate the difference equation relating the balance on successive months:

$$a(n + 1) = [1 + r]a(n) + P(n) + t(n), \tag{9}$$

where

$$P(n) := \begin{cases} P, & n = -1 \\ 0, & \text{otherwise}, \end{cases} \quad \text{and} \quad t(n) := \begin{cases} t, & n \ge 0, \\ 0, & n < 0. \end{cases}$$

Clearly $a(n) = 0$ for $n < 0$, so $a(n)$ is causal. Thus we assume that $a(n)$ has a $z$-transform $A(z)$ for $z$ sufficiently large. Taking the $z$-transform of both sides of Eq. (9), we use Theorem 8 and the fact that

$$\sum_{n=0}^{\infty} z^{-n} = \frac{1}{1 - 1/z} = \frac{z}{z - 1} \quad \text{for } |z| > 1$$

to obtain

$$z A(z) = [1 + r]A(z) + Pz + \frac{tz}{z - 1},$$

or

$$A(z) = \frac{Pz^2 + (t - P)z}{(z - 1)(z - 1 - r)} = z\frac{Pz + (t - P)}{(z - 1)(z - 1 - r)} \quad (|z| \text{ "large"}).$$

With a partial fraction decomposition this becomes

$$A(z) = z\left(\frac{P + t/r}{z - 1 - r} - \frac{t/r}{z - 1}\right),$$

which has the following Laurent series expansion for large $|z|$ (for $|z| > 1 + r$, in fact):

$$A(z) = \left(P + \frac{t}{r}\right) \frac{1}{1 - (1 + r)/z} - \frac{t}{r} \frac{1}{1 - 1/z}$$

$$= \left(P + \frac{t}{r}\right) \sum_{n=0}^{\infty} (1 + r)^n z^{-n} - \frac{t}{r} \sum_{n=0}^{\infty} z^{-n}.$$

The corresponding sequence is therefore given by

$$a(n) = \begin{cases} \left(P + \dfrac{t}{r}\right)(1 + r)^n - \dfrac{t}{r}, & n \geq 0, \\ 0, & n < 0. \end{cases} \quad \blacksquare$$

The solution of difference equations with initial conditions is most conveniently accomplished with the *unilateral z-transform* $A^+(z)$, which omits the negatively indexed terms in the series (5):

$$A^+(z) := \sum_{n=0}^{\infty} a(n) z^{-n}. \tag{10}$$

Clearly if $a(n)$ is causal, then $A^+(z) = A(z)$, and the region of convergence of $A^+(z)$ is the exterior of a circle ($|z| > \limsup\limits_{n>0} \sqrt[n]{|a(n)|}$). The shifting property for the unilateral z-transform is similar to that described in Theorem 8, modified along the lines of the Laplace transform formula for derivatives [Eq. (8), Sec. 8.3, page 402].

---

**Theorem 9.** Let $A^+(z)$ be the unilateral z-transform of the sequence $\{a(n): -\infty < n < \infty\}$ in the region $a < |z|$. Then the corresponding unilateral z-transform of the shifted sequence $\{b(n) = a(n + 1): -\infty < n < \infty\}$ is given by $z[A^+(z) - a(0)]$. More generally, the unilateral z-transform of the sequence $\{c(n) = a(n + N): -\infty < n < \infty\}$ equals

$$z^N[A^+(z) - a(0) - a(1)z^{-1} - a(2)z^{-2} - \cdots - a(N - 1)z^{-(N-1)}]$$

for any positive $N$.

---

Again the proof is easy. The unilateral z-transform of $b(n)$ is

$$\sum_{n=0}^{\infty} b(n)z^{-n} = \sum_{n=0}^{\infty} a(n + 1)z^{-n} = z \sum_{n=0}^{\infty} a(n + 1)z^{-(n+1)}$$

$$= z \sum_{m=0}^{\infty} a(m)z^{-m} - za(0).$$

$$= z[A^+(z) - a(0)].$$

The generalization is left to the reader.

**Example 2**

Suppose the sequence $a(n)$ satisfies the difference equation

$$a(n + 2) - 3a(n + 1) + 2a(n) = 0 \qquad (11)$$

for $n \geq 0$ and that $a(0) = 1$ and $a(1) = -1$. Find a formula for $a(n)$ for $n \geq 0$.

**Solution.**   Since (11) holds for all $n \geq 0$ we are justified in applying the unilateral $z$-transform. Employing Theorem 9 we derive

$$z^2[A^+(z) - 1 - (-1)z^{-1}] - 3z[A^+(z) - 1] + 2A^+(z) = 0$$

or

$$A^+(z) = \frac{z^2 - 4z}{z^2 - 3z + 2} = z\frac{z - 4}{(z - 1)(z - 2)} = z\left(\frac{3}{z - 1} + \frac{-2}{z - 2}\right)$$

$$= 3\frac{1}{1 - 1/z} - 2\frac{1}{1 - 2/z}$$

$$= 3\sum_{n=0}^{\infty} z^{-n} - 2\sum_{n=0}^{\infty} 2^n z^{-n}.$$

Thus

$$a(n) = 3 - 2^{n+1}, \qquad \text{for } n \geq 0. \quad \blacksquare$$

Further properties of the $z$-transform and its engineering applications are developed in the references.

## EXERCISES 8.4

1. Show that the $z$-transform is a linear operator; i.e., if $A(z)$ and $B(z)$ denote the $z$-transforms of $\{a(n)\}$ and $\{b(n)\}$, respectively, then the transform of $\{\alpha a(n) + \beta b(n)\}$ is $\alpha A(z) + \beta B(z)$, in the common region of convergence.

2. If $A(z)$ is the $z$-transform of $\{a(n)\}$, show that the $z$-transform of the "linearly weighted" sequence $\{na(n)\}$ is $-zA'(z)$. Show that the annulus of convergence is unchanged. [HINT: $\lim_{n \to \infty} \sqrt[n]{n} = 1$.]

3. If $A(z)$ is the $z$-transform of $\{a(n)\}$, show that the $z$-transform of the "exponentially weighted" sequence $\{\alpha^n a(n)\}$ is $A(z/\alpha)$. How is the new annulus of convergence related to the old?

4. Verify the entries in the following table of *causal* $z$-transforms: for $n \geq 0$,

   (a) $a(n) = \begin{cases} 1, & n = 0 \\ 0, & n > 0 \end{cases}$    $A(z) = 1$

   (b) $a(n) = 1$    $A(z) = \dfrac{z}{z - 1}$

   (c) $a(n) = n$    $A(z) = \dfrac{z}{(z - 1)^2}$

   (d) $a(n) = \alpha^n$    $A(z) = \dfrac{z}{z - \alpha}$

**(e)** $a(n) = \sin n\omega$ $\qquad\qquad A(z) = \dfrac{z \sin \omega}{z^2 - 2z \cos \omega + 1}$

**(f)** $a(n) = \cos n\omega$ $\qquad\qquad A(z) = \dfrac{z(z - \cos \omega)}{z^2 - 2z \cos \omega + 1}$

**5.** Find inverse $z$-transforms for the following functions in the indicated annuli:

**(a)** $A(z) = \dfrac{1}{1 + 1/(3z)}, \qquad |z| < \dfrac{1}{3}$ $\qquad$ **(b)** $A(z) = \dfrac{1}{1 + 1/(3z)}, \qquad |z| > \dfrac{1}{3}$

**(c)** $A(z) = \dfrac{z^4}{z + 2}, \qquad |z| < 2$ $\qquad\qquad$ **(d)** $A(z) = \dfrac{z^4}{z + 2}, \qquad |z| > 2$

**(e)** $A(z) = \dfrac{z + 2}{2z^2 - 7z + 3}, \qquad \dfrac{1}{2} < |z| < 3$

**(f)** $A(z) = \dfrac{1 - 1/(2z)}{1 + 3/(4z) + 1/(8z^2)}, \qquad |z| > \dfrac{1}{2}$

**(g)** $A(z) = \dfrac{z}{(z - \frac{1}{2})(z - 1)^2}, \qquad |z| > 1$

**(h)** $A(z) = \dfrac{1 - \alpha/z}{\alpha - 1/z}, \qquad |z| > \dfrac{1}{|\alpha|}$

**6.** If $A(z)$ is the $z$-transform of a causal sequence $\{a(n)\}$, show that $a(0) = \lim_{z \to \infty} A(z)$.

**7.** Use unilateral $z$-transforms to solve the following difference equations.
   **(a)** $a(n + 1) = (0.5)a(n), \qquad a(0) = 2$
   **(b)** $a(n + 1) + 2a(n) = 1, \qquad a(0) = 1$
   **(c)** $a(n + 2) - 5a(n + 1) + 6a(n) = 1, \qquad a(0) = 2, \ a(1) = 3$

**8.** Derive the formula for the unilateral $z$-transform of the backward-shifted sequence $a(n - N), N > 0$.

## 8.5 CAUCHY INTEGRALS AND THE HILBERT TRANSFORM

An integral of the form

$$\int_\Gamma \frac{f(\zeta)}{\zeta - z} d\zeta$$

is known as a *Cauchy integral*. The study of Cauchy integrals is quite provocative and rewarding, and in this section we shall explore some of the theoretical and practical aspects of such forms.

If $\Gamma$ is a simple smooth closed curve as in Fig. 8.12, page 416, and $f$ is analytic inside and on $\Gamma$, then Cauchy's formula and theorem tell us that

$$\oint_\Gamma \frac{f(\zeta)}{\zeta - z} d\zeta = \begin{cases} 2\pi i f(z) & \text{if } z \text{ lies inside } \Gamma, \\ 0 & \text{if } z \text{ lies outside } \Gamma. \end{cases} \tag{1}$$

The question naturally arises as to how the values of the integral evolve as the point $z$ crosses $\Gamma$. To explore this, consider the indented contour $\Gamma'_\varepsilon$ in Fig. 8.13. The point $z_0$ lies on the original contour $\Gamma$, but since it falls inside $\Gamma'_\varepsilon$, Cauchy's formula tells us that

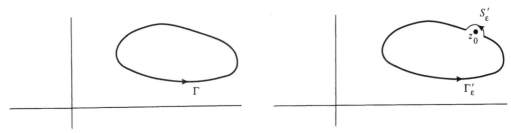

**Figure 8.12**    Contour for Cauchy integral.          **Figure 8.13**    Indented contour.

$$\oint_{\Gamma'_\varepsilon} \frac{f(\zeta)}{\zeta - z_0} \, d\zeta = 2\pi i f(z_0). \tag{2}$$

Here we have presumed that the radius $\varepsilon$ of the semicircular indentation $S'_\varepsilon$ is sufficiently small that $\Gamma'_\varepsilon$ remains within the domain of analyticity for $f$. Now as $\varepsilon$ goes to zero, the contribution from the semicircle $S'_\varepsilon$ to the integral along $\Gamma'_\varepsilon$ approaches $\pi i f(z_0)$ (Lemma 4, Sec. 6.5, page 275). This is half the value shown in Eq. (2); therefore, the remaining $\pi i f(z_0)$ must come from the rest of the contour, which is a facsimile of $\Gamma$ snipped "symmetrically" around $z_0$ in a manner generalizing the *principal value* concept (Sec. 6.5). To summarize: On the basis of Fig. 8.13 we express

$$\underbrace{\oint_{\Gamma'_\varepsilon} \frac{f(\zeta)}{\zeta - z_0} \, d\zeta}_{\substack{\| \\ 2\pi i f(z_0) \\ \text{(Cauchy's formula)}}} = \underbrace{\text{p.v.} \int_\Gamma \frac{f(\zeta)}{\zeta - z_0} \, d\zeta}_{} + \underbrace{\lim_{\varepsilon \to 0} \int_{S'_\varepsilon} \frac{f(\zeta)}{\zeta - z_0} \, d\zeta}_{\substack{\| \\ \pi i f(z_0) \\ \text{(Lemma 4)}}}, \tag{3}$$

from which we conclude

$$\text{p.v.} \int_\Gamma \frac{f(\zeta)}{\zeta - z_0} \, d\zeta = \pi i f(z_0). \tag{4}$$

Of course our Fig. 8.13 benignly sidesteps any topological complications; see Prob. 3, for example. We direct the reader to the references for a more rigorous statement and derivation of (4).

**Example 1**

Confirm Eq. (4) for the case when $\Gamma$ is the positively oriented unit circle centered around $z = 1$, $f(z) \equiv 1$, and $z_0 = 0$.

**Solution.**   Obviously $\int_\Gamma 1/(\zeta - z)\, d\zeta$ equals $2\pi i$ for $|z - 1| < 1$ and zero for $|z - 1| > 1$. For $z = z_0 = 0$ we refer to Fig. 8.14 to derive

$$\text{p.v.} \int_\Gamma \frac{1}{\zeta - 0} \, d\zeta = \lim_{z_1, z_2 \to 0} [\text{Log}\, z_1 - \text{Log}\, z_2]$$

with $|z_1| = |z_2|$ as shown. Thus we have

$$\text{p.v.} \int_\Gamma \frac{1}{\zeta - 0} \, d\zeta = \lim_{z_1, z_2 \to 0} [i\,\text{Arg}\, z_1 - i\,\text{Arg}\, z_2] = i\frac{\pi}{2} + i\frac{\pi}{2} = \pi i. \quad \blacksquare$$

**Figure 8.14**  Contour for Example 1.          **Figure 8.15**  Interior indentation.

What happens if we construct an indentation penetrating the *interior of* $\Gamma$, as in Fig. 8.15? Then the (clockwise-oriented) semicircle $S''_\varepsilon$ contributes *minus* $\pi i f(z_0)$ and the decomposition of the integral (3) takes the form

$$\oint_{\Gamma''_\varepsilon} \frac{f(\zeta)}{\zeta - z_0}\, d\zeta \;=\; \text{p.v.} \int_\Gamma \frac{f(\zeta)}{\zeta - z_0}\, d\zeta + \lim_{\varepsilon \to 0} \int_{S''_\varepsilon} \frac{f(\zeta)}{\zeta - z_0}\, d\zeta. \tag{5}$$

$$\begin{array}{ccc} \| & \| & \| \\ 0 & \pi i f(z_0) & -\pi i f(z_0) \\ \text{(Cauchy's theorem)} & [\text{Eq. (4)}] & \text{(Lemma 4)} \end{array}$$

From these considerations we can visualize what happens to the Cauchy integral (1) as the point $z$ crosses the contour $\Gamma$. In Fig. 8.16 the points $\{z_n^+ : n = 1, 2, 3, \ldots\}$ approach $z_0$ from the inside, and the Cauchy integrals equal $2\pi i f(z_n^+)$. Furthermore, if the contour is indented as in Fig. 8.13, these integrals are unchanged (by the deformation invariance theorem, page 136). They approach the limit $2\pi i f(z_0)$, and invoking Eqs. (3) and (4) we can attribute half of this limit to the principal value and the other half to the exterior indentation $S'_\varepsilon$.

Now for the sequence $\{z_n^-\}$ approaching $z_0$ from *outside* $\Gamma$ (Fig. 8.17), we use the interior indentation $S''_\varepsilon$ to argue that the contributions to the limit—zero—of the Cauchy integrals are $\pi i f(z_0)$ from the principal value and $-\pi i f(z_0)$ from the semicircle $S''_\varepsilon$. Thus as the point $z$ crosses the contour $\Gamma$, we can ascribe the jump in the Cauchy integral to the substitution of one indentation for the other; $S'_\varepsilon$ "opens the gate" and lets $z$ through; then $S''_\varepsilon$ closes the gate behind it. The difference between the interior and exterior limits is due to the opposing contributions of the semicircles:

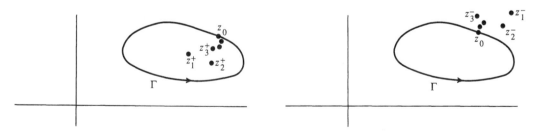

**Figure 8.16**  Approaching $z_0$ from inside $\Gamma$.          **Figure 8.17**  Approaching $z_0$ from outside $\Gamma$.

$$\lim_{z_n^+ \to z_0} \oint_\Gamma \frac{f(\zeta)}{\zeta - z_n^+} \, d\zeta - \lim_{z_n^- \to z_0} \oint_\Gamma \frac{f(\zeta)}{\zeta - z_n^-} \, d\zeta = \int_{S_\varepsilon'} \frac{f(\zeta)}{\zeta - z_0} \, d\zeta - \int_{S_\varepsilon''} \frac{f(\zeta)}{\zeta - z_0} \, d\zeta$$

$$= \pi i f(z_0) - [-\pi i f(z_0)]$$

$$= 2\pi i f(z_0). \tag{6}$$

Note that by similar accounting the *average* of the interior and exterior limits yields the principal value of the integral:

$$\lim_{z_n^+ \to z} \oint_\Gamma \frac{f(\zeta)}{\zeta - z_n^+} \, d\zeta + \lim_{z_n^- \to z} \oint_\Gamma \frac{f(\zeta)}{\zeta - z_n^-} \, d\zeta$$

$$= 2\left( \text{p.v.} \int_\Gamma \frac{f(\zeta)}{\zeta - z_0} \, d\zeta \right) + \int_{S_\varepsilon'} \frac{f(\zeta)}{\zeta - z_0} \, d\zeta + \int_{S_\varepsilon''} \frac{f(\zeta)}{\zeta - z_0} \, d\zeta$$

$$= 2\left( \text{p.v.} \int_\Gamma \frac{f(\zeta)}{\zeta - z_0} \, d\zeta \right). \tag{7}$$

If the contour $\Gamma$ is *not* closed we have no general theorem to tell us the values of $\int_\Gamma f(\zeta)/(\zeta - z) \, d\zeta$, but the argumentation motivated by Figs. 8.13 through 8.17 still validates Eqs. (6) and (7), as long as we interpret the sequence $\{z_n^+\}$ as approaching $z_0$ from the *left* of $\Gamma$ (as determined by its orientation) and $\{z_n^-\}$ as approaching from the right. The *Sokhotskyi-Plemelj formulas* (proved in the references) extend these considerations to more general (not necessarily analytic) functions and contours; they state that the difference between the limiting values of the Cauchy integral $\int_\Gamma f(\zeta)/(\zeta - z) \, d\zeta$ as $z$ approaches $z_0$ (on $\Gamma$) from the left and from the right is always equal to $2\pi i f(z_0)$, whereas their average equals the principal value.

A particularly useful identity results when $\Gamma$ encloses a half-plane, as in Fig. 8.18. If $f(z)$ is analytic in, say, the upper half-plane and goes to zero at infinity so rapidly that the contribution of the semicircle $S_R$ vanishes as $R \to \infty$, then Eq. (4) takes the form

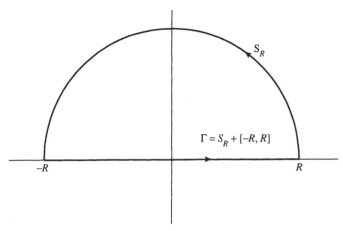

**Figure 8.18**   Contour enclosing half-plane (as $R \to \infty$).

$$\text{p.v.} \int_{-\infty}^{\infty} \frac{f(\xi)}{\xi - x} \, d\xi = \pi i f(x). \tag{8}$$

The integral over $S_R$ will disappear, for instance, if $|f(z)| < K/|z|$ (by the usual estimates) or if $|f(z)| < K|e^{imz}|$ for positive $m$ (by Jordan's lemma).

Expressing $f(x + iy) = u(x, y) + iv(x, y)$ and separating (8) into its real and imaginary parts, we obtain

$$v(x, 0) = -\frac{1}{\pi} \text{p.v.} \int_{-\infty}^{\infty} \frac{u(\xi, 0)}{\xi - x} \, d\xi, \qquad u(x, 0) = \frac{1}{\pi} \text{p.v.} \int_{-\infty}^{\infty} \frac{v(\xi, 0)}{\xi - x} \, d\xi. \tag{9}$$

The first of these formulas motivates the following definition.

---

**Definition 2.**   The **Hilbert transform** of an arbitrary real-valued function $\phi(x)$ $(-\infty < x < \infty)$ is defined by

$$\psi(x) := -\frac{1}{\pi} \text{p.v.} \int_{-\infty}^{\infty} \frac{\phi(\xi)}{\xi - x} \, d\xi \tag{10}$$

(when the integral exists).

---

Our derivation shows that whenever $\phi(x)$ and $\psi(x)$ are a pair of functions such that the combination $\phi + i\psi$ can be extended as an analytic function in the upper half-plane, suitably "dying off" at infinity therein, then $\psi$ is the Hilbert transform of $\phi$. Thus the Hilbert transform of $\cos x$ is $\sin x$, since $\cos x + i \sin x = e^{iz}$ for $z = x$, and $|e^{i(x+iy)}| = e^{-y} \to 0$ as $y \to +\infty$. Note that the transform of $\sin x$ is *minus* $\cos x$ (because $\sin x - i \cos x = -ie^{iz}$ for $z = x$). This is consistent with the appearance of the signs in Eqs. (9) (the second of which may be regarded as the formula for the *inverse Hilbert transform*).

Clearly, a collection of Hilbert transforms can be generated by writing down analytic functions with the requisite properties in the upper half-plane and separating their real and imaginary parts on the $x$-axis. In this manner one derives (see Prob. 1) the accompanying table.

### TABLE OF HILBERT TRANSFORMS

| Function $\phi(x)$ | | Hilbert transform $\psi(x)$ |
|---|---|---|
| $\cos \omega x$ | $(\omega > 0)$ | $\sin \omega x$ |
| $\cos \omega x$ | $(\omega < 0)$ | $-\sin \omega x$ |
| $\sin \omega x$ | $(\omega > 0)$ | $-\cos \omega x$ |
| $\sin \omega x$ | $(\omega < 0)$ | $\cos \omega x$ |
| $\dfrac{a}{a^2 + x^2}$ | $(a > 0)$ | $\dfrac{x}{a^2 + x^2}$ |
| $\dfrac{\sin ax}{x}$ | $(a > 0)$ | $\dfrac{1 - \cos ax}{x}$ |

**Example 2**

Verify Eq. (10) for the first entry in the transform table.

**Solution.** We break up the Cauchy integral into two parts:

$$\text{p.v.} \int_{-\infty}^{\infty} \frac{\cos \omega \xi}{\xi - x} \, d\xi = \text{p.v.} \int_{-\infty}^{\infty} \frac{e^{i\omega \xi}}{2(\xi - x)} \, d\xi + \text{p.v.} \int_{-\infty}^{\infty} \frac{e^{-i\omega \xi}}{2(\xi - x)} \, d\xi.$$

To evaluate the first integral we close the contour with a semicircle and an indentation in the upper half $\xi$-plane, as shown in Fig. 8.19(a). Using Jordan's lemma, Lemma 4 from Sec. 6.5 (page 275), and Cauchy's theorem, we obtain

$$\text{p.v.} \int_{-\infty}^{\infty} \frac{e^{i\omega \xi}}{2(\xi - x)} \, d\xi = \pi i \, \text{Res}(\xi = x) = \frac{\pi i e^{i\omega x}}{2}.$$

The contour for the second integral is closed as in Fig. 8.19(b), and we have

$$\text{p.v.} \int_{-\infty}^{\infty} \frac{e^{-i\omega \xi}}{2(\xi - x)} \, d\xi = -\pi i \, \text{Res}(\xi = x) = -\frac{\pi i e^{-i\omega x}}{2}.$$

The sum of these equals

$$\text{p.v.} \int_{-\infty}^{\infty} \frac{\cos \omega \xi}{\xi - x} \, d\xi = \pi i \, \frac{e^{i\omega x} - e^{-i\omega x}}{2} = -\pi \sin \omega x$$

and dividing by $-\pi$, we confirm the entry. ∎

The reader may have observed that a methodology for calculating the Hilbert transform is already at hand: starting from the function $\phi(x)$, one could use Poisson's formula for the half-plane (Prob. 14, Exercises 4.7, page 176) to extend it as a suitable harmonic function $u(x, y)$ in the upper half-plane; then Theorem 25 of that section (page 170) establishes the harmonic conjugate $v(x, y)$, whose restriction to the x-axis becomes the transform $\psi(x)$. Eq. (10), then, accomplishes all this directly. It also affords the extension of the transform to a wider class of functions. These generalizations can be found in the references.

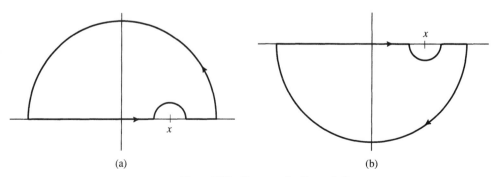

(a)                    (b)

**Figure 8.19**   Contours for Example 2.

Many applications of the Hilbert transform are based upon the way it interacts with the Fourier transform. Let us write the latter and its inversion formula as

$$\Phi(\omega) = \frac{1}{2\pi} \int_{-\infty}^{\infty} \phi(x)e^{-i\omega x}\, dx, \tag{11}$$

$$\phi(x) = \int_{-\infty}^{\infty} \Phi(\omega)e^{i\omega x}\, d\omega. \tag{12}$$

Now the Hilbert transform is a linear operation whose effect on cosines and sines has been determined. Therefore, if we adopt the point of view (espoused in Sec. 8.2) that Eq. (12) expresses $\phi$ as a superposition of sinusoids, then we rewrite (12) as

$$\phi(x) = \int_{-\infty}^{0} \Phi(\omega)[\cos \omega x + i \sin \omega x]\, d\omega$$
$$+ \int_{0}^{\infty} \Phi(\omega)[\cos \omega x + i \sin \omega x]\, d\omega \tag{13}$$

and take its Hilbert transform as follows:

$$\psi(x) = \int_{-\infty}^{0} \Phi(\omega)[-\sin \omega x + i \cos \omega x]\, d\omega + \int_{0}^{\infty} \Phi(\omega)[\sin \omega x - i \cos \omega x]\, d\omega$$
$$= i \int_{-\infty}^{0} \Phi(\omega)e^{i\omega x}\, d\omega + (-i) \int_{0}^{\infty} \Phi(\omega)e^{i\omega x}\, d\omega$$
$$= \int_{-\infty}^{0} \Phi(\omega)e^{i(\omega x + \pi/2)}\, d\omega + \int_{0}^{\infty} \Phi(\omega)e^{i(\omega x - \pi/2)}\, d\omega. \tag{14}$$

Comparing (14) with (13) we observe that positive frequencies are "phase-delayed" by $\pi/2$ radians with the Hilbert transform, and negative frequencies are phase-advanced by the same amount. The Hilbert transform is thus a 90° phase-shift operation.

This suggests an efficient numerical algorithm for the implementation of the Hilbert transform. First we use the fast Fourier transform algorithm to approximate $\Phi(\omega)$. Then we reverse the real and imaginary parts and change the sign of the former for $\omega < 0$ and the latter for $\omega > 0$. Finally, we take the inverse Fourier transform, again employing the FFT. This procedure is commonly used in signal-processing applications such as the following.

**Example 3**

The human ear can detect sound waves with frequencies up to a certain level $\Omega_0$; higher frequencies go unheard. (The frequency $\Omega_0$ is about 25,000 radians per second, or 4 kiloHertz.) Thus in practice when we perform a Fourier transform on a sound signal $h(t)$, we need to retain only the contributions for $|\omega|$ below $\Omega_0$:

$$h(t) = \int_{-\infty}^{\infty} H(\omega)e^{i\omega t}\, d\omega \approx \int_{-\Omega_0}^{\Omega_0} H(\omega)e^{i\omega t}\, d\omega.$$

In radio communications it is difficult to transmit at such low frequencies, so often the *amplitude-modulation* scheme is employed. For reasons to be seen, the *baseband* signal is first artificially biased so that $h(t) > 0$. Then the signal is electronically multiplied by a *carrier wave function* $\cos \Omega_1 t$ at a high frequency $\Omega_1$

(typically 1000 kiloHertz) so that the resulting Fourier transform resides in the high-frequency regions near $\pm\Omega_1$:

$$h(t)\cos\Omega_1 t \approx \int_{-\Omega_0}^{\Omega_0} H(\omega)e^{i\omega t}\cos\Omega_1 t\, d\omega$$

$$\approx \int_{-\Omega_0}^{\Omega_0} \frac{H(\omega)e^{i\omega t}[e^{i\Omega_1 t} + e^{-i\Omega_1 t}]}{2}\, d\omega$$

$$\approx \int_{\Omega_1-\Omega_0}^{\Omega_1+\Omega_0} \frac{H(\omega'-\Omega_1)e^{i\omega' t}}{2}\, d\omega' + \int_{-\Omega_1-\Omega_0}^{-\Omega_1+\Omega_0} \frac{H(\omega'+\Omega_1)e^{i\omega' t}}{2}\, d\omega'.$$

To extract $h(t)$ from the received signal $[h(t)\cos\Omega_1 t]$—i.e., to *demodulate* the signal—apparently we could merely divide by $\cos\Omega_1 t$; but this is not feasible because the receiver might not know exactly when the instant $t=0$ occurred in the sender's time frame. [Thus the receiver would effectively be dividing by $\cos(\Omega_1 t + \text{constant})$]. Use Hilbert transforms to design a better demodulation scheme.

**Solution.** The trick is to observe that when $|\omega| < \Omega_1$, the Hilbert transform of $(\cos\omega x\cos\Omega_1 x)$ is simply $(\cos\omega x\sin\Omega_1 x)$; after all, the function $f(z) = (\cos\omega z)e^{i\Omega_1 z}$ is analytic in the upper half-plane and satisfies the conditions at infinity there. Similarly, the transform of $(\sin\omega x\cos\Omega_1 x)$ is $(\sin\omega x\sin\Omega_1 x)$. By linearity, then, the Hilbert transform of $h(t)\cos\Omega_1 t$ is $h(t)\sin\Omega_1 t$, if we can neglect the Fourier transform of $h(t)$ for frequencies higher than $\Omega_1$. Thus by adding the squares of the modulated signal and its Hilbert transform and taking the square root, one demodulates the signal:

$$\sqrt{[h(t)\cos\Omega_1 t]^2 + [h(t)\sin\Omega_1 t]^2} = |h(t)| = h(t). \qquad \blacksquare$$

(In communications applications such as this the combination $\phi(x) + i\psi(x)$ is sometimes called the "analytic signal" associated with $\phi$.)

The Hilbert transform is also useful in other areas. Recall that in Sec. 3.4 we argued that *RLC* (resistor-inductor-capacitor) electrical circuits, when subjected to a sinusoidal external voltage, will eventually reach a steady state wherein all the internal voltages and currents oscillate sinusoidally at the same frequency as the driving voltage. In particular, we showed that for the circuit of Fig. 3.9 on page 45, the (complex) current $I_s$ resulting from the voltage $V_s(t) = Ae^{i\omega t}$ was given by

$$I_s = \frac{Ae^{i\omega t}}{R_{\text{eff}}},$$

where

$$R_{\text{eff}} = \frac{R/i\omega C}{R + 1/i\omega C} + i\omega L. \tag{15}$$

This "synchronous" behavior is characteristic of most closed (autonomous) physical systems, in that the steady-state response $y(t)$ to an input $e^{i\omega t}$ takes the form $k(\omega)e^{i\omega t}$; the *transfer function* $k(\omega)$ (in this case $1/R_{\text{eff}}$) almost always depends on the applied frequency.

An alternative description of the driver-response relationship for such physical systems is available from differential equation theory (see the references). Using a technique known traditionally as the *variation of parameters*, one can express the response $y(t)$ of the system as a *weighted sum* of the functional values of the input $u(t)$. The identity takes the form

$$y(t) = \int_{-\infty}^{\infty} G(t - \tau)u(\tau)\, d\tau. \tag{16}$$

$G(t - \tau)$, the *Green's function* for the system, thus measures the extent to which the values of $u$ at time $\tau$ affect the output at time $t$. $G$ is also called the *impulse response*.

Let's put these two observations together. If the input $u(t)$ is a sinusoid $e^{i\omega t}$, the transfer function description and the Green's function description of the output must agree. Therefore,

$$k(\omega)e^{i\omega t} = \int_{-\infty}^{\infty} G(t - \tau)e^{i\omega \tau}\, d\tau,$$

or

$$k(\omega) = \int_{-\infty}^{\infty} G(t - \tau)e^{-i\omega(t - \tau)}\, d\tau = \int_{-\infty}^{\infty} G(T)e^{-i\omega T}\, dT, \tag{17}$$

where $T = t - \tau$. Equation (17) has the form of an inverse Fourier transform, except for a missing factor of $2\pi$. Thus we can invert it and express the Green's function in terms of the transfer function:

$$G(T) = \frac{1}{2\pi} \int_{-\infty}^{\infty} k(\omega)e^{i\omega T}\, d\omega. \tag{18}$$

If the system is a true physical model (such as the *RLC* circuit), it must obey the *causality* principle: the value of the response at a given time $t$ cannot depend on the values of the input at *later* times $\tau$. Thus from Eq. (16) we see that for causal systems the Green's function $G(t - \tau)$ equals zero for $\tau > t$, and as a result Eq. (18) yields the equality

$$\int_{-\infty}^{\infty} k(\omega)e^{i\omega T}\, d\omega = 0 \qquad \text{for all} \quad T < 0. \tag{19}$$

As a criterion for causality, condition (19) is intriguing. For $T < 0$ the exponential factor, as a function of the complex variable $\omega$, decays in the *lower* half-plane. Thus Jordan's lemma (Sec. 6.4) would guarantee (19) if the analytic continuation of $k(\omega)$ into the lower half-plane turned out to be analytic there and bounded by (constant$/|\omega|$). In practice $k(\omega)$ is often a rational function (as in the case of the *RLC* circuit), and subject to a few assumptions about the nature of the physical system, condition (19) can be shown to be *necessary*, as well as sufficient, for causality.

### Example 4

Verify the causality condition (19) for the transfer function of the circuit in Fig. 3.9 on page 95.

**Solution.**  From Eq. (15) we find, with some algebra,

$$k(\omega) = \frac{1}{R_{\text{eff}}} = \frac{i\omega CR + 1}{-\omega^2 CRL + i\omega L + R},$$

whose poles are

$$\omega_{\pm} = \frac{i \pm \sqrt{-1 + 4CR^2/L}}{2CR}. \tag{20}$$

If $4CR^2/L > 1$, the radical is real and both poles lie in the upper half-plane. Otherwise, we have $0 < 4CR^2/L < 1$ and the radical is an imaginary number between 0 and $i$; thus the poles still stay in the upper half-plane. Therefore, $k(\omega)$ is analytic in the lower half-plane, and since it falls off like $|\omega|^{-1}$, the system is causal. ∎

How does the Hilbert transform come into play in this context? The causal transfer function $k(\omega)$ has all the properties that are necessary to validate identity (8); it goes to zero at infinity like $|\omega|^{-1}$ and it is analytic in a half-plane, *albeit the wrong one*. But if we modify the derivation of (8) by using a lower half-plane contour instead of Fig. 8.18, page 418, we conclude that the real and imaginary parts of $k(\omega)$ are related by the Hilbert transform equations, with a sign change to account for the clockwise orientation of the contour. Thus we have

$$\operatorname{Im} k(\omega) = \frac{1}{\pi} \text{ p.v.} \int_{-\infty}^{\infty} \frac{\operatorname{Re} k(\eta)}{\eta - \omega} \, d\eta,$$

$$\operatorname{Re} k(\omega) = -\frac{1}{\pi} \text{ p.v.} \int_{-\infty}^{\infty} \frac{\operatorname{Im} k(\eta)}{\eta - \omega} \, d\eta. \tag{21}$$

The first identity is very useful in circuit theory and atomic and electromagnetic scattering applications, because the real part of the transfer function can often be determined efficiently and accurately using power-loss measurements. Then (21) enables the calculation of the experimentally more elusive imaginary part. In optical applications the dielectric constant plays the role of the transfer function $k$, and Eqs. (21) are known as *dispersion relations* in this context.

### Example 5

Verify the relations (21) for the transfer function of the circuit shown in Fig. 8.20.

**Solution.** The net impedance is seen by the methods of Sec. 3.4 to be $R_{\text{eff}} = R + i\omega L$. Therefore,

$$I_s = \frac{V_s}{R_{\text{eff}}} = \frac{V_s}{R + i\omega L}$$

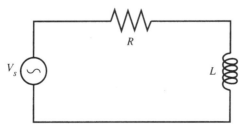

**Figure 8.20**   Circuit for Example 5.

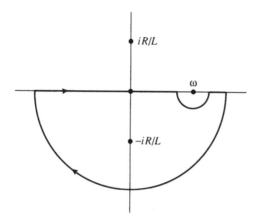

**Figure 8.21**  Contour for Example 5.

and $k(\omega) = 1/[R + i\omega L]$. This function is analytic in the lower half-plane and goes to zero like $|\omega|^{-1}$. We have

$$\text{Re } k(\omega) = \frac{R}{R^2 + \omega^2 L^2}, \qquad \text{Im } k(\omega) = \frac{-\omega L}{R^2 + \omega^2 L^2}.$$

From the contour $\Gamma$ in Fig. 8.21 we derive

$$\text{p.v.} \int_{-\infty}^{\infty} \frac{\text{Re } k(\eta)}{\eta - \omega} \, d\eta = \text{p.v.} \int_{-\infty}^{\infty} \frac{R}{R^2 + \eta^2 L^2} \frac{1}{\eta - \omega} \, d\eta$$

$$= -2\pi i \, \text{Res}(-iR/L) - \pi i \, \text{Res}(\omega)$$

and some algebra reveals this to be

$$\frac{-\pi \omega L}{R^2 + \omega^2 L^2} = \pi \, \text{Im } k(\omega),$$

in accordance with the first of Eqs. (21). The verification of the second equation is left as an exercise.  ■

## EXERCISES 8.5

1. Verify the entries in the Hilbert transform table. [HINT: For the last two entries consider the analytic functions $i/(z + ai)$ and $(e^{iaz} - 1)/iz$.]

2. Confirm Eq. (10) for some of the entries in the Hilbert transform table.

3. Argue that for the contour shown in Fig. 8.22, page 426, the contributions of the semicircles $S'_\varepsilon$ and $S''_\varepsilon$ to the integrals in Eqs. (3) and (5) will be $3\pi i f(z_0)/2$ and $-\pi i f(z_0)/2$, respectively. (Note that the Sokhotskyi-Plemelj prediction remains valid, however.)

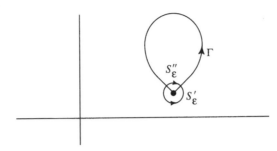

**Figure 8.22**   Contour for Prob. 3.

4. Use Euler's formula and the Hilbert transform table to show directly that if $|\omega| < \Omega_1$, the Hilbert transform of $\cos \omega x \cos \Omega_1 x$ is $\cos \omega x \sin \Omega_1 x$.

5. Verify the second dispersion relation (21) for the circuit in Fig. 8.20, page 424.

6. Verify that the value of the jump in the following integrals, as $z$ crosses the contour at $z_0$, is $2\pi i f(z_0)$, in accordance with the Sokhotskyi-Plemelj formula:

**(a)** $\oint_{|\zeta| = 1} \dfrac{1/\zeta}{\zeta - z}\, d\zeta$ 　　　　　　　　　　　　**(b)** $\displaystyle\int_{-\infty}^{\infty} \dfrac{1}{\xi - z}\, d\xi$

**(c)** $\displaystyle\int_{-\infty}^{\infty} \dfrac{\cos \xi}{\xi - z}\, d\xi$ 　　[HINT: Consult Prob. 10, Exercises 6.4, page 272.]

**(d)** $\displaystyle\int_{-\infty}^{\infty} \dfrac{\xi/(\xi^2 + 1)}{\xi - z}\, d\xi$ 　　[HINT: Consult Prob. 18, Exercises 4.7, page 178.]

7. The identity

$$\lim_{\varepsilon \downarrow 0} \frac{1}{x - x_0 - i\varepsilon} = \text{p.v.} \frac{1}{x - x_0} + i\pi\delta(x - x_0) \tag{22}$$

is frequently used by theoretical physicists. Here $\delta(x - x_0)$ is the *Dirac delta function*, an "idealized function" postulated to have the property that

$$\int_{-\infty}^{\infty} f(x)\delta(x - x_0)\, dx = f(x_0)$$

for any continuous function $f(x)$. Equation (22) is, strictly speaking, a crude abbreviation for the identity resulting when it is multiplied by $f(x)$ and integrated over $(-\infty, \infty)$, with the limits reversed:

$$\lim_{\varepsilon \downarrow 0} \int_{-\infty}^{\infty} \frac{f(x)}{x - x_0 - i\varepsilon}\, dx = \text{p.v.} \int_{-\infty}^{\infty} \frac{f(x)}{x - x_0}\, dx + i\pi f(x_0). \tag{23}$$

**(a)** Derive Eq. (23), assuming that $f(z)$ is analytic for Im $z \geq 0$ and approaches zero at infinity sufficiently rapidly that one can close the contour with a semicircle $C_\rho^+$ over which the integral goes to zero. [HINT: Let your analysis be guided by the sketches in Fig. 8.23.]

**(b)** What is $\displaystyle\lim_{\varepsilon \downarrow 0} \frac{1}{x - x_0 + i\varepsilon}$ [in the spirit of (22)]?

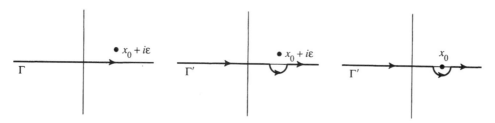

$$\int_\Gamma \frac{f(x)}{x - x_0 - i\varepsilon}\, dx \quad = \quad \int_{\Gamma'} \frac{f(x)}{x - x_0 - i\varepsilon}\, dx \quad \xrightarrow[\varepsilon \longrightarrow 0]{} \quad \int_{\Gamma'} \frac{f(x)}{x - x_0}\, dx$$

**Figure 8.23**   Contour for Prob. 7.

## SUMMARY

The chief value of sinusoidal analysis in applied mathematics lies in the fact that when a linear system is driven by a function $Ae^{i\omega t}$ its response takes the same form, under the proper circumstances. To take advantage of this we have to be able to express a given function of a real variable as a superposition of sinusoids.

When the function $F$ is periodic, say of period $L$, and satisfies certain continuity conditions, the appropriate decomposition is given by the Fourier series

$$F(t) = \sum_{n=-\infty}^{\infty} c_n e^{in2\pi t/L},$$

with coefficients defined by

$$c_n = \frac{1}{L} \int_{-L/2}^{L/2} F(t) e^{-in2\pi t/L}\, dt.$$

The fast Fourier transform is an efficient algorithm for approximating these coefficients.

If $F$ is not periodic but $|F|$ is integrable (and the continuity conditions still hold), then the decomposition takes the form

$$F(t) = \int_{-\infty}^{\infty} G(\omega) e^{i\omega t}\, d\omega,$$

where $G$ is the Fourier transform of $F$, defined by

$$G(\omega) = \frac{1}{2\pi} \int_{-\infty}^{\infty} F(t) e^{-i\omega t}\, dt.$$

"Direct" differentiation (i.e., termwise or under the integral sign) of both representations can be justified in many circumstances, and since this operation merely amounts to multiplication by $i\omega$, the solution of differential equations can often be greatly simplified by employing Fourier analysis.

To handle initial conditions and transients it is convenient to use the Laplace transform of $F$,

$$\mathcal{L}\{F\}(s) = \int_0^\infty F(t)e^{-st}\, dt.$$

The Laplace transform can be identified as the Fourier transform of a function which is "turned on" at $t = 0$ [i.e., $F(t) = 0$ for $t < 0$], with $i\omega$ replaced by $s$. The effect of the initial conditions is displayed in the formulas for the derivatives $F^{(n)}(t)$:

$$\mathcal{L}\{F^{(n)}\}(s) = s^n\mathcal{L}\{F\}(s) - s^{n-1}F(0) - s^{n-2}F'(0) - \cdots - F^{(n-1)}(0).$$

Hence the Laplace transform is the appropriate tool for solving initial-value problems.

Another advantage of the Laplace transform is its ability to handle certain non-integrable functions. There is an inversion formula for recovering $F(t)$ from $\mathcal{L}\{F\}(s)$, but the use of tables is frequently more convenient.

The tool that plays the role of the Fourier/Laplace transform in the cases where the data set is discrete is the $z$-transform. It can be related to the theory of Laurent series, from which many of its properties are derived.

The behavior of Cauchy integrals as their singularities cross the contours is discontinuous, with jumps predicted by the Sokhotskyi-Plemelj formulas. When the contour is the real axis the integrals yield the Hilbert transform formulas, which relate the real and imaginary parts of the integrand. This transform finds applications in the analysis of causal autonomous systems.

## SUGGESTED READING

The references mentioned at the end of Chapter 3 under the heading "Sinusoidal Analysis" may also be useful for a further study of the material in this chapter. In addition, the following texts deal with the subjects from a more rigorous viewpoint.

### Fourier Series

[1] Bachman, G., and Narici, L. *Functional Analysis*. Academic Press, Inc., New York, 1966.

[2] Edwards, R. E. *Fourier Series*, 2nd ed., Vols. I and II. Springer-Verlag, New York, 1979.

[3] Friedman, A. *Advanced Calculus*. Holt, Rinehart and Winston, Inc., New York, 1971.

### Fourier and Laplace Transforms

[4] Churchill, R. V. *Operational Mathematics*, 3rd ed. McGraw-Hill Book Company, New York, 1972.

[5] Jeffreys, H., and Jeffreys, B. *Methods of Mathematical Physics*, 3rd ed. Cambridge University Press, New York, 1972.

### z-Transforms

[6] Oppenheim, A. V., and Schafer, R. W. *Digital Signal Processing*. Prentice-Hall, Inc., Englewood Cliffs, N.J., 1975.

[7] Papoulis, A. *Signal Analysis*. McGraw-Hill Book Company, New York, 1977.

### Cauchy Integrals and Hilbert Transforms

[8] Bendat, J. S. *The Hilbert Transform and Applications to Correlation Measurements*. Bruel and Kjaer, Naerum, Denmark.

[9] Henrici, P. *Applied and Computational Complex Analysis*, Vol. 3. John Wiley & Sons, Inc., New York, 1986.

[10] Levinson, N., "Simplified Treatment of Integrals of Cauchy Type, the Hilbert Problem, and Singular Integral Equations," *SIAM Review*, 7, October 1965.

[11] Panofsky, W. K. H., and Phillips, M. *Classical Electricity and Magnetism*, 2nd ed. Addison-Wesley Publishing Company, Inc., Reading, Mass.; 1962.

# Appendix I:
# Numerical
# Construction
# of Conformal Maps

## by Lloyd N. Trefethen

In Chapter 7 we looked at conformal mapping and some of its applications, and we developed a formula to represent these maps in the case of arbitrary polygonal domains, the Schwarz-Christoffel transformation. It might seem that once we have this formula, applying it to construct a conformal map onto a given polygon should be just a matter of paperwork. But as we indicated in Chapter 7, this is not so. In fact, for nearly a century after Schwarz and Christoffel discovered their formula in the 1860s, the construction of conformal maps for all but the simplest polygons remained in practice impossible. The arrival of computers changed this situation.

The purpose of this appendix is to describe some of the mathematical techniques that can be applied on the computer to make the Schwarz-Christoffel transformation a practical reality. In doing this we can only scratch the surface of the issues of numerical analysis involved; this is a lively field in its own right, which the student is encouraged to explore at his or her leisure. But we hope at least to convey the flavor of how conformal mapping, and complex analysis in general, can be used in serious scientific problems. At the end we will also give an indication of how the Schwarz-Christoffel ideas are related to the broader problem of constructing conformal maps onto arbitrary domains.

## I.1 *THE SCHWARZ-CHRISTOFFEL PARAMETER PROBLEM*

Recall from Section 7.5 that in the Schwarz-Christoffel problem, our aim is to find an analytic function $f$ that maps the upper half $z$-plane onto the interior of a polygon $P$ in the $w$ plane defined by $n$ vertices $w_1, \ldots, w_n$. In Theorem 7 and Eq. (8) of Sec. 7.5, page 342, we showed that such a map can always be represented by the formula

$$f(z) = Ag(z) + B$$

$$= A \int_0^z (\zeta - x_1)^{\theta_1/\pi}(\zeta - x_2)^{\theta_2/\pi} \cdots (\zeta - x_{n-1})^{\theta_{n-1}/\pi} \, d\zeta + B \qquad (1)$$

for some suitable choice of complex numbers $A$ and $B$ and real numbers $x_1 < x_2 < \cdots < x_{n-1}$. Here $\theta_i$ is the right-turn angle of the polygon at the vertex $w_i$, as indicated in Fig. 7.40, page 341. The numbers $x_i$, the "prevertices," are the points along the real axis that are mapped to the vertices $w_i$ by $f$.

As we mentioned in Section 7.5, there are two reasons why applying Eq. (1) is not a trivial task. First, there is the problem of *numerical integration*. Even the simple triangular domain of Figure 7.43 (page 344) led to an integral (1) that could not be evaluated in closed form. For some complicated polygons, the evaluation of Schwarz-Christoffel integrals is almost invariably beyond the reach of exact formulas. Therefore one is forced to look for efficient numerical approximations. We will consider this problem in Sec. I.3. But until then, let us pretend that all integrals (1) can be evaluated effortlessly.

The second and more serious difficulty arises as soon as $n$, the number of vertices of $P$, is four or more. As we pointed out, only three of the prevertices $x_i$ can be specified arbitrarily, and the rest are then implicitly determined by the geometry of $P$. If $P$ is symmetrical, as was the rectangle of Fig. 7.50 for example, then the correct values of these missing parameters are often evident. But for general polygons they are emphatically unknown. What use then is Eq. (1)? Obviously the prevertices must first be determined somehow, and the problem of finding them is known as the Schwarz-Christoffel *parameter problem*.

To be precise, let us fix the following three values: $x_1 = -1$, $x_2 = 0$, and (implicitly) $x_n = \infty$. What remain are then $n - 3$ prevertices whose values are sought:

$$n - 3 \text{ unknowns:} \quad x_3, x_4, \ldots, x_{n-1} \qquad (2)$$

subject to the constraints

$$0 < x_3 < x_4 < \cdots < x_{n-1} < \infty. \qquad (3)$$

For any choice of these quantities, Eq. (1) defines a function $g$ that maps the real axis onto a polygon $P'$. Our problem is to find a set of values $x_i$ for which $P'$ is geometrically similar to $P$. Once we have accomplished this, as mentioned in Sec. 7.5, the constants $A$ and $B$ in Eq. (1) can be adjusted so that the two polygons actually coincide.

For example, suppose the right-turn angles in Fig. 7.40, page 341, are

$$\theta_1 = -146°, \qquad \theta_2 = -82°, \qquad \theta_3 = 160°, \qquad \theta_4 = -150°, \qquad \theta_5 = -142°,$$

and suppose we try the prevertex values

$$x_3 = 1, \qquad x_4 = 2.$$

Figure I.1(a) shows the resulting polygon $P'$, computed numerically and scaled and rotated so that the side $[w_1', w_2']$ is the same as $[w_1, w_2]$. As expected, the angles are correct but the side lengths are far off. To lengthen the side $[w_2', w_3']$, let us try the new guess

$$x_3 = 3, \qquad x_4 = 4.$$

The result shown in Fig. I.1(b) looks better, but $[w_2', w_3']$ is still too short, and $[w_3', w_4']$ has shrunk ominously. If we now try

$$x_3 = 10, \qquad x_4 = 15,$$

we get the polygon of Fig. I.1(c), which is beginning to look about right. But it is obvious that it is going to take a long time to make $P'$ precisely similar to $P$—say, to six digits of accuracy—by trial and error. If $n$ were 10 or 20 instead of 5, the process would be impossibly slow.

To make speedier progress, we need to formulate the condition of similarity algebraically. Consider the following set of equations:

$$n - 3 \text{ equations:} \quad \frac{|g(x_i) - g(x_{i-1})|}{|g(x_2) - g(x_1)|} - \frac{|w_i - w_{i-1}|}{|w_2 - w_1|} = 0, \quad i = 3, 4, \ldots, n - 1. \tag{4}$$

Notice that each absolute value in the right-hand quotient of this formula represents a side length in $P$ while, by Eq. (1), each absolute value in the left-hand quotient represents a side length in $P'$. Thus these identities assert that the sides $[w_1', w_2']$ through $[w_{n-2}', w_{n-1}']$ of $P'$ have the same lengths as the corresponding sides of $P$,

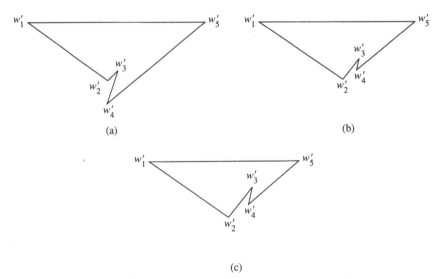

(a)

(b)

(c)

**Figure I.1** Schwarz-Christoffel approximations to the polygon $P$ of Fig. 7.40 based on different choices of prevertices $0 < x_3 < x_4 < \infty$.

except for a uniform scale factor $C = |g(x_2) - g(x_1)|/|w_2 - w_1|$. It may appear that this leaves the two sides $[w'_{n-1}, w'_n]$ and $[w'_n, w'_1]$ unaccounted for, but suppose $P'$ has all but these two side lengths equal to $C$ times the corresponding side lengths in $P$. Then, since the angles are fixed by the exponents in Eq. (1), and since any two nonparallel lines have a unique point of intersection, the remaining two side lengths of $P$ and $P'$ automatically have to be related in the same way too. In other words, Eq. (4) fully expresses the desired condition that $P'$ and $P$ are *similar*.

Equations (2) to (4) constitute a *constrained system of $n - 3$ nonlinear equations in $n - 3$ unknowns*. The reason we have put both quotients on the left-hand side in Eq. (4) is so that the system can be written symbolically in the form

$$F_i(\mathbf{x}) = 0, \quad i = 3, 4, \ldots, n - 1,$$

where $\mathbf{x}$ denotes the vector $(x_3, \ldots, x_{n-1})$ of unknowns, of length $n - 3$, and each $F_i$ is a real-valued function of $\mathbf{x}$. If we go further and let $\mathbf{F}$ denote the vector-valued function $\mathbf{F}(\mathbf{x}) = (F_3(\mathbf{x}), \ldots, F_{n-1}(\mathbf{x}))$, and $\mathbf{0}$ the vector $(0, \ldots, 0)$, then the system takes the even more compact form

$$\mathbf{F}(\mathbf{x}) = \mathbf{0}. \tag{5}$$

We have now translated the Schwarz-Christoffel parameter problem into the following algebraic task: Find a root $\mathbf{x}^*$ of Eq. (5), subject to (3). It can be proved that there is a unique solution to this problem, and that this solution is precisely the vector of prevertices for the Schwarz-Christoffel mapping (1) onto $P$.

Of course, in getting to Eq. (5) we have not really simplified the mathematics, just the notation. Written out in full, for example, the component $i = 3$ of Eq. (5) looks like this for our polygon with $n = 5$:

$$\frac{\left| \int_{x_2}^{x_3} (\zeta - x_1)^{-146°/\pi}(\zeta - x_2)^{-82°/\pi}(\zeta - x_3)^{+160°/\pi}(\zeta - x_4)^{-150°/\pi} \, d\zeta \right|}{\left| \int_{x_1}^{x_2} (\zeta - x_1)^{-146°/\pi}(\zeta - x_2)^{-82°/\pi}(\zeta - x_3)^{+160°/\pi}(\zeta - x_4)^{-150°/\pi} \, d\zeta \right|} - \frac{|w_3 - w_2|}{|w_2 - w_1|} = 0. \tag{6}$$

(The degree symbols represent $\pi/180$, so that $146°/\pi$ is equivalent to $146/180$, and so on.) Nevertheless, a simplification of notation is often highly useful in formulating a problem for numerical computation, because it may reveal that the problem has a standard form for which numerical methods are already in existence.

Such is exactly the case here. In fact, solving a nonlinear system of equations of the form (5) is one of the fundamental, extensively studied problems in numerical analysis. In general, there is no hope of finding the root $\mathbf{x}^*$ exactly, so one has to approximate it by iteration. First, some initial guess vector $\mathbf{x}^{(0)}$ is chosen. Then a sequence of new guesses $\mathbf{x}^{(1)}, \mathbf{x}^{(2)}, \ldots$ are successively computed, which it is hoped will converge to $\mathbf{x}^*$. Most of the commonly used algorithms for generating these guesses are variants of *Newton's method*, with which the reader may be familiar for the case of a single scalar nonlinear equation. Many computer programs have been written to implement this kind of iteration, and they can be found in the software libraries of most large computer systems and also in interactive numerical problem-solving environments such as MATLAB [3]. Under favorable circumstances, such a computer program will often converge astonishingly fast. It is not unusual that after a few

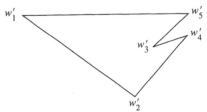

**Figure I.2** An approximation to $P$ obtained with prevertices out of order: $x_3 > x_4$.

dozen iterations, $\mathbf{x}^{(k)}$ will be so near $\mathbf{x}^*$ that $\mathbf{F}(\mathbf{x}^{(k)})$ has magnitude $10^{-16}$ or smaller; in other words, $\mathbf{x}^{(k)}$ has converged to "machine accuracy" in double precision on a typical computer with a 32-bit word length.

If we blindly apply a program for solving systems of equations to our Schwarz-Christoffel problem, however, the correct solution will usually not be found. The reason is that we have ignored the constraints (3). Even if $\mathbf{x}^{(0)}$ satisfies these constraints, it is highly likely that a later iterate $\mathbf{x}^{(k)}$ will violate them, unless we take pains to prevent this. In such a situation Eq. (1) will define a polygon $P'$ in which the vertices are out of order. For example, suppose we take

$$x_3 = 2, \qquad x_4 = 1$$

in the pentagonal example considered above. Figure I.2 shows the result. It is a perfectly good polygon, but it belongs to a different family from the shape for which we are looking, and adjusting its side lengths to satisfy Eq. (4) will be a waste of time.

A simple but effective way to get around the problem of constraints is by introducing a change of variables. If $\mathbf{x}$ is a vector of positive numbers satisfying (3), and $x_2 = 0$ as usual, then the formula

$$\hat{x}_i = \log(x_i - x_{i-1}) \in \mathbf{R}, \quad i = 3, 4, \ldots, n-1 \tag{7}$$

defines a new vector $\hat{\mathbf{x}} = (\hat{x}_3, \ldots, \hat{x}_{n-1})$ of unconstrained real numbers. Conversely, given an arbitrary such vector $\hat{\mathbf{x}}$, the inverse formula

$$x_i = x_{i-1} + e^{\hat{x}_i}, \quad i = 3, 4, \ldots, n-1 \tag{8}$$

defines a vector $\mathbf{x}$ that satisfies (3). To solve the parameter problem (2), (3), and (5), we can rewrite Eq. (5) as a new nonlinear system in the new unconstrained variables,

$$\hat{\mathbf{F}}(\hat{\mathbf{x}}) = \mathbf{0}. \tag{9}$$

The parameter problem becomes the unconstrained nonlinear system of Eqs. (7) and (9). Now we can apply a numerical algorithm based on Newton's method without worrying about constraints.

## I.2 EXAMPLES

Let us look at a few examples to see how these ideas perform in practice. First, we should finish off the polygon of Fig. 7.40, but before we can do this, we have to specify its dimensions. Let us assume that

$$|w_2 - w_1| = 8, \qquad |w_3 - w_2| = 3, \qquad |w_4 - w_3| = 1;$$

the other two side lengths are then determined implicitly, as discussed above. Beginning with the initial guess $\hat{x}_3^{(0)} = \hat{x}_4^{(0)} = 0$ in Eq. (7), that is, $x_3^{(0)} = 1$ and $x_4^{(0)} = 2$ by Eq. (8), we now iterate numerically with our computer program to solve the parameter problem of Eq. (9). Figure I.3(a) shows the polygon $P'$ at some of the first few steps (ignoring $n - 3$ uninteresting initial steps required to estimate a *Jacobian matrix* needed for the approximate Newton method). Figure I.3(b) shows the final polygon, obtained to 15-digit accuracy in 14 iterations. The shapes of the intermediate polygons, as well as the exact number of iterations, would differ depending on the details of the programming.

The line above each polygon in Fig. I.3(a) represents the segment $[-1, 15]$ of the real axis, and the vertical cross-segments indicate the points $x_1, \ldots, x_{n-1}$ for the given step. At the beginning, the four prevertices are evenly spaced. The points $x_1 = -1$ and $x_2 = 0$ remain fixed throughout the iteration, but $x_3$ and $x_4$ change.

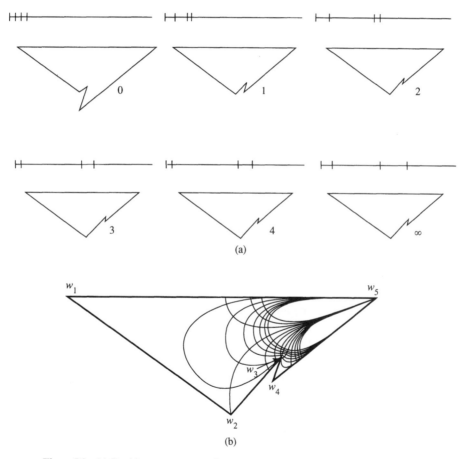

**Figure I.3**   (a) Rapid convergence to the correct prevertices and the correct $P$ by a variant of Newton's method. (b) The polygon $P$ successfully mapped onto the upper half-plane. The curves inside $P$ are the conformal images of vertical and horizontal lines in the half-plane.

One can see the rapid convergence of these parameters as the final polygon is approached. The converged values turn out to be

$$x_3 = 8.22544, \qquad x_4 = 11.07029.$$

Notice that these numbers are surprisingly far from the guess $x_3 = 10$, $x_4 = 15$ that led to Fig. I.1(c).

In Fig. I.3(b), the final polygon $P$ is plotted with additional curves drawn on the inside. These curves are the images of the horizontal and vertical lines

$$\text{Im } z = 0.2, 1, 2, 3, 4, 5, 6 \quad \text{and} \quad \text{Re } z = -6, -4, -2, 0, \ldots, 18$$

in the $z$-plane. All 20 curves meet at $w_5$, which corresponds to $z = \infty$. But inside the polygon, as they must, the curves meet everywhere at right angles. In fact, they form a grid in which each individual cell looks like a distorted rectangle of length-to-width ratio 2. This confirms what we know about conformal maps in general: locally, they change the scale but not the shape.

Figures I.4 and I.5 (page 438) show analogous plots for two other polygons, again beginning with the initial guess $x_i^{(0)} = i - 2$ for $1 \le i \le n - 1$. In Fig. I.4, $P$ is a six-sided $L$ shape. Since this polygon happens to be symmetrical about the line through the corners $w_3$ and $w_6$, the converged prevertices end up symmetrically located around $x_3$ (and around $x_6 = \infty$). In Fig. I.5, $P$ is a seven-sided polygon shaped like a claw. In this latter case, the preliminary steps shown represent polygons that fold over themselves. This means that the function $g$ is not one-to-one.

Notice in Fig. I.5(a) that although it is easy to distinguish $x_4$, $x_5$, and $x_6$ early in the iteration, their final values lie nearly on top of each other. Here are the converged values:

$$x_1 = -1, \qquad x_2 = 0, \qquad x_3 = 2.48883,$$

$$x_4 = 3.89729, \qquad x_5 = 3.91233, \qquad x_6 = 3.96037.$$

Thus $|x_4 - x_5| \approx .015$, for example. Such uneven distribution of prevertices is not the exception but the rule in conformal mapping, and indeed, the unevenness is often far more pronounced. The root of the phenomenon is that the two ends of our "claw" have very little influence on each other. For example, if we think of the claw as a heat conductor, it is intuitively clear that a change in the temperature applied along the boundary near one end will have almost no effect near the other.

To carry these observations further, perhaps the most valuable insight to be gained from looking at pictures of numerical conformal maps is an appreciation of the great distortions they often introduce. Consider Figs. I.3(b), I.4(b), and I.5(b). One kind of distortion that appears in all three cases is near $w_n$, where the curves bunch up tightly. Since $w_n$ corresponds to $\infty$ under the conformal map, this is not surprising. Another kind of distortion appears near outward-pointing or "salient" corners, which the interior image curves tend to avoid. This is especially pronounced in Fig. I.5(b), where the lower claw is entirely bare. Salient corners are "deadwater" regions, little affected by what goes on in the rest of the polygon. In contrast, at an inward-pointing or "reentrant" corner, such as $w_3$ in Fig. I.3(b), the image curves bunch up closely. Physically, at such a corner the speed of flow, the electric field strength, or the temperature gradient will be infinite.

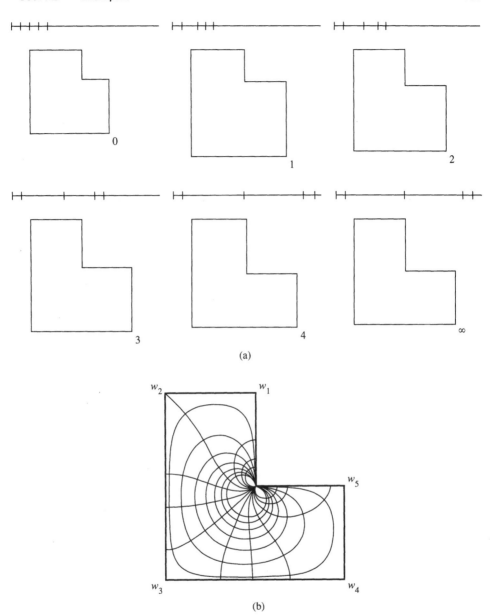

(a)

(b)

**Figure I.4**  Another example of numerical Schwarz-Christoffel mapping.

For a final computed example, recall that Fig. 7.55, page 354, showed the streamlines for the flow over a polygonal streambed. To obtain that solution, the domain was mapped analytically onto the upper half-plane, and the desired streamlines were then the conformal images of the lines Im $z$ = constant in the upper half-plane. For any more complicated streambed, we have to resort to the numerical

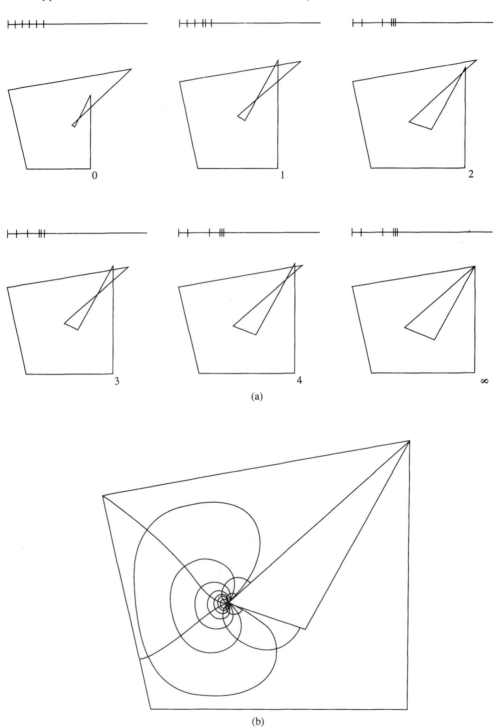

(a)

(b)

**Figure I.5** A third example.

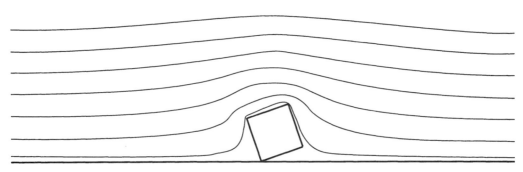

**Figure I.6** "Flow over a square boulder," the idealized potential flow streamlines computed by a Schwarz-Christoffel map to the upper half-plane.

techniques described in this appendix. Figure I.6 shows a plot generated by the computer for the flow over an idealized square boulder. Now the vertex $x_n$ lies at infinity, but this is no problem since $x_n$ does not appear in Eq. (1). Schwarz and Christoffel would have liked this picture!

## I.3 NUMERICAL INTEGRATION

Having shown these successful results, we must now return to the first question mentioned at the start of this appendix. How were these Schwarz-Christoffel integrals computed? Notice that every time the vector-valued function $F(x)$ of Eqs. (4) and (5) is evaluated, the computer program must compute $n - 2$ integrals of the form

$$g(x_i) - g(x_{i-1}) = \int_{x_{i-1}}^{x_i} (\zeta - x_1)^{\theta_1/\pi} \cdots (\zeta - x_{n-1})^{\theta_{n-1}/\pi} \, d\zeta. \tag{10}$$

Since $F(x)$ will be evaluated many times during the course of the iterative solution of the parameter problem, this may add up to hundreds or thousands of numerical integrations. Moreover, once the parameter problem is solved, further integrations will be necessary to produce plots like Figs. I.4 to I.6. Clearly, it is imperative to be able to compute Eq. (10) efficiently.

Numerical integration, also called *quadrature*, is another old and well-understood problem of numerical analysis. Suppose we want to compute the integral from $a$ to $b$ of a function $\phi(z)$. (It does not matter in what we say here whether $a$, $b$, and $\phi$ are real or complex.) For a first attempt, an obvious idea is to calculate a Riemann sum based on a division of the interval $[a, b]$ into subintervals. Let $N$ be a positive integer, and set

$$\Delta z = \frac{b - a}{N}, \quad z_i = a + \left(i - \frac{1}{2}\right)\Delta z, \quad i = 1, \ldots, N.$$

Then the integral can be approximated by

$$\int_a^b \phi(z) \, dz \approx \sum_{i=1}^{N} \phi(z_i) \, \Delta z. \tag{11}$$

This method of approximate integration is called the *midpoint rule*, and it is not hard to show that if $\phi$ is smooth, then the difference between the numerical estimate and the exact integral will decrease at least in proportion to $N^{-2}$ as $N \to \infty$. We write this as

$$\int_a^b \phi(z) \, dz - \sum_{i=1}^N \phi(z_i) \, \Delta z = O(N^{-2}).$$

Thus the midpoint rule is a usable numerical method in principle; but its performance can at best be described as so-so. To get eight significant digits, we will need $N \approx 10,000$.

For greater efficiency, we can generalize (11) to a formula

$$\int_a^b \phi(z) \, dz \approx \sum_{i=1}^N \omega_i \phi(z_i), \tag{12}$$

where the numbers $\{\omega_i\}$ are a set of real *weights*, not necessarily equal, and the numbers $\{z_i\}$ are a set of *points* or *nodes* in $[a, b]$, not necessarily equally spaced. Of course, everything will depend on how the weights and nodes are chosen. In *Simpson's rule*, with which the reader may be familiar, the points are equally spaced but the weights alternate in magnitude in a pattern 1-4-2-4-2-$\cdots$-4-2-4-1, and this leads to an error estimate $O(N^{-4})$. Now only roughly 100 points are needed to get eight significant digits. There are also formulas of higher order based on equally spaced points, called *Newton-Cotes formulas*. But the king of them all is the method known as *Gaussian quadrature*. In this method, the points are unevenly distributed in an "optimal" way (they depend on the zeros of the so-called *Legendre polynomial* of order $N$), and the weights are irrational numbers. The result is a formula under which the errors decrease faster than $O(N^{-K})$ for any fixed $K$. In fact, if $\phi$ is analytic in an open region of the complex plane containing the interval $[a, b]$, then the error in Gaussian quadrature decreases at a geometric rate $O(\rho^N)$ for some $\rho < 1$, as $N \to \infty$.

Now what if we go ahead and use Gaussian quadrature to integrate Eq. (10) for our Schwarz-Christoffel problem? The results will be virtually useless! The reason is that the integrand of Eq. (10), far from being analytic on $[x_{i-1}, x_i]$, has singularities at both endpoints. For example, there are singularities with exponents $-82°/\pi$ and $+160°/\pi$ in the numerator on the left in Eq. (6). As a consequence, it turns out that both Gaussian quadrature and the midpoint rule reduce to thoroughly unacceptable rates of convergence for this integral, not even always as good as $O(N^{-1})$. Simpson's rule is in general not even defined, since it insists on evaluating the integrand at the endpoints, where it may be infinite. Certainly the presence of endpoint singularities is a key issue in Schwarz-Christoffel integration.

Once this difficulty has been recognized, there are many ways to cope with it. One of these is to use a modified procedure called *Gauss-Jacobi quadrature*. A Gauss-Jacobi formula has the usual form (12), but the weights and nodes are specially chosen on the basis of the assumption that $\phi$ has singularities of type $(z - a)^\alpha$ at one endpoint and of type $(z - b)^\beta$ at the other, for some numbers $\alpha, \beta > -1$. (The nodes now depend on the zeros of the so-called *Jacobi polynomials*.) In fact, if $\phi$ can be written as a product

$$\phi(z) = (z - a)^\alpha (z - b)^\beta \psi(z),$$

where $\psi$ is analytic in an open region containing $[a, b]$, then the Gauss-Jacobi formula will also have errors decreasing geometrically with $N$, as the Gaussian formula did in the case where $\phi$ was analytic. See the notes for references on how these nodes and weights can be computed. Obviously, Gauss-Jacobi quadrature is made to order for our Schwarz-Christoffel application, where we have singularities of type $\alpha = \theta_{i-1}/\pi$, $\beta = \theta_i/\pi$.

For an illustration of the power of Gauss-Jacobi quadrature, consider again the integral that came up in Example 1 of Sec. 7.5, page 343, for the conformal map onto a triangle, which reduces to $(1 + i)/\sqrt{2}$ times

$$\int_0^1 (1 - x^2)^{-3/4}\, dx \approx 2.622058. \tag{13}$$

Here we have $a = 0$, $\alpha = 0$, $b = 1$, and $\beta = -3/4$. With $N = 4$, for example, the Gauss-Jacobi nodes and weights turn out to be approximately

$$z_1 = .02867, \qquad z_2 = .38609, \qquad z_3 = .75428, \qquad z_4 = .98366,$$

$$\omega_1 = .13109 \qquad \omega_2 = .31759, \qquad \omega_3 = .56220, \qquad \omega_4 = 1.36754.$$

Inserting these values in (11), we get the approximate integral 2.622057, which is accurate to five places! The table below shows more systematically how fast the approximation converges as $N$ increases.

| $N$ | Approx. integral |
|---|---|
| 1 | 2.57 |
| 2 | 2.6208 |
| 3 | 2.62202 |
| 4 | 2.622057 |

Simpson's rule and its relatives, even if we took care of the singularity somehow, could not come close to this performance.

In applying Gauss-Jacobi quadrature to more complicated Schwarz-Christoffel maps, there is one more difficulty we have to address before we can count on accurate results at low cost. Recall that in mapping the "claw" of Fig. I.5, we found that the prevertices ended up quite unevenly distributed. Now consider in that problem, for example, the integral in Eq. (1) from $x_3$ to $x_4$, which we will have to compute repeatedly during the iteration to determine the length of the corresponding side $[w'_3, w'_4]$. By using the appropriate Gauss-Jacobi formula, we can avoid any ill effects due to the singularity at $x_4$. But $x_5$, with its own singularity, lies only a distance .015 away. It turns out that, as a result, Gauss-Jacobi quadrature will not begin to give accurate answers until $N$ is at least as large as $1/.015 \approx 10^2$—quite unacceptable! This problem of singularities near endpoints comes up in almost every practical Schwarz-Christoffel mapping problem.

Fortunately, this difficulty too can be circumvented once it is recognized. One way to do so is to use *compound Gauss-Jacobi quadrature*, in which troublesome intervals of integration are automatically subdivided near the endpoints into small subintervals whose length is comparable to the distance to the nearest singularity. With this improvement, the remarkable behavior of geometric convergence—errors $O(\rho^N)$—can be maintained for the conformal mapping of arbitrary polygons.

## I.4 CONFORMAL MAPPING OF SMOOTH DOMAINS

The Schwarz-Christoffel integral of Eq. (1) contains a product of terms $(\zeta - x_i)^{\theta_i/\pi}$, one for each corner $w_i$ of the polygon $P$. Now what could we do if, instead of a polygon, we wanted to map the upper half-plane conformally onto a more general domain $D$ bounded by a smooth curve $C$? There is an obvious way to try to go about this. Let us attempt to treat $C$ as a polygon with infinitely many corners $w_i$, at each of which the right turn angle $\theta_i$ is infinitesimal.

Making this idea precise is easier than one might expect. From Eq. (1) here or Eq. (3) of Sec. 7.5,

$$f'(z) = A(z - x_1)^{\theta_1/\pi}(z - x_2)^{\theta_2/\pi} \cdots (z - x_{n-1})^{\theta_{n-1}/\pi}$$

for a conformal map onto a polygon. Let us rewrite this as

$$f'(z) = A \exp\left[\frac{1}{\pi} \sum_{i=1}^{n-1} \theta_i \log(z - x_i)\right]. \tag{14}$$

Each term $\log(z - x_i)$ here cannot be defined as a single-valued function in a neighborhood of $x_i$, but fortunately, we only need to evaluate $f'(z)$ in the upper half-plane. Thus let $\log(z - x_i)$ be defined t. be single-valued for Im $z \geq 0$, $z \neq x_i$, and then Eq. (14) becomes a well-defined analytic function where we need it. And now it should be obvious what formula to use for curved boundaries—just replace the sum in Eq. (14) by an integral!

$$f'(z) = A \exp\left[\frac{1}{\pi} \int_{-\infty}^{\infty} \theta(x) \log(z - x)\, dx\right]. \tag{15}$$

In this equation, $\theta(x)$ is a function that can be identified as the amount of turning of $C$ per unit length along the real $x$-axis. Integrating, we get the following analog of Eq. (1) for smooth domains:

$$f(z) = A \int_0^z \exp\left[\frac{1}{\pi} \int_{-\infty}^{\infty} \theta(x) \log(\zeta - x)\, dx\right] d\zeta + B. \tag{16}$$

This is a *continuous Schwarz-Christoffel formula*.

The reader is entitled at this point to be skeptical. If applying the usual Schwarz-Christoffel integral, with its finite number of vertices, turned out to be such a major undertaking, what hope can there be of solving Eq. (16)? In particular, we now face a seemingly much harder *continuous parameter problem*. Before, we had to determine the point $x_i$ corresponding to each corner $w_i$. Now, we need to determine *for every* $x$ the "amount of corner" $\theta(x)$ present there.

Nevertheless, it turns out that Eq. (16) can be solved numerically with a reasonable computational effort. This equation is a (somewhat irregular) example of an *integral equation,* a term for an equation in which an unknown function—$\theta(x)$—appears inside an integral. Several methods are known for solving integral equations numerically. This particular equation has been applied by a number of computational aerodynamicists for determining maps onto complicated domains (see [5]). In addition, there are a dozen or more other well-known integral equations that can also be used to compute conformal maps. Determining how these equations can be efficiently implemented, and which of them is best in practice, is an area of active research today, 150 years after Riemann first proposed his mapping theorem.


## SUGGESTED READING


Some references on the material in this appendix are given below. Many textbooks of numerical analysis touch on the questions of integration and solution of nonlinear systems we have discussed; [2] is one example and [4] is another that is more advanced. Most of the standard numerical algorithms have been implemented in the form of robust and widely distributed computer programs, and a good way to find out about available software is to send the electronic mail message SEND INDEX to the address NETLIB@ORNL.GOV [1]. In particular, software is available for solving nonlinear systems of equations and for numerical integration, both by Gauss-Jacobi quadrature and by other methods. Numerical conformal mapping, on the other hand, is a research topic that is still mentioned more often in journal articles than in books.

### Numerical Analysis and Software

[1]  Dongarra, J. J., and Grosse, E. "Distribution of Mathematical Software via Electronic Mail," *Communications of the Assoc. of Computing Machinery* 30 (1987), 403–7.

[2]  Kahaner, D., Moler, C., and Nash, S. *Numerical Methods and Software.* Prentice-Hall, Inc., Englewood Cliffs, N.J., 1989.

[3]  The Mathworks. *Matlab User's Guide.* The MathWorks, Inc., 24 Prime ParkWay, Natick, Mass. 01760.

[4]  Stoer, J., and Bulirsch, R. *Introduction to Numerical Analysis.* Springer-Verlag, New York, 1980.

### Numerical Conformal Mapping

[5]  Davis, R. T. "Numerical Methods for Coordinate Generation based on Schwarz-Christoffel Transformations," *4th AIAA Comp. Fluid Dynamics Conf. Proc.,* 1978.

[6]  Henrici, P. *Applied and Computational Complex Analysis,* vol. III. John Wiley & Sons, Inc., New York, 1986.

[7]  Trefethen, L. N. "Numerical Computation of the Schwarz-Christoffel Transformation," *SIAM Journal of Scientific and Statistical Computing* 1, 1980, 82–102.

[8]  Trefethen, L. N., ed. *Numerical Conformal Mapping.* North-Holland Publishing Co., Amsterdam, 1986.

# Appendix II: Table of Conformal Mappings

The following five pages list some conformal mappings that arise in applications. Notice that corresponding points are labeled by the same letter; thus the point $a$ is mapped to the point $A$, etc. A more extensive tabulation appears in Ref. [5] at the end of Chapter 7.

## *II.1 MÖBIUS TRANSFORMATIONS*

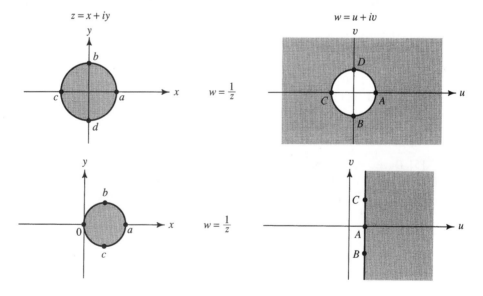

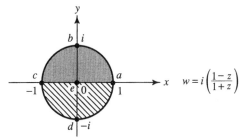

$$w = i\left(\frac{1-z}{1+z}\right)$$

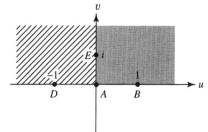

$z = x + iy$ 　　　　　　　　　 $w = u + iv$

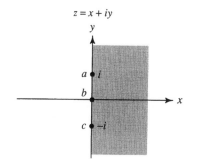

$$w = \frac{z-1}{z+1}$$

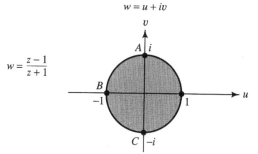

$$w = \frac{z-a}{1-\bar{a}z}, \ (|a| < 1)$$

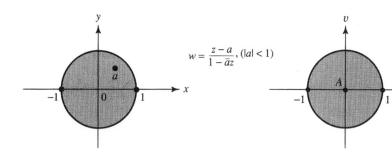

$$w = \frac{z-\lambda}{\lambda z - 1}$$

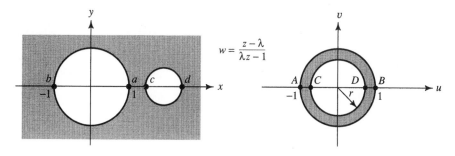

$$\lambda = \frac{1 + cd + \sqrt{(c^2-1)(d^2-1)}}{c+d}, \ (1 < c < d)$$

$$r = \frac{cd - 1 - \sqrt{(c^2-1)(d^2-1)}}{d-c}, \ (0 < r < 1)$$

## II.2 OTHER TRANSFORMATIONS

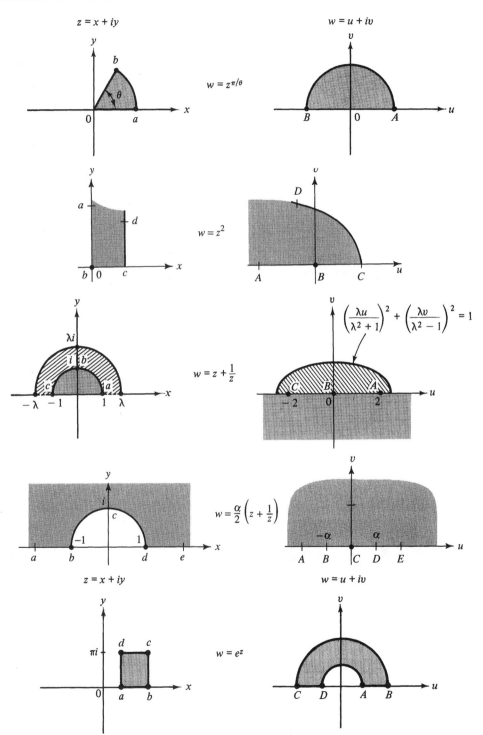

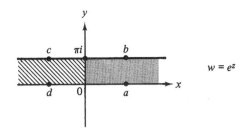

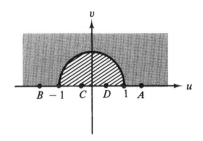

$$w = e^z$$

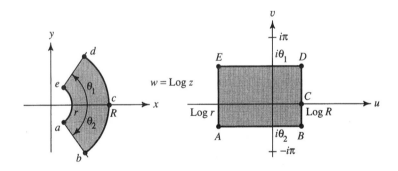

$$w = z + e^z$$

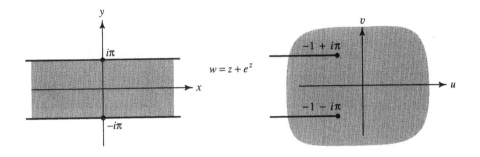

$$w = \text{Log } z$$

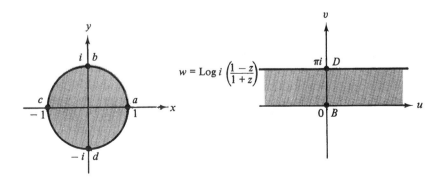

$$w = \text{Log } i \left( \frac{1-z}{1+z} \right)$$

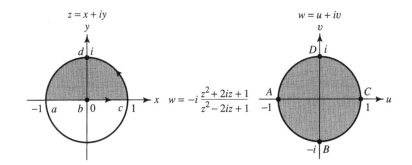

$$z = x + iy$$

$$w = u + iv$$

$$w = -i\frac{z^2 + 2iz + 1}{z^2 - 2iz + 1}$$

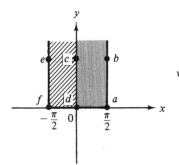

$$w = \sin z$$

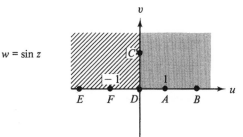

$$z = x + iy$$

$$w = u + iv$$

$$w = 2\sqrt{z+1} + \mathrm{Log}\,\frac{\sqrt{z+1} - 1}{\sqrt{z+1} + 1}$$

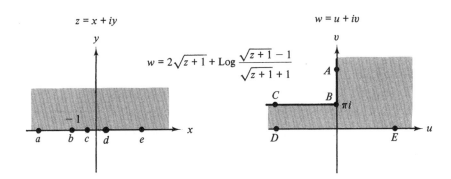

$$w = \frac{i\sigma_0}{\pi}\left\{\mathrm{Log}\left(\frac{1+\zeta}{1-\zeta}\right) - i\alpha\,\mathrm{Log}\left(\frac{1+i\alpha\,\zeta}{1-i\alpha\,\zeta}\right)\right\}$$

$$\alpha = \frac{\tau_0}{\sigma_0}\,;\quad \zeta^2 = \frac{z+1}{z-\alpha^2}$$

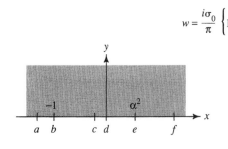

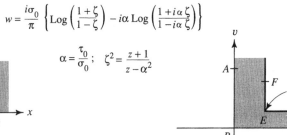

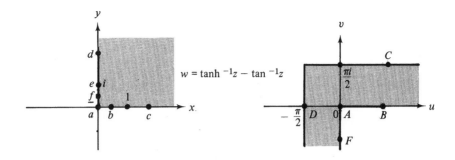

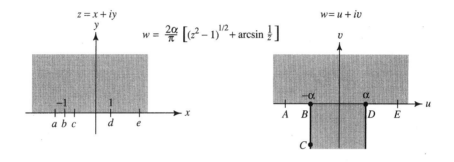

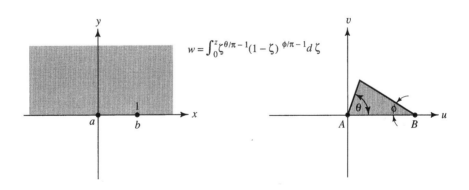

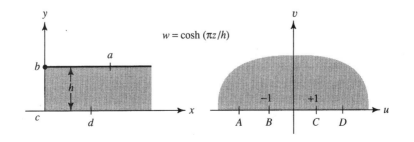

# Answers
# to Odd-Numbered
# Problems

## CHAPTER 1

### Exercises 1.1, Page 4

**5.** **(a)** $0 + (-3/2)i = -3i/2$ **(b)** $3 + 0i = 3$ **(c)** $0 + (-2)i = -2i$

**7.** **(a)** $8 + i$ **(b)** $1 + i$ **(c)** $0 + (-8/3)i = -8i/3$

**9.** $\frac{61}{185} - \frac{107}{185}i$

**11.** $2 + 0i = 2$

**13.** $6 + 5i$

**17.** $8 - 10i$

**19.** $a^3 - 3ab^2 + 5a^2 - 5b^2 = a$, $3a^2b - b^3 + 10ab = b + 3$, where $z = a + bi$

**21.** $z_1 = 1 + i$, $z_2 = -i$

### Exercises 1.2, Page 11

**3.** $-3$

**7.** **(a)** horizontal line $y = -2$ **(b)** circle, center $= 1 - i$, rad $= 3$

**(c)** circle, center $= i/2$, rad $= 2$

**(d)** perpendicular bisector of segment joining $z = 1$ and $z = -i$

**(e)** parabola $y^2 = 4(x + 1)$ with vertex $-1$, focus 0

**(f)** ellipse with foci at $\pm 1$ **(g)** circle, center $= \frac{9}{8}$, rad $= \frac{3}{8}$

**(h)** half-plane consisting of all points on or to the right of the vertical line $x = 4$

**(i)** all points inside the circle with center $i$ and rad $= 2$ (open disk)

**(j)** all points outside the circle $|z| = 6$

## Exercises 1.3, Page 19

**3.** $|z_1 + z_2|^2 + |z_1 - z_2|^2 = 2|z_1|^2 + 2|z_2|^2$

**5.** (a) 1 (b) $5\sqrt{26}$ (c) $5\sqrt{5}/2$ (d) 1

**7.** (a) $\arg = \pi + 2k\pi$, polar form: $\frac{1}{2}\operatorname{cis}(\pi)$

   (b) $\arg = 3\pi/4 + 2k\pi$, polar form: $3\sqrt{2}\operatorname{cis}(3\pi/4)$

   (c) $\arg = -\dfrac{\pi}{2} + 2k\pi$, polar form: $\pi\operatorname{cis}(-\pi/2)$

   (d) $\arg = 7\pi/6 + 2k\pi$, polar form: $4\operatorname{cis}(7\pi/6)$

   (e) $\arg = 7\pi/12 + 2k\pi$, polar form: $2\sqrt{2}\operatorname{cis}(7\pi/12)$

   (f) $\arg = 5\pi/3 + 2k\pi$, polar form: $4\operatorname{cis}(5\pi/3)$

   (g) $\arg = 5\pi/12 + 2k\pi$, polar form: $\dfrac{\sqrt{2}}{2}\operatorname{cis}(5\pi/12)$

   (h) $\arg = 13\pi/12 + 2k\pi$, polar form: $\dfrac{\sqrt{14}}{2}\operatorname{cis}(13\pi/12)$

**9.** rotation of vector $z$ about origin through an angle $\phi$ in the counterclockwise direction

**13.** (b), (d)

**21.** $r = \sqrt{r_1^2 + r_2^2 + 2r_1 r_2 \cos(\theta_1 - \theta_2)}$, and $\theta$ is determined by the pair of equations: $\cos\theta = (r_1 \cos\theta_1 + r_2 \cos\theta_2)/r$, $\sin\theta = (r_1 \sin\theta_1 + r_2 \sin\theta_2)/r$

**23.** The center of mass of three particles that lie inside or on the unit circle also lies inside or on the circle.

## Exercises 1.4, Page 25

**1.** (a) $\dfrac{\sqrt{2}}{2} - \dfrac{\sqrt{2}}{2}i$ (b) $e^2 i$ (c) $e^{\cos 1}\cos(\sin 1) + ie^{\cos 1}\sin(\sin 1)$

**3.** (a) $\dfrac{\sqrt{2}}{3}e^{-i\pi/4}$ (b) $16\pi e^{i4\pi/3}$ (c) $8e^{i3\pi/2}$

**11.** (a), (c), (d)

**17.** (a) circle $|z| = 3$ traversed counterclockwise

   (b) circle $|z - i| = 2$ traversed counterclockwise

   (c) upper half of circle $|z| = 2$ traversed counterclockwise

   (d) circle $|z - (2 - i)| = 3$ traversed clockwise

**21.** $\left|\dfrac{1 - z^n}{1 - z}\right| = |1 + z + \cdots + z^{n-1}| \le 1 + 1 + \cdots + 1 = n$ for $z = e^{i\theta} \ne 1$

## Exercises 1.5, Page 31

**5.** (a) $2(\cos \pi/4 + i \sin \pi/4)$, $2(\cos 3\pi/4 + i \sin 3\pi/4)$, $2(\cos 5\pi/4 + i \sin 5\pi/4)$, $2(\cos 7\pi/4 + i \sin 7\pi/4)$

   (b) $1$, $\cos 2\pi/5 + i \sin 2\pi/5$, $\cos 4\pi/5 + i \sin 4\pi/5$, $\cos 6\pi/5 + i \sin 6\pi/5$, $\cos 8\pi/5 + i \sin 8\pi/5$

   (c) $\cos \pi/8 + i \sin \pi/8$, $\cos 5\pi/8 + i \sin 5\pi/8$, $\cos 9\pi/8 + i \sin 9\pi/8$, $\cos 13\pi/8 + i \sin 13\pi/8$

   (d) $\sqrt[3]{2}[\cos(-\pi/9) + i \sin(-\pi/9)]$, $\sqrt[3]{2}(\cos 5\pi/9 + i \sin 5\pi/9)$, $\sqrt[3]{2}(\cos 11\pi/9 + i \sin 11\pi/9)$

   (e) $\sqrt[4]{2}(\cos 3\pi/8 + i \sin 3\pi/8)$, $\sqrt[4]{2}(\cos 11\pi/8 + i \sin 11\pi/8)$

   (f) $\sqrt[12]{2}\exp[i\pi(1 + 8k)/24]$, $k = 0, 1, 2, 3, 4, 5$

7. **(a)** $-\dfrac{1}{4} \pm i\dfrac{\sqrt{23}}{4}$   **(b)** $2 - i,\ 1 - i$   **(c)** $1 \pm \sqrt[4]{2}e^{-i\pi/8}$

9. $1,\ 1 + i\sqrt{3},\ 1 - i\sqrt{3}$

11. $z = 1/(w - 1),\ w = e^{i2k\pi/5},\ k = 1, 2, 3, 4$

15. $\sqrt[4]{8}(\cos 5\pi/8 + i \sin 5\pi/8),\ \sqrt[4]{8}(\cos 13\pi/8 + i \sin 13\pi/8)$

### Exercises 1.6, Page 35

3. (b), (c), (f)

5. (a), (c)

7. (a), (b), (c), (d), (e)

11. all points of $S$ and 0

17. No, $S \cap T$ might not be connected.

19. $D$ is not connected.

### Exercises 1.7, Page 39

3. Elevation $= 3.59 \times 10^7$ m, velocity $= 3.07 \times 10^3$ m/sec

5. $l = .0732$ m, $dl/dt = -.1155$ m/sec

## CHAPTER 2

### Exercises 2.1, Page 44

1. **(a)** $3x^2 - 3y^2 + 5x + 1 + i(6xy + 5y + 1)$   **(b)** $x/(x^2 + y^2) - iy/(x^2 + y^2)$
   **(c)** $x/[x^2 + (y - 1)^2] + i(1 - y)/[x^2 + (y - 1)^2]$
   **(d)** $(2x^2 - 2y^2 + 3)/\sqrt{(x - 1)^2 + y^2} + i4xy/\sqrt{(x - 1)^2 + y^2}$
   **(e)** $e^{3x}\cos 3y + ie^{3x}\sin 3y$   **(f)** $2\cos y \cosh x + i2 \sin y \sinh x$

3. **(a)** half-plane Re $w > 5$   **(b)** upper half-plane Im $w \geq 0$   **(c)** $|w| \geq 1$
   **(d)** circular sector $|w| < 2,\ -\pi < \operatorname{Arg} w < \pi/2$

5. **(a)** domain of definition $= \mathbf{C}$, range $= \mathbf{C}\backslash\{0\}$   **(c)** circle $|w| = e$
   **(d)** ray (half-line) $\operatorname{Arg} w = \pi/4$   **(e)** infinite sector $0 \leq \operatorname{Arg} w \leq \pi/4$

13. **(a)** $\left|\dfrac{1}{\bar{z} - \bar{z}_0}\right| = \dfrac{1}{|z - z_0|},\ \arg\left(\dfrac{1}{\bar{z} - \bar{z}_0}\right) = \arg(z - z_0)$   **(b)** $\sqrt{2}/3 + 2i/3;\ -\sqrt{2}/3 + 2i/3$

### Exercises 2.2, Page 49

1. spirals to 0

5. **(a)** converges to 0   **(b)** does not converge   **(c)** converges to $\pi$
   **(d)** converges to $2 + i$   **(e)** converges to 0   **(f)** does not converge

9. **(a)** $-8i$   **(b)** $-7i/2$   **(c)** $6i$   **(d)** $-\frac{1}{2}$   **(e)** $2z_0$   **(f)** $4\sqrt{2}$

11. limits exist except at $z = -1$; continuous except at $z = -1$ and $z = 0$; removable
    discontinuity at $z = 0$

15. no

17. $-\frac{1}{2} - i$

19. **(a)** 1   **(b)** 0   **(c)** $-\dfrac{\pi}{2} + i$   **(d)** 1

### Exercises 2.3, Page 56

7. (a) $18z^2 + 16z + i$ (b) $-12z(z^2 - 3i)^{-7}$ (c) $\dfrac{(iz^3 + 2z + \pi)2z - (z^2 - 9)(3iz^2 + 2)}{(iz^3 + 2z + \pi)^2}$

(d) $(z + 2)^2[-5z^2 + (-16 - i)z + 3 - 8i]/(z^2 + iz + 1)^5$

(e) $24i(z^3 - 1)^3(z^2 + iz)^{99}(53z^4 + 28iz^3 - 50z - 25i)$

9. (a) $2 - 3i$ (b) $\pm i$ (c) $\frac{1}{2}(-1 \pm i\sqrt{15})$ (d) $\frac{1}{2}, 1$

11. (a) nowhere analytic (b) nowhere analytic (c) analytic except at $z = 5$
(d) everywhere analytic (e) nowhere analytic (f) analytic except at $z = 0$
(g) nowhere analytic (h) nowhere analytic

13. (a), (b), (d), (f), (g)

15. $\frac{3}{5}$

### Exercises 2.4, Page 62

3. $g(z) = 3z^2 + 2z - 1$

5. $f'(z) = 2e^{x^2-y^2}(x + iy)[\cos(2xy) + i\sin(2xy)]$

7. Hint: If $f' = g'$ in $D$, consider $h := f - g$.

9. If $F$ were analytic, then $\underline{\mathrm{Im}}\, F(z) \equiv 0$ implies $F(z)$ is constant, which is a contradiction.

11. Hint: Consider $f(z) + \overline{f(z)}$.

13. Hint: $|f(z)| \equiv$ constant by condition (8). Now use the result of Prob. 12.

### Exercises 2.5, Page 68

3. (a) $v = -x + a$ (b) $v = -e^x \cos y + a$ (c) $v = y^2/2 - y - x^2/2 - x + a$
(d) $v = \cos x \sinh y + a$ (e) $v = \mathrm{Tan}^{-1}(y/x) + a = \mathrm{Arg}\, z + a$
(f) $v = -\mathrm{Re}(e^{z^2}) + a = -e^{x^2-y^2}\cos(2xy) + a$

7. $\phi(x, y) = x + 1$

9. $\phi(x, y) = xy - 1$

13. (one example) $\phi(r, \theta) = r^4 \sin 4\theta = 4x^3y - 4xy^3$

15. $\phi(x, y) = \ln|z/3|$

17. (a) $\phi(x, y) = \mathrm{Re}(z^2 + 5z + 1)$ (b) $\phi(x, y) = 2\,\mathrm{Re}\left(\dfrac{z^2}{z + 2i}\right)$

### Exercises 2.6, Page 73

3. See Fig. 2.6(a).

## CHAPTER 3

### Exercises 3.1, Page 79

3. sum $= \dfrac{1 - e^{101z}}{1 - e^z}$ for $z \neq 2k\pi i$; sum $= 101$ for $z = 2k\pi i$

5. (a) $e^2\dfrac{\sqrt{2}}{2} + ie^2\dfrac{\sqrt{2}}{2}$ (b) $ie^2$ (c) $i\sinh 2$ (d) $\cos(1)\cosh(1) + i\sin(1)\sinh(1)$
(e) $-\sinh(1)$ (f) $0$

9. (a) $2\pi z \exp(\pi z^2)$ (b) $-2\sin(2z) - (i/z^2)\cos(1/z)$ (c) $2\cos(2z)\exp[\sin(2z)]$
(d) $3\tan^2 z \sec^2 z$ (e) $2(\sinh z + 1)\cosh z$ (f) $1 - \tanh^2 z = \mathrm{sech}^2 z$

**17. (a)** $z = ik\pi/2$, $k = 0$, $\pm 1$, $\pm 2, \ldots$     **(b)** $z = 2k\pi - i \ln 3$, $k = 0$, $\pm 1$, $\pm 2, \ldots$
    **(c)** no solution
**19.** Hint: If $z_1 \neq z_2$ and $e^{z_1} = e^{z_2}$, then $|z_1 - z_2| = |2k\pi i| \geq 2\pi$.

### Exercises 3.2, Page 86

**1. (a)** $i(\pi/2 + 2k\pi)$, $k = 0$, $\pm 1$, $\pm 2, \ldots$     **(b)** $\frac{1}{2} \text{Log } 2 + i(7\pi/4 + 2k\pi)$, $k = 0$, $\pm 1$, $\pm 2, \ldots$
    **(c)** $-i\pi/2$   **(d)** $\text{Log } 2 + i(\pi/6)$
**5. (a)** $\text{Log } 2 + i(\pi/2 + 2k\pi)$, $k = 0$, $\pm 1$, $\pm 2, \ldots$     **(b)** $\pm \sqrt[4]{2} \exp(i\pi/8)$
    **(c)** $i(2\pi/3 + 2k\pi)$, $i(4\pi/3 + 2k\pi)$, $k = 0$, $\pm 1$, $\pm 2, \ldots$
**9.** cut plane: $\mathbf{C} \setminus \{z = x + i: x \geq 4\}$, $f'(z) = -1/(4 + i - z)$
**11.** (one example) $f(z) = \text{Log}(z^2 + 2z + 3)$, $f'(-1) = 0$
**13. (a)** $\text{Log } (2z - 1)$   **(b)** $\mathcal{L}_0(2z - 1)$, where $\mathcal{L}_0(re^{i\theta}) = \text{Log } r + i\theta$, $0 < \theta < 2\pi$
    **(c)** $\mathcal{L}_{\pi/2}(2z - 1)$, where $\mathcal{L}_{\pi/2}(re^{i\theta}) = \text{Log } r + i\theta$, $\pi/2 < \theta < 5\pi/2$
**15.** $w = (1/\pi)\text{Log } z$
**19.** Choose a branch of $\arg z$ that is continuous on the complement of the half-parabola;
    say, $\arg(re^{i\theta}) = \theta$, where $g(r) < \theta < g(r) + 2\pi$, $g(r) = \text{Tan}^{-1}\sqrt{2/(\sqrt{1 + 4r^2} - 1)}$.

### Exercises 3.3, Page 93

**1. (a)** $\exp(-\pi/2 - 2k\pi)$, $k = 0$, $\pm 1$, $\pm 2, \ldots$     **(b)** $+1$, $-\dfrac{1}{2} + i\dfrac{\sqrt{3}}{2}$, $-\dfrac{1}{2} - i\dfrac{\sqrt{3}}{2}$
    **(c)** $\exp[-2k\pi^2 + \pi i \text{ Log } 2]$, $k = 0$, $\pm 1$, $\pm 2, \ldots$
    **(d)** $(1 + i) \exp[2k\pi + \pi/4 - (i/2) \text{ Log } 2]$, $k = 0$, $\pm 1$, $\pm 2, \ldots$     **(e)** $-2 + 2i$
**3. (a)** $2$   **(b)** $e^{-\pi}$   **(c)** $(1 + i) \exp[(i/2) \text{ Log } 2 - \pi/4]$
**5.** Take, for example, $z_1 = -1 + i$, $z_2 = i$, $\alpha = 1/2$.
**7.** $(1 + i) \exp(-\pi/2)$
**11.** $z = \pi/4 + k\pi$, $k = 0$, $\pm 1$, $\pm 2, \ldots$
**15. (a)** $i \exp[\frac{1}{2} \text{ Log}(1 - z^2)]$   **(b)** $z \exp[\frac{1}{2} \text{ Log}(4/z^2 + 1)]$
    **(c)** $z^2 \exp[\frac{1}{2} \text{ Log}(1 - 1/z^4)]$   **(d)** $z \exp[\frac{1}{3} \text{ Log}(1 - 1/z^3)]$
**17.** $0 < \text{Sec}^{-1} x < \pi/2$ for $x > 1$; $\pi/2 < \text{Sec}^{-1} x < \pi$ for $x < -1$

### Exercises 3.4, Page 99

**1. (a)** $I_s = \dfrac{A}{R_0} \cos(\omega t - \phi_0)$

  where $R_0 = \left| R + \dfrac{1}{i\omega C} \right| = \sqrt{R^2 + \dfrac{1}{\omega^2 C^2}}$

  and  $\phi_0 = \text{Arg}\left( R + \dfrac{1}{i\omega C} \right) = -\tan^{-1}\left( \dfrac{1}{\omega RC} \right)$

**(b)** $I_s = \dfrac{A\sqrt{\omega^2 L^2 + (\omega^2 RCL - R)^2}}{\omega RL} \cos[\omega(t + \beta) + \phi_0]$

  where $\phi_0 = \tan^{-1}\left( \dfrac{\omega^2 RCL - R}{\omega L} \right)$

(c) $I_s = \dfrac{A\omega_1}{\omega_1^4 + 4} \sqrt{16 + 4\omega_1^2 + 4\omega_1^4 + \omega_1^6} \cos(\omega_1 t + \phi_1)$

$\quad + \dfrac{B\omega_2}{\omega_2^4 + 4} \sqrt{16 + 4\omega_2^2 + 4\omega_2^4 + \omega_2^6} \sin(\omega_2 t + \phi_2)$

where $\phi_1 = \tan^{-1}\left(\dfrac{4}{\omega_1^3 + 2\omega_1}\right)$, $\phi_2 = \tan^{-1}\left(\dfrac{4}{\omega_2^3 + 2\omega_2}\right)$

# CHAPTER 4

## Exercises 4.1, Page 113

1. (a) $z(t) = (1 + i) + t(-3 - 4i)$, $0 \le t \le 1$   (b) $z(t) = 2i + 4e^{-it}$, $0 \le t \le 2\pi$
   (c) $z(t) = Re^{it}$, $\pi/2 \le t \le \pi$   (d) $z(t) = t + it^2$, $1 \le t \le 3$
3. $z(t) = a \cos t + ib \sin t$, $0 \le t \le 2\pi$
5. yes

7. $z(t) = \begin{cases} -1 - i + 8t & 0 \le t \le \frac{1}{4} \\ 1 - i + 8i(t - \frac{1}{4}) & \frac{1}{4} \le t \le \frac{1}{2} \\ 1 + i - 8(t - \frac{1}{2}) & \frac{1}{2} \le t \le \frac{3}{4} \\ -1 + i - 8i(t - \frac{3}{4}) & \frac{3}{4} \le t \le 1 \end{cases}$

   length $= 8$

9. $z(t) = \begin{cases} -2 + \exp(-6\pi i t) & 0 \le t \le \frac{1}{3} \\ -1 + 6(t - \frac{1}{3}) & \frac{1}{3} \le t \le \frac{2}{3} \\ 2 - \exp[6\pi i(t - \frac{2}{3})] & \frac{2}{3} \le t \le 1 \end{cases}$

11. $15\pi$
13. (a) instantaneous velocity at time $t$   (b) instantaneous speed at time $t$
    (c) infinitesimal (differential) distance traveled during time interval $dt$
    (d) distance traveled from time $t = a$ to time $t = b$

## Exercises 4.2, Page 122

1. yes
3. (a) $1 + i/3$   (b) $(1 + i) \sinh 2$
   (c) $\dfrac{i}{12}[1 - (1 + 2i)^6] = \dfrac{11}{3} - \dfrac{29i}{3}$   (d) $\dfrac{1}{2i} - \dfrac{1}{8 + 2i} = -\dfrac{2}{17} - \dfrac{8i}{17}$
5. $4\pi i$
7. $\frac{1}{2} + i$
9. $\dfrac{13}{10} + \dfrac{i}{6}$
11. (a) $3 + i$   (b) $3 + i$   (c) $3 + i$
13. $-2i$

### Exercises 4.3, Page 128

1. (a) $-3 + 2i$  (b) $-2 \sinh 1$  (c) $i\pi$  (d) $0$  (e) $-\dfrac{i}{3} \sinh^3 1$

   (f) $\dfrac{e^\pi}{2} + \dfrac{e^i}{2} (\cosh 1 + i \sinh 1)$  (g) $-\dfrac{\sqrt{2}}{3} - \dfrac{2}{3}(\sqrt{\pi})^3 + \dfrac{i\sqrt{2}}{3}$

   (h) $\pi - 2 + i\left(2 - \dfrac{\pi^2}{4}\right)$  (i) $\dfrac{\pi}{4} - \dfrac{1}{2} \arctan 2 + \dfrac{i}{4} \operatorname{Log} 5$

5. Hint: Consider Theorem 7.

7. Hint: Consider a branch of $\log(z - z_0)$ whose branch cut does not intersect $C$.

### Exercises 4.4, Page 148

1. (a), (c)
3. (a), (b), (d), (e)
5. $z(s, t) = (2 - s) \cos 2\pi t + i(3 - 2s) \sin 2\pi t,\ 0 \le s,\, t \le 1$
9. (a), (c), (d), (f)
11. Since the whole plane $\mathbf{C}$ is simply connected, Theorem 10 (or Theorem 13) applies.
13. (a) $\pi$  (b) $0$  (c) $-\pi$
15. $-4\pi i$
17. $0$

### Exercises 4.5, Page 160

1. $0$
3. (a) $-2\pi i$  (b) $\dfrac{3\pi i e^{3/2}}{2}$  (c) $\dfrac{2\pi i}{9}$  (d) $10\pi i$  (e) $-2e\pi i$  (f) $-i\pi/2$
5. $G(1) = 4\pi i,\ G'(i) = -2\pi(2 + i),\ G''(-i) = 4\pi i$
7. $\dfrac{-2\pi i}{9}$
9. $|f^{(n)}(0)| \le Mn!$
11. $\dfrac{\partial^2 u}{\partial x^2} = \operatorname{Re} f''$
13. $g(z)$ is not analytic inside $\Gamma$; note $G(z) \equiv 0$.

### Exercises 4.6, Page 167

3. Hint: Apply the Cauchy estimates to the disk $\{\zeta : |\zeta - z| \le r - |z|\}$.
5. $|e^{f(z)}| \le e^M$, so $e^{f(z)} \equiv$ constant, which implies $f'(z)e^{f(z)} \equiv 0$. Thus $f'(z) \equiv 0$.
7. If $f$ is entire and $|f(z)| \le M|z|^n$ for $|z| > r_0$, where $n$ is a nonnegative integer, then $f$ must be a polynomial of degree at most $n$.
15. Hint: Suppose $f$ does not vanish and apply the maximum modulus principle as well as the minimum modulus principle (Prob. 14).
17. $\dfrac{9\sqrt{2}}{8}$

**Exercises 4.7, Page 174**

1. $\phi(z) \equiv -5$
5. Consider $\phi_1(x, y) = y$ and $\phi_2(x, y) \equiv 0$ in the upper half-plane.
11. 3

15. $\dfrac{1}{\pi}\left[\tan^{-1}\left(\dfrac{1-x}{y}\right) - \tan^{-1}\left(\dfrac{-1-x}{y}\right)\right]$ for $y > 0$.

# CHAPTER 5

**Exercises 5.1, Page 185**

1. (a) $\frac{9}{10} + i\frac{3}{10}$  (b) $3(1 - i)$  (c) $\frac{3}{5}$  (d) $\dfrac{-2+i}{5 \cdot 2^{13}}$  (e) $\frac{9}{8}$  (f) $-1$

3. Hint: If $z_n, z_{n+1}, z_{n+2}, \ldots$ are within $\varepsilon$ of their limit $L$, how far apart can any two $z_j$ be?
5. Apply Prob. 3.
7. (a) diverges  (b) converges  (c) diverges  (d) diverges  (e) converges  (f) diverges
9. Hint: How does $|z|$ compare with $|x| + |y|$?
11. (a) $|z| < 1$  (b) $|z - i| < 2$  (c) all $z$  (d) $|z + 5i| < 1$
17. Hint: Apply Log to the inequality $x^n < \frac{1}{2}$.

**Exercises 5.2, Page 195**

3. Hint: Use the chain rule to find the derivatives of the composite functions.

5. (a) $\displaystyle\sum_{j=0}^{\infty} (-z)^j, |z| < 1$  (b) $\displaystyle\sum_{j=0}^{\infty} \dfrac{(-1)^j z^{2j}}{j!}$, all $z$  (c) $\displaystyle\sum_{j=0}^{\infty} \dfrac{(-1)^j 3^{2j+1} z^{2j+4}}{(2j+1)!}$, all $z$

(d) $\displaystyle\sum_{j=0}^{\infty} \dfrac{i^j[1 + (-1)^j] - i}{j!} z^j$, all $z$  (Note that $[1 + (-1)^j]$ vanishes for odd $j$.)

(e) $i + \displaystyle\sum_{j=1}^{\infty} \dfrac{2}{(1 - i)^{j+1}} (z - i)^j, |z - i| < \sqrt{2}$

(f) $\dfrac{1}{\sqrt{2}}\left\{1 - \left(z - \dfrac{\pi}{4}\right) - \dfrac{1}{2!}\left(z - \dfrac{\pi}{4}\right)^2 + \dfrac{1}{3!}\left(z - \dfrac{\pi}{4}\right)^3 + \dfrac{1}{4!}\left(z - \dfrac{\pi}{4}\right)^4 - \cdots\right\}$, all $z$

(g) $\displaystyle\sum_{j=1}^{\infty} jz^j \left(= z\dfrac{d}{dz}\dfrac{1}{1 - z}\right), |z| < 1$

7. $2\displaystyle\sum_{j=0}^{\infty} \dfrac{z^{2j+1}}{2j + 1}, |z| < 1$

11. (a) $1 + z - \dfrac{z^3}{3} + \cdots$  (b) $-1 - 2z - \dfrac{5z^2}{2} - \cdots$  (c) $1 + \dfrac{z^2}{2} + \dfrac{5z^4}{24} + \cdots$

(d) $z - \dfrac{z^3}{3} + \dfrac{2z^5}{15} + \cdots$

13. $\dfrac{z(1 + z)}{(1 - z)^3}$

17. $f(z) = (1 - z)^{-1}$ is not analytic at $z = 1$.

19. nine terms ($n = 0$ to 8)

### Exercises 5.3, Page 203

3. (a) $|z| = 1$   (b) $|z - 1| = \frac{1}{2}$   (c) $|z| = 0$   (d) $|z - i| = 3$   (e) $|z + 2| = \dfrac{1}{\sqrt{10}}$   (f) $|z| = 2$

5. (a) $\dfrac{6!6^3}{3^6}$   (b) $2\pi i$   (c) $0$   (d) $0$

7. $z + \dfrac{z^3}{3} + \dfrac{z^5}{10}$

9. Hint: The polynomials are analytic inside and on $C$.

11. (b) Hint: Argue that two analytic functions that agree on a real interval must have identical derivatives.

13. (a) $\displaystyle\sum_{k=0}^{\infty} \frac{z^{2k}}{2^k k!} = e^{z^2/2}$

(b) $1 + z - \dfrac{4}{2!} z^2 - \dfrac{4}{3!} z^3 + \dfrac{4^2}{4!} z^4 + \dfrac{4^2}{5!} z^5 + \cdots$

$= \left[ 1 - \dfrac{(2z)^2}{2!} + \dfrac{(2z)^4}{4!} - \cdots \right] + \dfrac{1}{2}\left[ (2z) - \dfrac{(2z)^3}{3!} + \dfrac{(2z)^5}{5!} - \cdots \right]$

$= \cos 2z + \dfrac{1}{2}\sin 2z$

(c) $\displaystyle\sum_{k=0}^{\infty} (k + 1)z^{2k} = \frac{1}{(1 - z^2)^2}$

15. (a) $\displaystyle\sum_{k=0}^{\infty} \frac{(-1)^k}{(2k + 1)!} \left[ \int_{-1}^{2} t^{2k+1} g(t)\, dt \right] z^{2k+1}$   (b) Hint: Differentiate (a) termwise.

### Exercises 5.4, Page 209

1. (a) $2$   (b) $\infty$   (c) $0$   (d) $\infty$

3. (a) $2$   (b) $1$   (c) $\frac{1}{3}$   (d) $e$ (Hint: Use the ratio test)   (e) $1$   (f) $1$

5. (a) $R$   (b) $R^4$   (c) $\sqrt{R}$   (d) $R$   (e) $\infty$ (if $R > 0$)

9. $f(z) = a_0 + \displaystyle\sum_{j=0}^{\infty} z^{2^j}, \ |z| < 1$

11. $1, \zeta, \dfrac{(3\zeta^2 - 1)}{2}, \dfrac{(5\zeta^3 - 3\zeta)}{2}$

### Exercises 5.5, Page 217

1. (a) $\displaystyle\sum_{j=-1}^{\infty} (-1)^{j+1} z^j$   (b) $\displaystyle\sum_{j=2}^{\infty} (-1)^j z^{-j}$   (c) $-\displaystyle\sum_{j=-1}^{\infty} (z + 1)^j$   (d) $\displaystyle\sum_{j=2}^{\infty} (z + 1)^{-j}$

3. (a) $\dfrac{1}{3}\displaystyle\sum_{j=0}^{\infty} \left[ (-1)^j - \left(\dfrac{1}{2}\right)^j \right] z^j$   (b) $\dfrac{1}{3}\displaystyle\sum_{j=1}^{\infty} (-1)^{j-1} z^{-j} - \dfrac{1}{3}\displaystyle\sum_{j=0}^{\infty} \left(\dfrac{1}{2}\right)^j z^j$

(c) $\dfrac{1}{3}\displaystyle\sum_{j=1}^{\infty} [(-1)^{j-1} + 2^j] z^{-j}$

5. $\dfrac{\frac{5}{4}}{(z - 4)^3} + \displaystyle\sum_{j=-2}^{\infty} (-1)^{j+1} \left(\dfrac{1}{4}\right)^{j+4} (z - 4)^j$

7. **(a)** $\dfrac{1}{z^2} + \dfrac{1}{z^3} + \dfrac{3}{2z^4} + \cdots$ **(b)** $\dfrac{1}{z} - \dfrac{1}{2} + \dfrac{z}{12} + \cdots$ **(c)** $\dfrac{1}{z} + \dfrac{z}{6} + \dfrac{7z^3}{360} + \cdots$

**(d)** $\dfrac{1}{e}\left[ 1 + z + \dfrac{z^2}{2!} + \cdots \right]$

9. $\frac{1}{2} < |z| < 2$

11. $\displaystyle\sum_{j=n}^{\infty} \alpha^{j-n} \dfrac{(j-1)!}{(j-n)!(n-1)!} z^{-j}$

13. Hint: Use Eq. (1).

## Exercises 5.6, Page 225

1. **(a)** pole of order 2 at 0, removable singularity at $-1$
   **(b)** essential singularity at 0  **(c)** simple poles at $\pm i$
   **(d)** simple poles at $2n\pi i$ $(n = 0, \pm 1, \pm 2, \ldots)$
   **(e)** simple poles at $\dfrac{2n+1}{2}\pi$  $(n = 0, \pm 1, \pm 2, \ldots)$  **(f)** essential singularity at 0
   **(g)** removable singularity at 0
   **(h)** essential singularity at 0, simple poles at $\dfrac{1}{n\pi}$ $(n = \pm 1, \pm 2, \ldots)$

3. Possible answers are:
   **(a)** $\dfrac{(z-i)^2}{(z-2+3i)^5}$  **(b)** $ze^{1/(z-1)}$  **(c)** $\dfrac{(\sin z)e^{1/(z-i)}}{z(z-1)^6}$  **(d)** $\dfrac{e^{1/[z(z-1)]}}{(z-1-i)^2}$

5. **(a)** false  **(b)** true  **(c)** true  **(d)** false  **(e)** true
7. essential
9. Yes; $e^{1/z}$ is bounded on the negative real axis.
11. Hint: If, say, Re $f \le M$ then $e^{f(z)}$ would have a removable singularity. Now take the log.
13. Hint: Choose a tiny contour in the formula for $a_{-j}$.

## Exercises 5.7, Page 230

1. **(a)** essential singularity  **(b)** essential singularity  **(c)** analytic  **(d)** zero of order 2
   **(e)** pole of order 2  **(f)** essential singularity  **(g)** essential singularity
   **(h)** essential singularity  **(i)** analytic

3. **(a)** $1 + 2\displaystyle\sum_{j=1}^{\infty} (-1)^j/z^j, |z| > 1$  **(b)** $\displaystyle\sum_{j=0}^{\infty} (-1)^j/z^{2j}, |z| > 1$  **(c)** $\displaystyle\sum_{j=0}^{\infty} i^j/z^{3j+3}, |z| > 1$.

5. $(\deg Q) - (\deg P)$

7. Observe $\displaystyle\oint_{|z|=1} \dfrac{dz}{z} = 2\pi i$, page 219.

11. Hint: Exploit Theorem 16, page 219, and Lemma 7, page 222.
13. Hint: Write $z - z_0 = re^{i\theta}$ and count the sign changes in Re$(z - z_0)^{\pm m}$ as $z$ encircles $z_0$.
15. **(b)** Hint: Observe that Re $f(z)$ is large positive when $z$ lies in the right half-plane near an imaginary pole of the type considered.

### Exercises 5.8, Page 240

1. $z^2$

3. $\sin\dfrac{1}{1-z}$ vanishes at $z = 1 - \dfrac{1}{n\pi}$, $n = 1, 2, \ldots$.

5. all values except $0 \le \alpha < 1$

7. They both sum to $\dfrac{1}{1-z}$.

9. (a) no  (b) yes  (c) yes  (d) no  (e) yes  (f) yes

11. $f(z) = zg'(z)$

15. If $\phi(x, y) \to 0$, it can be harmonically extended as an odd function of $y$. If $\partial\phi/\partial y \to 0$, $\phi$ can be harmonically extended as an even function of $y$.

## CHAPTER 6

### Exercises 6.1, Page 251

1. (a) $\text{Res}(2) = e^6$  (b) $\text{Res}(1) = -2$, $\text{Res}(2) = 3$  (c) $\text{Res}(0) = 0$  (d) $\text{Res}(-1) = -6$
   (e) $\text{Res}(0) = 1$, $\text{Res}(-1) = -5/2e$  (f) $\text{Res}(0) = \frac{1}{3}$

   (g) $\text{Res}\left[\pm\dfrac{(2n+1)\pi}{2}\right] = -1$, $n = 0, 1, 2, \ldots$  (h) $\text{Res}(n\pi) = (-1)^n(n\pi - 1)$, $n = 0$,
   $\pm 1, \pm 2, \ldots$

   (i) $\text{Res}(1) = -2$

3. (a) $\pi i \sin 2$  (b) $\dfrac{\pi i(e^2 - 1)}{4}$  (c) $-8\pi i$  (d) $\pi i\left[\dfrac{(2 - 5i)e^{2i}}{58} - \dfrac{12 - 5i}{50}\right]$  (e) $\dfrac{\pi i}{3}$

   (f) 0  (g) 0

5. no; yes ($1/z^2$, for example)

7. $2\pi i$

### Exercises 6.4, Page 271

5. $\dfrac{\pi \sin 3}{2e^3}$

7. $\dfrac{\pi}{3e}\left(1 - \dfrac{1}{2e}\right)$

9. $\dfrac{i\pi}{e^6}$

11. $m > 0$, $\deg P < \deg Q$

### Exercises 6.5, Page 279

1. (a) $\dfrac{i\pi}{2}$  (b) $\dfrac{3\pi i e^{3i}}{8}$  (c) 0  (d) $-\pi i$

9. $\dfrac{3\pi}{4}$

11. $-\pi \cot(a\pi)$

### Exercises 6.7, Page 296

1. (a), (c), (e), and (f)
3. 1
7. Hint: Compare with $f(z) = 27$.
9. 4
13. Easy; let $h(z) = -f(z)$
15. Hint: $f$ and $f + h$ have the same poles; now apply Prob. 14.
21. A "zero" for $F(z)$ is a "minus 1" for $P(z)$; apply the argument principle.

## CHAPTER 7

### Exercises 7.1, Page 306

1. $\text{Log}|w| + \text{Arg}\,w;\quad e^x \cos y + e^x \sin y$

3. $\text{Arg}(z - 2) - \text{Arg}(z + 1);\quad a_4 + \dfrac{1}{\pi}\displaystyle\sum_{k=1}^{3}(a_k - a_{k+1})\,\text{Arg}(z - x_k)$

5. $\dfrac{1}{2} - \dfrac{1}{2}\dfrac{x^2 + y^2 - 1}{(1 + x)^2 + y^2}$

7. These are the Cauchy-Riemann equations for $f^{-1}(w)$.

### Exercises 7.2, Page 314

1. **(a)** $1;\ f(-1 + \zeta) = f(-1 - \zeta)$  **(b)** $1;\ f(n\pi + \zeta) = f(n\pi - \zeta)$
   **(c)** $2;\ f(r) = f(re^{i2\pi/3}) = f(re^{i4\pi/3})$
3. Angles increase (decrease) for $\alpha > 1$ ($\alpha < 1$).
5. pure imaginary constants
11. **(a)** the upper half-plane: $\text{Im}\,w > 0$
    **(b)** the whole plane minus the logarithmic spiral $\rho = e^\phi,\ -\infty \le \phi < \infty$
    **(c)** $\{w\colon |w| < 1,\ \text{Im}\,w > 0\}$  **(d)** $\{w\colon |w| > 1,\ \text{Im}\,w > 0\}$
    **(e)** the upper half-annulus $\{w\colon e < |w| < e^2,\ \text{Im}\,w > 0\}$  **(f)** $\{w\colon |w| > 1\}$ and $\{w\colon |w| < 1\}$
13. **(a)** the upper half-plane: $\text{Im}\,w > 0$  **(b)** fourth quadrant
    **(c)** the whole plane minus the real intervals $(-\infty, -1], [1, \infty)$
    **(d)** the interior of the ellipse $(u^2/\cosh^2 1) + (v^2/\sinh^2 1) = 1$ excluding the real segments
    $[-\cosh 1, -1], [1, \cosh 1]$

### Exercises 7.3, Page 325

1. $w = 3iz + 5$
3. **(a)** $\{w\colon |w - 2 + 2i| \le 1\}$  **(b)** $\{w\colon |w - 6i| \le 3\}$  **(c)** $\{w\colon \text{Re}(w) \le \tfrac{1}{2}\}$;
   **(d)** $\{w\colon \text{Re}(w) \ge \tfrac{3}{2}\}$  **(e)** $\{w\colon |w - \tfrac{2}{3}| \le \tfrac{1}{3}\}$

5. $w = e^{3\pi i/4}\left(\dfrac{z + i}{z - 1}\right)$

7. **(a)** $w = iz$  **(b)** $w = \dfrac{2z}{z + 1}$  **(c)** $w = \dfrac{z + i}{z - 1}$  **(d)** $w = \dfrac{z + 1}{z - 1}$

9. the region exterior to both of the circles $C_1\colon |w - (1 - i)/2| = 1/\sqrt{2}$ and
   $C_2\colon |w - (1 + i)/2| = 1/\sqrt{2}$

11. $w = \exp\left(4\pi\left[-i\left(\dfrac{1}{z-2}+\dfrac{1}{4}\right)\right]\right)$

### Exercises 7.4, Page 334

1. $z - 2$

3. (a) $\dfrac{4-3i}{25}$   (b) $\dfrac{7-i}{6}$   (c) $\dfrac{5-2i}{3}$

5. $\lambda > 0$

7. No. (This would violate the symmetry principle.)

9. 0

15. $w = \dfrac{i-2z}{2+iz}\, e^{i\theta}$ (any real $\theta$)

17. $w = \dfrac{az+b}{cz+d}$ with $a$, $b$, $c$, $d$ real and $ad - bc > 0$

19. (b) No. (A shift, for example, does not commute with an inversion.)

### Exercises 7.5, Page 346

1. $w = A(z-x_1)^2 - 1$, where $A < 0$

3. $w = \dfrac{i}{2} - \dfrac{i}{\pi}\sin^{-1} z$

5. $w = -\dfrac{2}{\pi}(\sin^{-1} z + z\sqrt{1-z^2})$

7. $w = \dfrac{i}{\pi}\sqrt{z^2-1} + \dfrac{\sin^{-1} z}{\pi} + \dfrac{1}{2}$

9. $w = \dfrac{1}{\pi}\mathrm{Log}\left(\dfrac{z-x_2}{x_2-x_1}\right)$, where $x_1 < x_2$

11. The analytic continuation of the S-C transformation across the interval $(x_{j-1}, x_j)$ maps the lower half-plane onto the figure obtained by reflecting the polygon $P$ through the mirror containing $f(x_{j-1})$ and $f(x_j)$.

### Exercises 7.6, Page 359

1. $\phi(x, y) = \dfrac{2}{\pi}\mathrm{Arg}\left(\dfrac{1+z}{1-z}\right)$

3. $T(x, y) = \dfrac{2}{\pi}\mathrm{Arg}\left(\dfrac{1+z^2}{1-z^2}\right) - 1$   5. $\phi(x, y) = \dfrac{1}{\pi}\mathrm{Arg}\left(\dfrac{\sqrt{z^2+1}-1}{\sqrt{z^2+1}+1}\right)$

7. (a) $T(x, y) = 1 - \dfrac{1}{\pi}[\mathrm{Arg}(\cos(\pi i z) + 1) + \mathrm{Arg}(\cos(\pi i z) - 1)]$

   (b) $T(x, y) = \dfrac{2}{\pi}\mathrm{Arg}(e^{\pi z} + 1) - \dfrac{1}{\pi}\mathrm{Arg}(e^{\pi z})$

9. $\phi(x, y) = (\mathrm{Log}\, s)^{-1}\mathrm{Log}\left|\dfrac{4-\lambda z}{4\lambda - z}\right|$, where $\lambda = \dfrac{19 + \sqrt{105}}{16}$ and $s = \dfrac{13 - \sqrt{105}}{8}$.

11. A parametric representation of the streamlines is obtained by holding $y$ constant in the S-C mapping equation $w = f(x + iy)$.

### Exercises 7.7, Page 366

1. isotherms: $z(t) = g(a + it)$, $t \geq 0$, $-\dfrac{\pi}{2} < a < \dfrac{\pi}{2}$, where $g(w) = \dfrac{1}{2} + \dfrac{w - \cos w}{\pi}$ maps the half-strip $-\dfrac{\pi}{2} < u < \dfrac{\pi}{2}$, $v > 0$ onto the given region.

3. $T(z) = \dfrac{2}{\pi} \operatorname{Re}[\sin^{-1}(z^2)]$

5. $T(z) = -\dfrac{20}{\pi} \operatorname{Re}[\sin^{-1}(-e^{-\pi z})]$

7. streamlines: $\operatorname{Im}(z^{\pi/\alpha} + z^{-\pi/\alpha}) = \text{constant}$

# CHAPTER 8

### Exercises 8.1, Page 384

1. (a) $\dfrac{i}{8}[-3e^{it} + 3e^{-it} + e^{3it} - e^{-3it}]$   (b) $\displaystyle\sum_{n=-\infty}^{\infty} \dfrac{12(-1)^n}{\pi(9 - 4n^2)(1 - 4n^2)} e^{i2n\pi t/3}$

   (c) $c_n = \dfrac{2(-1)^n}{n^2}$ for $n \neq 0$; $c_0 = \dfrac{\pi^2}{3}$   (d) $c_n = (-1)^n \left[\dfrac{i\pi}{n} - \dfrac{2i}{\pi n^3}\right] + \dfrac{2i}{\pi n^3}$ for $n \neq 0$; $c_0 = 0$

3. (a), (b)

5. $\bar{c}_n = c_{-n}$

7. (a) $\dfrac{1}{8} + \dfrac{\cos 2t}{2} - \dfrac{\cos 4t}{104}$   (b) $\dfrac{\pi^2}{3} + \displaystyle\sum_{\substack{n=-\infty \\ n \neq 0}}^{\infty} \dfrac{c_n e^{int}}{1 + in - n^2}$, $c_n$ as in Prob. 1(c)

   (c) $\displaystyle\sum_{\substack{n=-\infty \\ n \neq 0}}^{\infty} \dfrac{c_n e^{int}}{2 + 4in - n^2}$, $c_n$ as in Example 4.

11. $u(x, t) = \displaystyle\sum_{n=1}^{\infty} b_n \sin nx \cos nt + \sum_{n=1}^{\infty} c_n \sin nx \sin nt$

   $b_n = \dfrac{2}{\pi} \displaystyle\int_0^{\pi} f_1(\xi) \sin n\xi \, d\xi$

   $c_n = \dfrac{2}{\pi n} \displaystyle\int_0^{\pi} f_2(\xi) \sin n\xi \, d\xi$

### Exercises 8.2, Page 397

1. (a) $G(\omega) = \dfrac{1}{\pi}\left(\dfrac{1}{1 + \omega^2}\right)$   (b) $G(\omega) = \dfrac{1}{2\sqrt{\pi}} e^{-\omega^2/4}$   (c) $G(\omega) = \dfrac{-i\omega e^{-\omega^2/4}}{4\sqrt{\pi}}$

   (d) $G(\omega) = \begin{cases} \frac{1}{2} & \text{if } |\omega| < 1 \\ 0 & \text{if } |\omega| > 1 \\ \frac{1}{4} & \text{if } \omega = \pm 1 \end{cases}$   (e) $G(\omega) = \begin{cases} -\dfrac{i}{2} \sin \omega & \text{if } |\omega| \leq \pi \\ 0 & \text{if } |\omega| \geq \pi \end{cases}$

3. (a) $G(\omega) = \dfrac{e^{-|\omega|/\sqrt{2}}}{2\sqrt{2}}\left(\cos\dfrac{|\omega|}{\sqrt{2}} + \sin\dfrac{|\omega|}{\sqrt{2}}\right)$    (b) $G(\omega) = -\dfrac{ie^{-|\omega|/\sqrt{2}}}{2}\sin\dfrac{\omega}{\sqrt{2}}$

(c) $G(\omega) = \dfrac{1}{2\sqrt{\pi}}e^{-\omega^2/4}$

## Exercises 8.3, Page 407

1. (a) $\dfrac{3s}{s^2+4} - \dfrac{8}{s+2}$   (b) $\dfrac{2}{s} - \dfrac{\pi}{(4-s)^2+\pi^2}$   (c) $\dfrac{1}{s}(1-e^{-s})$

(d) $\dfrac{1}{s}(e^{-s} - e^{-2s})$   (e) $\dfrac{2}{s(s^2+4)}$   (f) $\sqrt{\dfrac{\pi}{s}}$

3. (a) $\tfrac{1}{2}\sin(2t)$   (b) $4te^t$   (c) $e^{-2t} - te^{-2t}$   (d) $\tfrac{1}{2}[1 - 2e^{-t} + e^{-2t}]$

(e) $e^{-2t}\left(\cos\sqrt{3}t + \dfrac{1}{\sqrt{3}}\sin\sqrt{3}t\right)$

5. (a) $f(t) = \tfrac{1}{2}e^{3t} + \tfrac{3}{2}e^t$   (b) $f(t) = 4e^{2t} - 3e^{3t}$

(c) $f(t) = \tfrac{28}{39}e^{2t} - \tfrac{5}{6}e^{-t} - \tfrac{1}{13}e^{-t}\sin 2t + \tfrac{3}{26}e^{-t}\cos 2t$

(d) $f(t) = \begin{cases} 0 & \text{if } 0 \le t \le 3 \\ \tfrac{1}{2} + \tfrac{1}{2}e^{2t-6} - e^{t-3} & \text{if } 3 \le t \le 6 \\ \tfrac{1}{2}e^{2t-6} - \tfrac{1}{2}e^{2t-12} + e^{t-6} - e^{t-3} & \text{if } t \ge 6 \end{cases}$

7. (b) Hint: Consider the partial fraction expansion of $F(s)$, and compare with the table entries.

9. (b)  (i) $x(t) = \cos\sqrt{3}t$, $y(t) = -\cos\sqrt{3}t$;

  (ii) $x(t) = y(t) = \cos t$;

  (iii) $x(t) = \tfrac{1}{2}(\cos t + \cos\sqrt{3}t)$, $y(t) = \tfrac{1}{2}(\cos t - \cos\sqrt{3}t)$.

(c) (i), (ii)

## Exercises 8.4, Page 414

3. Multiply the radii by $|\alpha|$.

5. (a) $a(n) = -(-3)^{-n}\,(n \le -1)$, 0 otherwise   (b) $a(n) = (-\tfrac{1}{3})^n\,(n \ge 0)$, 0 otherwise

(c) $a(n) = (-1)^n 2^{n+3}\,(n \le -4)$, 0 otherwise   (d) $a(n) = (-2)^{n+3}\,(n \ge -3)$, 0 otherwise

(e) $a(n) = -(\tfrac{1}{2})^n\,(n \ge 1)$, $-3^{n-1}\,(n \le -1)$, $-\tfrac{1}{3}(n=0)$

(f) $a(n) = 4(-\tfrac{1}{2})^n - 3(-\tfrac{1}{4})^n\,(n \ge 0)$, 0 otherwise

(g) $a(n) = 4(\tfrac{1}{2})^n + 2n - 4\,(n \ge 0)$, 0 otherwise

(h) $a(n) = \dfrac{1}{\alpha^n}[\alpha^{-1} - \alpha]\,(n \ge 1)$, $\dfrac{1}{\alpha}(n=0)$, 0 otherwise

7. (a) $2^{-n+1}$   (b) $\tfrac{1}{3} + \tfrac{2}{3}(-2)^n$   (c) $\dfrac{1}{2} + 2^{n+1} - \dfrac{3^n}{2}$

## Exercises 8.5, Page 425

3. Hint: Use Lemma 4, Section 6.5.

7. (b) p.v. $\dfrac{1}{x-x_0} - i\pi\delta(x-x_0)$

# Index